FOREST-BASED BIOMASS ENERGY
Concepts and Applications

ENERGY AND THE ENVIRONMENT

SERIES EDITOR

Abbas Ghassemi
New Mexico State University

PUBLISHED TITLES

Forest-Based Biomass Energy : Concepts and Applications
Frank R. Spellman

Introduction to Renewable Energy
Vaughn Nelson

Geothermal Energy: Renewable Energy and the Environment
William E. Glassley

Solar Energy: Renewable Energy and the Environment
Robert Foster, Majid Ghassemi, Alma Cota
Jeanette Moore, Vaughn Nelson

Wind Energy: Renewable Energy and the Environment
Vaughn Nelson

FOREST-BASED BIOMASS ENERGY
Concepts and Applications

Frank R. Spellman

CRC Press
Taylor & Francis Group
Boca Raton London New York

CRC Press is an imprint of the
Taylor & Francis Group, an **informa** business

CRC Press
Taylor & Francis Group
6000 Broken Sound Parkway NW, Suite 300
Boca Raton, FL 33487-2742

First issued in paperback 2017

ISBN-13: 978-1-4398-6019-9 (hbk)
ISBN-13: 978-1-138-07716-4 (pbk)

Library of Congress Cataloging-in-Publication Data

Spellman, Frank R.
   Forest-based biomass energy : concepts and applications / author, Frank R. Spellman.
      p. cm.
   "A CRC title."
   Includes bibliographical references and index.
   ISBN 978-1-4398-6019-9 (hardback)
   1. Forest biomass. 2. Biomass energy. I. Title.

SD387.B48S69 2011
333.95'39--dc23
2011025401

For Lynn Eubanks

Contents

Section II Forest-Based Biomass

Series Preface

By 2050, the demand for energy could double or even triple, as the global population rises and developing countries expand their economies. All life on Earth depends on energy and the cycling of carbon. Energy is essential for economic and social development but also poses an environmental challenge. We must explore all aspects of energy production and consumption, including energy efficiency, clean energy, global carbon cycle, carbon sources and sinks, and biomass, as well as their relationship to climate and natural resource issues. Knowledge of energy has allowed humans to flourish in numbers unimaginable to our ancestors. The world's dependence on fossil fuels began approximately 200 years ago. Are we running out of oil? No, but we are certainly running out of the affordable oil that has powered the world economy since the 1950s. We know how to recover fossil fuels and harvest their energy for operating power plants, planes, trains, and automobiles, which modify the carbon cycle and contribute to greenhouse gas emissions. This has resulted in debate on the availability of fossil energy resources, the peak oil era and timing for the anticipated end of the fossil fuel era, price and environmental impact versus various renewable resources and use, carbon footprints, emissions, and control, including cap and trade and emergence of "green power."

Our current consumption has largely relied on oil for mobile applications and coal, natural gas, and nuclear or water power for stationary applications. In order to address the energy issues in a comprehensive manner, it is vital to consider the complexity of energy. Any energy resource—including oil, coal, wind, biomass, and so on—is an element of a complex supply chain and must be considered in the entirety as a system from production through consumption. All of the elements of the system are interrelated and interdependent. Oil, for example, requires consideration for interlinking of all of the elements, including exploration, drilling, production, water, transportation, refining, refinery products and byproducts, waste, environmental impact, distribution, consumption or application, and finally emissions. Inefficiencies in any part of the system will impact the overall system, and disruption in any one of these elements would cause major interruption in consumption. As we have experienced in the past, interrupted exploration will result in disruption in production, restricted refining and distribution, and consumption shortages; therefore, any proposed energy solution requires careful evaluation, which may be one of the key barriers to implementing the proposed use of hydrogen as a mobile fuel.

Even though an admirable level of effort has gone into improving the efficiency of fuel sources for the delivery of energy, we are faced with severe challenges on many fronts. This includes population growth, emerging economies, new and expanded usage, and limited natural resources. All energy solutions include some level of risk, including technology snafus and changes in market demand and economic drivers. This is particularly true when proposing energy solutions that involve implementation of untested alternative energy technologies.

There are concerns that emissions from fossil fuels will lead to climate change with possible disastrous consequences. Over the past five decades, the world's collective greenhouse gas emissions have increased significantly even as efficiency has increased, resulting in extending energy benefits to more of the population. Many propose that we improve the efficiency of energy use and conserve resources to lessen greenhouse gas emissions and avoid a climate catastrophe. Using fossil fuels more efficiently has not reduced overall greenhouse gas emissions for various reasons, and it is unlikely that such initiatives will have a perceptible effect on atmospheric greenhouse gas content. Despite a debatable correlation between energy use and greenhouse gas emissions, there are effective means to produce energy, even from fossil fuels, while controlling emissions. There are also emerging technologies and engineered alternatives that will actually manage the makeup of the atmosphere but will require significant understanding and careful use of energy.

We need to step back and reconsider our role in and knowledge of energy use. The traditional approach of micromanagement of greenhouse gas emissions is not feasible or functional over a long period of time. More assertive methods to influence the carbon cycle are needed and will be emerging in the coming years. Modifications to the cycle mean we must look at all options in managing atmospheric greenhouse gases, including various ways to produce, consume, and deal with energy. We need to be willing to face reality and search in earnest for alternative energy solutions. There appear to be technologies that could assist; however, they may not all be viable. The proposed solutions must not be in terms of a "quick approach," but a more comprehensive, long-term (10, 25, and 50 plus years) approach that is science based and utilizes aggressive research and development. The proposed solutions must be capable of being retrofitted into our existing energy chain. In the meantime, we must continually seek to increase the efficiency of converting energy into heat and power.

To achieve sustainable development, long-term, affordable availability of resources, including energy, is necessary. There are many potential constraints to sustainable development. Foremost is the competition for water use in energy production, manufacturing, and farming, for example, versus a shortage of freshwater for consumption and development. Sustainable development is also dependent on the Earth's limited amount of soil; in the not too distant future, we will have to restore and build soil as a part of sustainable development. Hence, possible solutions must be comprehensive and based on integrating our energy use with nature's management of carbon, water, and life on Earth as represented by the carbon and hydrogeological cycles. Obviously, the challenges presented by the need to control atmospheric greenhouse gases are enormous and require "out of the box" thinking, innovative approaches, imagination, and bold engineering initiatives in order to achieve sustainable development. We will need to ingeniously exploit even more energy and integrate its use with control of atmospheric greenhouse gases. The continued development and application of energy are essential to the development of human society in a sustainable manner through the coming centuries. All alternative energy technologies are not equal and have risks and drawbacks. When evaluating our energy options, we must consider all aspects, including performance against known criteria, basic economics and benefits, efficiency, processing and utilization requirements, infrastructure requirements, subsidies and credits, waste, and ecosystems, as well as unintended consequences such as impacts to natural resources and the environment. Additionally, we must include the overall changes and emerging energy picture based on current and future efforts to modify fossil fuels and evaluate the energy return for the investment of funds and other natural resources such as water.

A significant motivation in creating this book series, which is focused on alternative energy and the environment, was brought about as a consequence of lecturing around the country and in the classroom on the subject of energy, environment, and natural resources such as water. Water is a precious commodity in the West, in general, and the Southwest, in particular, and has a significant impact on energy production, including alternative sources, due to the nexus between energy and water and the major correlation with the environment and sustainability-related issues. Although the correlation between these elements, how they relate to each other, and the impacts of one on the other are understood, these issues are not adequately considered when it comes time to integrate alternative energy resources into the energy matrix. Additionally, as renewable technology implementation grows by various states, nationally and internationally, the need for informed and trained human resources continues to be a significant driver in future employment resulting in universities, community colleges, and trade schools offering minors, certificate programs, and even, in some cases, majors in renewable energy and sustainability. As the field grows, the demand for trained operators, engineers, designers, and architects who would be able to incorporate these technologies into their daily activity is increasing. Additionally, we receive daily deluges of flyers, e-mails, and texts promoting various short courses in solar, wind, geothermal, biomass, and other forms of alternative energy under the umbrella of retooling an individual's career and providing the trained resources needed to interact with financial, governmental, and industrial organizations.

In all my interactions throughout the years in this field, I have conducted significant searches to locate integrated textbooks that explain alternative energy resources in a suitable manner and that would complement a syllabus for a potential course to be taught at the university while providing good reference material for people interested in this field. I have been able to locate a number of books related to energy; energy systems; resources such as fossil, nuclear, and renewable energy; and energy conversion, as well as specific books in the subjects of natural resource availability, use, and impact as related to energy and the environment. However, specific books that are correlated and present the various subjects in detail are few and far between. We have, therefore, initiated a series of texts addressing specific technology fields in the renewable energy arena. This series includes textbooks on wind, solar, geothermal, biomass, hydro, and other forms of energy yet to be developed. These texts are intended for upper-level undergraduate and graduate students and for informed readers who have a solid fundamental understanding of science and mathematics, as well as individuals and organizations involved in the development of renewable energy who are interested in making reference material available to their scientists and engineers, consulting organizations, and reference libraries. Each book presents fundamentals as well as a series of numerical and conceptual problems designed to stimulate creative thinking and problem solving.

The series author wishes to express his deep gratitude to his wife, Maryam, who has served as a motivator and intellectual companion and too often was victim of this effort. Her support, encouragement, patience, and involvement have been essential to the completion of this series.

Abbas Ghassemi, PhD
Las Cruces, New Mexico

Series Editor

Abbas Ghassemi, PhD, is director and chief operating officer of the Institute for Energy and Environment (IE&E) and a professor of chemical engineering at New Mexico State University. At IE&E, he is responsible for administering over $10 million in programs annually in education and research, as well as outreach in energy resources including renewable energy, water quality and quantity, and environmental issues. His administrative duties also include operations, budget, planning, and personnel supervision for the program. Dr. Ghassemi has authored and edited several textbooks and has many publications and papers in the areas of energy, water, carbon cycle (including carbon generation and management), process control, thermodynamics, transport, education management, and innovative teaching methods. His research areas of interest include risk-based decision making, renewable energy and water, carbon management and sequestration, energy efficiency, pollution prevention, and multiphase flows and process control. Dr. Ghassemi serves on a number of public and private boards, editorial boards, and peer review panels.

Preface

Why a text on forest-based biomass energy? Good question. A better question might be what is forest-based biomass energy? Simply, forest-based biomass energy and energy products are derived from forest biomass in forestlands. Forest biomass (trees, shrubs, limbs, and vegetation debris) can be converted to energy and energy products in a number of ways: direct combustion, pellets, gasification, or cofiring. Many of these uses are described in this text.

A follow-up question might be why study or care about forest-based biomass energy? Simply put, studying renewable energy without knowledge or an understanding of forest-based biomass energy is analogous to attempting to cook without a complete recipe. Moreover, studying renewable energy and its forest-based biomass feedstock contributors affords us the opportunity not only to grasp the pressing need for renewable energy but also to understand the complicated task ahead of us in developing viable forms of renewable energy for the future. Keep in mind that we need to replace hydrocarbon fuels because we are literally emptying the wells and because we need to clean up our environment—we need to produce a reliable renewable energy source that will not destroy or pollute our fragile environment. When you get right down to it, isn't sustaining our current lifestyle—the so-called good life—and maintaining a healthy environment on Earth goals we should all strive for?

We think they are. Moreover, many of us have come to realize that we pay a price (sometimes a high price) for the good life. Our consumption and use of the world's resources make us all at least partially responsible for the shortages and the resulting pollution of our environment. Simply, it is a matter of record that pollution and its ramifications are inevitable products of a good life maintained through the use of fossil fuels. But, obviously, fuel shortages and the impact of fossil fuel pollution are not something any single individual causes or can totally prevent or correct. To reduce fuel shortages, pollution, and its harmful effects, we must band together as an informed, knowledgeable group and pressure our elected decisionmakers to manage the problem now and in the future. We must replace fossil fuels with clean and efficient renewable energy.

Although this text addresses a pressing topic of interest to us all, we have to the best of our ability deliberately avoided the intrusion of media hype and political ideology; instead, this text attempts to relate facts. Political ideology and the media work to fog the facts, which in turn fogs both public perception of the issues and scientific assessment of environmental quality. In this text, we emphasize that every problem has a solution, and the solution to replacing fossil fuels with reliable and efficient nonpolluting renewable energy is out there. We simply need to find sources and develop them; the world's forests are a good place to begin. Why? Because in addition to traditional value-added products, such as lumber, paper, and composites, exciting new opportunities are on the horizon for forest-based biomass use in the production of electricity, transportation fuels, chemical feedstock, and cellulose nanofibers.

Throughout this text, we present common-sense approaches and practical (and sometimes poetic) examples. Because this is a science text, we have adhered to scientific principles, models, and observations; however, you need not be a scientist to understand the principles and concepts presented here—the text goes easy on the hard math and science and presents the material in a user-friendly manner. To gain an understanding of the principles and practices offered in this text you need an open mind, a love for the challenge of wading through the muck, an ability to decipher problems, and the patience to answer the questions relevant to each topic presented. We weave real-life situations throughout the fabric of this text, which is written in straightforward, plain English, to equip you with the facts, knowledge, and information necessary to understand the complex issues and to make your own informed decisions. Several basic and practical mathematical operations are illustrated with example problems and solutions.

This text is not an answer book; instead, it is designed to stimulate thought. Although answers to specific renewable energy and forest-derived feedstock questions are provided, we also point out the disadvantages and hurdles that must be overcome to make forest-based renewable energy a viable alternative to fossil fuels. Our goal herein is to provide a framework of principles you can use to understand the complexity of substituting renewables, such as forest-based biomass for fossil fuels.

Author

Frank R. Spellman, PhD, is retired assistant professor of environmental health at Old Dominion University, Norfolk, Virginia, and author of more than 70 books covering topics ranging from concentrated animal feeding operations (CAFOs) to all areas of environmental science and occupational health. Many of his texts are readily available online, and several have been adopted for classroom use at major universities throughout the United States, Canada, Europe, and Russia; two have been translated into Spanish for South American markets. Dr. Spellman has been cited in more than 400 publications. He serves as a professional expert witness for three law groups and as an incident/accident investigator for the U.S. Department of Justice and a northern Virginia law firm. In addition, he consults on homeland security vulnerability assessments for critical infrastructure, including water/wastewater facilities nationwide and conducts pre-Occupational Safety and Health Administration (OSHA)/Environmental Protection Agency (EPA) audits throughout the country. Dr. Spellman receives frequent requests to co-author with well-recognized experts in several scientific fields; for example, he is a contributing author for the prestigious text *The Engineering Handbook*, 2nd ed. (CRC Press). Dr. Spellman lectures on sewage treatment, water treatment, and homeland security and health and safety topics throughout the country and teaches water/wastewater operator short courses at Virginia Tech (Blacksburg, Virginia). He holds a BA in public administration, a BS in business management, an MBA, and an MS and PhD in environmental engineering.

Prologue: May I Borrow a Light?

Timeframe: 402,010 BCE

Inside the deep recesses of his clan's cave, located at what we today call Swartkrans in South Africa, a man and a few of his fellow clan members stood briefly (there was little time for appreciating anything other than the ongoing struggle for survival) looking at their brilliant artwork. For days, weeks, months, years, and even decades he and his brother and sister clan mates had contributed to the massive depictions of animals, such as bison, horses, aurochs (ancestors of domestic cattle), and deer, intermixed with tracings of human hands as well as finger flutings (lines fingers leave on soft surfaces). Certainly these distinctive naturalistic animal figures depicted the animals necessary for their survival and fitness, but they also illustrated the challenge of the hunt and the rewards of slaughter. Many of the drawings were silhouettes of animals incised in the rock first, and then filled in with red and yellow ochre, hematite, manganese oxide, and charcoal. Line drawings covered almost every square foot of the reachable wall and ceiling area within the inner cave. The outer cave area had been decorated in the same or like manner many decades earlier. In fact, in those ancient times (relative to what today we would describe as ancient), most cave art was located as close to cave entrances as possible; having enough daylight to see made this a necessity. In this particular cave, a chronological history of early to late cave drawings was clearly displayed. This timeline reflected the initial need to make use of available daylight at the cave entrance, but these early drawings eventually gave way to a steady progression of artwork deeper and deeper within the recesses of the cave, made possible by shining artificial light on the walls and ceiling.

Modern humans look upon prehistoric cave paintings today and are amazed by the creativity of our early ancestors. Intuitively, some of us can imagine the major purpose of cave art: To take the edge off the drudgery and dangers associated with daily life—with survival of the fittest, so to speak. But, right now many readers of this text, although appreciative of our early ancestors' accomplishments, are probably asking themselves what do caveman and cave paintings have to do with the topic of this text, forest-based biomass energy? Good question. The simple answer is everything. The complex answer is that the use of forest-based biomass energy began somewhere, sometime, and someplace. We should not have to grope a lot for imaginative license to discern that it was a caveman who first discovered the practical applications of wood for fire—heat, light, and scaring off wild beasts. It was cavemen who first discovered that you don't confront a bunch of sabertoothed cats without a fire-stick in hand. Surely it was continuous exposure to the natural world that made them aware of the awesome power and force of fire as exhibited by forest fires caused by lightning or the heat and fire created by molten magma erupting from the very bowels of the Earth. Certainly, cavemen knew of the hazards of fire—the potential for destructive, life-threatening conflagrations; these had to be a constant worry.

It is highly unlikely, of course, that cavemen used the energy derived from fire to power various and sundry propulsion mechanisms; however, it is likely that they used fire for cooking, generating heat, signaling, lighting the interior of their caves, and self-protection. Archeological evidence certainly supports this view.

Again, it is safe to speculate that cavemen observed lightning or volcanic eruptions that set many forests of that time aflame. Thus, not only did our early ancestors observe the terrible destructive power of uncontrolled fire but sooner or later they had to discern the advantages of capturing and harnessing fire. Initially, they probably accomplished this by carefully removing burning tree branches and carrying them to their caves. Once they had fire inside the cave, inside some type of fire pit, they next had to find something to sustain the burn. Again, when cavemen observed naturally generated forest fires they were learning; basically, they were students residing in Nature's classroom. An important thing they learned was that wood burns—wood burns very well; it burns extremely well, thank you very much! Eventually, caveman discovered through trial and error that some woods burn better than others. Certainly they learned which types of wood burned the longest and gave off the best heat. We can't say for sure, but perhaps they learned the following, a poem penned much later by John Estabrook ("Tree Farm"):

> Beech wood fires are bright and clear
> if the logs are kept a year.
> Chestnut's only good, they say
> if for along it's laid away.
> Make a fire of elder tree
> Death within your house will be.
> But ash new or ash old
> is fit for a Queen with a crown of gold.

But, have no doubt, caveman learned. Today some might argue that our ancient ancestors were not smart enough to learn, not smart enough to put such a string of ideas, concepts, and thoughts together in any coherent form. Really? We beg to differ with this view. For a group of people who had to constantly outmaneuver Mother Nature at almost every twist and turn of existence, who had to outthink, outsmart, and outwit myriad untold hungry, vicious, wild beasts, as well as other competitive and sometimes deadly clan members, cavemen had an intelligence level for the art of staying alive that many of us might envy today.

One thing seems certain; once cavemen harnessed fire and later found a way to transport it safely from place to place and learned how to use flint and damp moss to strike their own fires, their limited world expanded. Not only were they able to shift habitats and wander to other regions in pursuit of food, but they were also able to provide fire to those who needed to borrow some. Cavemen became very familiar with fire and its many uses, such as providing enough light to allow painting deep within a normally dark cave, as well as the charcoal necessary to draw. When a friendly visitor arrived at a cave and grunted, gestured, signed, or in some other way asked that familiar question "May I borrow a light?" the cave dweller generously passed the flame to that visitor … and to another … and to another … and to us.

The obvious question now is will we pass the light on to those who follow us?

Section I

The Basics

Nothing was the matter with the Spruce-firs, exactly; but the history of their excitement was as follows:—They, and a number of other trees, were growing together in a pretty wood. There were oaks, and elms, and beeches, and larches, and firs of many sorts; and, here and there, there was a silver-barked Birch. And there was one silver-barked Birch in the particular, who had been observing the spruce-firs all that spring; noticing how fast they were growing, and what a stupid habit (as he thought) they had, of always getting into everybody's way, and never bending to accommodate the convenience of others.

The Law of the Wood is to be as accommodating as possible.

—**Mrs. Alfred Gatty,** *Parables from Nature* **(1861)**

1

Introduction*

Fossil fuels are the lifeblood of an industrialized society, supplying most of its energy needs. In recent years, the problems with these fossil fuels—including environmental damage, unequal global distribution of fossil fuel resources, price instability, and ultimately supply constraints—have led to a reexamination of their use and a search for alternatives.

—**Paul Komor,** *Renewable Energy Policy* **(2004)**

Grasshoppers were likely to eat anything. Wheat was a favorite grain, but they enthusiastically tackled corn, oats, barley, Lucerne [alfalfa], and clover—even grass. They ate almost all garden crops—potatoes, onions, peppers, rhubarb, beets, cabbages, radishes, turnips, tomatoes.

—**Cecil Alter,** *In the Beginning* **(1927, p. 157)**

From Caveman to the Greatest to the Grasshopper Generation[†]

In an editorial appearing February 10, 2010, in *The New York Times*, Thomas Friedman stated that, "The fat lady has sung." Specifically, Friedman was speaking about America's transition from the "Greatest Generation" to what Kurt Anderson referred to as the "Grasshopper Generation." According to Friedman, we are "going from the age of government handouts to the age of citizen givebacks, from the age of companions fly free to the age of paying for each bag." Friedman goes on to say that we all accept that our parents were the greatest generation, but it is us that we are concerned about and that it is the "we" that comprise the Grasshopper Generation: "We have been eating through the prosperity that was bequeathed us like hungry locusts."

What we are eating through, among other things, is our readily available, relatively inexpensive source of fossil fuel energy—our insatiable demand is outpacing our dwindling, nonreplaceable supply. Like the grasshopper, we can gobble it all up until it is all gone or we can find alternatives—that is, alternative sources of energy, renewable energy, or, specifically for our purposes in this text, forest-based biomass energy.

* Much of the information and data in this chapter is from EIA, *Renewable Energy Trends*, U.S. Energy Information Administration, Washington, D.C., 2004 (http:www.eia.doe.gov/cneaf/solar.renewables/page/trends/rentrends04.html); EIA, *How Much of Our Electricity Is Generated from Renewable Sources?*, U.S. Energy Information Administration, Washington, D.C., 2007 (http://tonto.eia.doe.gov/energy_in_brief/renewable_energy.cfm); Spellman, F.R., *The Science of Renewable Energy*, CRC Press, Boca Raton, FL, 2011.
† Spellman, F.R., *The Science of Renewable Energy*, CRC Press, Boca Raton, FL, 2011.

World Energy Supplies: Current Status

In addition to the problems associated with the ever-increasing demand for energy and its dwindling supply, the current status of the worldwide use of fossil fuels, politics, and other persistent forces (e.g., a growing population with a growing need for energy) are pushing for the development of renewable energy sources. Moreover, the energy supply and usage problem is exacerbated by the current and future economic problems generated by $4+/gal gasoline and by the perceived crisis developing with high carbon dioxide emissions—the major contributing factor of global climate change.

Before proceeding with an introductory discussion of purpose-grown, forest-based (or derived) biomass energy sources, it is important to make a clear distinction between the two current buzzwords alternative and renewable energy. *Alternative energy* is an umbrella term that refers to any source of usable energy intended to replace fuel sources without the undesired consequences presented by the use of the current fuels. The term "alternative" presupposes an undesirable connotation (for many people, the term "fossil fuels" has joined that endless list of offensive four-letter words). Examples of alternative fuels include petroleum as an alternative to whale oil, coal as an alternative to wood, alcohol as an alternative to fossil fuels, and coal gasification as an alternative to petroleum. These alternative fuels *need not be* renewable.

Renewable energy is energy generated from natural resources—such as sunlight, wind, water (hydro), ocean thermal, wave and tide action, biomass, and geothermal heat—that are naturally replenished and thus renewable. Renewable energy resources are virtually inexhaustible—they are replenished at the same rate as they are used—but they are limited in the amount of energy that is available per unit time. If we have not come full circle in our cycling from renewable to nonrenewable energy, we are getting close. Consider, for example, that in 1850, about 90% of the energy consumed in the United States came from renewable energy resources (e.g., hydropower, wind, burning wood). Today, though, the United States is heavily reliant on nonrenewable fossil fuels (natural gas, oil, and coal). In 2009, about 8% of all energy consumed came from renewable energy resources (see Table 1.1).

TABLE 1.1

U.S. Energy Consumption of Renewable Energy Sources, 2009 (Quadrillion Btu)

Total energy consumption	94.820
Renewable energy consumption	7.745
Biomass	3.884
Biofuels	1.546
Waste	0.447
Wood-derived fuels	1.891
Geothermal	0.373
Hydroelectric conventional	2.682
Solar thermal/photovoltaic	0.109
Wind	0.697

Source: EIA, *U.S. Energy Consumption by Energy Source*, Energy Information Administration, Washington, D.C., 2007.

TABLE 1.2

Renewable Energy Consumption for Production
of Electricity, 2009 (Quadrillion Btu)

Total renewable energy consumption	7.745
Electric power	4.113
Biomass	0.426
Waste	0.253
Landfill gas	0.088
Municipal solid waste (MSW) biogenic	0.138
Other biomass	0.027
Wood and wood-derived fuels	0.173
Geothermal	0.320
Hydroelectric conventional	2.663
Solar thermal/photovoltaic	0.008
Wind	0.697

Source: EIA, *Renewable Energy Consumption and Electricity Preliminary Statistics*, U.S. Energy Information Administration, Washington, D.C., 2010.

Most of the renewable energy is used to generate electricity, provide heat for industrial processes, heat and cool buildings, and transport fuels. Electricity producers (utilities, independent producers, and combined heat and power plants) accounted for 53% of the total U.S. renewable energy consumed in 2009 (see Table 1.2). Most of the rest of the remaining renewable energy consumed was biomass, which was used for industrial applications (principally papermaking) by plants producing only heat and steam. Biomass is also used for transportation fuels (ethanol) and to provide residential and commercial space heating. The largest share of renewable-generated electricity comes from hydroelectric energy (65%), followed by wind (17%), biomass (10%), geothermal (8%), and solar (0.2%). Wind-generated electricity increased by almost 28% in 2009 over 2008, more than any other energy source.

From these tables it is obvious that currently there are five primary forms of renewable energy: solar, wind, biomass, geothermal, and hydroelectric. Each of theses holds promise and poses challenges regarding future development; however, as the title of this book suggests, it is biomass that we are most interested in here. Specifically, we are interested in forest-based biomass energy production because it is the author's view that the greatest hope for throwing off the chains of fossil fuel dependence lies in the production of liquid fuels from biomass.

DID YOU KNOW?

The greatest amounts of forest-based biomass per area are located in the Pacific Northwest. Moderate levels are located along the Appalachian Mountains, from northern Georgia into central Maine, encompassing much of the hardwood region of the United States. The rest of the United States is occupied by forests containing between 1 and 200 tons CO_2 per hectare on average, with infrequent extremely heavy biomass accumulations (350+ tons CO_2 per hectare). Overall, the conterminous U.S. forests contain 16 billion metric tons of carbon in aboveground live biomass (USFS, 2009).

In light of the pressing need to develop alternative energy sources to replace dwindling supplies of fossils fuels, the U.S. Department of Energy (DOE) and the U.S. Department of Agriculture (USDA) are both strongly committed to expanding the role of biomass as an energy source. In particular, they support biomass fuels and producers as a way to reduce the need for oil and gas imports; to support the growth of agriculture, forestry, and rural economies; and to foster major new domestic industries—biorefineries—making a variety of fuels, chemicals and other products (USDA, 2005).

Bioenergy to Feed the Grasshopper

A promising source of renewable energy is bioenergy. *Bioenergy* is a general term that refers to energy derived from materials such as straw, wood, or animals wastes, which, in contrast to fossil fuels, were living matter relatively recently. Such materials can be burned directly as solids (biomass) to produce heat or power, but they can also be converted into liquid biofuels. In the last few years, there has been increased interest in the use of bioenergy fuels as *biofuels* (biodiesel and bioethanol), which can be used for transport. At the moment, transport has taken center stage in our search for renewable, alternative fuels to eventually replace the hydrocarbon fuels that we are exhausting. Unlike liquid biofuels, *solid biomass fuel* is used primarily for electricity generation or heat supply.

Even though bioenergy is a promising source of energy in the future, it is rather ironic when the experts (or anyone else, for that matter) state this point without qualification. The reality? Simply, keep in mind that it was only a little over 100 years ago that our economy was based primarily on bioenergy from biomass, or carbohydrates, rather than from hydrocarbons. In the late 1800s, the largest selling chemicals were alcohols made from wood and grain, the first plastics were produced from cotton, and about 65% of the nation's energy came from wood (USDOE, 2003).

By the 1920s, the economy began to shift toward the use of fossil resources, and after World War II this trend accelerated as technology breakthroughs were made. By the 1970s, fossil fuel energy was established as the backbone of the U.S. economy, and all but a small portion of the carbohydrate economy remained. By 1989, plants accounted for about 16% of inputs in the industrial sector, compared with 35% in 1925.

Processing costs and the availability of inexpensive fossil energy resources continue to be driving factors in the dominance of hydrocarbon resources. In many cases, it is still more economical to produce goods from petroleum or natural gas than from plant matter. This trend is about to shift dramatically, though, as we reach peak oil and as the world continues to demand unprecedented amounts of petroleum supplies from an ever-dwindling supply. Assisting in this shift are technological advances being made in the biological sciences and engineering fields, political changes, and growing concern for the environment. These factors have begun to swing the economy back toward the use of carbohydrates on a number of fronts. Consumption of biofuels in American vehicles, for example, rose from zero in 1977 to nearly 6.8 billion gallons of bioethanol and 491 million gallons of biodiesel in 2007. The use of inks produced from soybeans in the United States increased by fourfold between 1989 and 2000 and is now at more than 22% of total use (ILSR, 2002).

Technological advances are also beginning to make an impact on reducing the costs of producing industrial products and fuels from biomass, making them more competitive with those produced from petroleum-based hydrocarbons. Developments in pyrolysis, ultracentrifuges, membranes, and the use of enzymes and microbes as biological factories are enabling the extraction of valuable components from plants at a much lower cost. As a result, industry is investing in the development of new bioproducts that are steadily gaining a share of current markets (USDOE, 2003).

New technology is also allowing the chemical and food processing industries to develop new processes for more cost-effective production of all kinds of industrial products from biomass. One example is a plastic polymer derived from corn that is now being produced at a 300-million-pound-per-year plant in Nebraska, a joint venture between Cargill, the largest grain merchant, and Dow Chemical, the largest chemical producer (Fahey, 2001).

It is interesting to note that, at the time of this writing, the national average for a gallon of regular gasoline topped $4. The recent rising price of crude oil has been pushing up the price of gasoline, and the question is what will be the price of a gallon of gasoline next year? Will it be $3, $4, $5, $10, $20, or more a gallon? Well, we do not know for sure, do we? We can guess, though. Some say we are turning the corner on our current economic downturn and things are beginning to look up. Some are under the misguided notion that some great White Knight will ride into this country, wave his magic sword in the air, and solve not only our economic problems but also our energy problems. Wishful thinking? Absolutely. Our current economic woes have been caused by numerous factors. Keep in mind that, even if we could restore the 9 million high-paying jobs that left this country and are now in China, Mexico, or some other foreign locale, as soon as this country begins to make any progress toward economic recovery one thing is certain—foreign suppliers of crude oil will increase their prices and eventually this will put a lid on that recovery. Are we doomed? No, not necessarily, but, if we do not innovate and provide a homegrown source of energy to fuel our industries, homes, and vehicles, then, yes, failure is a distinct possibility—the consequences of which are difficult to fathom. Renewable energy sources, including forest-based biomass energy, are not only our way to future success but also the key that will unlock those chains that fetter us to those who despise us and who want to destroy us. Is there another alternative to moving ahead aggressively with renewable energy development? Yes—simply, move over, caveman.

Chapter Review Question

1.1 *Thought-provoking question:* When fossil fuels are depleted, will forest-based biomass converted to liquid fuel for transportation and other needs be enough to provide the energy needs of future generations? Explain.

References and Recommended Reading

Fahey, J. (2001). Shucking petroleum, *Forbes Magazine*, November 1, p. 206.

Friedman, T. (2010). The fat lady has sung, *The New York Times*, December 12, p. WK8 (www.nytimes.com/2010/02/21/opinion/21friedman.html).

ILSR. (2002). *Accelerating the Shift to a Carbohydrate Economy: The Federal Role: Executive Summary of the Minority Report of the Biomass Research and Development Technical Advisory Committee*, Institute for Local Self-Reliance, Washington, D.C.

Komor, P. (2004). *Renewable Energy Policy*, iUniverse, Inc., New York.

Spellman, F.R. (2011). *The Science of Renewable Energy*, CRC Press, Boca Raton, FL.

USDA. (2005). *Biomass as Feedstock for a Bioenergy and Bioproducts Industry: The Technical Feasibility of a Billion-Ton Annual Supply*. U.S. Department of Agriculture, Washington, D.C.

USDOE. (2003). *Industrial Bioproducts: Today and Tomorrow*, U.S. Department of Energy, Washington, D.C.

USFS. (2009). *Forest Inventory and Analysis National Program*, U.S. Forest Service, Washington, D.C. (http://fia.fs.fed.us).

2

Basic Math Operations

> If you can not measure it, you can not improve it.
> —**Lord Kelvin**

Introduction

In the course of practice, most mathematical calculations made by biomass technicians and practitioners in various branches of biomass production and usage, such as forest-based biomass energy personnel (among many others), begin with the basics, such as addition, subtraction, multiplication, division, and sequence of operations. Biomass industry personnel should master basic math definitions and the formation of problems. Daily forest-based biomass operations require calculations of many types, so a foundation in basic and higher mathematics is important. In this chapter, we present some basic definitions and math operations that will aid the reader in understanding the technical material that follows.

Basic Math Terminology and Definitions

- An *integer*, or an *integral number*, is a whole number; thus, 1, 2, 3, 4, 5, 6, 7, 8, 9, 10, 11, and 12 are the first 12 positive integers.
- A *factor*, or *divisor*, of a whole number is any other whole number that exactly divides it; thus, 2 and 5 are factors of 10.
- A *prime number* in math is a number that has no factors except itself and 1; examples of prime numbers include 1, 3, 5, 7, and 11.
- A *composite number* is a number that has factors other than itself and 1; examples of composite numbers include 4, 6, 8, 9, and 12.
- A *common factor*, or *common divisor*, of two or more numbers is a factor that will exactly divide each of them. If this factor is the largest factor possible, it is called the *greatest common divisor*; thus, 3 is a common divisor of 9 and 27, but 9 is the greatest common divisor of 9 and 27.
- A *multiple* of a given number is a number that is exactly divisible by the given number. If a number is exactly divisible by two or more other numbers, it is a common multiple of them. The least (smallest) such number is called the *lowest common multiple*. Thus, 36 and 72 are common multiples of 12, 9, and 4; however, 36 is the lowest common multiple.

- An *even number* is a number exactly divisible by 2; thus, 2, 4, 6, 8, 10, and 12 are even integers.
- An *odd number* is an integer that is not exactly divisible by 2; thus, 1, 3, 5, 7, 9, and 11 are odd integers.
- A *product* is the result of multiplying two or more numbers together; thus, 25 is the product of 5 × 5, and 4 and 5 are factors of 20.
- A *quotient* is the result of dividing one number by another; for example, 5 is the quotient of 20 divided by 4.
- A *dividend* is a number to be divided, and a *divisor* is a number that divides; for example, in 100 ÷ 20 = 5, 100 is the dividend, 20 is the divisor, and 5 is the quotient.
- *Area* is the area of an object, measured in square units.
- *Base* is a term used to identify the bottom leg of a triangle, measured in linear units.
- *Circumference* is the distance around an object, measured in linear units. When such a distance is determined for shapes other than circles, it may be called the *perimeter* of the figure, object, or landscape.
- *Cubic units* are measurements used to express volume, such as cubic feet or cubic meters.
- *Depth* is the vertical distance from, for example, the bottom of a tank to the top. This is normally measured in terms of liquid depth and given in terms of sidewall depth (SWD), measured in linear units.
- *Diameter* is the distance from one edge of a circle to the opposite edge passing through the center; it is measured in linear units.
- *Height* is the vertical distance from the base or bottom of a unit to the top or surface.
- *Linear units* are measurements used to express distances, such as feet, inches, meters, or yards.
- *Pi* (π) is a number used in calculations involving circles, spheres, or cones: $\pi = 3.14$.
- *Radius* is the distance from the center of a circle to the edge, measured in linear units.
- A *sphere* is a container shaped like a ball.
- *Square units* are measurements used to express area, such as square feet or square meters.
- *Volume* is the capacity of a unit (how much it will hold), measured in cubic units (e.g., cubic feet, cubic meters) or in liquid volume units (e.g., gallons, liters, million gallons).
- *Width* is the distance from one side of a tank, for example, to the other, measured in linear units.

DID YOU KNOW?

Of means to multiply.	*And* means to add.
Per means to divide.	*Less than* means to subtract.

Sequence of Operations

Mathematical operations such as addition, subtraction, multiplication, and division are usually performed in a certain order or sequence. Typically, multiplication and division operations are done prior to addition and subtraction operations. In addition, mathematical operations are also generally performed from left to right using this hierarchy. The use of parentheses is also common to set apart operations that should be performed in a particular sequence.

Note: It is assumed that the reader has a fundamental knowledge of basic arithmetic and math operations. Thus, the purpose of the following text is to provide only a brief review of the mathematical concepts and applications frequently employed by forest-based biomass energy specialists.

Rules

Rule 1.   In a series of additions, the terms may be placed in any order and grouped in any way. Thus, $4 + 3 = 7$ and $3 + 4 = 7$. Similarly, $(4 + 3) + (6 + 4) = 17$, $(6 + 3) + (4 + 4) = 17$, and $[6 + (3 + 4)] + 4 = 17$.

Rule 2.   In a series of subtractions, changing the order or grouping of the terms may change the result. Thus, $100 - 30 = 70$, but $30 - 100 = -70$. Similarly, $(100 - 30) - 10 = 60$, but $100 - (30 - 10) = 80$.

Rule 3.   When no grouping is given, the subtractions are performed in the order written, from left to right. Thus, $100 - 30 - 15 - 4 = 51$. This operation can also be written as steps: $100 - 30 = 70$, $70 - 15 = 55$, $55 - 4 = 51$.

Rule 4.   In a series of multiplications, the factors may be placed in any order and in any grouping. Thus, $[(2 \times 3) \times 5] \times 6 = 180$ and $5 \times [2 \times (6 \times 3)] = 180$.

Rule 5.   In a series of divisions, changing the order or the grouping may change the result. Thus, $100 \div 10 = 10$, but $10 \div 100 = 0.1$. Similarly, $(100 \div 10) \div 2 = 5$, but $100 \div (10 \div 2) = 20$. Again, if no grouping is indicated, the divisions are performed in the order written, from left to right; thus, $100 \div 10 \div 2$ is understood to mean $(100 \div 10) \div 2$.

Rule 6.   In a series of mixed mathematical operations, the convention is as follows: Whenever no grouping is given, multiplications and divisions are to be performed in the order written, then additions and subtractions in the order written.

Examples

- In a series of *additions*, the terms may be placed in any order and grouped in any way:

$$4 + 6 = 10 \text{ and } 6 + 4 = 10$$

$$(4 + 5) + (3 + 7) = 19, \ (3 + 5) + (4 + 7) = 19, \text{ and } [7 + (5 + 4)] + 3 = 19$$

- In a series of *subtractions*, changing the order or the grouping of the terms may change the result:

$$100 - 20 = 80, \text{ but } 20 - 100 = -80$$

$$(100 - 30) - 20 = 50, \text{ but } 100 - (30 - 20) = 90$$

When no grouping is given, the subtractions are performed in the order written, from left to right:

$$100 - 30 - 20 - 3 = 47$$

or by steps:

$$100 - 30 = 70, \ 70 - 20 = 50, \ 50 - 3 = 47$$

- In a series of *multiplications*, the factors may be placed in any order and in any grouping:

$$[(3 \times 3) \times 5] \times 6 = 270 \ \text{ and } \ 5 \times [3 \times (6 \times 3)] = 270$$

- In a series of *divisions*, changing the order or the grouping may change the result:

$$100 \div 10 = 10, \text{ but } 10 \div 100 = 0.1$$

$$(100 \div 10) \div 2 = 45, \text{ but } 100 \div (10 \div 2) = 20$$

If no grouping is indicated, the divisions are performed in the order written—from left to right:

$$100 \div 5 \div 2 \text{ is understood to mean } (100 \div 5) \div 2$$

Note: In a series of mixed mathematical operations, the rule of thumb is that, whenever no grouping is given, multiplications and divisions are to be performed in the order written, then additions and subtractions in the order written.

■ *Example 2.1*

Problem: Perform the following mathematical operations to solve for the correct answer:

$$(2+4)+(2\times6)+\left(\frac{6+2}{2}\right)=$$

Solution:

1. Mathematical operations are typically performed going from left to right within an equation and within sets of parentheses.
2. Perform all math operations within the sets of parentheses first:

$$(2+4)=6, \quad (2\times6)=12, \quad \frac{6+2}{2}=\frac{8}{2}=4$$

Note that the addition of 6 and 2 was performed prior to dividing.

3. Perform all math operations outside of the parentheses. In this case, go from left to right.

$$6 + 12 + 4 = 22$$

■ *Example 2.2*

Problem: Solve the following equation:

$$(4 - 2) + (3 \times 3) - (15 \div 3) - 8 =$$

Solution:

1. First perform the mathematical operations inside each set of parentheses:

$$(4 - 2) = 2$$
$$(3 \times 3) = 9$$
$$(15 \div 3) = 5$$

2. Now perform the addition and subtraction operations from left to right:

$$2 + 9 - 5 - 8 = -2$$

There may be cases where several operations will be performed within multiple sets of parentheses. In these cases, we must perform all operations within the innermost set of parentheses first and move outward. We must continue to observe the hierarchical rules throughout the problem. Brackets, [ ], may be used to indicate additional sets of parentheses.

■ *Example 2.3*

Problem: Solve the following equation:

$$[2 \times (3 + 5) - 5 + 2] \times 3 =$$

Solution:

1. First perform the operations in the innermost set of parentheses:

$$(3 + 5) = 8$$

2. Rewrite the equation:

$$[2 \times 8 - 5 + 2] \times 3 =$$

3. Perform multiplication prior to addition and subtraction within the bracket:

$$[16 - 5 + 2] \times 3 = [11 + 2] \times 3 = [13] \times 3 = 13 \times 3 = 39$$

■ *Example 2.4*

Problem: Solve the following equation:

$$7 + [2(3 + 1) - 1] \times 2 =$$

Solution:

$$7 + [2(4) - 1] \times 2 =$$
$$7 + [8 - 1] \times 2 =$$
$$7 + [7] \times 2 =$$
$$7 + 14 = 21$$

■ *Example 2.5*

Problem: Solve the following equation:

$$[(12 - 4) \div 2] + [4 \times (5 - 3)] =$$

Solution:

$$[(8) \div 2] + [4 \times (2)] =$$
$$[4] + [8] =$$
$$4 + 8 = 12$$

Rounding and Significant Digits

Rounding Numbers

When rounding numbers, the following key points are important:

1. Numbers are rounded to reduce the number of digits to the right of the decimal point. This is done for convenience, not for accuracy.
2. *Rule:* A number is rounded off by dropping one or more numbers from the right and adding zeroes if necessary to place the decimal point. If the last figure dropped is 5 or more, increase the last retained figure by 1. If the last digit dropped is less than 5, do not increase the last retained figure.

■ *Example 2.6*

Problem: Round off 10,546 to 4, 3, 2, and 1 significant figures.
Solution:

$$10{,}546 = 10{,}550 \text{ to 4 significant figures}$$

$$10{,}546 = 10{,}500 \text{ to 3 significant figures}$$

$$10{,}546 = 11{,}000 \text{ to 2 significant figures}$$

$$10{,}547 = 10{,}000 \text{ to 1 significant figure}$$

Determining Significant Figures

In determining significant figures, the following key points are important:

1. The concept of significant figures is related to rounding.
2. It can be used to determine where to round off.
3. *Rule:* Significant figures are those numbers that are known to be reliable. The position of the decimal point does not determine the number of significant figures.

> **DID YOU KNOW?**
>
> No answer can be more accurate than the least accurate piece of data used to calculate the answer.

■ *Example 2.7*

Problem: How many significant figures are in a measurement of 1.35 in?
Solution: Three significant figures: 1, 3, and 5.

■ *Example 2.8*

Problem: How many significant figures are in a measurement of 0.000135?
Solution: Again, three significant figures: 1, 3, and 5. The three zeros are used only to place the decimal point.

■ *Example 2.9*

Problem: How many significant figures are in a measurement of 103,500?
Solution: Four significant figures: 1, 0, 3, and 5. The remaining two zeros are used to place the decimal point.

■ *Example 2.10*

Problem: How many significant figures are in 27,000.0?
Solution: There are six significant figures: 2, 7, 0, 0, 0, 0. In this case, the .0 means that the measurement is precise to 1/10 unit. The zeros indicate measured values and are not used solely to place the decimal point.

Powers and Exponents

When working with powers and exponents, the following key points are important:

1. Powers are used for *area* (as in square feet, ft^2) and *volume* (as in cubic feet, ft^3).
2. In multiplication, when all of the factors are alike, such as $4 \times 4 \times 4 \times 4 = 256$, the product is called a *power*. Thus, 256 is a power of 4, and 4 is the *base* of the power. A *power* is a *product* obtained by using a base a certain number of times as a factor.

3. Powers can be used to indicate that a number should be squared, cubed, etc.—in other words, the number of times a number must be multiplied by itself. Instead of writing $4 \times 4 \times 4 \times 4$, it is more convenient to use an *exponent* to indicate that the factor 4 is used as a factor four times. This exponent, a small number placed above and to the right of the base number, indicates how many times the base is to be used as a factor. Using this system of notation, the multiplication $4 \times 4 \times 4 \times 4$ can be written as 4^4. The ^{4} is the *exponent*, showing that 4 is to be used as a factor 4 times.

4. These same considerations apply to letters (a, b, x, y, etc.), as well:

$$z^2 = z \times z$$

$$z^4 = z \times z \times z \times z$$

Key Point: When a number or letter does not have an exponent, it is considered to have an exponent of one.

The powers of 1:

$1^0 = 1$ $1^3 = 1$

$1^1 = 1$ $1^4 = 1$

$1^2 = 1$

The powers of 10:

$10^0 = 1$ $10^3 = 1000$

$10^1 = 10$ $10^4 = 10{,}000$

$10^2 = 100$

■ Example 2.11

Problem: How is the term 2^3 written in expanded form?
Solution: The power (exponent) of 3 means that the base number (2) is multiplied by itself three times:

$$2^3 = 2 \times 2 \times 2$$

Key Point: When parentheses are used, the exponent refers to the entire term within the parentheses.

■ Example 2.12

Problem: How is the term $(3/8)^2$ written in expanded form?
Solution:

$$(3/8)^2 = 3/8 \times 3/8$$

When a negative exponent is used with a number or term, that number can also be expressed using a positive exponent:

$$6^{-3} = 1/6^3$$

Another example is

$$11^{-5} = 1/11^5$$

■ *Example 2.13*

Problem: How is the term 8^{-3} written in expanded form?
Solution:

$$8^{-3} = \frac{1}{8^3} = \frac{1}{8 \times 8 \times 8}$$

Note: A number or letter written as 3^0 or x^0 does not equal 3×1 or $x \times 1$, but simply 1.

Solving for the Unknown

Many forest-based biomass energy calculations involve the use of formulas and equations; for example, operations related to calculations used in biorefinery process control operations may require the use of equations to solve for an unknown quantity. To make these calculations, you must first know the values for all but one of the terms of the relevant equation. The obvious question now is "What is an equation?" Simply, an equation is a mathematical statement that tells us that what is on one side of an equal sign (=) is equal to what is on the other side.

As an example, $4 + 5 = 9$ is an equation. Now suppose we decide to add 4 to the left side, so we have $4 + 5 + 4$. What must we do now? To keep the other side equal, we must add 4 to the right side, so now we have $4 + 5 + 4 = 9 + 4$. Let's try that again by subtracting 3 from the left side of $6 + 2 = 8$, so we obtain $6 + 2 - 3$. What comes next? We must also subtract 3 from the right side, so we now have $6 + 2 - 3 = 8 - 3$.

It follows that if the right side were multiplied by a certain number, we would also have to multiply the left side by that same number. Finally, if one side is divided by 4, for example, then the other side must also be divided by 4.

The bottom line: What we do to one side of the equation, we must do to the other side. This is the case, of course, because the two sides, by definition, are always equal.

Equations

An *equation* is a statement indicating that two expressions or quantities are equal in value. Here is an example of an equation that operators are likely to encounter:

$$W = F \times D$$

where W is work, F is force, and D is distance. Thus, this equation can also be written as

Work (ft-lb or in.-lb) = Force (lb) × Distance (ft or in.)

The statement of equality $6x + 4 = 19$ is also an equation; that is, it is algebraic shorthand for "the sum of 6 times a number plus 4 is equal to 19." It can be seen that the equation $6x + 4 = 19$ is much easier to work with than the equivalent sentence. When thinking about equations, it is helpful to consider an equation as being similar to a balance. The equal sign tells you that two quantities are in balance (i.e., they are equal). Returning to our equation, $6x + 4 = 19$, the solution to this problem may be summarized in three steps:

Step 1. $6x + 4 = 19$

Step 2. $6x = 15$ (subtract 4 from both sides)

Step 3. $x = 2.5$ (divide both sides by 6)

Key Point: An equation is kept in balance (both sides of the equal sign are kept equal) by subtracting the same number from both sides, by adding the same number to both, or by dividing or multiplying by the same number.

The expression $6x + 4 = 19$ is called a *conditional equation*, because it is true only when x has a certain value. The number to be found in a conditional equation is called the *unknown number*, the *unknown quantity*, or, more briefly, the *unknown*. By following the axioms presented below, solving for unknowns (e.g., solving for x in $60 = x \times 2$) can be quite simple.

Key Point: Solving an equation is finding the value or values of the unknown that will make the equation true.

Note: It is important to point out that the following discussion includes only what the axioms are and how they work; it does not provide the arguments for why they are true.

Axioms

1. If equal numbers are added to equal numbers, the sums are equal.
2. If equal numbers are subtracted from equal numbers, the remainders are equal.
3. If equal numbers are multiplied by equal numbers, the products are equal.
4. If equal numbers are divided by equal numbers (except zero), the quotients are equal.
5. Numbers that are equal to the same number or to equal numbers are equal to each other.
6. Like powers of equal numbers are equal.
7. Like roots of equal numbers are equal.
8. The whole of anything equals the sum of all its parts.

Axioms 2 and 4 were used to solve the equation $6x + 4 = 19$.

Solving Equations

■ *Example 2.14*

Problem: Find the value of x if $x - 8 = 2$.

Solution: Here it can be seen by inspection that $x = 10$, but inspection does not help in solving more complicated equations. In this case, mentally we added 8 to each side of the equation to determine that $x = 10$; thus, we have acquired a method or procedure that can be applied to similar but more complex problems:

$$x - 8 = 2$$

Add 8 to each member (axiom 1):

$$x - 8 + 8 = 2 + 8$$

Collecting the terms (that is, adding 2 and 8), we obtain:

$$x = 10$$

■ Example 2.15

Problem: Solve for x if $4x - 4 = 8$.
Solution: Move the term -4 to the right of the equal sign as $+4$:

$$4x = 8 + 4$$
$$4x = 12$$
$$\frac{4x}{4} = \frac{12}{4} \quad \text{(divide both sides by 4)}$$
$$x = 3 \quad (x \text{ is alone on the left and is equal to the value on the right})$$

■ Example 2.16

Problem: Solve for x, if $x + 10 = 15$.
Solution: Subtract 10 from each side (axiom 2):

$$x + 10 = 15$$
$$x + 10 - 10 = 15 - 10$$

Collecting the terms, we obtain:

$$x = 5$$

■ Example 2.17

Problem: Solve for x, if $5x + 5 - 7 = 3x + 6$.
Solution: Collect the terms $(+5)$ and (-7):

$$5x - 2 = 3x + 6$$

Add 2 to both sides (axiom 2):

$$5x - 2 + 2 = 3x + 6 + 2$$
$$5x = 3x + 8$$

Subtract $3x$ from both sides (axiom 2):

$$5x - 3x = 3x - 3x + 8$$
$$2x = 8$$

Divide both members by 2 (axiom 4):

$$\frac{2x}{2} = \frac{8}{2}$$
$$x = 4$$

Checking the Answer

After obtaining a solution to an equation, we should always check it. This is an easy process. All we need to do is substitute the solution for the unknown quantity in the given equation. If the two members of the equation are then identical, the number substituted is the correct answer.

■ *Example 2.18*

Problem: Solve and check your solution for $4x + 5 - 7 = 2x + 6$.
Solution:

$$4x + 5 - 7 = 2x + 6$$
$$4x - 2 = 2x + 6$$
$$4x = 2x + 8$$
$$2x = 8$$
$$x = 4$$

Substitute the answer $x = 4$ in the original equation:

$$4x + 5 - 7 = 2x + 6$$
$$4(4) + 5 - 7 = 2(4) + 6$$
$$16 + 5 - 7 = 8 + 6$$
$$14 = 14$$

Because the statement $14 = 14$ is true, the answer $x = 4$ must be correct.

Setting up Equations

The equations discussed to this point were expressed in *algebraic* language. It is important to learn how to set up an equation by translating a sentence into an equation (into algebraic language) and then solving this equation. In setting up an equation properly, the following suggestions and examples should help:

1. Always read the statement of the problem carefully.
2. Select the unknown number and represent it by some letter. If more than one unknown quantity exists in the problem, try to represent those numbers in terms of the same letter—that is, in terms of one quantity.
3. Develop the equation, using the letter or letters selected, and then solve.

■ *Example 2.19*

Problem: We are given that one number is eight more than another. The larger number is two less than three times the smaller. What are the two numbers?

Solution: If we let n represent the smaller number, then $n + 8$ represents the larger number, and

$$n + 8 = 3n - 2$$
$$n = 5 \text{ (smaller number)}$$
$$n + 8 = 13 \text{ (larger number)}$$

■ *Example 2.20*

Problem: If five times the sum of a number and six is increased by three, the result is two less than ten times the number. Find the number.

Solution: Let n represent the number; thus, we can write the problem as

$$5(n + 6) + 3 = 10n - 2$$
$$5n + (5 \times 30) + 3 = 10n - 2$$
$$5n + 33 = 10n - 2$$
$$33 = 5n - 2$$
$$n = 7$$

■ *Example 2.21*

Problem: If $2x + 5 = 10$, solve for x.

Solution:

$$2x + 5 = 10$$
$$2x = 5$$
$$x = 5/2 = 2.5$$

■ *Example 2.22*

Problem: If $.5x - 1 = -6$, find x.

Solution:

$$.5x - 1 = -6$$
$$.5x = -5$$
$$x = -10$$

Dimensional Analysis

Dimensional analysis is a problem-solving method that uses the fact that any number or expression can be multiplied by 1 without changing its value. It is a useful technique used to check if a problem is set up correctly. When using dimensional analysis to check a math setup, we work with the dimensions (units of measure) only—not with numbers. Unit factors may be made from any two terms that describe the same or equivalent amounts of what we are interested in; for example, we know that 1 inch = 2.54 centimeters. In order to use the dimensional analysis method, we must know how to perform three basic operations.

Basic Operation: Horizontal to Vertical

To complete a division of units, always ensure that all units are written in the same format; it is best to express a horizontal fraction (such as gal/ft^2) as a vertical fraction:

$$\text{gal/ft}^3 \text{ to } \frac{\text{gal}}{\text{ft}^3}$$

$$\text{psi to } \frac{\text{lb}}{\text{in.}^2}$$

The same procedure is applied in the following examples:

$$\text{ft}^3/\text{min becomes } \frac{\text{ft}^3}{\text{min}}$$

$$\text{s/min becomes } \frac{\text{s}}{\text{min}}$$

Basic Operation: Divide by a Fraction

We must know how to divide by a fraction; for example,

$$\frac{\left(\dfrac{\text{lb}}{\text{day}}\right)}{\left(\dfrac{\text{min}}{\text{day}}\right)} \text{ becomes } \frac{\text{lb}}{\text{day}} \times \frac{\text{day}}{\text{min}}$$

In the above, notice that the terms in the denominator were inverted before the fractions were multiplied. This is a standard rule that must be followed when dividing fractions.
   Another example is

$$\frac{\text{mm}^2}{\left(\dfrac{\text{mm}^2}{\text{m}^2}\right)} \text{ becomes } \text{mm}^2 \times \frac{\text{m}^2}{\text{mm}^2}$$

Basic Operation: Cancel or Divide Numerators and Denominators

We must know how to cancel or divide terms in the numerator and denominator of a fraction. After fractions have been rewritten in the vertical form and division by the fraction has been reexpressed as multiplication as shown above, the terms can be canceled (or divided) out. For every term that is canceled in the numerator of a fraction, a similar term must be canceled in the denominator and *vice versa*, as shown below:

$$\frac{kg}{\cancel{d}} \times \frac{\cancel{d}}{min} = \frac{kg}{min}$$

$$\cancel{mm}^{\,2} \times \frac{m^2}{\cancel{mm}^{\,2}} = m^2$$

$$\frac{\cancel{gal}}{min} \times \frac{ft^3}{\cancel{gal}} = \frac{ft^3}{min}$$

Question: How are units that include exponents calculated?

Answer: When written with exponents, such as ft^3, a unit can be left as is or put in expanded form, (ft)(ft)(ft), depending on other units in the calculation. The point is that it is important to ensure that square and cubic terms are expressed uniformly (e.g., sq ft, ft^2, cu ft, ft^3). For dimensional analysis, use of the superscript is preferred.

Let's say we wish to convert 1400 ft^3 to gallons. We will use 7.48 gal/ft^3 in the conversion. The question becomes do we multiply or divide by 7.48? In this instance, it is possible to use dimensional analysis to answer this question of whether we multiply or divide by 7.48.

To determine if the math setup is correct, only the dimensions are used. First, try dividing the dimensions:

$$\frac{ft^3}{gal/ft^3} = \frac{ft^3}{\left(\frac{gal}{ft^3}\right)}$$

Multiply the numerator and denominator to get:

$$\frac{ft^6}{gal}$$

So, by dimensional analysis we determine that if we divide the two dimensions (ft^3 and gal/ft^3) then the units of the answer are ft^6/gal, not gal. It is clear that division is not the right approach to making this conversion.

What would have happened if we had multiplied the dimensions instead of dividing?

$$ft^3 \times \left(gal/ft^3\right) = ft^3 \times \left(\frac{gal}{ft^3}\right)$$

Multiply the numerator and denominator to obtain:

$$\frac{ft^3 \times gal}{ft^3}$$

and cancel the common terms to obtain:

$$\frac{\cancel{ft}^{\,3} \times gal}{\cancel{ft}^{\,3}}$$

Obviously, by multiplying the two dimensions (ft^3 and gal/ft^3), the answer will be in gallons, which is what we want. Thus, because the math setup is correct, we would then multiply the numbers to obtain the number of gallons:

$$(1400 \text{ ft}^3) \times (7.48 \text{ gal/ft}^3) = 10{,}472 \text{ gal}$$

Now let's try another problem with exponents. We wish to obtain an answer in square feet. If we are given the two terms—70 ft^3/s and 4.5 ft/s—is the following math setup correct?

$$(70 \text{ ft}^3/\text{s}) \times (4.5 \text{ ft/s})$$

First, only the dimensions are used to determine if the math setup is correct. By multiplying the two dimensions, we get:

$$(\text{ft}^2/\text{s}) \times (\text{ft/s}) = \frac{\text{ft}^3}{\text{s}} \times \frac{\text{ft}}{\text{s}}$$

Then multiply the terms in the numerators and denominators of the fraction:

$$\frac{\text{ft}^3}{\text{s}} \times \frac{\text{ft}}{\text{s}} = \frac{\text{ft}^4}{\text{s}^2}$$

Obviously, the math setup is incorrect because the dimensions of the answer are not square feet; therefore, if we multiply the numbers as shown above, the answer will be wrong.
  Let's try division of the two dimensions instead:

$$(\text{ft}^3/\text{s}) = \frac{\left(\dfrac{\text{ft}^3}{\text{s}}\right)}{\left(\dfrac{\text{ft}}{\text{s}}\right)}$$

Invert the denominator and multiply to get:

$$= \frac{\text{ft}^3}{\text{s}} \times \frac{\text{s}}{\text{ft}} = \frac{(\text{ft} \times \text{ft} \times \text{ft}) \times \text{s}}{\text{s} \times \text{ft}} = \frac{(\text{ft} \times \text{ft} \times \cancel{\text{ft}}) \times \cancel{\text{s}}}{\cancel{\text{s}} \times \cancel{\text{ft}}} = \text{ft}^2$$

Because the dimensions of the answer are square feet, this math setup is correct; therefore, by dividing the numbers as was done with the units, the answer will also be correct.

$$\frac{70 \text{ ft}^3/\text{s}}{4.5 \text{ ft/s}} = 15.56 \text{ ft}^2$$

■ *Example 2.23*

Problem: We are given two terms, 5 m/s and 7 m², and the answer to be obtained is in cubic meters per second (m³/s). Is multiplying the two terms the correct math setup?
Solution:

$$(m/s) \times (m^2) = \frac{m}{s} \times m^2$$

Multiply the numerators and denominator of the fraction:

$$\frac{m \times m^2}{s} = \frac{m^3}{s}$$

Because the dimensions of the answer are cubic metes per second (m³/s), the math setup is correct; therefore, multiply the numbers to get the correct answer:

$$5 \text{ m/s} \times 7 \text{ m}^2 = 35 \text{ m}^3/\text{s}$$

■ *Example 2.24*

Problem: The flow rate in a water line is 2.3 ft³/s. What is the flow rate expressed as gallons per minute?
Solution: Set up the math problem and then use dimensional analysis to check the math setup:

$$(2.3 \text{ ft}^3/\text{s}) \times (7.48 \text{ gal/ft}^3) \times (60 \text{ s/min})$$

Dimensional analysis can be used to check the math setup:

$$(\text{ft}^3/\text{s}) \times (\text{gal/ft}^3) \times (\text{s/min}) = \frac{\text{ft}^3}{\text{s}} \times \frac{\text{gal}}{\text{ft}^3} \times \frac{\text{s}}{\text{min}} = \frac{\cancel{\text{ft}^3}}{\cancel{\text{s}}} \times \frac{\text{gal}}{\cancel{\text{ft}^3}} \times \frac{\cancel{\text{s}}}{\text{min}} = \frac{\text{gal}}{\text{min}}$$

The math setup is correct as shown above; therefore, this problem can be multiplied out to get the answer in correct units:

$$(2.3 \text{ ft}^3/\text{s}) \times (7.48 \text{ gal/ft}^3) \times (60 \text{ s/min}) = 1032.24 \text{ gal/min}$$

Chapter Review Questions

2.1 If $x - 6 = 2$, what is the value of x?

2.2 If $x + 3 = 4$, what is the value of x?

2.3 If $x - 8 = 17$, what is the value of x?

2.4 If $x + 4 = 12$, what is the value of x?

2.5   If $2x = 50$, what is the value of x?

2.6   If $4x = 8$, what is the value of x?

2.7   If $6x = 15$, what is the value of x?

2.8   If $.4x = 9$, what is the value of x?

2.9   If $3.5x = 15$, what is the value of x?

2.10   If $x + 10 = 2$, what is the value of x?

2.11   An observation is recorded as 8.600 inches, which implies a measurement that is precise and to the nearest _____ inch.

3

Units of Measurement and Conversions

The trouble with measurement is its seeming simplicity.

—**Old adage**

The only man who behaved sensibly was my tailor; he took my measurement anew every time he saw me, while all the rest went on with their old measurements and expected them to fit me.

—**George Bernard Shaw**

Introduction

Management of forest-based biomass feedstock areas (forested land) requires knowledge of the location and current volume of timber resources. The role of measurements is to supply the numerical data required to make produce management decisions. Most of the calculations made in the forest-based biomass energy industry have *units* connected or associated with them. Whereas the number tells us how many, the units tell us what we have. When we measure something, we always have to specify what units we are measuring in. In its broadest sense, measurement consists of the assignment of numbers to measurable properties (Husch et al., 2003).

Symbols and Abbreviations

As in most scientific disciplines, there have been periodic attempts to standardize the nomenclature, symbols, and abbreviations associated with various quantities. Despite this, symbols adopted for one discipline may have entirely different connotations in another scientific field. In this book, we have attempted to employ abbreviations and symbols commonly found in forestry literature in the United States and to abide by typical publisher's standards for abbreviations and symbols of measurement in compliance with SI unit rules and style conversions. Thus, only units of the SI and those units recognized for use with the SI are used in the text to express the values of quantities. Equivalent values in other units are given in parentheses following values in acceptable units only when deemed necessary for the intended audience. In the following, we present some SI unit rules and style conventions typically used today, as well as their standardized abbreviations or symbols and relevance to forest-based biomass energy topics discussed within this text.

SI Unit Rules and Style Conventions*

1. **Abbreviations**—Abbreviations such as sec, cc, or mps are avoided, and only standard unit symbols, prefix symbols, unit names, and prefix names are used.

   Proper: s or second; cm^3 or cubic centimeter; m/s or meter per second

   Improper: sec; cc; mps

2. **Plurals**—Unit symbols are unaltered in the plural.

   Proper: $l = 75$ cm

   Improper: $l = 75$ cms

3. **Punctuation**—Unit symbols are not followed by a period unless at the end of a sentence (with the exception of in. for inch).

   Proper: The length of the bar is 75 cm.

                        The bar is 75 cm long.

   Improper: The bar is 75 cm. long.

4. **Multiplication and division**—A space or center dot is used to signify the multiplication of units. A slash, horizontal line, or negative exponent is used to signify the division of units.

   Proper: The speed of sound is about 344 $m \cdot s^{-1}$ (meters per second).

                        The decay rate of ^{113}Cs is about 21 ms^{-1} (reciprocal milliseconds).

                        m/s

   Improper: The speed of sound is about 344 ms^{-1} (reciprocal milliseconds).

                        The decay rate of ^{113}Cs is about 21 $m \cdot s^{-1}$ (meters per second).

                        m ÷ s

5. **Typeface**—Variables and quantity symbols are in italic type. Unit symbols are in Roman type. Numbers should generally be written in Roman type. These rules apply irrespective of the typeface used in the surrounding text.

   Proper: She exclaimed, *"That dog weighs* 10 kg!"

                        $t = 3$ s (where t is time and s is second)

                        $T = 22$ K (where T is thermodynamic temperature, and K is kelvin)

   Improper: He exclaimed, *"That dog weighs 10 kg!"*

                        t = 3 s (where t is time and s is second)

                        T = 22 K (where T is thermodynamic temperature, and K is kelvin)

6. **Typeface**—Superscripts and subscripts are in italic type if they represent variables, quantities, or running numbers; however, they are in Roman type if they are descriptive.

   Quantity (italic): c_p (specific heat capacity at constant pressure)

   Descriptive (Roman): m_p (mass of a proton)

---

* Adapted from NIST, *SI Unit Rules and Style Conventions*, National Institute of Standards and Technology, Washington, D.C., 2004 (physics.nist.gov/cuu/Units/checklist.html).

7. **Information and units**—Information is not mixed with unit symbols or names.

Proper:	The water content is 20 mL/kg
Improper:	20 mL H_2O/kg
	20 mL of water/kg

8. **Unit symbols and names**—Unit symbols and unit names are not mixed and mathematical operations are not applied to unit names.

Proper:	kg/m^3, $kg \cdot m^{-3}$, *or* kilogram per cubic meter
Improper:	kilogram/m^3, kg/cubic meter, kilogram/cubic meter, kg per m^3, kilogram per $meter^3$

9. **Weight vs. mass**—When the word "weight" is used, the intended meaning must be clear. In science and technology, weight is a force, for which the SI unit is the newton; in commerce and everyday use, weight is usually a synonym for mass, for which the SI unit is the kilogram.

> **DID YOU KNOW?**
>
> The term *weight* is commonly used for *mass*, although this is, strictly speaking, incorrect. Weight refers to the force on an object due to gravity, which varies in time and space and which differs according to location on Earth. Mass refers to the amount of matter in an object. Because it is important to know whether mass or force is being measured, the SI has established two units: the kilogram for mass and the newton for force.

10. **Object and quantity**—An object and any quantity describing the object are distinguished.

Proper:	A body of mass 5 g
Improper:	A mass of 5 g

With these provisions, rules, and conventions in mind, the standardized abbreviations and symbols used herein are presented in Table 3.1. SI units are divided into two kinds: base and derived. Examples of *base units* include length (meter, m), mass (kilogram, kg), time (second, s), and electric current (ampere, A). The symbols for *derived* units are obtained by means of the mathematical operations of multiplication and division. Examples of derived units include area (square meter, m^2), volume (cubic meter, m^3), velocity (meter per second, m/s), density (kilogram per cubic meter, kg/m^3), specific volume (cubic meter per kilogram, m^3/kg), and concentration (mol/m^3). Table 3.2 shows non-SI units accepted for use with the International System of Units.

> **DID YOU KNOW?**
>
> The *nominal scale* is used for numbering tree species or forest types in a stand. The *ordinal scale* is used to express degree, quality, or position in a series, such as first, second, and third. The *interval scale* includes a series of graduations marked off at uniform intervals to form a reference point of fixed magnitude. The *ratio scale*, unlike the interval scale, always has an absolute zero of origin (Husch et al., 2003).

TABLE 3.1

Abbreviations and Symbols

Abbreviation or Symbol	Meaning
B, b	Cross-sectional areas of logs or bolts
BA, ba	Basal area
BAF, baf	Basal-area factor (point sampling)
bd ft	Board foot
cd	Cord
CFI	Continuous forest inventory
cu ft or ft^3	Cubic foot
D, d	Tree or log diameter (at any specified point)
DBH	Diameter breast height
DIB	Diameter inside bark
DOB	Diameter outside bark
t	Frequency (statistical notation)
GIS	Geographic information system
GPS	Global positioning system
H, h	Height
L, l	Log or bolt length
M	Thousand
MBF	Thousand board feet
N, n	Number of (statistical notation)
RF	Representative fraction
sp gr	Specific gravity
V, v	Volume

Scales of Measurement

Measurement is the assignment of numerals to objects or events according to rules. Scales of measurement are expressions that typically refer to the theory of scale types; that is, all measurement in science has been conducted using four different types of scales known as *nominal, ordinal, interval,* and *ratio* (Stevens, 1946). These classifications are shown in Table 3.3.

TABLE 3.2

Non-SI Units Accepted for Use with the International System of Units

Unit	Symbol	Equivalence in SI Units
Minute	min	1 min = 60 s
Hour	hr	1 hr = 60 min = 3600 s
Day	d	1 d = 24 hr = 86,400 s
Degree (angular)	°	$1° = (\pi/180)$ rad
Minute (angular)	′	$1′ = (1/60)° = (\pi/10,800)$ rad
Second (angular)	″	$1″ = (1/60)′ = (\pi/648,000)$ rad
Liter	L	$1 L = 1 dm^3 = 10^{-3} m^3$
Metric ton	t	$1 t = 10^3 kg$

TABLE 3.3

Classification of Scale of Measurement

Scale Type	Permissible Statistics	Admissible Scale Transformation	Mathematical Structure
Nominal (also known as categorical)	Mode, chi-square, counting, number of cases	One to one; equality (=)	Standard set structure (unordered)
Ordinal	Median, percentiles, order correlation	Monotonic increasing; order (<)	Totally ordered set
Interval	Mean, standard deviation, regression, analysis of variance	Positive linear (affine)	Affine line
Ratio	All statistics permitted for interval scales: geometric mean, harmonic mean, coefficient of variation, logarithms	Positive similarities	Field

Units of Measurement

A basic knowledge of units of measurement and how to use them and convert them is essential. Forest-based biomass energy specialists should be familiar both with the U.S. Customary System (USCS), or English System, and the International System of Units (SI). Some of the important units are summarized in Table 3.4. In the study of forest-based biomass energy operations (and in actual practice), it is quite common to encounter both extremely large quantities and extremely small ones. To describe quantities that may take on such large or small values, it is useful to have a system of prefixes for the units. Some of the more important prefixes are presented in Table 3.5.

Conversion Factors

Sometimes we have to convert between different units. Suppose a 60-inch piece of wood is attached to a 6-foot piece of wood. How long are they together? Obviously, we cannot find the answer to this question by adding 60 to 6. The reason is that the two figures are given in different units. Before we can add the two numbers, we have to convert one of

TABLE 3.4

Commonly Used Units

Quantity	SI Units	USCS Units
Length	Meter	Feet (ft)
Mass	Kilogram	Pound (lb)
Temperature	Celsius	Fahrenheit (F)
Area	Square meter	Square foot (ft^2)
Volume	Cubic meter	Cubic foot (ft^3)
Energy	Kilojoule	British thermal unit (Btu)
Power	Watt	Btu/hr
Velocity	Meter/second	Mile/hour (mi/hr)

TABLE 3.5

Common Prefixes

Quantity	Prefix	Abbreviation
10^{-12}	pico	p
10^{-9}	nano	n
10^{-6}	micro	μ
10^{-3}	milli	m
10^{-2}	centi	c
10^{-1}	deci	d
10	deca	da
10^{2}	hecto	h
10^{3}	kilo	k
10^{6}	mega	M
10^{9}	giga	G
10^{12}	tera	T

them to the units of the other. Then, when we have two numbers in the same units, we can add them. In order to perform this conversion, we need a *conversion factor*. In this case, we have to know how many inches make up a foot: 12 inches = 1 foot. Knowing this, we can perform the calculation in two steps as follows:

1. 60 inches is really 60/12 = 5 feet
2. 5 feet + 6 feet = 11 feet

From the example above, it can be seen that a conversion factor changes known quantities in one unit of measure to an equivalent quantity in another unit of measure. When making the conversion from one unit to another, we must know two things:

1. The exact number that relates the two units
2. Whether to multiply or divide by that number

When making conversions, confusion over whether to multiply or divide is common; on the other hand, the number that relates the two units is usually known and, thus, is not a problem. Understanding the proper methodology (mechanics) to use for various operations requires practice and common sense. Along with using the proper mechanics (in addition to practice and common sense) when making conversions, probably the easiest and fastest way to convert units is to use a conversion table.

The simplest conversion requires that the measurement be multiplied or divided by a constant value. In a forest- or pulp-waste treatment drying bed, for example, if the depth of biosolids (organic waste materials) is 0.85 feet, multiplying by 12 inches per foot converts the measured depth to inches (10.2 inches). Likewise, if the depth of the waste residue is measured as 16 inches, dividing by 12 inches per foot converts the depth measurement to feet (1.33 feet). Table 3.6 lists many of the conversion factors used in water/wastewater treatment. Note that Table 3.6 is designed with a unit of measure in the left and right columns and a constant (conversion factor) in the center column.

Key Point: To convert in the opposite direction (e.g., inches to feet rather than feet to inches), divide by the factor rather than multiply.

TABLE 3.6

Conversion Table

Multiply	by	To Get
Feet	12	Inches
Yards	3	Feet
Yards	36	Inches
Inches	2.54	Centimeters
Meters	3.3	Feet
Meters	100	Centimeters
Meters	1000	Millimeters
Square yards	9	Square feet
Square feet	144	Square inches
Acres	43,560	Square feet
Cubic yards	27	Cubic feet
Cubic feet	1728	Cubic inches
Cubic feet (water)	7.48	Gallons
Cubic feet (water)	62.4	Pounds
Acre-feet	43,560	Cubic feet
Gallons (water)	8.34	Pounds
Gallons (water)	3.785	Liters
Gallons (water)	3785	Milliliters
Gallons (water)	3785	Cubic centimeters
Gallons (water)	3785	Grams
Liters	1000	Milliliters
Days	24	Hours
Days	1440	Minutes
Days	86,400	Seconds
Million gallons/day	1,000,000	Gallons/day
Million gallons/day	1.55	Cubic feet/second
Million gallons/day	3.069	Acre-feet/day
Million gallons/day	36.8	Acre-inches/day
Million gallons/day	3785	Cubic meters/day
Gallons/minute	1440	Gallons/day
Gallons/minute	63.08	Liters/minute
Pounds	454	Grams
Grams	1000	Milligrams
Pressure (psi)	2.31	Head, ft (water)
Horsepower	33,000	Foot-pounds/minute
Horsepower	0.746	Kilowatts
To Get	**by**	**Divide**

General Conversions

Several common equivalents used when converting English to metric units and *vice versa* are provided below:

Converting English units to metric units

1 in. or 1000 mils	=	2.5400 cm
1 ft or 12 in.	=	30.4800 cm

1 yd or 3 ft	=	0.9144 m
1 U.S. statute mile or 5280 ft	=	1.6093 km
1 acre or 43,560 sq ft	=	0.4047 ha
1 ft^3 or 1728 in.^{3}	=	0.0283 m^3

Converting metric units to English units

1 cm or 10 mm	=	0.3937 in.
1 dm or 10 cm	=	3.9370 in.
1 m or 10 dm	=	39.3700 in.
1 km or 1000 m	=	0.6214 U.S. statute mile
1 ha or 10,000 m^2	=	2.4710 acres
1 m^3 or 1,000,000 cm^3	=	35.3147 ft^3

Weight, Concentration, and Flow

Using Table 3.6 to convert from one unit expression to another and *vice versa* is good practice. When we make conversions to solve process problems in forest-based biomass energy operations involving black liquor and other required computations, we must be very familiar with the conversion calculations based on the relationships among weight, flow or volume, and concentration. The basic relationship is

$$Weight = Concentration \times (Flow\ or\ Volume) \times Factor$$

Table 3.7 summarizes the weight, volume, and concentration calculations. With practice, many of these calculations become second nature to specialists; the calculations are important relationships and are used often in biorefinery process control calculations, so on-the-job practice is highly likely.

Key Point: *Density* (also called *specific weight*) is mass per unit volume and may be indicated as lb/ft^3, lb/gal, g/mL, g/m^3. If we take a fixed volume container, fill it with a fluid, and weigh it, we can determine the density of the fluid (after subtracting the weight of the container).

Woody Biomass Utilization Conversion Factors

Here are a few woody biomass conversion factors and bone-dry ton conversions (see Table 3.8) that are commonly used by natural resource managers:

1. The gasoline market in the United States is about 118 billion gallons per year, or about 323 million gallons per day.
2. The theoretical limit of conversion of ethanol from wood is 120 gallons per ton. A high but achievable figure is about 80 gallons per ton.
3. If the 370 million tons of biomass available (dry weight) were all converted to ethanol, they would yield 29.6 billion gallons of ethanol.

TABLE 3.7

Weight, Volume, and Concentration Calculations

To Calculate	Formula
Pounds	Concentration (mg/L) × Tank Volume (MG) × 8.34 lb/MG/mg/L
Pounds/day	Concentration (mg/L) × Flow (MGD) × 8.34 lb/MG/mg/L
Million gallons/day	$\dfrac{\text{Quantity (lb/day)}}{\text{Concentration (mg/L)} \times 8.34 \text{ lb/MG/mg/L}}$
Milligrams/liter	$\dfrac{\text{Quantity (lb)}}{\text{Tank Volume (MG)} \times 8.34 \text{ lb/MG/mg/L}}$
Kilograms/liter	Concentration (mg/L) × Volume (MG) × 3.785 lb/MG/mg/L
Kilograms/day	Concentration (mg/L) × Flow (MGD) × 3.785 lb/MG/mg/L
Pounds/dry ton	Concentration (mg/kg) × 0.002 lb/dry ton/mg/kg

4. Ethanol is less energy dense than gasoline, as it requires 1.6 gallons of ethanol to produce the same energy as a gallon of gasoline (29.6 billion gallons ÷ 1.6 gallons = 18.5 billion gallons of "equivalent" gasoline).

5. It works out that 370 million tons of biomass could be converted to 57 days' worth of transportation fuel for the United States (18.5 billion gallons ÷ .323 billion gallons per day = 57.28 days).

> **DID YOU KNOW?**
>
> To put 370 million tons of biomass into perspective, the United States currently consumes about 300 million tons of wood per year.

Forest Fuel/Biomass Conversion Factors

1 green ton (GT) of wood chips = 2000 pounds (not adjusted for moisture)

1 bone-dry ton (BDT) of wood chips = 2000 dry pounds (assumes no moisture content)

1 bone-dry unit (BDU) of wood chips = 2400 dry pounds (assumes no moisture content)

1 unit of wood chips = 200 cubic feet

1 BDT of wood chips = 2.0 GT (assuming 50% moisture content)

1 unit of wood chips = 1.0 BDT wood chips

1 ccf (hundred cubic feet) roundwood = 1.0 BDU wood chips

1 ccf roundwood (logs) = 1.2 BDT wood chips

1 ccf roundwood (logs) = 1.2 units of wood chips

1 ccf roundwood (logs) = 1.2 cords roundwood (85 ft^3 wood per cord)

1 board foot (BF) = lumber measure equivalent to 12-in. by 12-in. by 1-in. thick

1 MBF (thousand board feet) = 1000 BF

1 GT of logs = 160 BF of lumber

6 GT of logs = 1 MBF

One standard wood chip van carries 25 green tons, or approximately 12.5 BDT at 50% moisture content.

TABLE 3.8

Bone-Dry Tons Conversion Table

Wood Density						
Diameter Breast Height (DBH) (in.)	Height (ft)	Stem Volume (ft³)	Stem Weight (bone-dry pounds)	Crown and Tip Weight (bone-dry pounds)	Total Weight	Bone-Dry Tons (BDT)
Douglas Fir						
2	10	0.1	2.5	—	2.5	0.00
4	30	1	25	40	65	0.03
6	50	4	100	64	164	0.08
8	70	10	250	97	347	0.17
10	90	20	500	137	637	0.32
12	110	35	875	184	1059	0.53
16	120	64	1600	301	1901	0.95
20	120	95	2375	482	2857	1.43
24	120	130	3250	725	3975	1.99
28	130	190	4750	1030	5780	2.8
Ponderosa Pine						
2	10	0.1	2.5	—	2.5	0.00
4	30	1	25	35	60	0.03
6	50	3.7	92.5	66	158.5	0.08
8	70	9	225	113	338	0.17
10	90	17.9	447.5	177	624.5	0.31
12	110	31.1	777.5	259	1036.5	0.52
16	120	58.3	1457.5	478	1935.5	0.97
20	120	88	2200	774	2974	1.49
24	120	123.1	3077.5	1150	4277.5	2.11
28	130	178.7	4467.5	1620	6087.5	3.0
White Fir						
2	10	0.1	2.5	—	2.5	0.00
4	30	1.2	30	45	75	0.04
6	50	4.5	112.5	77	189.5	0.09
8	70	11.1	277.5	120	397.5	0.20
10	90	22.3	557.5	175	732.5	0.37
12	110	39.2	980	242	1222	0.61
16	120	71.2	1780	422	2202	1.10
20	120	103.7	2592.5	637	3229.5	1.61
24	120	141	3525	852	4377	2.19
28	130	202.4	5060	1090	6150	3.08

Sources: Walters, D.K. et al., *Equations and Tables Predicting Gross Total Stem Volumes in Cubic Feet for Six Major Conifers of Southwest Oregon*, Res. Bull. 50, Forest Research Laboratory, Oregon State University, Corvallis, OR, 1985; Snell, J.A. and Brown, J.K., *Handbook for Predicting Residue Weights of Pacific NW Conifers*, General Technical Report PNW-103, USDA Forest Service, Pacific Northwest Forest and Range Experiment Station, Portland, OR, 1980; Hartman, D.A. et al., *Conversion Factors for the Pacific Northwest Forest Industry: Converting Forest Growth to Forest Products*, Institute of Forest Products, Seattle, WA, 1976.

When woody biomass is utilized in a commercial-scale (10+ megawatt electrical output) power generation facility, the following energy output rules of thumb apply:

1 BDT woody fuel = 10,000 pounds of steam

1 megawatt hour (MWh) = 10,000 pounds of steam

1 MW = 1000 horsepower

1 MW = power for approximately 750 to 1000 homes

Conversion Examples

Note: Use Table 3.6 and Table 3.7 to make the conversions indicated in the following example problems. Other conversions are presented in appropriate sections of the text.

■ *Example 3.1*

Convert cubic feet to gallons:

$$\text{Gallons} = \text{Cubic feet } (\text{ft}^3) \times 7.48 \text{ gal/ft}^3 \tag{3.1}$$

Problem: How many gallons of black liquor can be pumped to a holding tank that has 3600 cubic feet of volume available?
Solution:

$$\text{Gallons} = 3600 \text{ ft}^3 \times 7.48 \text{ gal/ft}^3 = 26,928 \text{ gal}$$

■ *Example 3.2*

Convert gallons to cubic feet:

$$\text{Cubic feet} = \frac{\text{Gallons}}{7.48 \text{ gal/ft}^3} \tag{3.2}$$

Problem: How many cubic feet of biosolids are removed when 18,200 gallons are withdrawn?
Solution:

$$\text{Cubic feet} = \frac{18,200 \text{ gal}}{7.48 \text{ gal/ft}^3} = 2433 \text{ ft}^3$$

■ *Example 3.3*

Convert gallons to pounds:

$$\text{Pounds} = \text{Gallons} \times 8.34 \text{ lb/gal} \tag{3.3}$$

Problem: If 1650 gallons of solids are removed from a primary settling tank, how many pounds of solids are removed?
Solution:

$$\text{Pounds} = 1650 \text{ gal} \times 8.34/\text{gal} = 13,761 \text{ lb}$$

■ *Example 3.4*

Convert pounds to gallons:

$$\text{Gallons} = \frac{\text{Pounds}}{8.34 \text{ lb/gal}} \tag{3.4}$$

Problem: How many gallons of water are required to fill a tank that holds 7540 pounds of water?
Solution:

$$\text{Gallons} = \frac{7540 \text{ lb}}{8.34 \text{ lb/gal}} = 904 \text{ gal}$$

Key Point: For some operations, concentrations in milligrams per liter or parts per million determined by laboratory testing must be converted to quantities of pounds, kilograms, pounds per day, or kilograms per day.

■ *Example 3.5*

Convert milligrams/liter to pounds:

$$\text{Pounds} = \text{Concentration (mg/L)} \times \text{Volume (MG)} \times 8.34 \text{ lb/MG/mg/L} \tag{3.5}$$

Problem: The solids concentration in the black liquor holding tank is 2580 mg/L. The secondary holding tank volume is 0.95 MG. How many pounds of solids are in the tank?
Solution:

$$\text{Pounds} = 2580 \text{ mg/L} \times 0.95 \text{ MG} \times 8.34 \text{ lb/MG/mg/L} = 20{,}441.3 \text{ lb}$$

■ *Example 3.6*

Convert million gallons per day (MGD) to gallons per minute (gpm):

$$\text{Flow (gpm)} = \frac{\text{Flow (MGD)} \times 1{,}000{,}000 \text{ gal/MG}}{1440 \text{ min/day}} \tag{3.6}$$

Problem: The current flow rate is 5.55 MGD. What is the flow rate in gallons per minute?
Solution:

$$\text{Flow} = \frac{5.5 \text{ MGD} \times 1{,}000{,}000 \text{ gal/MG}}{1440 \text{ min/day}} = 3854 \text{ gpm}$$

■ *Example 3.7*

Convert gallons per minute (gpm) to million gallons per day (MGD):

$$\text{Flow (MGD)} = \frac{\text{Flow (gpm)} \times 1440 \text{ min/day}}{1{,}000{,}000 \text{ gal/MG}} \tag{3.7}$$

Problem: The flow meter indicates that the current flow rate is 1,469 gpm. What is the flow rate in MGD?

Solution:

$$\text{Flow} = \frac{1469 \text{ gpm} \times 1440 \text{ min/day}}{1{,}000{,}000 \text{ gal/MG}} = 2.12 \text{ MGD (rounded)}$$

■ *Example 3.8*

Convert flow in cubic feet per second (cfs) to million gallons per day (MGD):

$$\text{Flow (MGD)} = \frac{\text{Flow (cfs)}}{1.55 \text{ cfs/MG}} \tag{3.8}$$

Problem: The flow in a channel is determined to be 3.89 cubic feet per second (cfs). What is the flow rate in million gallons per day (MGD)?

Solution:

$$\text{Flow} = \frac{3.89 \text{ cfs}}{1.55 \text{ cfs/MG}} = 2.5 \text{ MGD}$$

■ *Example 3.9*

Problem: The water in a tank weighs 675 pounds. How many gallons does the tank hold?

Solution: Water weighs 8.34 lb/gal; therefore,

$$\text{Gallons} = \frac{675 \text{ lb}}{8.34 \text{ lb/gal}} = 80.9 \text{ gal}$$

■ *Example 3.10*

Specific gravity is the ratio of the density of a substance to that of a standard material under standard conditions of temperature and pressure. Specific gravity of a chemical, for example, can be determined by dividing the weight of the chemical by the weight of water:

$$\text{Specific Gravity} = \frac{\text{Weight of Chemical}}{\text{Weight of Water}} \tag{3.9}$$

Problem: A liquid chemical weighs 62 lb/ft^3. How much does a 5-gallon can of it weigh?

Solution: Solve for specific gravity; get lb/gal; multiply by 5:

$$\text{Specific Gravity} = \frac{62 \text{ lb/ft}^3}{62.4 \text{ lb/ft}^3} = .99$$

$$.99 = \frac{\text{Weight of Chemical}}{8.34 \text{ lb/gal}}$$

$$.99 \times 8.34 \text{ lb/gal} = \text{Weight of Chemical} = 8.26 \text{ lb/gal}$$

$$8.26 \text{ lb/gal} \times 5 \text{ gal} = 41.3 \text{ lb}$$

■ *Example 3.11*

Problem: A wooden piling with a diameter of 16 inches and a length of 16 feet weighs 50 lb/ft^3. If it is inserted vertically into a body of water, what vertical force is required to hold it below the surface of the water?

Solution: If this piling had the same weight as water, it would rest just barely submerged. Find the difference between its weight and that of the same volume of water. That is the weight required to keep it down.

$$62.4 \text{ lb/ft}^3 \text{ (water)} - 50.0 \text{ lb/ft}^3 \text{ (piling)} = 12.4 \text{ lb/ft}^3$$

$$\text{Volume of piling} = .785 \times 1.332 \times 16 \text{ ft} = 22.21 \text{ ft}^3$$

$$12.4 \text{ lb/ft}^3 \times 22.21 \text{ ft}^3 = 275.4 \text{ lb to hold piling below surface of water}$$

■ *Example 3.12*

Problem: A liquid chemical with a specific gravity of 1.22 is pumped at a rate of 40 gpm. How many pounds per day are being delivered by the pump?

Solution: Solve for pounds pumped per minute; change to lb/day:

$$8.34 \text{ lb/gal water} \times 1.22 = 10.2 \text{ lb/gal}$$

$$40 \text{ gal/min} \times 10.2 \text{ lb/gal} = 408 \text{ lb/min}$$

$$408 \text{ lb/min} \times 1440 \text{ min/day} = 587,520 \text{ lb/day}$$

■ *Example 3.13*

Problem: A cinder block weighs 70 pounds in air. When immersed in water, it weighs 40 pounds. What are the volume and specific gravity of the cinder block?

Solution: The cinder block displaces 30 pounds of water; solve for cubic feet of water displaced (equivalent to volume of cinder block).

$$\frac{30 \text{ lb of water displaced}}{62.4 \text{ lb/ft}^3} = .48 \text{ ft}^3 \text{ of water displaced}$$

$$\text{Cinder Block Volume} = .48 \text{ ft}^3$$

$$\text{Cinder Block Weight} = 70 \text{ lb}$$

$$\text{Cinder Block Density} = \frac{70 \text{ lb}}{.48 \text{ ft}^3} = 145.8 \text{ lb/ft}^3$$

$$\text{Specific Gravity} = \frac{\text{Density of Cinder Block}}{\text{Density of Water}}$$

$$= \frac{145.8 \text{ lb/ft}^3}{62.4 \text{ lb/ft}^3} = 2.34$$

Temperature Conversions

The formulas used for Fahrenheit and Celsius temperature conversions are

$$°C = 5/9 \; (°F - 32) \tag{3.10}$$

$$°F = 9/5 \; (°C) + 32 \tag{3.11}$$

The difficulty arises when one tries to recall these formulas from memory. Probably the easiest way to recall these important formulas is to remember three basic steps for both Fahrenheit and Celsius conversions:

1. Add 40°.
2. Multiply by the appropriate fraction (5/9 or 9/5).
3. Subtract 40°.

Obviously, the only variable in this method is the choice of 5/9 or 9/5 in the multiplication step. To make the proper choice, you must be familiar with the two scales. The freezing point of water is 32° on the Fahrenheit scale and 0° on the Celsius scale. The boiling point of water is 212° on the Fahrenheit scale and 100° on the Celsius scale. What does all this mean? Notice that for the same temperature higher numbers are associated with the Fahrenheit scale and lower numbers with the Celsius scale. This important relationship will help you decide whether to multiply by 5/9 or 9/5. Let's look at a few conversion problems to see how the three-step process works.

■ *Example 3.14*

Problem: Convert 240°F to Celsius.
Solution: Use the three-step process as follows:

1. Add 40°:

$$240° + 40° = 280°$$

2. Multiply 280° by either 5/9 or 9/5. Because the conversion is to the Celsius scale, we will be moving to a number *smaller* than 280. Through reason and observation, obviously, if 280 were multiplied by 9/5, the result would be almost the same as multiplying by 2, which would double 280 rather than make it smaller. If we multiply by 5/9, the result will be about the same as multiplying by 1/2, which would cut 280 in half. Because in this problem we wish to move to a smaller number, we should multiply by 5/9:

$$(5/9) \times 380° = 156.0°C$$

3. Subtract 40°:

$$156.0°C - 40.0°C = 116.0°C$$

Thus, 240°F = 116.0°C.

■ *Example 3.15*

Problem: Convert 22°C to Fahrenheit.
Solution:

1. Add 40°:

$$22° + 40° = 62°$$

2. Multiply 280° by either 5/9 or 9/5. Because we are converting from Celsius to Fahrenheit, we are moving from a smaller to a larger number, so 9/5 should be used in the multiplication:

$$(9/5) \times 62° = 112°$$

3. Subtract 40°:

$$112° - 40° = 72°$$

Thus, 22°C = 72°F.

Obviously, knowing how to make these temperature conversion calculations is useful; however, in day-to-day operations you may wish to use a temperature conversion table.

Specific Gravity and Density

Earlier, we defined *specific gravity* as the ratio of the density of a substance to that of a standard material under standard conditions of temperature and pressure, and we presented a few specific gravity and density conversion example problems. The specific gravity of water is 1.0. Any substance with a density greater than that of water will have a specific gravity greater than 1.0, and any substance with a density less than that of water sill have a specific gravity less than 1.0. Specific gravity can be used to calculate the weight of a gallon of liquid chemical:

Chemical Weight (lb/gal) = Water Specific Gravity × Chemical Specific Gravity   (3.12)

■ *Example 3.16*

Problem: The label states that the ferric chloride solution has a specific gravity of 1.58. What is the weight of 1 gallon of ferric chloride solution?
Solution:

Ferric Chloride = 8.34 lb/gal × 1.58 lb/gal = 13.2 lb/gal

■ *Example 3.17*

Problem: If we say that the density of gasoline is 43 lb/ft³, what is the specific gravity of gasoline?

Solution: The specific gravity of gasoline is the comparison (or ratio) of the density of gasoline to that of water:

$$\text{Specific Gravity} = \frac{\text{Density of Gasoline}}{\text{Density of Water}} = \frac{43 \text{ lb/ft}^3}{62.4 \text{ lb/ft}^3} = 0.69$$

Key Point: Because the specific gravity of gasoline is less than 1.0 (lower than the specific gravity of water), it will float in water. If the specific gravity of gasoline were greater than the specific gravity of water, it would sink.

Flow

Flow is expressed in many different terms. The most commonly used flow units are as follows:

gpm = gallons per minute
cfs = cubic feet per second
gpd = gallons per day
MGD = million gallons per day

When converting flow rates, the most common flow conversions are 1 cfs = 448 gpm and 1 gpm =1440 gpd. To convert gallons per day to MGD, divide the gpd by 1,000,000; for example, let's convert 150,000 gallons to MGD.

$$\frac{150,000 \text{ gpd}}{1,000,000} = 0.150 \text{ MGD}$$

In some instances, flow is given in MGD but is needed in gpm. To make the conversion from MGD to gpm, two steps are required:

1. Convert the gpd by multiplying by 1,000,000.
2. Convert to gpm by dividing by the number of minutes in a day (1440 min/day).

■ *Example 3.18*

Problem: Convert 0.135 MGD to gpm.
Solution: First convert the flow in MGD to gpd:

$$0.135 \text{ MGD} \times 1,000,000 = 135,000 \text{ gpd}$$

Now convert to gpm by dividing by the number of minutes in a day:

$$\frac{135,000 \text{ gpd}}{1440 \text{ min/day}} = 93.8, \text{ or } 94 \text{ gpm}$$

When determining flow through a pipeline, channel, or stream, we use the following equation:

$$Q = V \times A \tag{3.13}$$

where Q = quantity (cubic feet per second, cfs), V = velocity (feet per second, ft/s), and A = area (square feet, ft^2).

■ *Example 3.19*

Problem: Find the flow in cubic feet per second in an 8-inch line if the velocity is 3 ft/s.
Solution:

1. Determine the cross-sectional area of the line in square feet. Start by determining the radius of the line. The diameter is 8 inches; therefore, the radius is 4 inches, which is 4/12 of a foot, or 0.33 feet.
2. Find the area in square feet:

$$A = \pi r^2$$

$$A = \pi \times (0.33 \text{ ft})^2 = \pi \times 0.109 \text{ ft}^2 = 0.342 \text{ ft}^2$$

3. Now determine the flow:

$$Q = V \times A$$

$$Q = 3 \text{ ft/s} \times 0.342 \text{ ft}^2 = 1.03 \text{ cfs}$$

■ *Example 3.20*

Problem: Find the flow in gpm when the total flow for the day is 75,000 gpd.
Solution:

$$\frac{75,000 \text{ gpd}}{1440 \text{ min/day}} = 52 \text{ gpm}$$

■ *Example 3.21*

Problem: Find the flow in gpm when the flow is 0.45 cfs.
Solution:

$$0.45 \times \frac{\text{cfs}}{1} \times \frac{448 \text{ gpm}}{1 \text{ cfs}} = 202 \text{ gpm}$$

Horsepower and Energy Costs

Horsepower is a common expression for power. One horsepower is equal to 33,000 foot-pounds (ft-lb) of work per minute. This value is determined, for example, when selecting the pump or combination of pumps necessary to ensure adequate pumping capacity. Pumping capacity depends on the flow rate desired and the feet of head against which the pump must pump (also known as the effective height). The basic concept from which the horsepower calculation is derived is the concept of work. *Work* involves the operation of a force (lb) over a specific distance (ft). The *amount of work* accomplished is measured in foot-pounds:

$$ft \times lb = ft\text{-}lb \tag{3.14}$$

The *rate of doing work* (*power*) involves a time factor. Originally, the rate of doing work or power compared the power of a horse to that of a steam engine. The rate at which a horse could work was determined to be about 550 ft-lb/s (or 33,000 ft-lb/min). This rate has become the definition of the standard unit known as horsepower:

$$\text{Horsepower (hp)} = \frac{\text{Power (ft-lb/min)}}{33,000 \text{ ft-lb/min/hp}} \tag{3.15}$$

The horsepower calculation can be modified as shown below.

Water Horsepower

The amount of power required to move a given volume of water a specified total head is known as water horsepower (Whp):

$$\text{Water Horsepower (Whp)} = \frac{\text{Pump Rate (gpm)} \times \text{Total Head (ft)} \times 8.34 \text{ lb/gal}}{33,000 \text{ ft-lb/min/hp}} \tag{3.16}$$

■ *Example 3.22*

Problem: A pump must deliver 1210 gpm to a total head of 130 feet. What is the required water horsepower?
Solution:

$$\text{Water Horsepower (Whp)} = \frac{1210 \text{ gpm} \times 130 \text{ ft} \times 8.34 \text{ lb/gal}}{33,000 \text{ ft-lb/min/hp}} = 40$$

Brake Horsepower

Brake horsepower (bhp) refers to the horsepower supplied to the pump from the motor. As power moves through the pump, additional horsepower is lost from slippage and friction of the shaft and other factors; thus, pump efficiencies range from about 50 to 85%, and pump efficiency must be taken into account.

$$\text{Brake Horsepower (bhp)} = \frac{\text{Whp}}{\text{Pump Efficiency (\%)}} \tag{3.17}$$

■ *Example 3.23*

Problem: Under the specified conditions, the pump efficiency is 73%. If the required water horsepower is 40 hp, what is the required brake horsepower?
Solution:

$$\text{Brake Horsepower (bhp)} = \frac{40 \text{ Whp}}{0.73} = 55$$

Motor Horsepower

Motor horsepower (mhp) is the horsepower the motor must generate to produce the desired brake and water horsepower:

$$\text{Motor Horsepower (mhp)} = \frac{\text{Brake Horsepower (bhp)}}{\text{Motor Efficiency (\%)}} \tag{3.18}$$

■ *Example 3.24*

Problem: A motor is 93% efficient. What is the required motor horsepower when the required brake horsepower is 49.0 bhp?
Solution:

$$\text{Motor Horsepower (mhp)} = \frac{49 \text{ bhp}}{0.93} = 53$$

Electrical Power

On occasion, electrical power calculations must be made—especially regarding electrical energy required or consumed during a period of time. To accomplish this, horsepower is converted to electrical energy (kilowatts), then multiplied by the hours of operation to obtain kilowatt-hours (kWh):

$$\text{Kilowatt-Hours} = \text{Horsepower} \times 0.746 \text{ kW/hp} \times \text{Operating Time (hr)} \tag{3.19}$$

■ *Example 3.25*

Problem: A 60-horsepower motor operates at full load 12 hours per day, 7 days a week. How many kilowatts of energy does it consume per day?
Solution:

$$\text{Kilowatt-Hours/day} = 60 \text{ hp} \times 0.746 \text{ kW/hp} \times 12 \text{ hr/day} = 537 \text{ kWh/day}$$

The cost per kilowatt-hour may be determined for any given period of operation:

$$\text{Cost} = \text{Power Required/day} \times \text{kWh/day} \times \text{Days/Period} \times \text{Cost/kWh} \tag{3.20}$$

■ *Example 3.26*

Problem: A 60-horsepower motor requires 458 kWh/day. The pump is in operation every day. The current cost of electricity is $0.0328/kWh. What is the yearly electrical cost for this pump?
Solution:

$$\text{Cost} = 458 \text{ kWh/day} \times 365 \text{ days/yr} \times \$0.0328/\text{kWh} = \$5483.18/\text{yr}$$

Fundamental Measurements

Forest-based biomass feedstock is generally derived from large parcels of land. Moreover, biorefinery operating systems and other processes are described in terms of length, time, weight, circumference, area, and volume measurements. Parameters such as the circumference or perimeter, area, or volume of the tank or channel are often part of information necessary to determine the result of various operational calculations. Many biorefinery process calculations require computation of surface areas. To aid in performing these calculations, the following definitions are provided:

Area is the area of an object, measured in square units.

Base is a term used to identify the bottom leg of a triangle, measured in linear units.

Circumference is the distance around an object, measured in linear units. When determined for shapes other than circles, it may be called the *perimeter* of the figure, object, or landscape.

Cubic units are measurements used to express volume, cubic feet, cubic meters, etc.

Depth is the vertical distance from the bottom of the tank to the top. Depth is normally measured in terms of liquid depth and given in terms of sidewall depth (SWD), measured in linear units.

Diameter is the distance from one edge of a circle to the opposite edge passing through the center; it is measured in linear units.

Height is the vertical distance from one end of an object to the other, measured in linear units.

Length is the distance from one end of an object to the other, measured in linear units.

Linear measurements consist of determining the length of a line extending from one point to another.

Linear units are measurements used to express distances, such as feet, inches, meters, or yards.

Pi (π) is a number used in calculations involving circles, spheres, or cones; $\pi = 3.14$.

Radius is the distance from the center of a circle to the edge, measured in linear units.

Sphere is a container shaped like a ball.

Square units are measurements used to express area, such as square feet, square meters, or acres.

Volume is the capacity of a unit (how much it will hold), measured in cubic units (cubic feet, cubic meters) or in liquid volume units (gallons, liters, million gallons).

Width is the distance from one side of the tank to the other, measured in linear units.

Linear Measurements

Linear measurement, or length measurement, consists of determining the length of a line extending from one point to another. Because the configurations of objects vary, a length might be the straight-line distance between two points on an object or the curved or irregular line distance between two points on an object (e.g., the periphery of the cross-section of a tree stem). The various length measurements taken in the practice of forest-based biomass operations include determination of the diameter or height of a tree, the length of a log, the width of the image of a tree crown on an aerial photograph, the length of the boundary of a tract of land, and so on.

Time Measurements

In forest-based biomass operations, time is utilized to denote position in a continuum of time at which some event took place, to measure the duration of a given event, or to determine the speed or rate at which an event or physical change occurred. One of the most practical time measurements is the determination of tree and stand age.

Perimeter and Circumference Measurements

On occasion, it may be necessary to determine the distance around grounds or landscapes. To measure the distance around property, buildings, and basin-like structures, it is necessary to determine either perimeter or circumference. The *perimeter* is the distance around an object; it is the border or outer boundary of the object. *Circumference* is the distance around a circle or circular object, such as a clarifier. Distance is linear measurement, which defines the distance (or length) along a line. Standard units of measurement, such as inches, feet, yards, and miles, and metric units, such as centimeters, meters, and kilometers, are used.

Perimeter

The perimeter of a rectangle (a four-sided figure with four right angles) is obtained by adding the lengths of the four sides (see Figure 3.1):

$$\text{Perimeter} = L_1 + L_2 + L_3 + L_4 \tag{3.21}$$

■ *Example 3.27*

Problem: Find the perimeter of the rectangle shown in Figure 3.2.
Solution:

$$\text{Perimeter} = 35 \text{ ft} + 8 \text{ ft} + 35 \text{ ft} + 8 \text{ ft} = 86 \text{ ft}$$

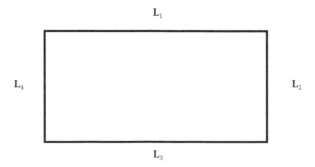

FIGURE 3.1
Perimeter.

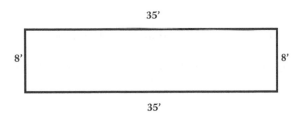

FIGURE 3.2
Example 3.27.

■ *Example 3.28*

Problem: What is the perimeter of a rectangular field that is 100 feet long and 50 feet wide?
Solution:

$$\text{Perimeter} = (2 \times \text{Length}) + (2 \times \text{Width})$$

$$\text{Perimeter} = (2 \times 100 \text{ ft}) + (2 \times 50 \text{ ft}) = 200 \text{ ft} + 100 \text{ ft} = 300 \text{ ft}$$

■ *Example 3.29*

Problem: What is the perimeter of a square whose sides are 8 inches?
Solution:

$$\text{Perimeter} = (2 \times \text{Length}) + (2 \times \text{Width})$$

$$\text{Perimeter} = (2 \times 8 \text{ in.}) + (2 \times 8 \text{ in.}) = 16 \text{ in.} + 16 \text{ in.} = 32 \text{ in.}$$

Circumference

The circumference is the length of the outer border of a circle. The circumference is found by multiplying *pi* (π) times the diameter (*D*). The diameter is a straight line from one side of a circle to the other that passes through the center of the circle, or the distance across the circle (see Figure 3.3). Use this calculation if, for example, the circumference of a circular tank must be determined:

$$\text{Circumference} = \pi \times D \tag{3.22}$$

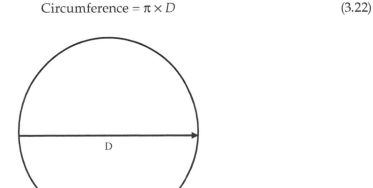

FIGURE 3.3
Diameter of circle.

■ *Example 3.30*

Problem: Find the circumference of a circle that has a diameter of 25 feet.
Solution:

$$\text{Circumference} = \pi \times 25 \text{ feet} = 3.14 \times 25 \text{ feet} = 78.5 \text{ feet}$$

■ *Example 3.31*

Problem: A circular chemical holding tank has a diameter of 18 meters. What is the circumference of this tank?
Solution:

$$\text{Circumference} = \pi \times 18 \text{ meters} = 3.14 \times 18 \text{ meters} = 56.52 \text{ meters}$$

■ *Example 3.32*

Problem: An influent pipe inlet opening has a diameter of 6 feet. What is the circumference of the inlet opening in inches?
Solution:

$$\text{Circumference} = \pi \times 6 \text{ feet} = 3.14 \times 6 \text{ feet} = 18.84 \text{ feet}$$

Area Measurements

For area measurements in forest-based biomass operations, three basic shapes are particularly important—namely circles, rectangles, and triangles. Area is the amount of surface an object contains or the amount of material it takes to cover the surface. The area on top of a chemical tank is called the *surface area*. The area of the end of a ventilation duct, for example, is called the *cross-sectional area* (the area at a right angle to the length of ducting). Area is usually expressed in square units, such as square inches (in.²) or square feet (ft²). Land may also be expressed in terms of square miles (sections) or acres (43,560 ft²) or, in the metric system, as *hectares* (10,000 m²).

Area of a Rectangle

A rectangle is a two-dimensional box. The area of a rectangle is found by multiplying the length (*L*) times width (*W*); see Figure 3.4.

$$\text{Area} = L \times W \tag{3.23}$$

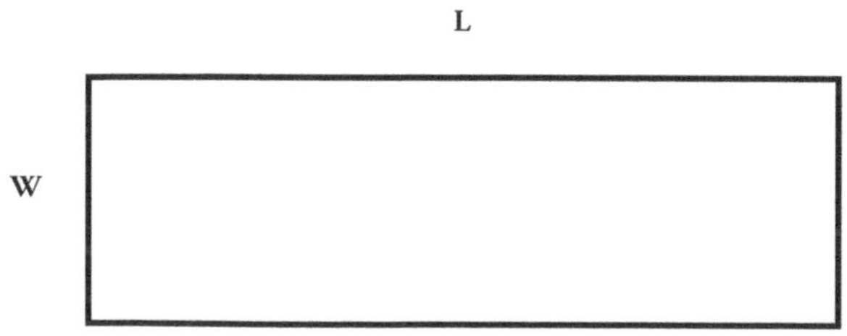

FIGURE 3.4
Area of rectangle.

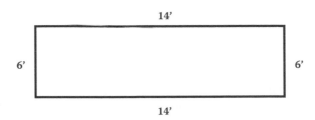

FIGURE 3.5
Example 3.33.

■ *Example 3.33*

Problem: Find the area of the rectangle shown in Figure 3.5.
Solution:

$$\text{Area} = L \times W = 14 \text{ feet} \times 6 \text{ feet} = 84 \text{ ft}^2$$

Area of a Circle

To find the area of a circle, we need to introduce a new term, *radius,* which is represented by *r*. Figure 3.6 shows a circle with a radius of 6 feet. The radius is any straight line that radiates from the center of the circle to some point on the circumference. By definition, all radii (plural of radius) of the same circle are equal. The surface area of a circle is determined by multiplying π by the radius squared (r^2):

$$\text{Area of Circle} = \pi \times r^2 \qquad (3.24)$$

■ *Example 3.34*

Problem: What is the area of the circle shown in Figure 3.6?
Solution:

$$\text{Area of Circle} = \pi \times r^2 = 3.14 \times (6)^2 = 3.14 \times 36 = 113 \text{ ft}^2$$

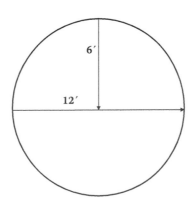

FIGURE 3.6
Example 3.34.

Area of a Circular or Cylindrical Tank

If we were assigned to paint a black liquor storage tank, we must determine the surface area of the walls of the tank so we know how much paint is required. To determine the surface area of the tank, we need to visualize the cylindrical walls as a rectangle wrapped around a circular base. The area of a rectangle is found by multiplying the length by the width; in this case, the width of the rectangle is the height of the wall, and the length of the rectangle is the distance around the circle (circumference). Thus, the area of the side walls of the circular tank is found by multiplying the circumference of the base ($\pi \times D$) times the height of the wall (H):

$$\text{Area} = \pi \times D \times H \tag{3.25}$$

For a tank with diameter of 20 feet and a height of 25 feet, the area can be determined as follows:

$$\text{Area} = \pi \times D \times H = 3.14 \times 20 \text{ ft} \times 25 \text{ ft} = 1570.8 \text{ ft}^2$$

To determine the amount of paint needed, remember to add the surface area of the top of the tank, which in this case is 314 ft^2. Thus, the paint must cover 1570.8 ft^2 + 314 ft^2 = 1884.8, or 1885 ft^2. If the tank floor should be painted, add another 314 ft^2.

Volume

The amount of space occupied by or contained in an object (see Figure 3.7), volume is expressed in cubic units, such as cubic inches (in.^{3}), cubic feet (ft^3), or acre-feet (1 acre-foot = 43,560 ft^3).

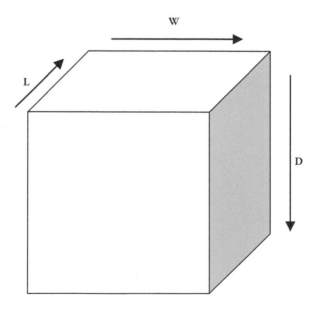

FIGURE 3.7
Volume.

TABLE 3.9

Volume Formulas

Sphere volume	=	$(\pi/6) \times$ (Diameter)^3
Cone volume	=	$1/3 \times$ Volume of Cylinder
Rectangular tank volume	=	Area of Rectangle $\times$ (Depth or Height)
	=	(Length $\times$ Width) $\times$ (Depth or Height)
Cylinder volume	=	Area of Cylinder $\times$ (Depth or Height)
	=	$\pi^2 \times$ (Depth or Height)
Pyramid volume	=	Area of triangle $\times$ (Depth or Height)
	=	[(Base $\times$ Height)/2] $\times$ (Depth or Height)

Volume of a Rectangular Basin

The volume (V) of a rectangular object is obtained by multiplying the length (L) times the width (W) times the depth or height (H):

$$V = L \times W \times H \tag{3.26}$$

■ Example 3.35

Problem: If a unit rectangular process basin has a length of 15 feet, width of 7 feet, and depth of 9 feet, what is the volume of the basin?
Solution:
$$V = L \times W \times H = 15 \text{ ft} \times 7 \text{ ft} \times 9 \text{ ft} = 954 \text{ ft}^3$$

Practical volume formulas are given in Table 3.9.

Volume of Round Pipe and Round Surface Areas

■ Example 3.36

Problem: Find the volume of a 3-inch-diameter round pipe that is 300 feet long.
Solution:

1. Change the diameter of the pipe from inches to feet by dividing by 12:

$$\text{Diameter} = 3 \div 12 = 0.25 \text{ ft}$$

2. Find the radius by dividing the diameter by 2:

$$\text{Radius } (r) = 0.25 \div 2 = 0.125 \text{ ft}$$

3. Find the volume:

$$V = L \times \pi \times r^2 = 300 \times 3.14 \times (0.125)^2 = 300 \times 3.14 \times 0.016 = 15.07 \text{ ft}^3$$

■ *Example 3.37*

Problem: Find the volume of a smokestack that is 24 inches in diameter (entire length) and 96 inches tall. Find the radius of the stack.

Solution: First find the radius of the stack. The radius is one-half the diameter:

$$24 \text{ in.} \div 2 = 12 \text{ in.}$$

Now determine the volume:

$$V = H \times \pi \times r^2 = 96 \times 3.14 \times (12)^2 = 96 \times 3.14 \times 144 = 43,407 \text{ in.}^3$$

Crown Volume

There are several attributes or parameters that could be measured for trees that are related to biomass. For example, the canopy dimensions of a tree would be related to biomass—the larger the tree, the greater the mass. The crown area of a tree is often highly correlated to the current season's growth biomass; therefore, crown area is a factor that can be used in sampling. Crown volume of a tree (an indirect measurement of tree biomass) and of various process tanks can be calculated using various volume formulas. To make an indirect estimate of tree biomass, the tree is first looked at in its natural state, then the viewer envisions a geometric shape that describes the shape of the tree. In this case, for our purposes, we have determined that the shape of the tree is a cone. Appropriate measurements, including diameter 1, diameter 2, and height, are made. Diameter 1 (D1) is the widest dimension of the crown of the tree, and the dimension perpendicular to D1 is diameter 2 (D2). Crown volume of a cone-shaped tree is found by

$$V = \frac{\pi(D1/2)(D2/2)(\text{Height})}{3} \tag{3.27}$$

Volume of a Cone

$$\text{Volume of Cone} = \frac{\pi}{12} \times \text{Diameter} \times \text{Diameter} \times \text{Height} \tag{3.28}$$

$$\text{where } \frac{\pi}{12} = \frac{3.14}{12} = 0.262$$

Key Point: The diameter used in the formula is the diameter of the base of the cone.

■ *Example 3.38*

Problem: The bottom section of a circular settling tank has the shape of a cone. How many cubic feet of water are contained in this section of the tank if the tank has a diameter of 120 feet and the cone portion of the unit has a depth of 6 feet?

Solution:

$$\text{Volume} = 0.262 \times 120 \text{ ft} \times 120 \text{ ft} \times 6 \text{ ft} = 22,637 \text{ ft}^3$$

Volume of a Sphere

$$\text{Volume of Sphere} = \frac{\pi}{6} \times \text{Diameter} \times \text{Diameter} \times \text{Diameter} \qquad (3.29)$$

$$\text{where } \frac{\pi}{6} = \frac{3.14}{6} = 0.524$$

■ *Example 3.39*

Problem: What is the volume in cubic feet of a gas storage container that is spherical and has a diameter of 60 feet?

Solution:

$$\text{Volume} = 0.524 \times 60 \text{ ft} \times 60 \text{ ft} \times 60 \text{ ft} = 113,184 \text{ ft}^3$$

Volume of a Circular or Cylindrical Tank

A circular tank consists of a circular floor surface with a cylinder rising above it (see Figure 3.8). The volume of a circular tank is calculated by multiplying the surface area by the height of the tank walls.

■ *Example 3.40*

Problem: If a tank is 20 feet in diameter and 25 feet deep, how many gallons of water will it hold? (*Hint:* In this type of problem, calculate the surface area first, multiply by the height, and then convert to gallons.)

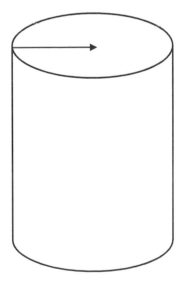

FIGURE 3.8
Circular tank.

Solution:

$$\text{Radius } (r) = \text{Diameter} \div 2 = 20 \text{ ft} \div 2 = 10 \text{ ft}$$

$$\text{Area } (A) = \pi \times r^2$$

$$A = \pi \times (10 \text{ ft})^2 = 3.14 \times 100 = 314 \text{ ft}^2$$

$$\text{Volume } (V) = A \times H$$

$$V = 314 \text{ ft}^2 \times 25 \text{ ft} = 7850 \text{ ft}^3$$

$$V = 7850 \text{ ft}^3 \times 7.5 \text{ gal/ft}^3 = 58{,}875 \text{ gal}$$

Weight Measurements

On Earth, the *weight* of a body is the result of the pull of gravity on that body. Weight is often used interchangeably with the term *mass*; however, the two are not the same. Mass is the measure of the amount of matter present in a body and thus has the same value at different locations. Weight varies depending on location of the body in Earth's gravitational field. Because the gravitational effect varies from place to place on Earth, the weight of a given mass also varies. "The distinction between weight and mass is confused by the use of the same units of measure: the gram, the kilogram, and the pound" (Husch et al., 2003, p. 25).

Chapter Review Questions

3.1   Convert the following measurements from English to metric units:

   a. 338 feet to kilometers

   b. 625 acres to hectares

   c. 58 square miles to square kilometers

   d. 310 cubic feet to cubic meters

3.2   Convert the following measurements from metric to English units:

   a. 7.4 kilometers to feet

   b. 69 hectares to acres

   c. 55 cubic meters to cubic feet

   d. 560 cubic meters per hectare to cubic feet per acre

3.3   Define the following abbreviations:

   a. GPS

   b. BA

   c MBF

   d. DBH

References and Recommended Reading

Hartman, D.A., Atkinson, W.A., Bryant, B.S., and Woodfin, R.O., Jr. (1976). *Conversion Factors for the Pacific Northwest Forest Industry: Converting Forest Growth to Forest Products*, Institute of Forest Products, Seattle, WA.

Husch, B., Beers, T.W., and Kershaw, J.A., Jr. (2003). *Forest Mensuration*, 4th ed., John Wiley & Sons, New York.

Snell, J.A. and Brown, J.K. (1980). *Handbook for Predicting Residue Weights of Pacific NW Conifers*, General Technical Report PNW-103, U.S. Department of Agriculture, Forest Service, Pacific Northwest Forest and Range Experiment Station, Portland, OR, 44 pp.

Stevens, S.S. (1946). On the theory of scales of measurement, *Science*, 103(2684):677–680.

Walters, D.K., Hann, D.W., and Clyde, M.A. (1985). *Equations and Tables Predicting Gross Total Stem Volumes in Cubic Feet for Six Major Conifers of Southwest Oregon*, Research Bulletin 50, Forest Research Laboratory, Oregon State University, Corvallis, OR, 36 pp.

4

Statistics

To the uninitiated it may often appear that the statistician's primary function is to prevent or at least impede the progress of research. And even those who suspect that statistical methods may be more boon than bane are at times frustrated in their efforts to make use of the statistician's wares.

—Freese (1967)

There are three kinds of lies: lies, damned lies, and statistics.

—Benjamin Disraeli

Statistical Concepts

Despite Disraeli's protestation, forest-based biomass energy production leans heavily on statistical applications. The principal concept of statistics is that of variation. When conducting typical forest-based biomass monitoring, sampling, and processing, variation is commonly found. Variation comes from the methods that were employed in the sampling process or, in this example, in the distribution of forest-based biomass. Several complex statistical tests can be used to determine the accuracy of data results; however, it is impossible to treat all aspects of statistics in one chapter. Instead, the aim of this chapter is simply to provide an overview of statistical concepts utilized in forest-based biomass and data analysis. Moreover, the intent is to provide a brief summary of and calculation procedures for the basic, fundamental statistical parameters. Specifically, this chapter provides a survey of the basic statistical and data analysis techniques that can be applied to solving many problems encountered on a daily basis. It covers the data analysis process: research design, data collection, analysis, drawing conclusions, and, most importantly, presentation of findings. Finally, it is important to point out that statistics can be used to justify the implementation of a program or process, identify areas that need to be addressed, or illustrate the impact that various environmental parameters, such as lightning-sparked forest fires, insect infestation, and so forth, have on losses and accidents. A set of forest data (or other data) is only useful if it is analyzed properly. Better decisions can be made when the nature of the data is properly characterized—that is, when precision, accuracy, and bias are considered.

Precision, Accuracy, and Bias

Precision, accuracy, and bias are terms with which most people are familiar, but it is important here to define the terms from a statistical viewpoint. *Precision* (or variance) as used here and in forest-based biomass energy production means the degree of agreement in a series of

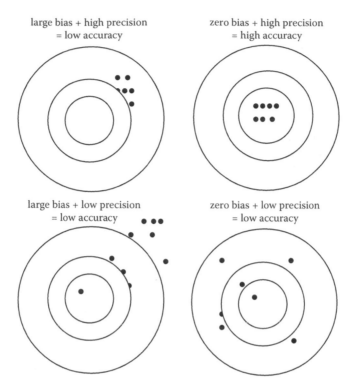

FIGURE 4.1
Precision, bias, and accuracy of a target shooter. The bull's-eye of the target is analogous to the unknown true population parameter, and the holes represent parameter estimates based on different samples. The goal is accuracy, which is the precise, unbiased target. (Adapted from Shiver, B.D. and Borders, B.E., *Forest Science*, 6, 42–50, 1996.)

measurements. *Accuracy*, on the other hand, is the closeness of a measurement or estimate to the true value; the ultimate objective, of course, is to obtain accurate measurements. *Bias* refers to the tendency of measures to systematically shift in one direction from the true value; this tendency is referred to as a *systematic error*. Such errors are often caused by poorly calibrated instruments, faulty measurement procedures, flaws in the sampling procedure, instrumental errors, mistakes in recording, and errors in the computations. A biased estimate may be precise, but it cannot be accurate; thus, it is evident that accuracy and precision are not synonymous or interchangeable terms. As an example, a forest-based biomass technician might make a series of careful measurements of a single tree with an instrument that is improperly calibrated or out of adjustment. If the measurements closely cluster about their average value, they are precise; however, because the instrument was not properly calibrated and was out of adjustment, the measured values will be biased—considerably off the true value. The resulting estimate will not be accurate. Although the results may not be accurate because of the presence of bias, the lack of precision may also be a factor (see Figure 4.1).

Measure of Central Tendency

The range of statistical applications in forest-based biomass energy is immense. As mentioned, it is impossible to treat all aspects of statistics in a single chapter. The practice of forest-based biomass energy production requires a much more complete understanding of statistics than can be presented in this book. Serious students of forest-based biomass energy production should keep in mind that the proper application of statistical methods in this field requires an understanding of statistics that must be covered in specialized courses on statistics.

If statistics are involved, it is usually because we are trying to estimate something with incomplete knowledge. Maybe we can only afford to test 1% of the items we are interested in, but we want to be able to say something about the properties of the entire lot. It is possible that we must destroy the sample in the process of testing it; in this case, 100% sampling is not feasible if someone is supposed to get the items after we are done with them.

The two primary objectives of statistical methods are (1) estimation of population parameters, and (2) testing of hypotheses about these parameters (Freese, 1967). Moreover, the questions we are usually trying to answer are "What is the central tendency of the item of interest?" and "How much dispersion about this central tendency can we expect?" Simply, the average or averages that can be compared are measures of *central tendency* or central location of the data.

Basic Statistical Terms

To speak effectively about forest-based biomass measurements, one must know the language. Many readers will be familiar with several of the following basic statistical terms, such as mean or average, median, mode, and range (Freese, 1967).

1. *Mean* is one of the most familiar and commonly estimated population parameters. It is the total of the values of a set of observations divided by the number of observations.
2. *Median* is the value of the central item when the data are arrayed by size.
3. *Mode* is the observation that occurs with the greatest frequency and thus is the most "fashionable" value.
4. *Range* is the difference between the values of the highest and lowest terms.
5. *Standard deviation* characterizes the dispersion of individual values about the mean. It gives us some idea about whether most of the individual values in a population are close to the mean or are spread out. The standard deviation of individuals in a population is frequently symbolized by σ (sigma). On the average, about two-thirds of the unit values of a normal population will be within 1 standard deviation of the mean. About 95% will be within 2 standard deviations, and about 99% will be within 2.6 standard deviations. Keep in mind that we will seldom know or be able to determine σ exactly; however, given a sample of individual values from the population we can often make an estimate of σ, which is commonly symbolized by s. We discuss standard deviation in more detail and provide examples later in this chapter.

6. *Coefficient of variation* is the ratio of the standard deviation to the mean. In nature, populations with large means often show more variation than populations with small means. The coefficient of variation facilitates comparison of variability about different sized means. A standard deviation of 2 for a mean of 10 indicates the same relative variability as a standard deviation of 16 for a mean of 80. The coefficient of variation would be 0.20, or 20%, in each case.

7. *Standard error of the mean* is a measure of the variation among sample means. There is usually variation among the individual units of a population, and the standard deviation is a measure of this variation. Because the individual units vary, variation may also exist among the means (or any other estimates) computed from samples of these units.

8. *Covariance* is a measure of the association between the magnitudes of two characteristics. Very often, each unit of a population will have more than a single characteristic. Trees, for example, may be characterized by their height, diameter, and class. If there is little or no association, the covariance will be close to zero. If the large values of one characteristic tend to be associated with the small values of another characteristic, then the covariance will be negative. If the large values of one characteristic tend to be associated with the large values of another characteristic, then the covariance will be positive.

9. *Simple correlation coefficient* is a measure of the degree of linear association between two variables. It is free of the effects of scale of measurement. It can vary from –1 to +1. A correlation of 0 indicates that there is no linear association (although there may be a very strong nonlinear association). A correlation of +1 or –1 would suggest a perfect linear association.

10. *Populations* are the aggregate of all arbitrarily defined, non-overlapping sample units; for example, if individual trees are the sample unit, then all trees on a given area of land could be considered the population.

11. *Parameters* are constants that describe the population as a whole.

12. *Variables* are characteristics that may vary from one sample unit to another.

13. *Samples* are used to provide the information necessary to determine the mean, mode, median, and range of a population. In statistics, we most often obtain data from a sample and use the results from the sample to describe an entire population. Measuring an entire population is difficult, if not impossible; for example, if we wanted to measure a characteristic of the entire population defined as "tree species," we would have to obtain a measure of that characteristic for every tree species possible.

14. *Subject or case* refers to a member of a population or sample. There are statistical methods for determining how many cases must be selected in order to have a credible study. *Data* is a term used to represent the measurements taken for the purposes of statistical analysis. Data can be classified as either qualitative or quantitative. *Qualitative* data describe a characteristic of an individual or subject (e.g., gender of a person or the color or a car). *Quantitative* data describe a characteristic in terms of a number (e.g., the age of a horse or the number of lost-time injuries an organization had over the previous year).

15. *Statistical symbols* are commonly used in the field of statistics. Statistical notation uses Greek letters and algebraic symbols to convey meaning about the statistical procedures to be followed during a particular study or test. Greek letters are used

TABLE 4.1

Commonly Used Statistical Symbols

Statistical Value	Population Symbol	Sample Notation
Mean	μ	$\bar{x}$
Standard deviation	σ	s
Variance	σ^2	s^2
Number of cases	N	n
Raw number or value	X	x
Correlation coefficient	R	r

Procedures	Symbol
Sum of	Σ
Absolute value of x	$|x|$
Factorial of n	$n!$

as statistical notation for a population, while English letters are used for statistical notation for a sample. Table 4.1 summarizes some of the more common statistical symbols, terms, and procedures used in statistical operations.

16. *Valid sampling methods* are essential, and their importance cannot be overemphasized. Several different methodologies are available, and a careful review of these methods (with an emphasis on designing appropriate sampling procedures) should be made before computing analytic results. Using appropriate sampling procedures along with careful sampling techniques will provide basic data that are accurate.

The need for statistics in forest-based biomass energy production is driven by the discipline itself. As mentioned, forest-based biomass studies often deal with entities that are variable. If there were no variation in collected data, there would be no need for statistical methods.

Frequency Distributions

Organized data are referred to *distributions*. The *frequency distribution* defines the relative frequency with which different values of a variable occur in a population. When a forest-based biomass practitioner conducts a research study, data are collected and a group of raw data is obtained. To make sense out of the data, the data must be organized into a meaningful format. The first step is to put the data into some kind of logical order and then group the data; to be compared to other data, the data must be organized. When confronted with masses of ungrouped data (i.e., lists of figures), it is difficult to generalize about the information they contain; however, if a frequency distribution of the figures is determined, many features become readily discernible. A frequency distribution records the number of cases that fall in each class of the data.

■ *Example 4.1*

Problem: A forest-based-biomass technician gathers data on the medical costs of 24 on-the-job injury claims filed by forest workers for a given year. The raw data collected were as follows:

$60	$1500	$85	$120
$110	$150	$110	$340
$2000	$3000	$550	$560
$4500	$85	$2300	$200
$120	$880	$1200	$150
$650	$220	$150	$4600

Solution: To develop a frequency distribution, the technician takes the values of the claims, places them in order, and counts the frequency of occurrences for each value:

Value	Frequency
$60	1
$85	2
$110	2
$120	2
$150	3
$200	1
$220	1
$340	1
$550	1
$560	1
$650	1
$880	1
$1200	1
$1500	1
$2000	1
$2300	1
$3000	1
$4500	1
$4600	1
Total	24

TABLE 4.2

Frequency Distribution for Example 4.1

Range	Frequency
$0–$999	17
$1000–$1999	2
$2000–$2999	2
$3000–$3999	1
$4000–$4999	2
Total	24

To develop a frequency distribution, the values in the table must be grouped, with each group having an equal range. The forest-based-biomass technician grouped the data into ranges of 1000. The lowest range and highest range are determined by the data. Because it was decided to group by thousands, values will fall in the ranges of $0 to $4999, and the distribution will end with this. The frequency distribution for these data appears in Table 4.2.

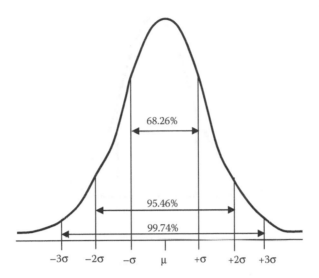

FIGURE 4.2
Normal distribution curve showing the frequency of a measurement.

Normal Distribution

When large mounts of data are collected on certain characteristics, the data and subsequent frequency can follow a distribution that is bell shaped—the *normal distribution*. Normal distributions are a very important class of statistical distributions. As stated, all normal distributions are symmetric and have bell-shaped curves with a single peak (see Figure 4.2). When speaking about a normal distribution, two quantities have to be specified: the mean, μ (mu), where the peak of the density occurs, and the standard deviation, σ (sigma). Different values of μ and σ yield different normal density curves and hence different normal distributions. Although there are many normal curves, they all share an important property that allows us to treat them in a uniform fashion. All normal density curves satisfy the following properties, which are often referred to as the *empirical rule*:

- 68% of the observations fall within 1 standard deviation of the mean—that is, between $\mu - \sigma$ and $\mu + \sigma$.
- 95% of the observations fall within 2 standard deviations of the mean—that is, between $\mu - 2\sigma$ and $\mu + 2\sigma$.
- 98% of the observations fall within 3 standard deviations of the mean—that is, between $\mu - 3\sigma$ and $\mu + 3\sigma$.

Thus, for a normal distribution, almost all values lie within 3 standard deviations of the mean (see Figure 4.3). It is important to stress that the rule applies to all normal distributions. Also remember that it applies *only* to normal distributions.

Key Point: Before applying the empirical rule it is a good idea to identify the data being described and the value of the mean and standard deviation. A sketch of a graph summarizing the information provided by the empirical rule should also be made.

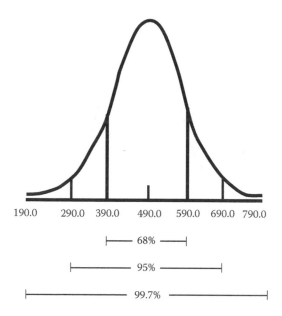

FIGURE 4.3
Sample Scholastic Aptitude Test (SAT) math percentages given by the empirical rule.

■ *Example 4.2*

Problem: The scores for all high-school seniors taking the math section of the Scholastic Aptitude Test (SAT) in a particular year had a mean of 490 and a standard deviation of 100. The distribution of SAT scores is bell shaped.

1. What percentage of seniors scored between 390 and 590 on this SAT test?
2. One student scored 795 on this test. How did this student do compared to the rest of the scores?
3. A rather exclusive university admits only students who score among the highest 16% on this test. What score would a student need on this test to be qualified for admittance to this university?

The data being described are the math SAT scores for all seniors taking the test one year. Because the data describe a population, we denote the mean and standard deviation as $\mu = 490$ and $\sigma = 100$, respectively. A bell-shaped curve summarizing the percentages given by the empirical rule is shown in Figure 4.3.

Solution:

1. From Figure 4.3, about 68% of seniors scored between 390 and 590 on this SAT test.
2. Because about 99.7% of the scores are between 10 and 790, a score of 795 is excellent. This is one of the highest scores on this test.
3. Because about 68% of the scores are between 390 and 590, this leaves 32% of the scores outside the interval. Because a bell-shaped curve is symmetric, one-half of the scores, or 16%, are on each end of the distribution.

Statistical Computations

Mean

The mean or average is the sum of all the data divided by the number of data points. Mathematically, it can be represented by the following equation:

$$\bar{x} = \frac{\sum x}{n}$$ (4.1)

where:

$\bar{x}$ = mean or average

$\sum x$ = sum of all the data

n = number of data points

■ *Example 4.3*

Problem: Let's assume that the following 12 numbers are the diameters of trees in a plot we sampled.

10.3, 6.6, 8.1, 9.4, 25.0, 16.1, 13.2, 5.1, 6.6, 30.9, 22.9, 11.3

For the 12 tree diameters under consideration, what is the average (mean) tree diameter?

Solution:

165.5 ÷ 12 = 13.8 in.

DID YOU KNOW?

The mean or average is a good indication of the center of the data if the data have a symmetric distribution. However, if the distribution of the data is not symmetric or the data include extremely high or low values, the mean may not accurately describe the center of the data.

Median

The median is used to describe the center or middle of the distribution of the data when they have been arranged in order of magnitude—that is, where half of the values are above and half are below. Let's refer back to our tree diameter example, but for simplicity assume that we added another sampling point with a diameter of 13.3.

■ *Example 4.4*

Problem: Following are the 13 tree diameters we sampled. What is the median?

10.3, 6.6, 8.1, 9.4, 25.0, 16.1, 13.2, 5.1, 6.6, 30.9, 22.9, 11.3, and now 13.3.

Solution: The first step in calculating the median is to rearrange the data so they are in numerical order. Your data should now look like the following:

5.1, 6.6, 6.6, 8.1, 9.4, 10.3, 11.3, 13.2, 13.3, 16.1, 22.9, 25.0, 30.9

Because we have an odd number of samples, we can simply find the number where half of the sample points fall above and half below. In this case, the median is equal to 11.3.

Mode

The mode is defined as the most frequently appearing value or class of values in a set of observations. It is the most commonly occurring value in a frequency distribution.

■ *Example 4.5*

Problem: Let's assume that the following data represent the number of willow trees per acre in 22 tracts. What is the mode?

110, 110, 110, 114, 116, 116, 116, 205, 205, 205, 205, 205,

215, 215, 216, 220, 220, 220, 225, 230, 240, 295

Solution: We can see that the value of 205 willow trees per acre occurred more frequently than any other result; therefore, the mode for this dataset would be 205.

Range (Spread of Data)

The easiest way to describe the spread of data is to calculate the range. Simply, the range is the difference between the highest and lowest values from a sample.

■ *Example 4.6*

Problem: Let's assume that the following data points represent the number of fires over 400 acres on a forest during the last 12 years. What is the range?

3, 3, 3, 5, 7, 9, 9, 10, 11, 11, 11, 12,

Solution: The range of fires over 400 acres during the last 12 years is equal to $12 - 3 = 9$ fires.

Variance

Variance (s^2) is one of several descriptors of a distribution. It describes how spread out the data are. The variance is computed as the average squared deviation of each number from its mean. Mathematically speaking, the formula is

$$s^2 = \frac{\sum (x - \bar{x})^2}{n - 1} \qquad (4.2)$$

where:

s^2 = variance

x = data point

$\bar{x}$ = mean of the data

n = number of data points

■ *Example 4.7*

Problem: Let's assume that the following five data points represent trees per acre of our sample points for a potential harvest area. What is the variance?

$$160, 166, 176, 184, 210$$

Solution: We would all agree that the range of these data is 50 trees per acre and that the mean is equal to about 179 trees per acre. We now need to calculate the variance. You can begin by subtracting the mean from each value and then squaring the result:

$$(160 - 179)^2 = 361$$
$$(166 - 179)^2 = 169$$
$$(176 - 179)^2 = 9$$
$$(184 - 179)^2 = 25$$
$$(210 - 179)^2 = 961$$

The next step is to add together the results to get 1525. Last, we divide this number by the number of sample points minus 1:

$$s^2 = 1525 \div (5 - 1) = 381.3$$

Standard Deviation

The standard deviation, s or σ, is often used as an indicator of precision. The standard deviation is a measure of the variation (the spread in a set of observations) in the values. To gain a better understanding and perspective of the benefits to be derived from using statistical methods in forest-based biomass energy production, it is appropriate to consider some of the basic theory of statistics. As mentioned, in any set of data, the true value (mean) will lie in the middle of all the measurements taken; however, this is true only when the sample size is large and only random error is present in the analysis. In addition, the measurements will show a normal distribution as shown in Figure 4.2. Figure 4.2 shows that 68.26% of the results fall between $M + s$ and $M - s$, 95.46% of the results lie between $M + 2s$ and $M - 2s$, and 99.74% of the results lie between $M + 3s$ and $M - 3s$. Therefore, if the values are precise, then 68.26% of all the measurements should fall between the true value estimated by the mean plus the standard deviation and the mean minus the standard deviation. The following equation is used to calculate the sample standard deviation:

$$s = \sqrt{\frac{\sum (x - \bar{x})^2}{(n - 1)}}$$

where:

    s = standard deviation

    x = each value in the dataset

    $\bar{x}$ = mean of the data

    n = number of samples

■ *Example 4.8*

Problem: Given the following data for percent solids in a black liquor, what is the standard deviation?

$$9.5, 10.5, 10.1, 9.9, 10.6, 9.5, 11.5, 9.5, 10.0, 9.4$$

Solution:

$$\bar{x} = 10.0$$

x	$x - \bar{x}$	$(x - \bar{x})^2$
9.5	−0.5	0.25
10.5	0.5	0.25
10.1	0.1	0.01
9.9	−0.1	0.01
10.6	0.6	0.36
9.5	−0.5	0.25
11.5	1.5	2.25
9.5	−0.5	0.25
10.0	0.0	0.00
9.4	−0.6	0.36
Total		3.99

$$s = \sqrt{\frac{3.99}{10 - 1}} = 0.67$$

Statistical Inference*

Statistical inference is the process of drawing conclusions from data that are subject to random variation (Upton and Cook, 2008). An understanding of statistical inference is important to forest science and to discussing the role of sampling in the inferential process. Scientific inference becomes statistical inference when the connection between the unknown "state of nature" and the observed information is expressed in probabilistic

---

* Much of the information presented here is from Schreuder, H.T. et al., *Statistical Techniques for Sampling and Monitoring Natural Resources*, USDA Forest Service RMRS-GTR-126, U.S. Department of Agriculture, Washington, D.C., 2004.

terms (Dawid, 1983). Statistical inference encompasses the entire field of statistics, as its focus is on what is logically implied by the information available (Fraser, 1983). Cramer (1946) summarized the role of statistical inference as having three functions: description, analysis, and prediction. Description is the reduction of datasets to as small a set of numbers as possible, such as the mean, variance, or skewness of a distribution. This allows describing a population as concisely and briefly as possible and making comparisons between populations. Analysis is the summarization of data for a particular purpose or objective: What are the estimates of certain population characteristics? Did the sample arise from a given distribution? Given two samples, did they arise from the same population or not? Statistics provides methods for doing such analyses. Statistical methods are used to predict and explain phenomena, often a very challenging task.

Ideally, statistical inference would always be based on Bayes' theorem, which combines prior information with information from surveys or experiments and would be acceptable to many statisticians if the prior belief is objective. The problem is that usually prior information is subjective, as the information available varies from person to person. Objective prior information indicates that people would normally agree on it. As an example of subjective prior information, a forest industry person could believe that there is plenty of old growth distributed nicely over the forest habitat for endangered species, whereas an environmentalist could believe equally strongly that the old growth in the forest is limited and badly distributed. People willing to accept prior subjective information are called Bayesians and rely on Bayes' theorem for inference. Non-Bayesians, or frequentists, a majority, use classical inference procedures relying only on objective data often based on normality assumptions and large sample theory based on the central limit theorem and related statistical properties. (Frequentist statistics is the inference framework on which the well-established methodologies of statistical hypothesis testing and confidence intervals are based.) Many believe that Bayesian procedures should be used when immediate, logically defensible decisions need to be made, and classical ones should be used when building a body of scientific knowledge. A forest manager who has to make decisions about whether or not to cut old growth and where it should be cut for management purposes may well choose to use all his prior information to construct a (subjective) prior distribution to combine with actual sample data in order to use Bayes' theorem to make such decisions. Such decisions can be defended at least on the basis of a systematic approach. Scientific databases can be used by different users applying different priors to make their decisions.

Statistical inference from sample surveys can be either model based or design based. In model-based sampling, inference relies on a statistical model to describe how the probability structure of the observed data depends on uncontrollable change variables and, frequently, on other unknown nuisance variables. Such modes can be based on a theoretical understanding of the process by which the data were generated, experimental techniques used, or past experience with similar processes. For inference in design-based sampling, reliance is placed on probabilistic sampling. It is the most widely accepted approach now.

In regard to which single method of inference is best, Smith (1994, p. 17) stated: "My view is that there is no single right method of inference. All inferences are the product of man's imagination and there can be no absolutely correct method of inductive reason. Different types of inference are relevant for different problems and frequently the approach recommended reflects the statistician's background such as science, industry, social sciences or government. ... I now find the case for hard-line randomization inference based on the unconditional distribution to be acceptable. ... Complete reconciliation is neither possible nor desirable. Vive la difference."

Chapter Review Questions

4.1 A number calculated from a _____ is a characteristic of a statistic.

4.2 A _____ is the set of all values of a single measurement from a random variable collected under certain environmental conditions.

4.3 A _____ is any number calculated from a population.

References and Suggested Reading

Cramer, H. (1963). *Mathematical Methods of Statistics*, Princeton University Press, Princeton, NJ.

Dawid, P.P. (1983). Inference, statistical: I, in Kotz, S., Johnson, N.L., and Read, C.B., Eds., *Encyclopedia of Statistical Sciences*, Vol. 4, John Wiley & Sons, New York.

Frazer, D.A.S. (1983). Inference, statistical: II, in Kotz, S., Johnson, N.L., and Read, C.B., Eds., *Encyclopedia of Statistical Sciences*, Vol. 4, John Wiley & Sons, New York.

Freese, F. (1967). *Elementary Statistical Methods for Foresters*, Agriculture Handbook 317, U.S. Department of Agriculture, Washington, D.C.

Schreuder, H.T., Ernst, R., and Ramirez-Maldonado, H. (2004). *Statistical Techniques for Sampling and Monitoring Natural Resources*, USDA Forest Service RMRS-GTR-126, U.S. Department of Agriculture, Washington, D.C.

Shiver, B.D. and Borders, B.E. (1996). Systematic sampling with multiple random starts. *Forest Science*, 6:42–50.

Smith, T.M.F. (1994). Sample surveys: 1975–1990; an age of reconciliation? *International Statistical Review*, 62:5–34.

Upton, G. and Cook, I. (2008). *Oxford Dictionary of Statistics*, Oxford University Press, New York.

Wadsworth, H.M. (1990). *Handbook of Statistical Methods for Engineers and Scientists*, McGraw-Hill, New York.

5

Forest-Based Biomass: Heat Energy, Weight Considerations, and Estimations

Wood is still the largest biomass energy resource today.

—NREL (2010)

Introduction

Wise utilization of the forest resource relates to awareness of its value (Ince, 1979). The amount of heat energy that can be recovered from wood or bark determines its fuel value. The amount of recoverable heat energy varies with moisture content and chemical composition. Recoverable heat energy varies among tree species and even within a species. In this chapter, we provide a summary of information that may be used to estimate recoverable heat energy in wood or bark fuel, biomass weight considerations, biomass weight examples, and methods for estimating stand-level biomass.

Lower and Higher Heating Values

The net calorific value, or *lower heating value* (LHV), of a fuel is defined as the amount of heat released by combusting a specified quantity (initially at 25°C) and returning the temperature of the combustion products to 150°C; it is assumed that the latent heat of vaporization of water in the reaction produces is not recovered. The LHV is the useful calorific value in boiler combustion plants and is frequently used in Europe. The *higher heating value* (HHV), the gross calorific value or gross energy, of a fuel is defined as the amount of heat released by a specified quantity (initially at 25°C) once it is combusted and the products have returned to a temperature of 25°C, which takes into account the latent heat of vaporization of water in the combustion products. The HHV is derived only under laboratory conditions and is frequently used in the United States for solid fuels. The lower and higher heating values of solid fuels are expressed in British thermal units (Btu). Table 5.1 lists the values for solid forest-based biomass fuels. Table 5.2 illustrates the variation in reported heat content values (on a dry weight basis) in the U.S. and European literature based on values in the ECN Phyllis database (http://www.ecn.nl/phyllis/), the U.S. Department of Energy Biomass Feedstock Composition and Property database (http://www1.eere.energy.gov/biomass/feedstock_databases.html), and selected literature sources.

TABLE 5.1

Lower and Higher Heating Values of Solid Fuels

	Lower Heating Value (LHV)			Higher Heating Value (HHV)		
Fuel	Btu/ton	Btu/lb	MJ/kg	Btu/ton	Btu/lb	MJ/kg
Farmed trees	16,811,000	8406	19,551	17,703,170	8852	20,589
Forest residue	13,243,490	6622	15,402	14,164,160	7082	16,473

Source: *Transportation Fuel Cycle Analysis Model*, Greet 1.8b, Argonne National Laboratory, Argonne, IL, 2008.

Effect of Fuel Moisture on Wood Heat Content

Because recently harvested wood fuels usually contain 30 to 55% water, it is useful to understand the effect of moisture content on the heating value of wood fuels. Table 5.3 shows the effect of percent moisture content (MC) on the higher heating value as-fired (HHV-AF) of a wood sample starting at 8500 Btu/lb (oven-dry). Fuel moisture content is usually reported as the wet weight basis moisture content. Moisture content expressed on a wet weight basis (also called "green" or "as-fired" moisture content) is the decimal fraction of the fuel that consists of water; for example, a pound of wet wood fuel at 50% moisture content contains 0.50 pound of water and 0.50 pound of wood. Note that the wet weight basis differs from the total dry weight basis method of expressing moisture content which is more commonly used for describing moisture content of finished wood products. The dry weight basis is the ratio of the weight of water in wood to the oven-dry weight of the wood. The formulas used require that moisture content be expressed on the wet weight basis (Ince, 1979).

TABLE 5.2

Heat Content Ranges for Forest Biomass Fuels (Dry Weight Basis)

	English		Metric					
	Higher Heating Value		Higher Heating Value		Lower Heating Value			
Fuel	Btu/lb	MBtu/ton	kJ/kg	MJ/kg	kJ/kg	MJ/kg		
Woody Crops								
Black locust	8409–8582	16.8–17.2	19,547–19,948	19.5–19.9	18,464	18.5		
Eucalyptus	8174–8432	16.3–16.9	19,000–19,599	19.0–19.6	17,963	18.0		
Hybrid poplar	8183–8491	16.4–17.0	19,022–19,737	19.0–19.7	17,700	17.7		
Willow	7983–8497	16.0–17.0	18,556–19,750	18.6–19.7	16,734–18,419	16.7–18.4		
Forest Residues								
Hardwood wood	8017–8920	16.0–17.5	18,635–20,734	18.6–20.7	—	—		
Softwood wood	8000–9120	16.0–18.24	18,595–21,119	18.6–21.1	17,514–20,768	17.5–20.8		

Sources: http://www1.eere.energy.gov/biomass/feedstock_databases.html; Bushnell, D., *Biomass Fuel Characterization: Testing and Evaluating the Combustion Characteristics of Selected Biomass Fuels*, Bonneville Power Administration, Portland, OR, 1989; Jenkins, B., *Properties of Biomass, Appendix to Biomass Energy Fundamentals*, EPRI Report TR-102107, Electric Power Research Institute, Palo Alto, CA, 1993; Jenkins, B.L. et al., *Fuel Processing Technology* 54, 17–46, 1998; Tillman, D., *Wood as an Energy Resource*, Academic Press, New York, 1978.

TABLE 5.3

Effect of Fuel Moisture on Wood Heat Content

	Moisture Content (MC) Wet Basis (%)										
	0	15	20	25	30	35	40	45	50	55	60
Higher heating value as fired (HHV-AF) (Btu/lb)	8500	7275	6800	6375	5950	5525	5100	4575	4250	3825	3400

Sources: Borman, G.L. and Ragland, K.W., *Combustion Engineering*, McGraw-Hill, New York, 1998; Kluender, R.A., *The Forester's Wood Energy Handbook*, Publ. No. 80-A-12, American Pulpwood Association, Washington, D.C., 1980; Maker, T.M., *Wood-Chip Heating Systems: A Guide for Institutional and Commercial Biomass Installations*, Biomass Energy Resource Center, Montpelier, VT, 2004.

Moisture Content Wet- and Dry-Weight Basis Calculations

Moisture content (MC) on a wet- or dry-weight basis is calculated as follows:

$$\text{MC (dry basis)} = 100 \times \frac{(\text{wet weight} - \text{dry weight})}{\text{dry weight}} \tag{5.1}$$

$$\text{MC (wet basis)} = 100 \times \frac{(\text{wet weight} - \text{dry weight})}{\text{wet weight}} \tag{5.2}$$

To convert MC (wet basis) to MC (dry basis):

$$\text{MC (dry basis)} = \frac{100 \times \text{MC (wet basis)}}{100 - \text{MC (wet basis)}} \tag{5.3}$$

To convert MC (dry basis) to MC (wet basis):

$$\text{MC (wet basis)} = \frac{100 \times \text{MC (dry basis)}}{100 + \text{MC (dry basis)}} \tag{5.4}$$

Some sources report the heat contents of fuels "as-delivered" rather than at 0% moisture for practical reasons. Because most wood fuels have bone-dry (oven-dry) heat contents in the range of 7600 to 9600 Btu/lb (15,200,000 to 19,200,000 Btu/ton or 18 to 22 GJ/Mg), lower values will always mean that some moisture is included in the delivered fuel.

Forestry Volume Unit to Biomass Weight Considerations

Biomass is frequently estimated from forestry inventory merchantable-volume data, particularly for purposes of comparing regional and national estimates of aboveground biomass and carbon levels. Making such estimations can be done several ways, but it always involves the use of conversion factors or biomass expansion factors (or both combined).

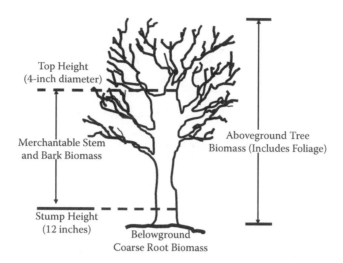

FIGURE 5.1
U.S. Forest inventory data. (Adapted from Jenkins, J.C. et al., *A Comprehensive Database of Diameter-Based Biomass Regressions for North American Tree Species*, General Technical Report NE-319, U.S. Department of Agriculture, Forest Service, Northeastern Research Station, Newtown Square, PA, 2004.)

Figure 5.1 defines what is included in each category of volume or biomass units, and Figure 5.2 illustrates indirect methods of large-scale biomass estimation. Total volume or biomass includes stem, bark, stump, branches, and foliage, especially if evergreen trees are being measured. When estimating biomass available for bioenergy, the foliage is not included, and the stump may or may not be appropriate to include depending on whether harvest occurs at ground level or higher. Both conversion and expansion factors can be used together to translate directly between merchantable volumes per unit area and total biomass per unit area, as demonstrated by the simple volume to weight conversion process shown below.

Estimation of Biomass Weights from Forestry Volume Data

Equation 5.5 is used to estimate merchantable biomass from merchantable volume assuming that the specific gravity and moisture content are known and the specific gravity basis corresponds to the moisture content of the volume involved (Briggs, 1994). Specific gravity (SG) is a critical element of the volume to biomass estimation equation. The SG content should correspond to the moisture content of the volume involved. SG varies considerably from species to species, differs for wood and bark, and is closely related to the moisture content as explained in graphs and tables in Briggs (1994). The wood specific gravity of species can be found in several references, although the moisture content basis is not generally given. Briggs (1994) suggested using a moisture content of 12% as the standard upon which many wood properties measurements should be based.

$$\text{Weight} = \text{Volume} \times \text{Specific Gravity} \times \text{Density of } H_2O \times (1 + MC_{od}/100) \qquad (5.5)$$

where the volume is expressed in cubic feet or cubic meters, the density of H_2O is 62.4 lb/ft^3 or 1000 kg/m^3, and MC_{od} is the oven-dry moisture content.

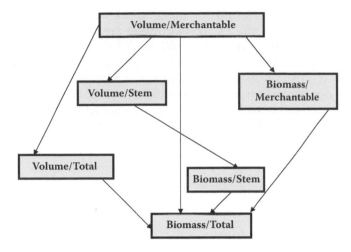

FIGURE 5.2
Indirect methods of large-scale biomass estimation. (Adapted from Somogyi, Z. et al., *European Journal of Forest Research*, 126(2), 197–207, 2007.)

■ *Example 5.1*

Problem: What is the weight of fiber in a 44-ft³ oven-dry log with a specific gravity of 0.40?
Solution:

Weight = 44 ft³ × 0.40 × 62.4 lb ft³ × (1 + 0/100) = 1098 lb, or 9.549 dry ton

Biomass Expansion Factors

Schroeder et al. (1997) described methods for estimating total aboveground dry biomass per unit area from growing stock volume data in the U.S. Forest Service Forest Inventory and Analysis (FIA) database. The growing stock volume data are limited to trees with diameters greater than or equal to 12.7 cm. It is highly recommended that this paper be studied for details of how the biomass expansion factors (BEFs) for oak–hickory and beech–birch were developed.

Stand-Level Biomass Estimation

At the individual field or stand level, biomass estimation is relatively straightforward, especially if it is being done for plantation-grown trees that are relatively uniform in size and other characteristics. The procedure involves first developing a biomass equation that predicts individual tree biomass as a function of diameter at breast height (DBH) or of DBH plus height. Second, the equation parameters (DBH and height) must be measured on a sufficiently large sample size to minimize variation around the mean values. Finally, the mean individual tree weight results are scaled to the area of interest based on percent survival or density information (trees per ace or hectare). Regression estimates are developed by directly sampling and weighing enough trees to cover the range of sizes being included in the estimation.

Regression equations can be found for many species in a wide range of literature. Examples for trees common to the Pacific Northwest are provided in Briggs (1994). The equations will differ depending on whether foliage or live branches are included, so care must be taken in interpreting the biomass data. For plantation trees grown on cropland or marginal cropland, it is usually assumed that tops and branches are included in the equations but foliage is not. For trees harvested from forests on lower quality land, it is usually recommended that tops and branches should not be removed (Pennsylvania DCNR, 2007) to maintain nutrient status and reduce erosion potential, thus biomass equations should assume regressions based on the stem weight only.

Biomass Equations*

The intensity of forest utilization has increased in recent years because of whole-tree harvesting and the use of wood for energy. Actually, estimating tree biomass (weight) based on parameters that are easily measured in the field is becoming a fundamental task in forestry and forest-based biomass technology. Traditionally, the cubic-foot or board-foot volume of merchantable products, such as sawlogs or pulpwood, adequately described forest stands; however, the intensity of forest utilization has increased in recent years because of whole-tree harvesting and the use of wood for energy. All aboveground branches, leaves, bark, small trees, and trees of poor form or vigor are now commonly included in the harvested product and are listed as biomass of whole trees (WT) or individual components. With this increasing emphasis on complete tree utilization and use of wood as a source of energy, tables and equations have been developed to show the whole tree biomass as weights of total trees and their components.

Numerous equations for estimating tree biomass from dry weight in kilograms and DBH (tree diameter at breast height, in centimeters) have been developed by researchers based on local and tree species. For example, Landis and Mogren (1975) developed an equation for estimating the biomass of individual Engelmann spruce trees employing the following model:

$$Y = b_0 + b_1 \times DBH^2 \tag{5.6}$$

where Y is tree component dry weight in kilograms, b_0 and b_1 are regression coefficients, and DBH is the tree diameter at breast height in centimeters.

Similar sets of equations have been developed for other species and locations. Regression equations are used to estimate tree biomass in both forestry and ecosystem studies. Examples of many common equations used in the northeastern United States are presented below. These equations are typically developed in the following way: Samples of major tree species are chosen for study, selected dimensions of each tree are recorded, the tree is felled and weighed either whole or in pieces, and subsamples are oven-dried and weighed again to determine tree moisture content. (Tree green weights are converted to dry weights by using moisture content values.) Because biomass is related to tree dimensions,

---

* Based on material contained in Tritton, L.M. and Hornbeck, J.W., *Biomass Equation for Major Tree Species of the Northeast*, U.S. Department of Agriculture, Washington, D.C., 1982.

regression analysis is used to estimate the constants or regression coefficients required for the actual calculation of biomass. The resultant regression equations may be used to estimate the biomass, by species, of all trees for which dimensional data are available. The equations shown below are of several different forms; they can be used to predict biomass from the DBH or the DBH and height.

Key to Abbreviations Used in Biomass Equations

Br	Branch biomass
DBH	Diameter at breast height (1.37 m) measured in inches (in.), millimeters (mm), or centimeters (cm)
DdBr	Dead branch biomass
ht	Tree height
Lf	Leaf biomass
Lf + Tw	Leaf and twig biomass
ln	Natural logarithm to the base e
log	Logarithm to the base 10
St	Stem biomass
St + Br	Steam and branch biomass but not foliage
weight	Weight measured in pounds (lb), grams (gm), or kilograms (kg)
WT	Whole tree biomass (all above-ground components, including leaves, branches, and stem)

Tree Species/Biomass Example Equations

- **Balsam fir** (*Abies balsams*) (Young et al., 1980)

  WT: $\ln (\text{weight}) = 0.5958 + 2.4017 \times \ln (\text{DBH})$

- **Red maple** (*Acer rubrum*) (Young et al., 1980)

  WT: $\ln (\text{weight}) = 0.9392 + 2.3804 \times \ln (\text{DBH})$

- **Sugar maple** (*Acer saccharum*) (Whittaker et al., 1974)

  St: $\log (\text{weight}) = 2.0877 + 2.3718 \times \log (\text{DBH})$

  Br: $\log (\text{weight}) = 0.6266 + 2.9740 \times \log (\text{DBH})$

  DdBr: $\log (\text{weight}) = 0.0444 + 2.2803 \times \log (\text{DBH})$

  Lf + Tw: $\log (\text{weight}) = 1.0975 + 1.9329 \times \log (\text{DBH})$

- **Yellow birch** (*Betula alleghaniensis* Britt.) (Ribe, 1973)

  Lf: $\log (\text{weight}) = 1.9962 + 1.9683 \times \log (\text{DBH})$

  Br: $\log (\text{weight}) = 2.5345 + 1.6179 \times \log (\text{DBH})$

  St: $\log (\text{weight}) = 2.9670 + 2.5330 \times \log (\text{DBH})$

- **Black birch** (*Betula lenta*) (Brenneman et al., 1978)

  WT: $\text{weight} = 1.6542 \times \text{DBH}^{2.6606}$

- **Paper birch** (*Betula papyrifera* Marsh) (Kinerson and Bartholomew, 1977)
  St: ln (weight) = 3.720 + 2.877 × ln (DBH)
  Br: ln (weight) = −1.351 + 4.368 × ln (DBH)

- **Gray birch** (*Betula populifolia* Marsh) (Young et al., 1980)
  Wt: ln (weight) = 1.0931 + 2.3146 × ln (DBH)

- **Hickory** (*Carya* spp.) (Wiant et al., 1977)
  St + Br: weight = 1.93378 × DBH$^{2.62090}$

- **Beech** (*Fagus grandifolia* Ehrh.) (Ribe, 1973)
  Lf: log (weight) = 2.0660 + 1.8089 × log (DBH)
  Br: log (weight) = 2.5983 + 1.5402 × log (DBH)
  St: log (weight) = 3.0692 + 2.4868 × log (DBH)

- **White ash** (*Fraxinus americana*) (Brenneman et al., 1978)
  WT: weight = 2.3626 × DBH$^{2.4798}$

- **Aspen** (*Populus* spp.) (MacLean and Wein, 1976)
  WT: log (weight) = −0.7891 + 2.0673 × log (DBH)

- **Spruce** (*Picea* spp.) (MacLean and Wein, 1976)
  WT: log (weight) = −0.2112 + 1.5639 × log (DBH)

- **Red pine** (*Pinus resinosa* Ait.) (Dunlap and Shipman, 1967)
  St: weight = −113.954 + 35.265 × (DBH)

- **White pine** (*Pinus strobus*) (Swank and Schreuder, 1974)
  Lf: ln (weight) = 3.051 + 2.1354 × ln (DBH)
  Br: ln (weight) = 3.158 + e2.5328 × ln (DBH)
  St: ln (weight) = −2.788 + 2.1338 × ln (DBH)

- **Yellow poplar** (*Liriodendron tulipifera*) (Hitchcock, 1978)
  St + Br: log (weight) = 1.9167 + 0.7993 × log (DBH^2 × ht)

- **Pin cherry** (*Prunus pensylvanica*) (Young et al., 1980)
  WT: ln (weight) = 0.9758 + 2.1948 × ln (DBH)

- **Black cherry** (*Prunus serotina* Ehrh.) (Wiant et al., 1979)
  St + Br: weight = 0.12968 (DBH^2 × ht)$^{0.97028}$

- **White oak** (*Quercus alba*) (Reiners, 1972)
  Lf: log (weight) = 2.1426 + 1.6684 × log (DBH)

- **Scarlet oak** (*Quercus coccinea*) (Clark and Schroeder, 1977)
  St + Br: weight = 0.12161 × (DBH^2 × ht)$^{1.00031}$

- **Chestnut oak** (*Quercus prinus*) (Wiant et al., 1979)
  St + Br: weight = 0.06834 × (DBH^2 × ht)$^{1.06370}$

- **Northern red oak** (*Quercus rubra*) (Clark and Schroeder, 1977)
  WT: weight = 0.10987 × (DBH^2 × ht)$^{1.00197}$

- **Black oak** (*Quercus velutina*) (Bridge, 1979)
  WT: ln (weight) = −0.34052 + 2.65803 × ln (DBH)

- **Hemlock** (*Tsuga canadensis*) (Young et al., 1980)
  WT: ln (weight) = 0.6803 + 2.3617 × ln (DBH)

- **General hardwoods** (Monk et al., 1970)
  WT: log (weight) = 1.9757 + 2.5371 × log (DBH)

- **General softwoods** (Monteith, 1979)
  WT: weight = 4.5966 − (0.2364 × DBH) + (0.00411 × DBH^2)

Chapter Review Questions

5.1   What is also known as the gross calorific value?

5.2   _____ is the ratio of the weight of water in wood to the oven-dry weight of the wood.

5.3   What does the total volume of biomass include?

References and Recommended Reading

Borman, G.L. and Ragland, K.W. (1998). *Combustion Engineering*, McGraw-Hill, New York.

Brenneman, B.B., Frederick, D.J., Gardner, W.E., Schoenhofen, L.H., and Marsh, P.L. (1978). Biomass of species and stands of West Virginia hardwoods, in Pope, P.E., Ed., *Proceedings of Central Hardwood Forest Conference II*, November 14–16, 1978, Purdue University, West Lafayette, IN.

Bridge, J.A. (1979). Fuelwood Production of Mixed Hardwoods on Mesic Sites in Rhode Island, MS thesis, University of Rhode Island, Kingston.

Briggs, D. (1994). *Forest Products Measurements and Conversion Factors: With Special Emphasis on the U.S. Pacific Northwest*, College of Forest Resources University of Washington, Seattle, Chapter 1.

Bushnell, D. (1989). *Biomass Fuel Characterization: Testing and Evaluating the Combustion Characteristics of Selected Biomass Fuels*, Bonneville Power Administration, Portland, OR.

Clark III, A. and Schroeder, J.G. (1977). *Biomass of Yellow Poplar in Natural Stands in Western North Carolina*, Paper SE-165, U.S. Department of Agriculture Forest Service, Washington, D.C., 41 pp.

Clark III, A., Phillips, D.R., and Hitchcock, H.C. (1980). *Predicted Weights and Volumes of Scarlet Oak Trees on the Tennessee Cumberland Plateau*, Paper SE-214, U.S. Department of Agriculture Forest Service, Washington, D.C., 23 pp.

Dunlap, W.H. and Shipman, R.D. (1967). *Density and Weight Prediction of Standing White Oak, Red Maple, and Red Pine*, Research Brief, Pennsylvania State University, University Park, PA, pp. 66–69.

Hitchcock III, H.C. (1978). Aboveground tree weight equations for hardwood seedlings and saplings, *TAPPI Journal*, 61(10):119–120.

Ince, P.J. (1979). *How to Estimate Recoverable Heat Energy in Wood or Bark Fuels*, General Technical Report FPL-GTR-29, U.S. Department of Agriculture, Forest Service, Forest Products Laboratory, Madison, WI.

Jenkins, B. (1993). *Properties of Biomass, Appendix to Biomass Energy Fundamentals*, EPRI Report TR-102107, Electric Power Research Institute, Palo Alto, CA.

Jenkins, B.L. et al. (1998). Combustion properties of biomass, *Fuel Processing Technology*, 54:17–46.

Jenkins, J.C., Chojnacky, D.C., Heath, L.S., and Birdsey, R.A. (2004). *A Comprehensive Database of Diameter-Based Biomass Regressions for North American Tree Species*, General Technical Report NE-319, U.S. Department of Agriculture, Forest Service, Northeastern Research Station, Newtown Square, PA.

Kinerson, R.S. and Bartholomew, I. (1977). *Biomass Estimation Equations and Nutrient Composition of White Pine, White Birch, Red Maple, and Red Oak in New Hampshire*, Research Report No. 62, New Hampshire Agricultural Experiment Station, University of New Hampshire, Durham, 8 pp.

Kluender, R.A. (1980). *The Forester's Wood Energy Handbook*, Publ. No. 80-A-12, American Pulpwood Association, Washington, D.C.

Landis, T.D. and Mogren, E.W. (1975). Tree strata biomass of subalpine spruce-fir stands in southwestern Colorado, *Forest Science*, 21(1):9–14.

MacLean, D.A. and Wein, R.W. (1976). Biomass of jack pine and mixed hardwood stands in northeastern New Brunswick, *Canadian Journal of Forest Research*, 6(4):441–447.

Maker, T.M. (2004). *Wood-Chip Heating Systems: A Guide for Institutional and Commercial Biomass Installations*, Biomass Energy Resource Center, Montpelier, VT.

Monk, C.D., Child, G.I., and Nicholson, S.A. (1970). Biomass, litter and leaf surface area estimates of an oak–hickory forest, *Oikos*, 21:138-141.

Monteith, D.B. (1979). *Whole-Tree Weight Table for New York*, AFRI Research Report 40, University of New York, Syracuse, 67 pp.

NREL. (2010). *Learning about Renewable Energy*, National Renewable Energy Laboratory, Golden, CO (http://www.nrel.gov/learning/).

Pennsylvania DCNR. (2007). *Guidance on Harvesting Woody Biomass for Energy in Pennsylvania*. Pennsylvania Department of Conservation and Natural Resources, Harrisburg (www.dcnr. state.pa.us/PA_Biomass_guidance_final.pdf).

Reiners, W.A. (1972). Structure and energetics of three Minnesota forests, *Ecological Monographs*, 42(1):71–94.

Ribe, J.H. (1973). *Puckerbrush Weight Tables*, University of Maine, Orono.

Schroeder, P., Brown, S., Mo, J., Birdsey, R., and Cieszewski, C. (1997). Biomass estimation of temperate broadleaf forests of the U.S. using forest inventory data, *Forest Science*, 43:424–434.

Somogyi, Z. et al. (2007). Indirect methods of large-scale biomass estimation, *European Journal of Forest Research*, 126(2):197–207.

Swank, W.T. and Schreuder, H.T. (1974). Comparison of three methods of estimating surface area and biomass for a forest of young eastern white pine, *Forest Science*, 20:91–100.

Tillman, D. (1978). *Wood as an Energy Resource*, Academic Press, New York.

Whittaker, R.H., Bormann, F.H., Likens, G.E., and Siccama, T.G. (1974). The Hubbard book ecosystem study: forest biomass and production, *Ecological Monographs*, 4:233–254.

Wiant, Jr., H.V., Sheetz, C.E., Colaninno, A., DeMoss, J.C., and Castaneda, F. (1977). *Tables and Procedures for Estimating Weights of Some Appalachian Hardwoods*, West Virginia University, Agricultural and Forestry Experiment Station, Morgantown, 8 pp.

Wiant, Jr., H.V., Castaneda F., Sheetz, C.E., Colaninno, A., and DeMoss, J.C. (1979). Equations for predicting weights of some Appalachian hardwoods, *West Virginia Forestry Notes*, 7:21–28.

Young, H.E., Ribe, J.H., and Wainwright, K. (1980). *Weight Tables for Tree and Shrub Species in Maine*, University of Maine, Life Sciences and Agriculture Experiment Station, Orono.

6

Forest Biomass Sampling

> Most human decisions are made with incomplete knowledge. In daily life, a physician may diagnose disease from a single drop of blood or a microscopic section of tissue; a housewife judges a watermelon by its "plug" or by the sound it emits when thumped; and amid a bewildering array of choices and claims we select toothpaste, insurance, vacation spots, mates, and careers with but a fragment of the total information necessary or desirable for complete understanding. All of these we do with the ardent hope that the drop of blood, the melon plug, and the advertising claim give a reliable picture of the population they represent.
>
> —Frank Freese (1976)

Introduction

Freese (1976) observed that partial knowledge is a normal state of doing business in many professions; the same can be said for the practice of forestry. The complete census is rare, and the sample is commonplace. A forester must advertise timber sales with estimated volume, estimated grade yield and value, estimated cost, and estimated risk. The nurseryperson sows seed whose germination is estimated from a tiny fraction of the seedlot, and at harvest he or she estimates the seedling crop with sample counts in the nursery beds. Enterprising pulp companies, seeking a source of raw material in sawmill residue, may estimate the potential tonnage of chippable material by multiplying reported production by a set of conversion factors obtained at a few representative sawmills.

On the surface, and in many cases, it would seem better to measure and not to sample; however, there are several good reasons why sampling is often preferred. In the first place, complete measurement or enumeration may be impossible. That is, not all units in the population can be identified. For example, how does one accurately count each branch or twig on a tree? How do we test the quality of every drop of water in a reservoir? How do we weigh every fish in a stream or count all seedlings in a 1000-bed nursery, enumerate all the egg masses in a turpentine beetle infestation, or measure the diameter and height of all merchantable trees in a 20,000-acre forest? Moreover, the nurseryperson might be somewhat better informed if he or she knew the germinative capacity of all the seed to be sown, but the destructive nature of the germination test precludes testing every seed. For identical reasons, it is impossible to conduct tests that are destructive on every chainsaw without destroying every chainsaw. Likewise, it is impossible to measure the bending strength of all the timbers to be used in a bridge, the tearing strength of all the paper to be put into a book, or the grade of all the boards to be produced for a timber sale. If the tests were permitted, no seedlings would be produced, no bridges would be built, no books printed, and no stumpage sold. Clearly where testing is destructive some sort of sampling is inescapable. Obviously, the enormity of the counting task and the destructive effects of testing demand some sort of sampling procedure.

Sampling will frequently provide the essential information at a far lower cost and in less time than a complete enumeration. Surveying 100% of the lumber market is not going to provide information that is very useful to a seller if it takes 11 months to complete the job. In addition, it is often the case that sampling information may at times be more reliable than that obtained by a 100% inventory. There are several reasons why this might be true. With fewer observations to be made, measurement of the units in the sample can be and is more likely to be made with greater care. Moreover, a portion of the savings resulting from sampling could be used to buy better instruments and to employ or train higher caliber personnel. It is not difficult to see that good measurements on 5% of the units in a population could provide more reliable information than sloppy measurements on 100% of the units.

The bottom line to making sampling effective and accurate is obtaining reliable data from the population sampled and making correct inferences about that population. The quality of the sampling depends on such factors as the rule by which the sample was drawn, the care exercised in measurement, and the degree to which bias was avoided (Avery and Burkhart, 2002).

Terms and Concepts

Inventory—The systematic acquisition and analysis of information necessary to describe, characterize, or quantify vegetation. As might be expected, data for many different vegetation attributes can be collected. Inventories can be used not only for mapping and describing ecological sites but also for determining ecological status, assessing the distribution and abundance of species, and establishing baseline data for monitoring studies.

Population—A population (used here in the structural, not biological, sense) is a complete collection of objects (usually called *units*) about which one wishes to make statistical inferences. Population units can be individual plants, points, plots, quadrats, or transects.

Sample—A set of units selected from a population used to estimate something about the population (a procedure that statisticians refer to as making inferences about the population). In order to properly make inferences about a population, the units must be selected using some random procedure. The units selected are called *sampling units*.

Sampling—A means by which inferences about a plant community can be made based on information from an examination of a small proportion of that community. The most complete way to determine the characteristics of a population is to conduct a complete enumeration (or census). In a census, each individual unit in the population is sampled to provide the data for the aggregate. This process is both time consuming and costly, and it may also result in inaccurate values when individual sampling units are difficult to identify; therefore, the best way to collect vegetation data is to sample a small subset of the population. If the population is uniform, sampling can be conducted anywhere in the population, but most vegetation populations are not uniform. It is important that data be collected that can ensure that the sample represents the entire population. Sample design is an important consideration in collected representative data.

Sampling unit—One of a set of objects in a sample that is drawn to make inferences about a population of those same objects. A collection of sampling units is a sample. Sampling units can be individual plants, points, plots, quadrats, or transects.

Shrub characterization—This topic is addressed here because it is not covered in most of the techniques in this text. Shrub characterization is the collection of data on the shrub and tree component of a vegetation community. Attributes that could be important for shrub characterization are height, volume, foliage density, crown diameter, form class, age class, and total number of plants by species (density). Another important feature of shrub characterization is the collection of data on a vertical as well as horizontal plant. Canopy layering is also important. The occurrence of individual species and the extent of canopy cover of each species is recorded in layers. The number of layers chosen should represent the herbaceous layer, the shrub layer, and the tree layers, though additional layers can be added if needed.

Trend—Refers to the direction of change. Vegetation data are collected at different points in time on the same site, and the results are then compared to detect a change. Trend is described as *moving toward meeting objectives, moving away from meeting objectives, not apparent,* or *state*. Trend data are important in determining the effectiveness of on-the-ground management actions. Trend data indicate whether rangeland is moving toward or away from specific objectives. The trend of a rangeland area may be judged by noting changes in vegetation attributes such as species composition, density, cover, production, and frequency. Trend data, along with actual use, authorized use, estimated use, utilization, climate, and other relevant data, are considered when evaluating activity plans.

Simple Random Sampling

Many statistical procedures assume *simple random sampling*—the fundamental selection method (not to be confused with random sampling). In this approach, a simple random sample is a subset of individuals (a sample) chosen from a larger set (a population). Each individual is chosen randomly and entirely by chance (must be free from deliberate choice), such that each individual has the same probability of being chosen at any stage during the sampling process, and each subset of k individuals has the same probability of being chosen for the sample as any other subset of k individuals (Yates et al., 2008). In small populations and often in large ones, such sampling is typically done without replacement; that is, one deliberately avoids choosing any member of the population more than once. Although simple random sampling can be conducted with replacement instead, this is less common and would normally be described more fully as simple random sampling with replacement. Sampling done without replacement is no longer independent but still satisfies exchangeability, so many results still hold. Further, for a small sample from a large population, sampling without replacement is approximately the same as sampling with replacement, as the odds of choosing the same sample twice are low. Conceptually, simple random sampling is the simplest of the probability sampling techniques. Though simple in form and use, the goal of simple random sampling, of course, is to come up with an unbiased random selection of individuals that in the long run represents the population.

Introduction to Simple Random Sampling*

It may be difficult to visualize giving every possible combination of n units an equal chance of appearing in a sample of size n, but it can be easily accomplished. It is only necessary to be sure that at any stage of the sampling the selection of a particular unit is in no way influenced by the other units that have been selected. Stated differently, the selection of any given unit should be completely independent of the selection of all other units. One way to do this is to assign every unit in the population a number and then draw n numbers from a table of random digits. Alternatively, the numbers can be written on some equal-sized slips of paper which are placed in a bowl, thoroughly mixed, and then drawn one at a time. For units such as individual tree seeds or seedlings, the units themselves may be drawn at random. The units may be selected with or without replacement. If selection is with replacement, each unit is allowed to appear in the sample as often as it is selected. In sampling without replacement, a particular unit is allowed to appear in the sample only once. Most forest sampling is without replacement.

Sample Selection

The selection method and computations may be illustrated by the sampling of a 250-acre plantation. The objective of the survey was to estimate the mean cordwood volume per acre in trees more than 5 inches diameter at breast height (DBH) outside bark. The population and sample units were defined to be square, quarter-acre plots with the unit value being the plot volume. The sample was to consist of 25 units selected at random and without replacement. The quarter-acre units were plotted on a map of the plantation and assigned numbers from 1 to 1000. From a table of random digits, 25 three-digit numbers were selected to identify the units to be included in the sample (the number 000 was associated with the plot numbered 1000). No unit was counted in the sample more than once. Units drawn a second time were rejected and an alternative unit was randomly selected. The cordwood volumes measured on the 25 units were as follows:

7	10	7	4	7
8	8	8	7	5
2	6	9	7	8
6	7	11	8	8
7	3	8	7	7

Total = 175

Estimates

If the cordwood volume on the ith sampling unit is designated y_i, then the estimated mean volume ($\bar{y}$) per sampling unit is

$$\bar{y} = \frac{\sum_{i=1}^{n} y_i}{n} = \frac{7+8+2+\ldots+7}{25} = \frac{175}{25} = 7 \text{ cords per quarter-acre plot}$$

---

* Much of this section is adapted from Freese, F., *Elementary Forest Sampling*, Agriculture Handbook No. 232, U.S. Department of Agriculture, Washington, D.C., 1976.

The mean volume per acre would, of course, be 4 times the mean volume per quarter-acre plot, or 28 cords. As there is a total of $N = 1000$ quarter-acre units in the 250-acre plantation, the estimated total volume ($\hat{Y}$) in the plantation would be

$$\hat{Y} = N \times \bar{y} = 1000 \times 7 = 7000 \text{ cords}$$

Alternatively,

$$\hat{Y} = 28 \text{ cords per acre} \times 250 \text{ acres} = 7000 \text{ cords}$$

Standard Errors

A first step in computing the standard error of estimate is to make an estimate (s_y^2) of the variance of individual values of y:

$$s_y^2 = \frac{\sum\limits_{i=1}^{n} y_i^2 - \frac{\left(\sum\limits_{i=1}^{n} y_i\right)^2}{n}}{(n-1)} \tag{6.1}$$

In this example,

$$s_y^2 = \frac{(7^2 + 8^2 + \ldots + 7^2) - \frac{175^2}{25}}{(25-1)} = \frac{1317 - 1225}{24} = 3.8333 \text{ cords}$$

When sampling is without replacement, the standard error of the mean ($s_{\bar{y}}$) for a simple random sample is

$$s_{\bar{y}} = \sqrt{\frac{s_y^2}{n}\left(1 - \frac{n}{N}\right)} \tag{6.2}$$

where N = total number of sample units in the entire population, and n = number of units in the sample.

For the plantation survey,

$$s_{\bar{y}} = \sqrt{\frac{3.8333}{25}\left(1 - \frac{25}{1000}\right)} = \sqrt{.1533 \times .975} = 0.387 \text{ cord}$$

This is the standard error for the mean per quarter-acre plot. By the rules for the expansion of variances and standard errors, the standard error for the mean volume per acre will be $4 \times 0.887 = 1.548$ cords.

Similarly, the standard error for the estimated total volume will be

$$s_{\hat{Y}} = N \times s_{\bar{y}} = 1000 \times .387 = 387 \text{ cords}$$

Sampling with Replacement

In the formula for the standard error of the mean, the term $(1 - n/N)$ is known as the *finite population correction* (FPC). It is used when units are selected without replacement. If units are selected with replacement, the FPC is omitted and the formula for the standard error of the mean becomes

$$s_{\bar{y}} = \sqrt{\frac{s_y^2}{n}} \qquad (6.3)$$

Even when sampling is without replacement, the sampling fraction (n/N) may be extremely small, making the FPC very close to unity. If (n/N) is less than 0.05, the FPC is commonly ignored, and the standard error is computed from the shortened formula.

Confidence Limits for Large Samples

By itself, the estimated mean of 28 cords per acre does not tell us very much. Had the sample consisted of only 2 observations we might conceivably have drawn the quarter-acre plots having only 2 and 3 cords, and the estimated mean would be 10 cords per acre. Or, if we had selected the plots with 10 and 11 cords, the mean would be 42 cords per acre. To make an estimate meaningful it is necessary to compute confidence limits that indicate the range within which we might expect (with some specified degree of confidence) to find the parameter. The 95% confidence limits for large samples are given by

Estimate ± (2 × Standard Error of Estimate)

Thus, the mean volume per acre (28 cords) that had a standard error of 1.548 cords would have confidence limits of

28 ± (2 × 1.548) = 24.90 to 31.10 cords per acre

The total volume of 7000 cords that had a standard error of 387 cords would have 95% confidence limits of

7000 ± (2 × 387) = 6226 to 7774 cords

Unless a 1-in-20 chance has occurred in sampling, the population mean volume per acre is somewhere between 24.9 and 31.1 cords, and the true total volume is between 6226 and 7774 cords.

Because of sampling variation, the 95% confidence limits will, on the average, fail to include the parameter in 1 case out of 20. It must be emphasized, however, that these limits and the confidence statement take account of sampling variation only. They assume that the plot values are without measurement error and that the sampling and estimating procedures are unbiased and free of computational mistakes. If these basic assumptions are not valid, the estimates and confidence statements may be nothing more than a statistical hoax.

Confidence Limits for Small Samples

Ordinarily, large-sample confidence limits are not appropriate for samples of less than 30 observations. For smaller samples the proper procedure depends on the distribution of the unit values in the parent population, a subject that is beyond the scope of this book. Fortunately, many forest measurements follow the bell-shaped normal distribution or a distribution that can be made nearly normal by transformation of the variable. For samples of any size from normally distributed populations, Student's *t* value can be used to compute confidence limits. The general formula is

$$\text{Estimate} \pm (t \times \text{Standard Error of Estimate})$$

The values of *t* have been tabulated. The particular value of *t* to be used depends on the degree of confidence desired and on the size of the sample. For 95% confidence limits, the *t* values are taken from the column for a probability of .05. For 99% confidence limits, the *t* value would come from the .01 probability column. Within the specified columns, the appropriate *t* for a simple random sample of *n* observations is found in the row for $(n - 1)$ degrees of freedom. For a simple random sample of 25 observations, the *t* value for computing the 95% confidence limits will be found in the .05 column and the 24 degrees of freedom row. This value is 2.064. Thus, for the plantation survey that showed a mean per-acre volume of 28 cords and a standard error of the mean of 1.548 cords, the small-sample 95% confidence limits would be

$$28 \pm (2064 \times 1.548) = 24.80 \text{ to } 31.20 \text{ cords}$$

The same *t* value is used for computing the 95% confidence limits on the total volume. As the estimated total was 7000 cores with a standard error of 387 cores, the 95% confidence limits are

$$7000 \pm (2.064 \times 387) = 6201 \text{ to } 7799 \text{ cords}$$

DID YOU KNOW?

In this book, *degrees of freedom* refers to a parameter in the distribution of Student's *t*. When a tabular value of *t* is required, the number of degrees of freedom (dfs) must be specified. The expression is not easily explained in nonstatistical language. One definition is that the dfs are equal to the number of observations in a sample minus the number of independently estimated parameters used in calculating the sample variance. Thus, in a simple random sample of *n* observations, the only estimated parameter required when calculating the sample variance is the mean $(\bar{x})$, so the dfs would be $(n - 1)$.

Size of Sample

In the example illustrating simple random sampling, 25 units were selected. But why 25? Why not 100? Or 10? All too often the number depends on the sampler's view of what looks about right, but there is a somewhat more objective solution: Take only the number of observations required to give the desired precision. In planning the plantation survey, we could have stated that unless a 1-in-2 change occurs we would like our sample estimate of the mean to be within $\pm E$ cords of the population mean. Because the small-sample confidence limits are computed as $\bar{y} \pm t(s_{\bar{y}})$, this is equivalent to saying that we want

$$t\left(s_{\bar{y}}\right) = E \tag{6.4}$$

For a simple random sample,

$$s_{\bar{y}} = \sqrt{\frac{s_y^2}{n}\left(1 - \frac{n}{N}\right)} \tag{6.5}$$

Substituting for $s_{\bar{y}}$ in the first equation we get

$$(t)\sqrt{\frac{s_y^2}{n}\left(1 - \frac{n}{N}\right)} = E \tag{6.6}$$

Rewritten in terms of the sample size (n) this becomes

$$n = \frac{1}{\dfrac{E^2}{t^2 s_y^2} + \dfrac{1}{N}} \tag{6.7}$$

To solve this relationship for n, we must have some estimate (s_y^2) of the population variance. Sometimes the information is available from previous surveys. In the illustration, we found $s_y^2 = 3.88$, a value that might be taken as representative of the variation among quarter-acre plots in this or similar populations. In the absence of this information, a small preliminary survey might be made to obtain an estimate of the variance. When, as often happens, neither of these solutions is feasible, a very crude estimate can be made from the relationship:

$$s_y^2 = \left(\frac{R}{4}\right)^2 \tag{6.8}$$

where R is the estimated range from the smallest to the largest unit value likely to be encountered in sampling.

For the plantation survey we might estimate the smallest y value on quarter-acre plots to be 1 cord and the largest to be 10 cords. As the range is 9, the estimated variance would be

$$s_y^2 = \left(\frac{9}{4}\right)^2 = 5.06$$

This approximation procedure should be used only when no other estimate of the variance is available.

Having specified a value of E and obtained an estimate of the variance, the last piece of information we need is the value of t_2. Here, we hit a bit of a snag. To use t we must know the number of degrees of freedom. But, the number of dfs must be $(n - 1)$, and n is not known and cannot be determined without knowing t.

An iterative solution will give us what we need, and it is not as difficult as it sounds. The procedure is to guess at a value of n, use the guessed value to get the degrees of freedom for t, and then substitute the appropriate t value in the sample-size formula to solve for a first approximation of n. Selecting a new n somewhere between the guessed value and the first approximation, but closer to the latter, we compute a second approximation. The process is repeated until successive values of n are the same or only slightly different. Three trials usually suffice.

To illustrate the process, suppose that in planning the plantation survey we had specified that, barring a 1-in-100 change, we would like the estimate to be within 3.0 cords of the true mean volume per acre. This is equivalent to $E = 0.75$ cord per quarter-acre. From previous experience, we estimate the population variance among quarter-acre plots to be $s_y^2 = 4$, and we know that there is a total of $N = 1000$ units in the population. To solve for n, this information is substituted in the sample-size formula (Equation 6.7):

$$n = \cfrac{1}{\cfrac{(0.75)^2}{t^2(4)} + \cfrac{1}{1000}}$$

We will have to use the t value for the .01 probability level, but we do not know how many degrees of freedom t will have without knowing n. As a first guess, we can try $n = 61$, in which case the value of t with 60 degrees of freedom at the .01 probability level is $t = 2.66$. Thus, the first approximation will be

$$n_1 = \cfrac{1}{\cfrac{(0.75)^2}{(2.66)^2(4)} + \cfrac{1}{1000}}$$

$$= \cfrac{1}{\cfrac{0.5625}{(7.0756)(4)} + \cfrac{1}{1000}} = 47.9$$

A second guessed value for n would be somewhere between 61 and 48, but closer to the computed value. We might test $n = 51$, for which the value of t (50 dfs) at the 0.1 level is about 2.68, so

$$n_2 = \cfrac{1}{\cfrac{0.5625}{(7.1824)(4)} + \cfrac{1}{1000}} = 48.6$$

The desired value is somewhere between 51 and 48.6 but much closer to the latter. Because the estimated sample size is, at best, only a good approximation, it is rather futile to strain on the computation of n. In this case, we would probably settle on $n = 50$, a value that could have been easily guessed after the first approximation was computed.

If the sampling fraction (n/N) is likely to be small (say, less than 0.05), then the finite-population correction ($1 - n/N$) may be ignored in the estimation of sample size and the formula can be simplified as

$$n = \frac{t^2 s_y^2}{E^2}$$

This formula is also appropriate in sampling with replacement. In the previous example, the simplified formula gives an estimated sample size of $n = 51$. The short formula is frequently used to get a first approximation of n. Then, if the sample size indicated by the short formula is a considerable proportion (say, over 10%) of the number of units in the population and sampling will be without replacement, the estimated sample size is recomputed with the long formula.

Effect of Plot Size on Variance

When estimating sample size, the effect of plot size and the scale of the unit values on variance must be kept in mind. In the plantation survey, a plot size of one-quarter acre was selected and the variance among plot volumes was estimated to be $s^2 = 4$. This is the variance among volumes per quarter-acre. Because the desired precision was expressed on a per-acre basis it was necessary to modify either the precision specification or s^2 to get them on the same scale. In the example, s^2 was used without change and the desired precision was divided by 4 to put it on a quarter-acre basis. The same result could have been obtained by leaving the specified precision unchanged and putting the variance on a per-acre basis. Because the quarter-acre volumes should be multiplied by 4 to put them on a per-acre basis, the variance of quarter-acre volumes should be multiplied by 16. (Remember: If x is a variable with variance s^2, then the variance of the variable $z = kx$ is $k^2 s^2$).

Plot size has an additional effect on variance. At the same scale of measurement, small plots will almost always be more variable than large ones. The variance in volume per acre on quarter-acre plots would be somewhat larger than the variance in volume per acre on half-acre pots but slightly smaller than the variance in volume per acre of fifth-acre plots. Unfortunately, the relation of plot size to variance changes from one population to another. Large plots tend to have a smaller variance because they average out the effect of clumping and holes. In very uniform populations, changes in plot size have little effect on variance. In nonuniform populations, the relationship of plot size to variance will depend on how the sizes of clumps and holes compare to the plot sizes. Experience is the best guide as to the effect of changing plot size on variance. When neither experience nor advice is available, a very rough approximation can be obtained by the rule:

$$s_2^2 = s_1^2 \sqrt{P_1/P_2} \tag{6.9}$$

Thus, if the variance in cordwood volume *per acre* on quarter-acre plots is $s_1^2 = 61$, then the variance in cordwood volume *per acre* on tenth-acre plots will be roughly

$$61\sqrt{0.25/0.10} = 96$$

The same results can be obtained without worrying about the scale of measurement if the squared coefficients of variation (C^2) are used in place of the variances. The formula would then be

$$C_2^2 = C_1^2 \sqrt{P_1/P_2} \tag{6.10}$$

■ *Example 6.1*

Problem: A survey is to be made to estimate the mean board-foot volume per acre in a 200-acre tract. Barring a 1-in-20 chance, we would like the estimate to be within 500 board feet of the population mean. Sample plots will be one-fifth acre. A survey in a similar tract showed the standard deviation among quarter-acre plot volumes to be 520 board feet. What size sample will be needed?

Solution: The variance among quarter-acre plot volumes is $520^2 = 270,400$. For quarter-acre volumes expressed on a per-acre basis the variance would be

$$s_1^2 = (4^2)(270,400) = 4,326,400$$

The estimated variance among fifth-acre plot volumes expressed on a per-acre basis would then be

$$s_2^2 = s_1^2 \sqrt{P_1/P_2} = 4,326,400\sqrt{0.25/0.20} = (4,326,400)(1.118) = 4,836,915$$

The population size is $N = 1000$ fifth-acre plots. If, as a first guess, $n = 61$, then the t value at the .05 level with 60 degrees of freedom is 2.00. The first computed approximation of n is

$$n_1 = \cfrac{1}{\cfrac{(500)^2}{(4)(4,836,915)} + \cfrac{1}{1000}} = 71.8$$

The correct solution is between 61 and 71.8 but much closer to the computed value. Repeated trials will give values between 71.0 and 71.8. The sample size (n) must be an integral value; because 71 is too small, a sample of $n = 72$ observations would be required for the desired precision.

Stratified Random Sampling*

Stratified random sampling refers to a sampling method where the population consists of N elements compiled during, for example, forest inventory work. The purpose of stratification is to reduce the variation within the forest subdivision and improve the precision of the population estimate, thus increasing the usefulness of the sample. The population is divided into H groups (on the basis of similarity of some characteristic), called *strata*. Each element of the population can be assigned to one, and only one, stratum. The number of observations within each stratum N_h is known, and $N = N_1 + N_2 + N_3 + \ldots + N_{H-1} + N_H$. A probability sample is drawn from each stratum. Stratified random sampling in forest inventory has the following advantages over simple random sampling:

---

* Much of this section is adapted from Freese, F., *Elementary Forest Sampling*, Agriculture Handbook No. 232, U.S. Department of Agriculture, Washington, D.C., 1976.

1. A stratified sample can provide greater precision than a simple random sample of the same size.
2. A stratified sample uses a smaller sample to ensure greater precision; this often saves money.
3. A stratified sample can guard against obtaining an unrepresentative sample (e.g., a sample of all pine trees from a mixed-stand population).
4. Researchers can obtain sufficient sample points to support a separate analysis of any subgroup.

On the other hand, the disadvantages of stratification are that it may require more administrative effort than a simple random sample. Also, the size of each stratum must be known, and the sampling units must be taken in each stratum for which an estimate is needed. When sampling a forest, we might set up strata corresponding to the major timber types, make separate sample estimates of each type, and then combine the type data to give an estimate for the entire population. If the variation among units within types is less than the variation among units that are not in the same type, the population estimate will be more precise than if sampling had been at random points over the entire population.

The sampling and computational procedures can be illustrated with data from a cruise made to estimate the mean cubic-ft volume per acre on an 800-acre forest. On aerial photographs, the tract was divided into three strata corresponding to the three major forest types: pine, bottom-land hardwoods, and upland hardwoods. The boundaries and total acreage of each type were known. Ten one-acre plots were selected at random and without replacement in each stratum:

Stratum	Observations			
I. Pine	570	510	600	
	640	590	780	Total = 6100
	480	670	700	
	560			
II. Bottom-land hardwoods	520	630	810	
	710	760	580	Total = 7370
	770	890	860	
	840			
III. Upland hardwoods	420	540	320	
	210	180	270	Total = 3040
	290	260	200	
	350			

Estimates

The first step in estimating the population mean per unit is to compute the sample mean ($\bar{y}_h$) for each stratum. The procedure is the same as for the mean of a simple random sample:

$$\bar{y}_I = 6100/10 = 610 \text{ cubic feet per acre for the pine type}$$
$$\bar{y}_{II} = 7370/10 = 737 \text{ cubic feet per acre for the bottom-land hardwoods}$$
$$\bar{y}_{III} = 3040/10 = 304 \text{ cubic feet per acre for the upland hardwoods}$$

The mean of a stratified sample $(\bar{y}_{st})$ is then computed by

$$\bar{y}_{st} = \frac{\sum_{h=1}^{L} N_h \bar{y}_h}{N}$$

where:

L = The number of strata

N_h = The total size (number of units) of stratum h ($h = 1, ..., L$)

N = The total number of units in all strata:

$$N = \sum_{h=1}^{L} N_h$$

If the strata sizes are

1. Pine = 320 acres = N_I
2. Bottom-land hardwoods = 140 acres = N_{II}
3. Upland hardwoods = 340 acres = N_{III}
   Total = 800 acres = N

then the estimate of the population means is

$$\bar{y}_{st} = \frac{(320 \times 610) + (140 \times 737) + (340 \times 304)}{800} = 502.174 \text{ cubic feet per acre}$$

To estimate the population total $(\hat{Y}_{st})$, simply omit the divisor N:

$$\hat{Y}_{st} = \sum_{h=1}^{L} N_h \bar{y}_h = (320 \times 610) + (140 \times 737) + (340 \times 304) = 401,740$$

Alternatively,

$$\hat{Y}_{st} = N\bar{y}_{st} = 800 \times 502.175 = 401,740$$

Standard Errors

To determine standard errors, it is first necessary to obtain the estimated variance among individuals within each stratum (s_h^2). These variances are computed in the same manner as the variance of a simple random sample. Thus, the variance within Stratum I (Pine) is

$$s_I^2 = \frac{(570^2 + 640^2 + ...700^2) - \dfrac{6100^2}{10}}{(10 - 1)} = \frac{3,794,000 - 3,721,000}{9} = 8111.111$$

Similarly,

$$s_{II}^2 = 15,556.6667$$

$$s_{III}^2 = 12,204.4444$$

From these values, we can find the standard error of the mean of a stratified random sample $(s_{\bar{y}_{st}})$ by the formula:

$$s_{\bar{y}_{st}} = \sqrt{\frac{1}{N^2} \sum_{h=1}^{L} \left[\frac{N_h^2 s_h^2}{n_h} \right] \left(1 - \frac{n_h}{N_h} \right)}$$

where n_h is the number of units observed in stratum h.

For our timber cruising example, we would have

$$s_{\bar{y}_{st}} = \sqrt{\frac{1}{800^2} \left[\frac{320^2 \times 8111.1111}{10} \right] \left(1 - \frac{10}{320} \right) + \ldots + \left[\frac{340^2 \times 12,204.4444}{10} \right] \left(1 - \frac{10}{340} \right)}$$

$$= \sqrt{383.920659} = 19.594$$

As a rough rule we can say that, unless a 1-in-20 chance has occurred, the population mean is included in the range:

$$\bar{y}_{st} \pm 2 \left(s_{\bar{y}_{st}} \right) = 502.175 \pm 2(19.594) = 463 \text{ to } 541$$

If sampling is with replacement or if the sampling fraction within a particular stratum (n_h/N_h) is small, then we can omit the finite-population correction $(1 - n_h/N_h)$ for that particular stratum when calculating the standard error.

Because the population total is estimated by

$$\hat{Y}_{st} = N\bar{y}_{st}$$

then the standard error of $\hat{Y}_{st}$ is simply

$$s_{\hat{Y}_{st}} = N \times s_{\bar{y}_{st}} = 800 \times 19.594 = 15,675$$

Worth repeating: Stratified random sampling offers two primary advantages over simple random sampling. First, it provides separate estimates of the mean and variance of each stratum. Second, for a given sampling intensity, it often gives more precise estimates of the population parameters than would a simple random sample of the same size. For this latter advantage, however, it is necessary that the strata be set up so that the variability among unit values within the strata is less than the variability among units that are not in the same stratum. Some of the drawbacks are that each unit in the population must be assigned to one and only one stratum, that the size of each stratum must be known, and that a sample must be taken in each stratum. The most common barrier to the use of stratified random sampling is lack of knowledge of the strata sizes. If the sampling fractions are small in each stratum, it is not necessary to know the exact strata sizes; the population mean and its standard error can be computed from the relative sizes.

Estimation of Number of Sampling Units

Assuming we have decided on a total sample size of n observations, how do we know how many of these observations to make in each stratum? To estimate the number of sampling units needed, it is necessary to have preliminary information of the variability of the strata in the population and to choose an allowable error and probability level. With this information, two common solutions to this problem are known as *proportional* and *optimum allocation*.

Proportional Allocation

In this procedure, the proportion of the sample that is selected in the hth stratum is made equal to the proportion of all units in the population that fall in that stratum. If a stratum contains half of the units in the population, half of the sample observations would be made in that stratum. In equation form, if the total number of sample units is to be n, then for proportional allocation the number to be observed in stratum h is

$$n_h = (N_h/N) \times n$$

In our earlier example, the 30 sample observations were divided equally among the strata. For proportional allocation, we would have used

$$n_{\text{I}} = (320 / 800) \times 30 = 12$$
$$n_{\text{II}} = (140 / 800) \times 30 = 5.25, \text{ or } 5$$
$$n_{\text{III}} = (340 / 800) \times 30 = 12.75, \text{ or } 13$$

Optimum Allocation

In optimum allocation, the observations are allocated to the strata so as to give the smallest standard error possible with a total of n observations. If we wish to get the most precise estimate of the population mean for the expenditure of money, then optimum allocation should be used. Note that this allocation can be done either if the costs of sampling in all strata are equal or if they differ. For a sample of size n, the number of observations (n_h) to be made in stratum h under optimum allocation is

$$n_h = \left(\frac{N_h s_h}{\sum_{h=1}^{L} N_h s_h} \right) n$$

In terms of our earlier example, the value of $N_h s_h$ for each stratum is

$$N_{\text{I}} s_{\text{I}} = 320\sqrt{8111.1111} = 320 \times 90.06 = 28,819.20$$
$$N_{\text{II}} s_{\text{II}} = 140\sqrt{15,556.6667} = 140 \times 124.73 = 17,462.20$$
$$N_{\text{III}} s_{\text{III}} = 340\sqrt{12,204.4444} = 340 \times 110.47 = 37,559.80$$

$$Total = 83,841.20 = \sum_{h=1}^{\text{III}} N_h s_h$$

Applying these values in the formula, we would get

$$n_\text{I} = \left(\frac{28{,}819.20}{83{,}841.20} \right) \times 30 = 10.3, \text{ or } 10$$

$$n_\text{II} = \left(\frac{17{,}462.20}{83{,}841.20} \right) \times 30 = 6.2, \text{ or } 6$$

$$n_\text{III} = \left(\frac{37{,}559.80}{83{,}841.20} \right) \times 30 = 13.4, \text{ or } 14$$

Here, optimum allocation is not much different from proportional allocation, but sometimes the difference is great.

Optimum Allocation with Varying Sampling Costs

Optimum allocation as described above assumes that the sampling cost per unit is the same in all strata. When sampling costs vary from one stratum to another, the allocation giving the most information per dollar is

$$n_h = \left(\frac{\dfrac{N_h s_h}{\sqrt{c_h}}}{\sum \left(\dfrac{N_h s_h}{\sqrt{c_h}} \right)} \right) \times n$$

where c_h is the cost per sampling unit in stratum h.

The best way to allocate a sample among the various strata depends on the primary objectives of the survey and our information about the population. One of the two forms of optimum allocation is preferable if the objective is to get the most precise estimate of the population mean for a given cost. If we want separate estimates for each stratum and the overall estimate is of secondary importance, we may want to sample heavily in the strata having high-value material. Then we would ignore both optimum and proportional allocation and place our observations so as to give the degree of precision desired for the particular strata.

We can't, of course, use optimum allocation without having some idea about the variability within the various strata. The appropriate measure of variability within the stratum is the standard deviation (not the standard error), but we need not know the exact standard deviation (s_h) for each stratum. In place of actual s_h values, we can use relative values. In our earlier example, if we had known that the standard deviations for the strata were about in the proportions $s_\text{I}{:}s_\text{II}{:}s_\text{III} = 9{:}12{:}11$, we could have used these values and obtained about the same allocation. Where optimum allocation is indicated but nothing is known about the strata standard deviations, proportional allocation is often very satisfactory.

Caution: In some situations, the optimum allocation formula will indicate that the number of units (n_h) to be selected in a stratum is larger than the stratum (N_h) itself. The common procedure then is to sample all units in the stratum and to recompute the total sample size (n) needed to obtain the desired precision.

Sample Size in Stratified Random Sampling

Overall sample (n) size needed to achieve a desired degree of precision at a specified probability level can be computed; however, the exact form of the sample-size formula varies somewhat depending on the method of allocating the sample to the strata. To estimate the total size of sample (n) required in a stratified random sample, the following pieces of information are required:

- A statement of the desired size of the standard error of the mean. This will be symbolized by D.
- A reasonably good estimate of the variance (s_h^2) or standard deviation (s_h) among individuals within each stratum.
- The method of sample allocation. If the choice is optimum allocation with varying sampling costs, the sampling cost per unit for each stratum must also be known.

Given this hard-to-come-by information, we can estimate the size of the sample (n) with these formulas:

- For equal samples in each of the L strata,

$$n = \frac{L \sum_{h=1}^{L} N_h^2 s_h^2}{N^2 D^2 + \sum_{h=1}^{L} N_h^2 s_h^2}$$

- For proportional allocation,

$$n = \frac{N \sum_{h=1}^{L} N_h s_h^2}{N^2 D^2 + \sum_{h=1}^{L} N_h s_h^2}$$

- For optimum allocation with equal sampling costs among strata,

$$n = \frac{\left(\sum_{h=1}^{L} N_h s_h \right)^2}{N^2 D^2 + \sum_{h=1}^{L} N_h s_h^2}$$

- For optimum allocation with varying sampling costs among strata,

$$n = \frac{\left(\sum_{h=1}^{L} N_h s_h \sqrt{c_h} \right) \left(\sum_{h=1}^{L} \frac{N_h s_h}{\sqrt{c_h}} \right)}{N^2 D^2 + \sum_{h=1}^{L} N_h s_h^2}$$

When the sampling fractions (n_h/N_h) are likely to be very small for all strata or when sampling will be with replacement, the second term of the denominators of the above formula:

$$\sum_{h=1}^{L} N_h s_h^2$$

may be omitted, leaving only N^2D^2.

If the optimum allocation formula indicates a sample (n_h) greater than the total number of units (N_h) in a particular stratum, n_h is usually made equal to N_h; that is, all units in that particular stratum are observed. The previously estimated sample size (n) should then be dropped and the total sample size (n') and allocation for the remaining strata recomputed, omitting the N_h and s_h values for the offending stratum but leaving N and D unchanged.

■ Example 6.2

Problem: Assume a population of four strata with sizes (N_h) and estimated variances s_h^2 as follows:

Stratum	N_h	s_h^2	s_h	$N_h s_h$	$N_h s_h^2$
1	200	400	20	4000	80,000
2	100	900	30	3000	90,000
3	400	400	20	8000	160,000
4	20	19,600	140	2800	392,000
$N = 720$				17,800	722,000

Solution: With optimum allocation (same sampling cost per unit in all strata), the number of observations to estimate the population mean with a standard error of $D = 1$ is

$$n = \frac{(17,800)^2}{(720^2 \times 1^2) + 722,000} = 255.4, \text{ or } 256$$

The allocation of these observations according to the optimum formula would be

$$n_1 = \left(\frac{4000}{17,800}\right) \times 256 = 57.5, \text{ or } 58$$

$$n_2 = \left(\frac{3000}{17,800}\right) \times 256 = 43.1, \text{ or } 43$$

$$n_3 = \left(\frac{8000}{17,800}\right) \times 256 = 115.1, \text{ or } 115$$

$$n_4 = \left(\frac{2800}{17,800}\right) \times 256 = 40.3, \text{ or } 40$$

The number of units allocated to the fourth stratum is greater than the total size of the stratum. Thus, every unit in this stratum would be selected $(n_4 = N_4 = 20)$, and the sample size for the first three strata would be recomputed. For these three strata,

$$\sum N_h s_h = 15,000$$

$$\sum N_h s_h^2 = 330,000$$

Hence,

$$n = \frac{15,000^2}{(720^2 \times 1^2) + 330,000} = 265$$

And the allocation of these observations among the three strata would be

$$n_1 = \left(\frac{4000}{15,000}\right) \times 265 = 70.7, \text{ or } 71$$

$$n_2 = \left(\frac{3000}{15,000}\right) \times 265 = 53.0, \text{ or } 53$$

$$n_3 = \left(\frac{8000}{15,000}\right) \times 265 = 141.3, \text{ or } 141$$

Regression Estimation*

Regression estimators, like stratification, were developed to increase the precision or efficiency of sampling by making use of supplementary information about the population being studied. If we have exact knowledge of the basal area (i.e., cross-section of a tree at breast height; used to determine percent stocking) of a stand of timber, the relationship between volume and basal area may help us to improve out estimate of stand volume. The sample data provide information on the volume–basal area relationship, which is then applied to the known basal area, giving a volume estimate that may be better or less expensive than would be obtained by sampling volume alone. Note that in this method a *regression coefficient* is used to adjust an estimate of the mean volume of sampling units. The regression coefficient indicates the average change in volume per unit change in area between the sampling units in the sample and the population. The total number of sampling units in the population and their average size along with the total size of the forest area must be known.

■ *Example 6.3*

Problem: Suppose a 100% inventory of a 200-acre pine stand indicates a basal area of 84 square feet per acre in trees 3.6 inches in DBH and larger. Assume further that on 20 random plots, each one-fifth acre in size, measurements were made of the basal area (x) and volume (y) per acre.

---

* Much of this section is adapted from Freese, F., *Elementary Forest Sampling*, Agriculture Handbook No. 232, U.S. Department of Agriculture, Washington, D.C., 1976.

Basal area per acre (x) (ft²)	Volume per acre (y) (ft³)	Basal area per acre (x) (ft²)	Volume per acre (y) (ft³)
88	1680	88	1840
72	1460	80	1630
80	1590	82	1560
96	1880	76	1560
64	1240	86	1610
48	1060	73	1370
76	1500	79	1490
85	1620	85	1710
93	1880	84	1600
110	2140	75	1440
Total		1620	31,860
Mean		81	1593

Some values and parameters that will be needed later are

$$n = 20$$
$$\Sigma y = 31,860$$
$$\bar{y} = 1593$$
$$\Sigma y^2 = 51,822,600$$
$$\Sigma xy = 2,635,500$$
$$\Sigma x = 1620$$
$$\bar{x} = 81$$
$$\Sigma x^2 = 134,210$$
$$SS_y = \Sigma y - \frac{(\Sigma y)^2}{n} = 51,822,600 - \frac{31,860^2}{20} = 1,069,620$$
$$s_y^2 = \frac{SS_y}{(n-1)} = \frac{1,069,620}{19} = 56,295.79$$
$$SS_x = \Sigma x - \frac{(\Sigma x)^2}{n} = 134,210 - \frac{1620^2}{20} = 2990$$
$$SS_{xy} = \Sigma xy - \frac{(\Sigma x)(\Sigma y)}{n} = 2,635,500 - \frac{1620 \times 31,860}{20} = 54,840$$
$$N = \text{Total number of fifth-acre plots in the population } (= 1000)$$

The relationship between y and x may take one of several forms, but here we will assume that it is a straight line. The equation for the line can be estimated from

$$\bar{y}_R = \bar{y} + b(X - \bar{x})$$

where:

$\bar{y}_R$ = Mean value of y as estimated from X (a specified value of variable X)

$\bar{y}$ = Sample mean of y (= 1593)

b = Linear regression coefficient of y on x

$\bar{x}$ = Sample mean of x (= 81)

For the linear regression estimator used here, the value of the regression coefficient is estimated by

$$b = \frac{SP_{xy}}{SS_x} = \frac{54,840}{2990} = 18.34$$

from the y values only. Thus, the equation would be

$$\bar{y}_R = 1593 + 18.34(X - 81) = 107.4 + 18.34X$$

To estimate the mean volume per acre for the tract, we substitute the known mean basal area per acre for X:

$$\bar{y}_R = 107.46 + (18.34 \times 84) = 1648 \text{ cubic feet per acre}$$

Standard Error

The standard errors for simple random sampling and stratified random sampling can be estimated. To obtain the standard error for a regression estimator, we need an estimate of the variability of the individual y values about the regression of y on x. A measure of this variability is the standard deviation from regression ($s_{y,x}$), which is computed by

$$s_{y,x} = \sqrt{\frac{SS_y - \frac{(SP_{xy})^2}{SS_x}}{(n-2)}}$$

$$= \sqrt{\frac{1,069,620 - \frac{54,840^2}{2990}}{(20-2)}} = 59.53$$

The symbol $s_{y,x}$ bears a strong resemblance to the covariance symbol (S_{yx}) with which it must not be confused. Having the standard deviation from regression, the standard error of $\bar{y}_R$ is

$$s_{\bar{y}_R} = s_{y,x}\sqrt{\left(\frac{1}{n} + \frac{(X-\bar{x})^2}{SS_x}\right)\left(1 - \frac{n}{N}\right)}$$

$$= 59.53\sqrt{\left(\frac{1}{20} + \frac{(84-81)^2}{2990}\right)\left(1 - \frac{20}{1000}\right)} = 13.57$$

With such a small sampling fraction ($n/N = 0.02$), the finite-population correction ($1 - n/N$) could have been ignored, and the standard error would be 13.71.

Family of Regression Estimators

The regression procedure in the above example is valid only if certain conditions are met. One of these is, of course, that we know the population mean for the supplementary variable (x). As will be shown in the next section, an estimate of the population mean can often be substituted. Another condition is that the relationship of y to x must be reasonably close to a straight line within the range of x values for which y will be estimated. If the relationship departs very greatly from a straight line, our estimate of the mean value of y will not be reliable. Often a curvilinear function is more appropriate. A third condition is that the variance of y about its mean should be the same at all levels of x. This condition is difficult to evaluate with the amount of data usually available. Ordinarily the question is answered from our knowledge of the population or by making special studies of the variability of y. If we know the way in which the variance changes with changes in the level of x, a weighted regression procedure may be used. Thus, the linear regression estimator that has been described is just one of a large number of related procedures that enable us to increase our sampling efficiency by making use of supplementary information about the population. Two other members of this family are the ratio-of-means estimator and the mean-of-ratios estimator.

Ratio Estimation

The *ratio-of-means estimator* is appropriate when the relationship of y to x is in the form of a straight line passing through the origin and when the standard deviation of y at any given level of x is proportional to the square root of x. The ratio estimate $(\bar{y}_R)$ of mean y is

$$\bar{y}_R = \hat{R} \times X$$

where:

R = The ratio of means obtained from the sample = $\dfrac{\bar{y}}{\bar{x}}$ or $\dfrac{\Sigma y}{\Sigma x}$

X = The known population mean of x

For large samples (generally taken as $n > 30$) (Cochran, 1953), the standard error of the ratio-of-means estimator can be approximated by

$$s_{\bar{y}_R} = \sqrt{\left(\frac{s_y^2 + \hat{R}^2 s_x^2 - 2\hat{R}s_{xy}}{n}\right)\left(1 - \frac{n}{N}\right)}$$

where:

s_y^2 = The estimated variance of y

s_x^2 = The estimated variance of x

s_{xy} = The estimated covariance of x and y

■ *Example 6.4*

Problem: Assume that for a population of $N = 400$ units, the population mean of x is known to be 62 and that from this population a sample of $n = 10$ units is selected. The y and x values of these 10 units are found to be

Observation	y_i	x_i
1	8	62
2	13	81
3	5	40
4	6	46
5	19	123
6	9	74
7	8	52
8	11	96
9	5	36
10	12	70
Total	96	680

Solution: From this sample the ratio-of-means is

$$\hat{R} = \frac{9.6}{68} = 0.141$$

The ratio-of-means estimator is then

$$\bar{y}_R = \hat{R} \times X = 0.141 \times 62 = 8.742$$

To compute the standard error of the mean, we will need the variances of y and x and also the covariance. These values are computed by the standard formulas for a sample random sample:

$$s_y^2 = \frac{(8^2 + 13^2 + \ldots + 12^2) - \dfrac{96^2}{10}}{(10-1)} = 18.7111$$

$$s_x^2 = \frac{(62^2 + 81^2 + \ldots + 70^2) - \dfrac{680^2}{10}}{(10-1)} = 733.5556$$

$$s_{xy} = \frac{(8 \times 62) + (13 \times 81) + \ldots + (12 \times 70) - \dfrac{(96 \times 680)}{10}}{(10-1)} = 110.2222$$

Substituting these values in the formula for the standard error of the mean gives

$$s_{\bar{y}_R} = \sqrt{\left(\frac{18.7111 + (0.141^2 \times 733.5556) - 2(0.141 \times 110.2222)}{10}\right)\left(1 - \frac{10}{400}\right)}$$

$$= \sqrt{.215690} = 0.464$$

Mean-of-Ratios Estimator

The *mean-of-ratios estimator* is appropriate when the relation of y to x is in the form of a straight line passing through the origin and the standard deviation of y at a given level of x is proportional to x (rather that to $\sqrt{x}$). The ratio (r_i) of y_i to x_i is computed for each pair of sample observations. Then the estimated mean of y for the population is

$$\bar{y}_R = \hat{R} \times X$$

where $\hat{R}$ is the mean of the individual ratios (r_i):

$$\hat{R} = \frac{\sum_{i=1}^{n} r_i}{n}$$

To compute the standard error of this estimate, we must first obtain a measure (s_r^2) of the variability of the individual ratios (r_i) about their mean:

$$s_r^2 = \frac{\sum_{i=1}^{n} r_i^2 - \frac{\left(\sum_{i=1}^{n} r_i\right)^2}{n}}{(n-1)}$$

The standard error for the mean-of-ratios estimator or mean y is then

$$s_{\bar{y}_R} = X\sqrt{\left(\frac{s_r^2}{n}\right)\left(1 - \frac{n}{N}\right)}$$

■ Example 6.5

Problem: A set of $n = 10$ observations is taken from a population of $N = 100$ units having a mean x value of 40:

Observation	y_i	x_i	r_i
1	36	18	2.00
2	95	48	1.98
3	108	46	2.35
4	172	74	2.32
5	126	58	2.17
6	58	26	2.23
7	123	60	2.05
8	98	51	1.92
9	54	25	2.00
Total			21.18

Solution: The sample mean-of-ratios is

$$\hat{R} = \frac{21.18}{10} = 2.118$$

And this is used to obtain the mean-of-ratios estimator:

$$\bar{y}_R = \hat{R} \times X = 2.118 \times 40 = 84.72$$

The variance of the individual ratios is

$$s_r^2 = \frac{(2.00^2 + 1.98^2 + \ldots + 2.00^2) - \dfrac{21.18^2}{10}}{(10-1)} = 0.022484$$

Thus, the standard error of the mean-of-ratios estimator is

$$s_{\bar{y}_R} = 40\sqrt{\left(\frac{0.022484}{10}\right)\left(1 - \frac{10}{100}\right)}$$

Double Sampling

Double sampling permits the use of regression estimators when the population mean or total of the supplementary variable is unknown. A large sample is taken to obtain a good estimate of the mean or total for the supplementary variable (x). On a subsample of the units in this large sample, the y values are also measured to provide an estimate of the relationship of y to x. The large sample mean or total or x is then applied to the fitted relationship to obtain an estimate of the population mean or total of y.

■ *Example 6.6*

Problem: In a forest inventory done in 1950, a sample of 200 quarter-acre plots in an 800-acre forest showed a mean volume of 372 cubic feet per plot (1488 cubic feet per acre). A subsample of 40 plots, selected at random from the 200 plots, was marked for remeasurement in 1955. The relationship of the subsample was applied to the 1950 volume to obtain a regression estimate of the 1955 volume. The subsample was as follows:

1955 Volume (y)	1950 Volume (x)	1955 Volume (y)	1950 Volume (x)
370	280	420	330
290	240	530	390
520	410	550	430
490	360	550	460

1955 Volume (y)	1950 Volume (x)	1955 Volume (y)	1950 Volume (x)
530	390	520	400
330	220	420	390
310	270	490	340
400	340	500	420
450	360	610	470
430	360	460	350
460	400	430	340
480	380	510	380
430	350	450	370
500	390	380	300
640	480	430	290
660	520	460	340
490	400	490	370
510	430	560	440
270	230	580	480
380	270	540	420
Total		18,820	14,790
Mean		470.50	369.75

Solution:

$$\sum y^2 = 9,157,400$$

$$\sum x^2 = 5,661,300$$

$$\sum xy = 7,186,300$$

A plotting of the 40 pairs of plot values on coordinate paper suggested that the variability of y was the same at all levels of x and that the relationship of y to x was linear. The estimator selected on the basis of this information was the linear regression:

$$\bar{y}_{Rd} = a + bX$$

Values needed to compute the linear-regression estimate and its standard error were as follows:

- Large-sample data (indicated by the subscript 1):

    n_1 = Number of observations in large sample = 200

    N = Number of sample units in population = 3200

    x_1 = Large sample mean of x = 200

- Small-sample data (indicated by the subscript 2):

    n_2 = Number of observations in subsample = 40

    $\bar{y}_2$ = Small sample mean of y = 470.50

$\bar{x}_2 = $ Small sample mean of $x = 369.75$

$$SS_y = \left(\Sigma y^2 - \frac{(\Sigma y)^2}{n_2}\right) = \left(9,157,400 - \frac{(18,820)^2}{40}\right) = 802,590.0$$

$$SS_x = \left(\Sigma x^2 - \frac{(\Sigma x)^2}{n_2}\right) = \left(5,661,300 - \frac{(14,790)^2}{40}\right) = 192,697.5$$

$$SP_{xy} = \left(\Sigma xy - \frac{(\Sigma x)(\Sigma y)}{n_2}\right) = \left(7,186,300 - \frac{18,820 \times 14,790}{40}\right) = 227,605.0$$

$$s_y^2 = \frac{SS_y}{(n_2 - 1)} = \frac{302,590}{40 - 1} = 7758.72$$

The regression coefficient (b) and the squared standard deviation from regression ($s_{y.x}$) are

$$b = \frac{SP_{xy}}{SS_x} = \frac{227,605.0}{192,697.5} = 1.18$$

$$s_{y.x}^2 = \frac{\left(SS_y - \frac{(SP_{xy})^2}{SS_x}\right)}{(n_2 - 2)} = \frac{\left(302,590.0 - \frac{227,605.0^2}{192,697.5}\right)}{(40 - 2)} = 888.2617$$

And the regression equation is

$$\bar{y}_{Rd} = \bar{y}_2 + b(X - \bar{x}_2)$$
$$= 470.50 + 1.18(X - 369.75)$$
$$= 34.2 + 1.18X$$

Substituting the 1950 mean volume (372 cubic feet) for X gives the regression estimate of the 1955 volume:

$$\bar{y}_{Rd} = 34.2 + (1.18 \times 372) = 473.16 \text{ cubic feet per plot}$$

Sampling Protocols and Vegetation Attributes*

By now you should understand that attempting to count every live, down-tree or down-wood material, or snag in an entire forest (of more than 10,000 acres, for example) is not fun, not feasible, not practicable, not time or cost effective, and just not likely to be conducted at all. Instead, sampling is preferred. Why sample? We sample because it is not an

---

* Material in this section is based on Lutes, D.C. et al., *FIREMON: Fire Effects Monitoring and Inventory System*, General Technical Report RMRS-GTR-164-CD, U.S. Department of Agriculture, Forest Service, Rocky Mountain Research Station, Fort Collins, CO, 2006; Coulloudon, B. et al., *Sampling Vegetation Attributes*, Technical Reference 1734-4, U.S. Department of Interior, Bureau of Land Management, Denver, CO, 1999.

arduous undertaking and is not as time consuming as a total count; simply, it is feasible and if conducted properly provides a fairly accurate representative sample to work with. This sample to "work with" provides the sampler/researcher with valuable information that is unbiased and empirically derived, which aids in the analysis of fuel loads for forest fires, carbon pools, and the wildlife habitat of the ecosystem sampled.

Note: Although this text dedicates most of its space to describing the author's modified version of the U.S. Forest Service FIREMON (Fire Effects Monitoring and Inventory System) sampling methods (designed to characterize changes in ecosystem attributes over time), which the author has used extensively over the years in various forest locations throughout the Cascade Mountain Range of Washington State, it is important to discuss other sampling protocols and methods used in forest inventory, sampling, and management. Keep in mind, however, that the same vegetation attributes and many of the following sampling methods are also used in the author's modified version of FIREMON.

Vegetation Attributes

When we are sampling forest biomass we are in fact measuring. Measuring what? Well, for our purposes in this text we are measuring the features of interest (attributes). A forest biomass inventory is where we take measurements to obtain dimensional or physiological information about the resource. For example, we might be interested in quantifying the amount of carbon sequestered (carbon pool) by a forest in a given year or the amount of water used by surrounding undergrowth in a semi-arid landscape.

A typical forest inventory normally focuses on assessing the volume or value of standing trees to use for different forest products. A forest biomass inventory is somewhat different than the typical forest inventory in that it can be defined as the process involved in the estimation of any forest feature such as tree height, grass cover, and so forth as precisely as available time, money, personnel, etc. can permit. Again, we are talking about measuring the features of interest. Often, in practice, we refer to these features as attributes or plant attributes. Generally, we measure only six attributes. A typical forest inventory normally focuses on assessing the volume or value of standing trees to use for different forest products. Vegetation or plant attributes are characteristics of vegetation or plants that can be measured or quantified with reference to how many, how much, or what kind of plant species are present. The most commonly used attributes are listed and described in the following.

Frequency

Was the plant present or not? This is one of the easiest and fastest methods available for monitoring vegetation. It describes the abundance and distribution of species and is useful to detect changes in a plant community over time. Frequency has been used to determine rangeland conditions, but only limited work has been done in most communities. This makes the interpretation difficult. The literature has discussed the relationship between density and frequency but this relation is only consistent with randomly distributed plants (Greig-Smith, 1983). Frequency is the number of times a species is present in a given number of sampling units. It is usually expressed as a percentage.

Advantages and Limitations

- Frequency is highly influenced by the size and shape of the quadrats used. Quadrats or nested quadrats are the most common measurement used; however, point sampling and step point methods have also been used to estimate frequency. The size and shape of a quadrat required to adequately determine frequency depends on the distribution, number, and size of the plant species.

- To determine change, the frequency of a species must generally be at least 20% and no greater than 80%. Frequency comparison must be made with quadrats of the same size and shape. Although change can be detected with frequency, the extent to which the vegetation community has changed cannot be determined.

- High repeatability is obtainable.

- Frequency is highly sensitive to changes resulting from seedling establishment. Seedlings present one year may not be persistent the following year. This situation is problematic if data are collected only every few years. It is less of a problem if seedlings are recorded separately.

- Frequency is also very sensitive to changes in pattern of distribution in the sampled area.

- Rooted frequency data are less sensitive to fluctuations in climatic and biotic influences.

- Interpretation of changes in frequency is difficult because of the inability to determine the vegetation attribute that changed. Frequency cannot tell which of three parameters has changed: canopy cover, density, or pattern of distribution.

Appropriate Use of Frequency for Rangeland Monitoring

If the primary reason for collecting frequency data is to demonstrate that a change in vegetation has occurred, then on most sites the frequency method is capable of accomplishing the task with statistical evidence more rapidly and at less cost than any other method that is currently available (Hironaka, 1985). Frequency should not be the only data collected if time and money are available. Additional information on ground cover, plant cover, and other vegetation and site data would contribute to a better understanding of the changes that have occurred (Hironaka, 1985). West (1985) noted the following limitations: "Because of the greater risk of misjudging a downward than upward trend, frequency may provide the easiest early warning of undesirable changes in key or indicator species. However, because frequency data are so dependent on quadrat size and sensitive to non-random dispersion patterns that prevail on rangelands, managers are fooling themselves if they calculate percentage composition from frequency data and try to compare different sites at the same time or the same site over time in terms of total species composition. This is because the numbers derived for frequency sampling are unique to the choice of sample size, shape, number, and placement. For variables of cover and weight, accuracy is mostly what is affected by these choices and the variable can be conceived independently of the sampling protocol" (p. 97).

Cover

How much space did they cover? Cover is an important vegetation and hydrologic characteristic. It can be used in various ways to determine the contribution of each species to a plant community. Cover is also important in determining the proper hydrologic function of a site.

This characteristic is very sensitive to biotic and edaphic forces (i.e., soil factors). For watershed stability, some have tried to use a standard soil cover, but research has shown each edaphic site has its own potential cover. Cover is generally referred to as the percentage of ground surface covered by vegetation; however, numerous definitions exist. It can be expressed in absolute terms (square meters/hectares), but it is most often expressed as a percentage. The objective being measured will determine the definition and type of cover measured.

- Vegetation cover is the total cover of vegetation on a site.
- Foliar cover is the area of ground covered by the vertical projection of the aerial portions of the plants. Small openings in the canopy and intraspecific overlap are excluded.
- Canopy cover is the area of ground covered by the vertical projection of the outermost perimeter of the natural spread of foliage of plants. Small openings within the canopy are included. Canopy cover may exceed 100%.
- Basal cover is the area of ground surface occupied by the basal portion of the plants.
- Ground cover is the cover of plants, litter, rocks, and gravel on a site.

Advantage and Limitations

- Ground cover is most often used to determine the watershed stability of the site, but comparisons between sites are difficult to interpret because of the different potentials associated with each ecological site.
- Vegetation cover is a component of ground cover and is often sensitive to climatic fluctuations that can cause errors in interpretation. Canopy cover and foliar cover are components of vegetation cover and are the most sensitive to climatic and biotic factors. This is particularly true with herbaceous vegetation.
- Overlapping canopy cover often creates problems, particularly in mixed communities. If species composition is to be determined, the canopy of each species is counted regardless of any overlap with other species. If watershed characteristics are the objective, only the uppermost canopy is generally counted.
- For trend comparisons in herbaceous plant communities, basal cover is generally considered to be the most stable. It does not vary as much due to climatic fluctuation or current-year grazing.

Density

How many plants were there and how close are individual plants to one another? Density has been used to describe characteristics of plant communities; however, comparisons can only be based on similar life-form and size. This is why density is rarely used as a measurement by itself when describing plant communities. For example, the importance of a particular species to a community is very different if there are 1000 annual plants per acre vs. 1000 shrubs per acre. It should be pointed out that density was synonymous with cover in the earlier literature.

Advantages and Limitations

- Density is useful in monitoring threatened and endangered species or other special status plants because it samples the number of individuals per unit area.
- Density is useful when comparing similar life-forms (annuals to annuals, shrubs to shrubs) that are approximately the same size. For trend measurements, this parameter is used to determine if the number of individuals of a specific species is increasing or decreasing.
- The problem with using density is being able to identify individuals and comparing individuals of different sizes. It is often difficult to identify individuals of plants that are capable of vegetative reproduction (e.g., rhizomatous plants such as western wheatgrass or Gamble oak). Comparisons of bunchgrass plants to rhizomatous plants are often meaningless because of these problems. Similar problems occur when looking at the density of shrubs of different growth forms or comparing seedlings to mature plants. Density of rhizomatous or stoloniferous (i.e., grows horizontally above the ground) plants is determined by counting the number of stems instead of the number of individuals. Seedling density is directly related to environmental conditions and can often be interpreted erroneously as a positive or negative trend measurement. Because of these limitations, density has generally been used with shrubs and not herbaceous vegetation. Seedlings and mature plants should be recorded separately.
- If the individuals can be identified, density measurements are repeatable over time because there is small observer error. The type of vegetation and distribution will dictate the technique used to obtain the density measurements. In homogeneous plant communities, which are rare, square quadrats have been recommended, while heterogeneous communities should be sampled with rectangular or line strip quadrats. Plotless methods have also been developed for widely dispersed plants.

Biomass (Production)

How much did the plants weigh? Many believe that the relative production of different species in a plant community is the best measure of these species' roles in the ecosystems. The terminology associated with vegetation is normally related to production.

- Gross primary production is the total amount of organic material produced, both above ground and below ground.
- Biomass is the total weight of living organisms in the ecosystem, including plants and animals.
- Standing crop is the amount of plant biomass present above ground at any given point.
- Peak standing crop is the greatest amount of plant biomass above ground present during a given year.
- Total forage is the total herbaceous and woody palatable plant biomass available to herbivores.
- Allocated forage is the difference of desired amount of residual material subtracted from the total forage.
- Browse is the portion of woody plant biomass accessible to herbivores.

Advantages and Limitations

- Biomass and gross primary production are rarely used in rangeland trend studies because it is impractical to obtain the measurements below ground. In addition, the animal portion of biomass is rarely obtainable.

- Standing crop and peak standing crop are the measurements most often used in trend studies. Peak standing crop is generally measured at the end of the growing season; however, different species reach their peak standing crop at different times. This can be a significant problem in mixed-plant communities.

- Often, the greater the diversity of plant species or growth patterns, the larger the error if only one measurement is made.

- Other problems associated with the use of plant biomass are that fluctuations in climate and biotic influences can alter the estimates. When dealing with large ungulates, exclosures are generally required to measure this parameter. Several authors have suggested that approximately 25% of the peak standing crop is consumed by insects or trampled; this is rarely discussed in most trend studies.

- Collecting production data also tends to be time and labor intensive. Cover and frequency have been used to estimate plant biomass in some species.

Structure

How tall were the plants and how were branches and leaves arranged? The structure of vegetation primarily looks at how the vegetation is arranged in a three-dimension space. The primary use for structure measurements is to help evaluate a vegetation community's value in providing habitat for associated wildlife species. Vegetation is measured in layers on vertical planes. Measurements generally look at the vertical distribution by either estimating the cover of each layer or measuring the height of the vegetation.

Advantages and Limitations

- Structure data provide information that is useful in describing the suitability of the sites for screening and escape cover, which are important for wildlife. Methods used to collect these data are quick, allowing for numerous samples to be obtained over relatively large areas. Methods that use visual obstruction techniques to evaluate vegetation height have little observer bias. Those techniques that estimate cover require more training to reduce observer bias. Structure is rarely used by itself when describing trend.

Composition

What kind of plant was it? Composition is a calculated attribute rather than one that is directly collected in the field. It is the proportion of various plant species in relation to the total of a given area. It may be expressed in terms of relative cover, relative density, relative weight, etc. Composition has been used extensively to describe ecological sites and to evaluate rangeland condition. To calculate composition, the individual value (weight, density, percent cover) of a species or group of species is divided by the total value of the entire population.

Advantages and Limitations

- Quadrats, point sampling, and step point methods can all be used to calculate composition.
- The repeatability of determining composition depends on the attribute collected and the method used.
- Sensitivity to change is dependent on the attribute used to calculate composition. For example, if plant biomass is used to calculate composition, the values can vary with climatic conditions and the timing of climate events (e.g., precipitation, frost-free period). Composition based on basal cover, on the other hand, would be relatively stable.
- Composition allows the comparison of vegetation communities at various locations within the same ecological sites.

Matrix of Monitoring Techniques and Vegetation Attributes

The following is a matrix of commonly used monitoring techniques and vegetation attributes. The X indicates that this is the primary attribute that the technique collects. Some techniques have the capability of collecting other attributes; the • indicates the secondary attribute that can be collected or calculated.

Method	Frequency	Cover	Density	Biomass	Structure	Composition
Frequency	X	•				
Dry-weight rank	•			•		X*
Daubenmire	•	X				•
Line intercept		X				•
Step-point		X				•
Point intercept		X				•
Density			X			•
Double weight sampling				X		•
Harvest				X		•
Comparative yield				X		•
Cover board		X			X	
Robel pole				•	X	

* Species composition is calculated using production data. Frequency data should not be used to calculate species composition.

Pace Frequency, Quadrat Frequency, and Nested Frequency

Pace frequency, quadrat frequency, and nested frequency methods consist of observing quadrats along transects, with quadrats systematically located at specified intervals along each transect. The only differences in these techniques are the size and configuration of the quadrat frames and the layout of the transect. The following vegetation attributes are monitored with these methods:

- Frequency
- Basal cover and general cover categories (including litter)
- Reproduction of key species (if seedling data are collected)

It is important to establish a photo plot and take both close-up and general-view photographs. This allows the portrayal of resource values and conditions and furnishes visual evidence of vegetation and soil changes over time. These methods are applicable to a wide variety of vegetation types and are suited for use with grasses, forbs, and shrubs.

Note: Both close-up and general view photographs should be taken for each of these methods.

Advantages and Disadvantages

- Frequency sampling is highly objective, repeatable, rapid, and simple to perform, and it involves a minimum number of decisions. Decisions are limited to identifying species and determining whether or not species are rooted within the quadrats (presence or absence).
- Frequency data can be collected in different-sized quadrats with the use of the nested frame. When a plant of a particular species occurs within a plot, it also occurs in all of the successively large plots. Frequency of occurrence for various size plots can be analyzed even though frequency is recorded for only one size plot. This eliminates problems with comparing frequency data from different plot sizes. Use of the nested plot configuration improves the chances of selecting a proper size plot for frequency sampling.
- Cover data can also be collected at the same time that frequency data are gathered; however, cover data collected in this manner will greatly overestimate cover. Unless the tines are honed to a fine point, observer bias will come into play. Another limitation is that the use of one size quadrat will likely result in values falling outside the optimum frequency range (greater than 20% to less that 80%) for some of the species of interest.

Dry-Weight Rank Method

The dry-weight rank method is used to determine species composition. It consists of observing various quadrats and ranking the three species that contribute the most weight in the quadrat. This method has been tested in a wide variety of vegetation types and is generally considered suitable for grassland/small shrubs types or understory communities of large shrub or tree communities. It does not work well on large shrubs and trees.

Advantages and Disadvantages

- One advantage of the dry-weight rank method is that a large number of samples can be obtained very quickly. Another advantage is that it deals with estimates of production, which allows for better interpretation of the data to make management decisions. It can be done in conjunction with frequency, canopy cover, or comparative yield methods. Because it is easier to rank the top three species in a quadrat, there is less observer bias.
- The limitation with this technique is that, by itself, it will not give a reliable estimate of plant standing crop, and it assumes that there are few empty quadrats. In many large shrub or sparse desert communities, a high percentage of quadrats are empty or have only one species present. The quadrat size required to address these concerns is often impractical.

Daubenmire Method

The Daubenmire method consists of systematically placing a 20- × 50-cm quadrat frame along a rope on permanently located transects. The following vegetation attributes are monitored using the Daubenmire method:

- Canopy cover
- Frequency
- Composition of canopy cover

This method is applicable to a wide variety of vegetation types as long as the plants do not exceed waist height.

Advantages and Disadvantages

- This method is relatively simple and rapid to use.
- A limitation is that there can be large changes in canopy cover of herbaceous species between years because of climatic conditions, with no relationship to the effects of management. In general, quadrats are not recommended for estimating cover (Floyd and Anderson, 1987; Kennedy and Addison, 1987). This method cannot be used to calculate rooted frequency.

Line Intercept Method

The line intercept method consists of horizontal, linear measurements of plant intercepts along the course of the line (tape). It is designed for measuring grass or grass-like plants, forbs, shrubs, and trees. The following vegetation attributes are monitored with this method:

- Foliar and basal cover
- Composition (by cover)

This method is ideally suited for semi-arid bunchgrass–shrub vegetation types.

Advantages and Disadvantages

- This method is best suited where the boundaries of plant growth are relatively easy to determine. It can be adapted to sampling varying densities and types of vegetation. It is not well adapted, however, for estimating cover on single-stemmed species, dense grassland situations, litter, or gravel less than 1/2 inch in diameter. It is best suited to estimating cover on shrubs.

Step-Point Method

The step-point method involves making observations along a transect at specified intervals, using a pin to record cover "hits." It measures cover for individual species, total cover, and species composition by cover. This method is best suited for use with grasses and forbs, as well as low shrubs. The greater the structure to the community, the more difficult it becomes to determine "hits" due to parallax, observer bias, wind, etc. This method is good for an initial overview of an area not yet subjected to intensive monitoring.

Advantages and Limitations

- This method is relatively simple and easy to use as long as careful consideration is given to the vegetation type to which it is applied. It is suitable for measuring major characteristics of the ground and vegetation cover of an area. Large areas can easily be sampled, particularly if the cover is reasonably uniform. It is possible to collect a fairly large number of samples within a relatively short time.

- A limitation of this method is that there can be extreme variation in the data collected among examiners when sample sizes are small. Tall or armored vegetation reduces the ability to pace in a straight line, and the offset for obstructions described in the procedures adds bias to the data collection by avoiding certain components of the community. Another limitation is that less predominant plant species may not be hit on the transects and therefore do not show up in the study records. The literature contains numerous studies utilizing point intercept procedures that required point densities ranging from 300 to 39,000 to adequately sample for minor species. One major consideration in the use of this method is to ensure that a sharpened pin is used and that only the point is used to record hits. Pins have finite diameters and therefore overestimate cover (Goodall, 1952). Another limitation of this method is that statistical analysis of the data is suspect unless two and preferably more transects are run per site.

Point Intercept Method: Sighting Devices, Pin Frames, and Point Frames

The point intercept method consists of employing a sighting device or pin/point frame along a set of transects to arrive at an estimate of cover. It measures cover for individual species, total cover, and species composition by cover. This method is suited to all vegetation types less than about 1.5 meters in height. This is because sighting devices and pin/point frames require the observer to look down on the vegetation from above in a vertical line with the ground. If the sighting device allows upward viewing, the method can also be used to estimate the canopy cover of large shrubs and trees.

Advantages and Limitations

- Point interception measurements are highly repeatable and lead to more precise measurements than cover estimates using quadrats. The method is more efficient than line intercept techniques, at least for herbaceous vegetation, and it the best method of determining ground cover and the cover of the more dominant species. Given the choice between sighting devices and pin/point frames, the optical sighing device is preferable.

- A limitation of point intercept sampling is the difficulty in picking up the minor species in the community without using a very large number of points. In addition, wind will increase the time required to complete a study because of the need to view a stationary plant.

- One limitation that is specific to the use of point frames is that a given number of points grouped in frames gives less precise estimates of cover than the same number of points distributed individually (Goodall, 1952; Greig-Smith, 1983). In fact, single-pin measurements require only one-third as many points as when point frames are used (Bonham, 1989). Another problem with frames is that they can overestimate the cover of large or clumped plants because the same plant is intercepted by different points on the same frame (Bonham, 1989). This problem is overcome with the method described here by treating the frames as the sampling units (rather than using the individual points as sampling units). However, this approach doesn't change the fact that more points must be read than when the points are independent.

- Use of a pin frame device (as opposed to a grid frame made of crossing strings) will result in overestimation of cover because the pins have finite diameter. The use of a sharpened pin will greatly reduce overestimation when only the point of the pin is used to record a hit or a miss.

Cover Board Method

The cover board method uses a profile board or density board to estimate the vertical area of a board covered by vegetation from a specified distance away. This technique is designed to evaluate changes in the vegetation structure over time. Quantifying the vegetation structure for statistical comparison was described by Nudds (1977). The following vegetation attributes are monitored using this method:

- Vertical cover
- Structure

This method is applicable to a wide variety of vegetation types. It should be used with those that show potential for changes, such as woody riparian vegetation.

Advantages and Disadvantages

- The cover board technique is a fast and easily duplicated procedure. The size of the board can be modified to meet the purpose of the study.

Density Method

Density is the number of individuals of a species in a given unit of area (e.g., plants per m^2). The term consequently refers to the closeness of individual plans to one another. For rhizomatous and other species for which the delineation of separate individual plants is difficult, density can also mean the number of stems, inflorescences, culm groups, or the plant parts per unit area. This method has wide applicability and is suited for use with grasses, forbs, shrubs, and trees.

Advantages and Disadvantages

- Generally, the density of mature perennial plant is not affected as much by annual variations in precipitation as are other vegetation attributes such as canopy cover of herbage production.
- Density is a quantitative and absolute attribute.
- Density is sensitive to changes in the adult population caused by long-term climatic conditions or resource uses.

- Density provides useful information on seedling emergence, survival, and mortality.
- Sampling is often quick and easy with certain life-forms (e.g., trees, shrubs, bunchgrasses).
- Plant communities on the same ecological sites can be compared using density estimates on specific species or life-forms.
- Density can be useful in estimating plant response to management actions.
- It can often be difficult to delineate an individual, especially when sampling sod-forming plants (stoloniferous or rhizomatous plants) and multistemmed grasses of closely spaced shrubs. Although in these cases a surrogate plant part (e.g., upright stems, inflorescences, culm groups) can be counted, the usefulness of such estimates is limited to the biological significance of changes in these surrogates.
- Sampling may be slow and tedious in dense populations; this also raises the risk of non-sampling errors.
- There is no single quadrat size and shape that will efficiently and adequately sample all species and life forms. For this reason, density estimations are usually limited to one or a few key species.

Double-Weight Sampling

Double-weight sampling has been referred to by some as the *calibrated weight estimate* method. The objective of this method is to determine the amount of current-year above-ground vegetation production on a defined area. The following vegetation is monitored:

- Peak standing crop, which is the aboveground annual production of each plant species
- Species composition by weight

This method can be used for a wide variety of vegetation types. It is best suited to grass-lands and desert shrubs. It can also be used in large shrub and tree communities, but the difficulties increase.

Advantages and Limitations

- Double-weight sampling measures the attribute historically used to determine capabilities of an ecosystem.
- It provides the basic data currently used for determining ecological status.
- Seasonal and annual fluctuations in climate can influence plant biomass.
- Measurements can be time consuming.
- Current year's growth can be difficult to separate from previous years' growth.
- Accurate measurements require collecting production data at peak production periods, which are usually short, or using utilization and phenology adjustment factors.
- Green weights require conversion to air-dry weights.
- In most areas, the variability in production between quadrats and the accuracy of estimating production within individual quadrats require the sampling of large numbers of quadrats to detect reasonable levels of change.

Harvest Method

The concept of the harvest method is to determine the amount of current-year above-ground vegetation production on a defined area. The following vegetation attributes are monitored:

- Peak standing crop, which is the aboveground annual production of each plant species
- Species composition by weight

This method can be used in a wide variety of vegetation types. It is best suited for grasslands and desert shrubs. It is not well suited to large shrub and tree communities.

Advantages and Limitations

- The harvest method measures the attribute historically used to determine the capabilities of an ecosystem.
- It provides the basic data currently used for determining ecological status.
- Seasonal and annual fluctuations in climate can influence plant biomass.
- Measurements can be time consuming.
- Current year's growth can be difficult to separate from previous years' growth.
- Accurate measurements require collecting production data at peak production periods, which are usually short, or using utilization and phenology adjustment factors.
- Green weights require conversion to air-dry weights.
- In most areas, the variability in production between quadrats requires the sampling of large numbers of quadrats to detect reasonable levels of change.

Comparative Yield Method

The comparative yield method is used to estimate total standing crops or production of a site. The total production in a sample quadrat is compared to one of five reference quadrats; relative ranks are recorded rather than estimating the weight directly. This method works best for herbaceous vegetation but can also be used successfully with small shrubs and half-shrubs. As with most production estimates, the comparative yield method can be used to compare relative production between different sites.

Advantages and Limitations

- The advantage of the comparative yield method is that a large sample can be obtained quickly. Total production is evaluated, so clipping calibration on a species basis is not needed. The process of developing reference quadrats for ranking purposes reduces both sampling and training time. This technique can be done in conjunction with the frequency, canopy cover, or dry-weight rank methods. Identification of individual species is not required.
- Large shrub communities are not well suited for this technique. If used in conjunction with other techniques (frequency and dry-weight rank), the quadrat size may have to be different. This technique can detect only large changes in production.

Visual Obstruction Method (Robel Pole)

This method is used for determining standing plant biomass on an area. It has primarily been used to determine the quality of nesting cover for birds on the Great Plains and is commonly referred to as the Robel pole method. This method is applicable to other ecosystems throughout the western United States where height and vertical obstruction of cover are important. The following vegetation attributes are monitored using this method:

- Vertical cover
- Production
- Structure

The Robel pole method is most effective in upland and riparian areas where perennial grasses, forbs, and shrubs less than 4 feet tall are the predominant species.

Advantages and Disadvantages

- Robel pole measurements are simple, quick, and accurate. This method can be used to monitor the height and density of standing vegetation over large areas quickly. Statistical reliability improves because numerous measurements can be taken in a relatively short time. Limitations of the method may stem from infrequent application in a variety of rangeland ecosystems. Although the Robel pole method has been used with great success on the Great Plains, more research in a variety of plant communities is necessary.

Modified FIREMON Sampling Procedures

The need to sample forest inventories is plain. The question is how do we accomplish the sampling? The U.S. Department of Agriculture, Forest Service, has developed the Fire Effects Monitoring and Inventory System (FIREMON), which uses the best estimate of resources that the manager can provide to help design the plot-level and landscape-level sampling strategy of a fire effects monitoring project. A sampling method is different from a sampling strategy in that a sampling strategy describes where, when, and how the sampling method (procedure for measuring things) is implemented across the landscape (Lutes and Keane, 2006). This system (although designed for fire service applications), along with other traditional sampling methods, has been proven effective, in general, for inventorying forests. Moreover, during personal field work involving forest plot surveys, sampling, and monitoring, the author has found that simple adaptations to FIREMON protocols can provide effective and practical tools to use in such endeavors. For this reason, FIREMON protocols are used in this text as examples of how to effectively monitor and inventory forest materials; again, these methods could just as easily be applied for inventorying or monitoring forest biomass. Moreover, keep in mind that with increasing focus on global, regional, and local carbon cycles, the ability to accurately predict forest biomass has taken on a new urgency. Thus, any proven methodology for inventorying and monitoring forest biomass is beneficial to those conducting such surveys, and FIREMON procedures have much to offer when estimating forest biomass quantities.

DID YOU KNOW?

The root of the word *monitoring* means "to warn," and an essential purpose of monitoring is to raise a warning flag that the current course of action is not working. Monitoring is a powerful tool for identifying problems in the early stages before they become dramatically obvious or turn into crises. If identified early, problems can be addressed while cost-effective solutions are still available. For example, an invasive species that threatens a rare plant population is much easier to control at the initial stages of invasion compared to eradication once it is well established at a site. Monitoring is also critical for measuring management success. Good monitoring can demonstrate that the current management approach is working and provide evidence supporting the continuation of current management (Elzinga et al., 1998).

FIREMON contains several different sampling procedures for monitoring many ecosystem characteristics:

- *Plot Description (PD)*—A generalized sampling scheme used to describe site characteristics on the FIREMON macroplot with biophysically based measurement.
- *Species Composition (SC)*—Used for making ocular estimates of vertically projected canopy cover of all or a subset of vascular and nonvascular species by DBH and height classes using a wide variety of sampling frames and intensities. This procedure is more appropriate for inventory than monitoring.
- *Cover/Frequency (CF)*—A microplot sampling scheme to estimate vertically projected canopy cover and nested rooted frequency for all or a subset of vascular and nonvascular species.
- *Point Intercept (PO)*—A microplot sampling scheme to estimate vertically projected canopy cover for all or a subset of vascular and nonvascular species. Allows more precise estimation of cover than the CF methods because it removes sampler error.
- *Line Intercept (LI)*—Primarily used when the fire manager wants to monitor changes in plant species cover and height of plant species with solid crowns or large basal areas where the plants are about 3 feet tall or taller.
- *Density (DE)*—Primarily used when the fire manager wants to monitor changes in plant species numbers. This method is best suited for grasses, forbs, shrubs, and small trees, which are easily separated into individual plants or counting units, such as stems. For trees and shrubs over 6 feet tall, the TD method may be more appropriate.
- *Rare Species (RS)*—Used specifically for monitoring rare plants such as threatened and endangered species.
- *Tree Data (TD)*—Trees and large shrubs are sampled on a fixed-area plot. Trees and shrubs less than 4.5 feet tall are counted on a subplot. Live and dead trees greater than 4.5 feet tall are measured on a larger plot.
- *Fuel Load (FL)*—The planar intercept (or line transect) technique is used to sample dead and down woody debris in the 1-hour, 10-hour, 100-hour, and 1000-hour and greater size classes. Litter and duff depths are measured at two points along the base of each sampling plane. Cover and height of live and dead woody and nonwoody vegetation are estimated at two points along each sampling plane.

- *Landscape Assessment (LA)*—This sampling methodology is only mentioned briefly in this text because it has no practicable application for use in forest-based biomass energy production; however, it is useful for mapping fire severity over large areas by combining a satellite-derived normalized burn ratio (BR) with a ground-based indicator of fire severity, the composition burn index (BI). The LA methodology will assist in determining landscape-level management actions where fire severity is a determining factor.

- *Composite Burn Index (BI)*—The BI methodology is a subset of the LA methods. It provides users with a ground-based fire severity index derived from a number of plot measurements.

- *Normalized Burn Ratio (BR)*—The BR method is a subset of the LA methods. It describes how to derive remotely sensed spatial information on burn severity, using Landsat satellite data.

Some of these procedures are presented in greater detail below. With regard to actual tree measurement field techniques used to describe density and basal area estimation for trees, the works by Avery and Burkhart (2002), Dilworth (1989), Dilworth and Bell (1973), Husch et al. (1982), Schreuder et al. (1993), and Shivers and Borders (1996) are all excellent resources that can be used as stand-alone models or can be used in conjunction with FIREMON sampling techniques and others discussed in this text. The reader will notice that many of the individual steps within each sampling protocol are repeated, again and again, with each protocol; this deliberate redundant treatment is presented because the steps are important for safety, accuracy, and other reasons.

Tree Data (TD) Sampling Method*

The Tree Data (TD) methods are used to sample individual live and dead trees on a fixed-area plot to estimate tree density, size, and age class distributions before and after fire (and for general inventory purposes) to assess tree survival and mortality rates. These methods were designed to measure important characteristics about individual trees so tree populations can be quantitatively described by various dimensional and structural classes before and after fires. A significant benefit of this method is that it can also be used to sample individual shrubs if they are over 4.5 ft tall. When trees are larger than the user-specified breakpoint diameter, the following are recorded: diameter at breast height, height, age, growth rate, crown length, insect/disease/abiotic damage, crown scorch height, and bole char height. Trees less than the breakpoint diameter and taller than 4.5 ft are tallied by species–diameter–status class. Trees less than 4.5 ft tall are tallied by species–height–status class and sampled on a subplot within the macroplot. Snag measurements are made on a snag plot that is usually larger than the macroplot. Snag characteristics include species, height, decay class, and cause of mortality. The fixed-area plot used to describe tree characteristics in the TD methods is different from the standard timber inventory techniques that use plotless point sampling implemented with the prism. We use the fixed-area technique

---

* Material in this section is based on Keane, R.E., *Tree Data (TD) Sampling Method*, General Technical Report RMRS-GTR-164-CD, U.S. Department of Agriculture, Forest Service, Rocky Mountain Research Station, Fort Collins, CO, 2006.

for a number of reasons. First, the plotless method was designed to quantify stand characteristics using many point samples across a large area (stand). This means that the sampling strategy is more concerned with conditions across the stand than within the plot (conditions within the plot are more meaningful to us in this text). Second, plotless sampling was designed for inventorying large, merchantable trees and is not especially useful for describing tree populations—especially within a plot—because the sampling distribution for the plotless methods undersamples small- and medium-diameter trees (for forest biomass use and production, we are not interested in undersamples). These small trees are the ones many fire managers (and biomass producers) are interested in monitoring, such as for a restoration project or fuel load estimate. Similarly, canopy fuels are not adequately sampled using plotless techniques because there is an insufficient number of trees in all classes to obtain realistic vertical fuel loadings and distributions. Finally, there are many ecosystem characteristics recorded at each biomass macroplot, and the origin and factors that control these characteristics are highly interrelated. Shrub cover, for example, is often inversely related to tree cover on productive sites.

Plotless sampling does not adequately allow the sampling of tree characteristics that influence other ecosystem characteristics, at the plot level, such as loading and regeneration. The expansion of trees-per-area estimates from fixed-area plots is much less variable than density estimation made using plotless methods. The contrast between point and fixed-area plot sampling is really a matter of scale. Point or prism sampling is an efficient means to obtain stand-level estimations, but it is inadequate for describing tree characteristics within a plot—for biomass to energy purposes and using materials in a mixed stand, it is important to know tree characteristics.

When using this protocol, many characteristics are recorded for each tree. The species and health status are recorded for each tree, then the structural characteristics of diameter at breast height (DBH), height, live crown percent, and crown position are measured to describe physical dimensions of the trees. Age and growth rate describe life state and productivity. Insect and disease evidence is recorded in the damage codes. A general description of dead trees is recorded in the snag codes. Fire severity is assessed by estimates of downhill bole char height and percent of crown scorched. User-defined codes can be developed for each tree, if needed. Each tree above the breakpoint diameter gets a tree tag that permanently identifies it for further measurements.

Besides being used to inventory general tree characteristics on the macroplot, the TD method can be used to determine tree survival or fire-caused tree mortality after burn and to describe the pre- and post-burn tree population characteristics by species, size, and age classes. Values in many TD fields can be used to compute a host of ecological characteristics of the tree; for example, the DBH, height, and live crown percent of a tree can be used to compute stand bulk density for modeling crown fire potential.

Sampling Procedure

This method assumes that the sampling strategy has already been selected and the macroplot has already been located. A set of general criteria recorded on the TD Data Form serves as the basis of the user-specified design of the TD sampling method. Each general TD field must be designed so the sampling captures the information necessary to successfully satisfy the management objective within the time, money, and personnel constraints. These general fields must be decided before the crews go into the field and should reflect a thoughtful analysis of the expected problems and challenges in the fire or biomass monitoring project.

TD Data Form Recordings

Mature Trees

Field 1	Macroplot Size
Field 2	Microplot Size
Field 3	Snag Plot Size
Field 4	Breakpoint Diameter
Field 5	Tag Number
Field 6	Species
Field 7	Tree Status
Field 8	DBH (ft/m)
Field 9	Height (ft/m)
Field 10	Live Crown Percent
Field 11	Crown Fuel Base Height (ft/m)
Field 12	Crown Class
Field 13	Age
Field 14	Growth Rate
Field 15	Decay Class
Field 16	Mortality Code
Field 17	Damage Code 1
Field 18	Severity Code 1
Field 19	Damage Code 2

Field 20 Severity Code 2

Field 21 Char Height (ft/m)

Field 22 Crown Scorch Percentage (%)

Field 23 Local Code

Saplings

Field 24 Diameter Class (in. or cm)

Field 25 Species

Field 26 Status

Field 27 Count

Field 28 Average Height (ft or m)

Field 29 Average Live Crown Percentage (%)

Field 30 Local Code

Seedlings

Field 31 Height Class (ft/m)

Field 32 Species

Field 33 Status

Field 34 Count

Field 35 Local Code

Plot Size Selection

There may be as many as three nested fixed plots in the TD methods protocol presented here. First, the macroplot is the primary sampling plot, and it is the plot where all live tree population characteristics are taken. Nest is the snag plot, which may be larger than the macroplot, and it is used to record a representative sample of snags. Often the incidence of snags on the landscape is so low that the macroplot area is not large enough to reliably describe snag populations. The snag macroplot allows more snags to be sampled. Last is the subplot where all seedlings are counted. This is smaller than the macroplot and is provided to streamline the counting of densely packed seedlings. All of the plots are con-centrically located around the macroplot center.

Macroplot Area

The size of the macroplot ultimately dictates the number of trees that will be measured, so large plot sizes usually take longer to sample because of the large number of trees on the plot; however, some ecosystems have widely spaced trees scattered over large areas so large plot sizes are needed to obtain statistically significant estimates. Many studies have been done to determine the optimum plot size for different ecosystems with mixed results. Table 6.1 helps to determine the plot size that best matches the monitoring application. Unless the project objectives dictate otherwise, use the median tree diameter and median tree height—for trees greater than breakpoint diameter—to determine the size of the macroplot. When filling out the TD Data Form, enter the macroplot size of the TD method in Field 1. In general, the 0.1-ac (0.04-ha) circular plot will be sufficient for most forest ecosystems and should be used if no other information is available. A general rule-of-thumb is

TABLE 6.1

Median Tree Diameter and Median Tree Height Determine Size of Sampling Macroplot

Median Tree Diameter (Trees Greater Than Breakpoint Diameter)	Median Tree Height	Suggested Plot Radius	Suggested Plot Size
<20 inches (<50 cm)	<100 ft (<30 m)	37.24 ft (11.28 m)	0.1 ac (0.04 ha)
20 to 40 inches (<100 cm)	100 to 130 ft (<40 m)	52.66 ft (12.6 m)	0.2 ac (0.05 ha)
>40 inches (<200 cm)	>130 ft (<50 m)	74.47 ft (17.84 m)	0.4 ac (0.1 ha)

that the plot should be big enough to capture at least 20 trees above breakpoint diameter (see definition in next section) on the average across all plots in your project. Though it is not absolutely necessary, extra measures should be taken so plot sizes are the same for all plots in a project.

Subplot Area

All seedlings—trees less than 4.5 ft (1.37 m) tall—are measured on a subplot nested within the macroplot (see Figure 6.1). Use seedling density to determine the subplot size (see Table 6.2) unless the project objectives dictate otherwise. Again, make an effort to keep the subplot radius constant across all plots in the project. The area of the subplot is entering Field 2 on the TD Data Form.

Snag Plot Area

Snags are dead trees greater than breakpoint diameter. Snags can be measured within the macroplot, but often their numbers are so low that a larger plot is needed to detect changes in snag populations. A suitable snag plot size is difficult to determine because snags are nonuniformly distributed across the landscape. A good rule-of-thumb for sizing the snag plot is to double the macroplot diameter, which will increase the snag plot area by a factor of four. Enter the snag plot area in Field 3 of the TD Data Form even if it is the same size as the macroplot. If snags are important to the project objectives, you should choose a plot radius that will provide you with at least 20 snags, meaning the plot could be quite large.

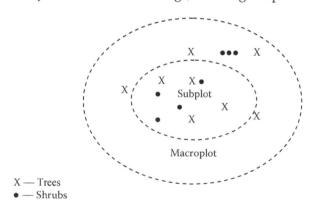

X — Trees
● — Shrubs

FIGURE 6.1

The subplot is nested inside the macroplot. The recommended circular plot shape is shown, but rectangular plots may also be used.

TABLE 6.2

Seedling Density Determines Subplot Radius

Seeding Density	Subplot Radius	Area
Typical	11.7 ft (3.57 m)	0.01 ac (0.004 ha)
<2 seedlings per species	18.62 ft (5.64 m)	0.025 ac (0.01 ha)
>100 seedlings per species	5.89 ft (1.78 m)	0.0025 ac (0.001 ha)

Breakpoint Diameter

The breakpoint diameter is the tree diameter at breast height (DBH) above which all trees are tagged and measured individually and below which trees are tallied to species–DBH classes. Choose a breakpoint diameter that allows at least 20 trees to be measured on each macroplot; see Figure 6.2). The same breakpoint diameter should be used for all the macroplots in the study. Selection of the breakpoint diameter must account for fire monitoring objectives as well as sampling limitations and efficiency. For example, a large breakpoint diameter (>8 in.) will exclude many small trees from individual measurements and reduce the sampling time. As long as the project objectives are broad (e.g., 70 to 80% mortality of saplings) then a tally of trees below breakpoint will be sufficient. If the objectives are specific (e.g., determine the effect of crown scorch and char height on fire-caused mortality of trees 2 in. and greater DBH), then a lower breakpoint diameter, and an associated increase in sampling effort, will be necessary. Individual measurement of small trees might be unrealistic if there are high sampling densities (>1000 trees) on the macroplot. Selection of an appropriate breakpoint diameter requires some field experience and knowledge of the resources available to complete the fire monitoring project. We suggest using a 4-inch (10-cm) breakpoint diameter if no other information is available. The breakpoint diameter is entered in Field 4 of the TD Data Form.

Preliminary Sampling Tasks

Before setting out for your field sampling, lay out a practice macroplot and subplot in an area with easy access. Even if there are just a few trees on your practice plot, getting familiar with the plot layout and the data that will be collected before heading out will make the first day or two of field sampling less frustrating. This will also let you determine if any pieces of equipment must be ordered.

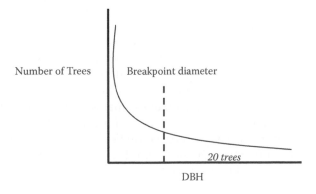

FIGURE 6.2
Choose a breakpoint diameter that leaves at least 20 trees in the tail of the distribution.

When you are ready to go into the field, gather the necessary materials. You will probably spend most of your day hiking from plot to plot, so it is important that supplies and equipment are placed in a comfortable daypack or backpack. Be sure you pack spare equipment so an entire day of sampling is not lost if something breaks. Spare equipment can be stored in the vehicle rather than the backpack. Be sure all equipment is well maintained and that you have plenty of extra supplies, such as Data Forms, map cases, and pencils.

All TD Data Forms should be copied onto waterproof paper because inclement weather can easily destroy valuable data recorded on standard paper. Data Forms should be transported into the field using a plastic, waterproof map protector or plastic bag. The day's sample forms should always be stored in a dry place (office or vehicle) and not be taken back into the field for the next day's sampling.

If the sampling project is resampling previously installed plots, then it is highly recommended that plot sheets from the first measurement be copied and brought to the field for reference. These data can be extremely valuable for help in identifying sample trees that have lost their tree tags or have fallen over, and the data can provide an excellent reference for verifying measurements.

One person on the field crew, preferably the crew boss, should have a waterproof, lined field notebook for recording logistic and procedural problems encountered during sampling. This helps with future remeasurements and future field campaigns. All comments and details not documented in the sampling methods should be written in this notebook; for example, snow on the plot might be described in the notebook, which would be helpful in plot remeasurement.

It is beneficial to have plot locations for several days of work in advance in case something happens, such as if the road to one set of plots is washed out by flooding. Locations or directions to the plots you will be sampling should be readily available to reduce travel times. If the plots were randomly located within the sampling unit, it is critical that the crew be provided plot coordinates before going into the field. Plots should be referenced on maps and aerial photos using pin-pricks or dots to make navigation easy for the crew and to provide a check of the referenced coordinates. It is easy to misrecord latitude and longitude coordinates, and marked maps can help identify any erroneous plot positions.

A field crew of three people is probably the most efficient for implementation of the TD sampling method. For safety reasons there should never be a one-person crew. A crew of two people will require excessive walking and trampling on the plot, and more than three people will probably result in some people waiting for critical tasks to be done and unnecessary physical damage to the plot.

For simplicity, we refer to the people in the three-person crew as the crew boss, note taker, and technician. The crew boss is responsible for all sampling logistics, including the vehicle, plot directions equipment, supplies, and safety. The note taker is responsible for recording the data on Data Forms or onto a laptop. The technical will perform most individual tree measurements while the note taker estimates tree heights. Of course, the crew boss can be the note taker, and probably should be for most situations. The initial sampling tasks of the field crew should be assigned based on field experience, physical capacity, and sampling efficiency. Sampling tasks should be modified as the field crew gains experience and tasks should also be shared to limit monotony.

Sampling Tasks

Define Macroplot Boundary

The snag and subplot boundaries usually do not require the same extensive marking and flagging as required for the macroplot. For the snag plot, experience suggests fixing a cloth or fiberglass tape to the macroplot center and proceeding out to the diameter of the snag plot radius. Again, based on experience, a cloth or fiberglass tape is suggested because your diameter tape will probably not be long enough. Leave the tape pulled out and start traversing the snag plot to search for snags. You will be able to determine whether most of the snags you find are "in" or "out" just by looking; however, you will have to use the tape to measure snags that are on the border. If you have a second cloth or fiberglass tape, this task will go more quickly. You probably will not need to flag the seedling subplot because it is so small. Seedling sampling is describing detail below.

Define Initial Sample Position

Once the macroplot has been defined and all perimeter flags have been hung, flag the tree inside the macroplot, farthest from plot center and greater than breakpoint diameter, that is closest to due north (360 degrees azimuth) from plot center. Mark it near the base so flagging will not be confused with the plot boundary marking. This tree will be the first tree measured and will also indicate when the sampling has been completed. If tree density is high, then you may want to flag the closest tree to the left of your "first" tree as the "last" tree (you will be sampling clockwise around the plot). Tie a flag at the base of the trees so they will not be confused with plot boundary trees. You may want to use another flag color or use tree chalk instead of a flag (see Figure 6.3). Subplot sampling is done before individual tree measurements because repeated walking around the macroplot center by field crews may trample some seedlings and bias the sample.

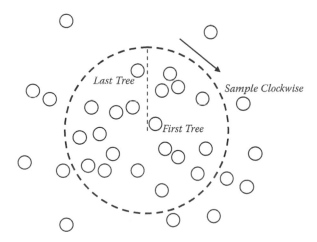

FIGURE 6.3
Sample trees clockwise around the plot starting with the first tree located clockwise from an azimuth of 360 degrees true north.

TABLE 6.3

Midpoint Height Classes for Seedlings

Feet		Meters	
Class	Height Range	Class	Height Range
0.2	>0.0–0.5	0.1	>0.0–0.2
1	>0.5–1.5	0.3	>0.2–0.5
2	>1.5–2.5	0.6	>0.5–0.8
3	>2.5–3.5	0.9	>0.8–1.0
4	>3.5–4.5	1.2	>1.0–1.4

Sampling Seedlings

To start your seedling counts, attach the zero end of a logger's tape at plot center and walk away headed due north until you are at the distance (corrected for slope) you selected for your subplot radius from Table 6.2. Once at the required distance, hold the tape just above the seedlings, then sweep clockwise around the plot, tallying seedlings by species into their respective status and height classes as you go. Use the following status classes:

H—Healthy (tree with little biotic or abiotic damage)

U—Unhealthy (tree with some biotic or abiotic damage that will reduce growth, although it appears that the tree will not immediately die from the damage)

S—Sick (tree with extensive biotic or abiotic damage that will ultimately cause death within the next 5 to 10 years)

D—Dead (tree or snag with no living tissue visible)

The height classes in Table 6.3 are suggested, but you may use any classes you choose. The FIREMON Analysis Tools program assumes that the height class value recorded is the midpoint of the class. Make sure you note any size class changes in the FIREMON Metadata table. Estimating height using a stick marked with the height classes will make counting quicker and easier. Experience has shown that using two crew members to sample is prudent, with one person tallying all of the seedlings on the inner part of the plot and the other person counting all the seedlings on the outer part. When the tape encounters a tree as it is swept clockwise, the sampler closest to the plot center can keep track of which seedlings have been counted so the sampler holding the tape can move around the obstructing tree and return the tape to the proper point before the tally is resumed. Be sure to carefully search the subplot for all seedlings, no matter how small. Use the dot tally method for seedlings, and when you have finished enter the counts in the Seedlings table of the TD Data Form by midpoint class (Fields 31 to 34). Sampling seedlings by height class allows a compromise between sampling efficiency and the detail required for describing fire effects. If a tree is broken below 4.5 ft (1.37 m), but you believe that the tree would be taller than 4.5 ft if unbroken, you should still sample it as a seedling.

Saplings and Mature Trees

Divide trees taller than breast height (4.5 ft) into two groups: saplings and mature trees. Both groups are measured on the macroplot. Saplings are trees taller than 4.5 ft (1.39 m) but smaller than the breakout diameter, and mature trees are greater than breakpoint

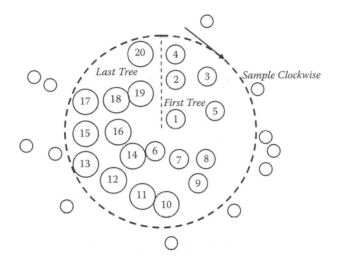

FIGURE 6.4
Sample trees by first measuring the tree you flagged due north of the plant center, then move in a clockwise direction. It is easiest to sample all trees above DBH at the same time.

diameter. Trees above the breakpoint diameter are tagged, measured, and recorded in the TD Data Form. Trees below the breakpoint, but taller than 4.5 ft (1.37 m), are measured and tallied in the TD Data Form. It is most efficient if both sampling tasks are done at the same time as the sampling proceeds around the macroplot because, especially in dense stands, it will be difficult to determine which trees have been measured.

As with seedlings, sample saplings and mature trees in a clockwise direction starting with your "first" tree. The best way to do a repeatable sample is to measure trees in the order that the second hand of a watch would hit them as it moved around the plot (see Figure 6.4). Sometimes this method means that you travel back and forth between the middle and outer portions of the plot, which may seem inefficient, but the benefit is that you will be able to relocate trees easier at a later date if something happens to the trees or the tree tags. For example, if tagged trees are blown down by heavy winds or knocked down by other falling trees, you will be able to account for their mortality by locating them, in order, around the plot.

Once a tree is identified for measurement, the sampler must decide if it is above or below the breakpoint diameter. Usually you will be able to do this visually, but use a ruler or diameter tape if the tree is borderline. Experience shows that it is usually less time consuming to initially estimate DBH using a clear plastic rule rather than with a diameter tape; however, if the ruler estimate is within 1 in. (2 cm) of the breakpoint diameter, we recommend using a diameter tape because the diameter measurement will be more reliable.

Measuring DBH

The diameter of a tree or shrub is conventionally measured at exactly 4.5 ft (1.37 m) above the ground surface, measured on the uphill side of the tree if it is on a slope. Wrap a diameter tape around the bole or stem of the plant, without twists or bends and without dead or live branches caught between the tape and the stem. Pull the tape tight and record the diameter. If you are only determining the diameter class, then the measurement only has to identify the class in which the tree falls. If you are measuring the diameter of a mature

TABLE 6.4

Midpoint Height Classes for Saplings

Inches		Centimeters	
Class	Height Range	Class	Height Range
0.5	>0–1	1.2	>0–2.5
1.5	>1–2	3.8	>2.5–5
2.5	>2–3	6.2	>5–7.5
3.5	>3–4	8.8	>7.5–10

tree, then measure to the nearest 0.10 in. (0.5 cm). Large-diameter trees are difficult to measure while standing at one point, so you have to hook the zero end of the tape to the plant bole (bark) and then walk around the tree, being sure to keep the tape exactly perpendicular to the tree stem and all foreign objects from under the tape.

Forked trees should be noted in the damage codes discussed in later sections. Forked trees often occur in tall shrublands or woodlands (pinyon juniper, whitebark pine) where trees are clumped due to bird caches or morphological characteristics. Many tree-based sampling techniques suggest that diameter be measured at the base rather than at breast height, but this may bias an estimate of tree mortality by not counting survival of individual stems. Moreover, basal diameters may not adequately portray canopy fuel for fire modeling.

Trees are "live" if they have any green foliage on them regardless of the angle at which they are leaning. If a tree has been tipped or is deformed by snow, make all the measurements but use the damage codes described below to record the nature of the damage.

Sampling Saplings

Saplings are recorded in species-diameter class-status groups rather than as individuals. As mentioned above a 4-ich (10-cm) break point diameter should be used. Tally trees by the diameter classes shown in Table 6.4. The diameter class value you record should be the midpoint of the class. Make sure any size class changes are listed in the Metadata table. Note that no age or growth characteristics are recorded for trees below breakpoint diameter. The only fire severity attribute assessed from these data is tree mortality by species–diameter–status class.

When a tree has been determined to be below the breakpoint diameter, the sample must (1) estimate diameter to the appropriate diameter class, (2) determine the species of the tree, (3) identify the tree status, and (4) estimate height to the nearest 1 ft (0.3 m). Tree diameter is most easily estimated using a clear plastic ruler. A diameter tape can be used, but it is often cumbersome when diameters are less than 4 inches (10 cm). Once the diameter class, species, status, counts, and average heights have been determined, record them in Fields 24 to 28, respectively. Optionally, average live crown percent can be recorded in Field 29 using the classes in Table 6.5. Record the class based on the percent of the tree that has live foliage growing from it, from the top of the live foliage to the ground.

Height and live crown percent are challenging to estimate because their values must represent all trees within the species–diameter–status class combination. The best way to do this is to look at the first tree, assign it to a species–diameter–status class, visually estimate height to the nearest 1 ft (0.5 m), estimate live crown percent class, and record

TABLE 6.5

Classes Used to Record Live Crown and Crown Scorch

Class	Live Crown (%)	Class	Live Crown (%)
0	0	50	>45–55
0.5	>0–1	60	>55–65
3	>1–5	70	>65–75
10	>5–15	80	>75–85
20	>15–25	90	>85–95
30	>25–35	98	>95–100
40	>35–45		

the estimates in the Saplings part of the TD Data Form. As more trees in the same species–diameter–status class are found, the note taker must adjust the average height and average crown ratio of the class. Methods for measuring height and live crown percent are discussed in detail in the next section.

Sampling Mature Trees

Mature trees are those with a diameter at breast height that is greater than or equal to the breakpoint diameter. If you are doing a monitoring project you should tag all of the mature trees on your macroplot so they can be identified in the future. (If you are simply doing an inventory, they do not need to be tagged. In this case, sequentially number the trees as you sample. The tree numbers are needed in the database.) There are two methods for tagging the sample trees depending on whether you are using steel or aluminum tags. The tag number is recorded on the TD Data Form. Experience suggests that steel casket tabs should be used because they will not melt due to the heat of a fire. Nail the casket tags to the tree using high-grade nails that also will not melt during a fire. Each tag should be tightly nailed to the tree bole with just enough pressure to prevent it from moving but not so tight that the tag is driven into the bark. If the tag is allowed to move or twirl in the wind, the movement might wear through the nail and the tag could fall off. Nail the tags at breast height with the tag facing toward plot center. This is done so each tree can be identified while standing in one place and will make relocating the plot center easier. It will also reduce macroplot travel, which can cause compaction and seedling trampling. If the trees are going to be cut, nail the tag less than 1 ft from the ground, facing plot center. This will leave the tags available for posttreatment sampling and will keep them out of the sawyer's way if the trees are cut with a chain saw.

Unfortunately, casket tags are expensive and the steel tags and nails can damage saws, so they should not be used if the trees being tagged will eventually be harvested. As an alternative to steel tags, nail aluminum tags at DBH on the downhill side of all the mature trees using aluminum nails. Putting tags on the downhill side of the trees will help keep them out of the hottest part of the flame and help prevent the tags and nails from melting. Pound the nails in at a downward angel, until the tags are tight but not driven into the bark. Typically, some of the tags will be melted by the fire, but by using tree characteristics (species and DBH, primarily) and the clockwise sampling scheme you should be able to relocate all the mature trees. Occasionally, the head of the nail will melt off and the wind will blow the tag off the angled nail, but you may be able to find the tag lying at the base of the tree. Although it is not recommended, sometimes it may be necessary to remove some branches with a hatchet or bow saw so the tag can be firmly attached to the tree. If necessary, remove just the problem branches. Removing too many branches may influence the health of the tree or modify the fuelbed around the tree.

Measuring General Attributes of Mature Trees

Record the general characteristics of the tree (species and health) on the TD Data Form. These characteristics allow the stratification of results and provide some input values required to compute crown biomass and potential tree mortality. Enter the species of the tagged tree and the tree status on the TD Data Form. Tree status describes the general health of the tree. Use the following tree status codes:

H—Healthy (tree with little biotic or abiotic damage)

U—Unhealthy (tree with some biotic or abiotic damage that will reduce growth, although it appears that the tree will not immediately die from the damage)

S—Sick (tree with extensive biotic or abiotic damage that will ultimately cause death within the next 5 to 10 years)

D—Dead (tree or snag with no living tissue visible)

Tree status is purely a qualitative measure of tree health, but it does provide an adequate characteristic for stratification of preburn tree health or determining postburn survival. Remember that marking trees as dead (code D) indicates that the tree was sampled on the snag plot.

Measuring Structural Characteristics of Mature Trees

Five important structural characteristics are measured for each mature tree: DBH, height, live crown percent, crown fuel base height, and crown class. These structural characteristics are used to assess a number of fire-related properties such as canopy bulk density, vertical fuel ladders, height to the base of the canopy, and potential fire-caused mortality. Tree DBH is measured using a diameter or logger's tape. Be sure to measure DBH on the uphill side of the tree with the diameter tape perpendicular to the tree stem and directly against the bark. Measure DBH to the nearest 0.1 in. (0.2 cm) and record the measurement in Field 8 on the TD Data Form.

Measure tree height from the ground level to the top of the bole or highest living foliage, whichever is higher, to the nearest 1 ft (0.5 m), and enter the height in Field 9 of the TD Data Form. Tree height is commonly measured with a clinometer, but it can be measured with laser technology or other surveying techniques if that equipment is available and the crew has adequate training in using that technology.

Live crown percent (LCP) is the percent of the tree bole that is supporting live crown. LCP is estimated by visually redistributing the live tree crown evenly around the tree so the branches are spaced at about the same branch density as seen along the bole and from the typical conical crown. In the instance of recent abiotic or biotic damage, use the damage and severity codes described below to improve the crown fuel estimates. Record the LCP in Field 10 on the TD Data Form using the classes shown in Table 6.5. Sometimes, a lone live branch at the bottom of the crown doesn't appear as part of the crown; in these cases, we ignore the branch's contribution to the LCP.

Crown fuel base height (CFBH) is important for assessing the risk of crown fire. In Field 11 record either the height of the dead material that is sufficient to carry a fire from the lower to the upper part of the tree crown or, if the dead fuel is insufficient, the height of the lowest live foliage. The dead material may include dead branches associated with mistletoe infection, lichens, dead needles, and so forth. This is probably the most subjective field of the TD assessment. Generally, this information is not collected unless a knowledgeable crew member is available who can consistently estimate CFBH. This person should also record in the Metadata table the characteristic used for the assessment so it can be made the same way in future sampling.

Optionally, users can record height to live crown (ignore the dead component) to limit some subjectivity. Height to live crown is defined as the height of the lowest branch whorl that has branching in two quadrants—excluding epicormic branches and whorls not part of the main crown—measured from the ground line on the uphill side of the tree. If you decide to estimate height to live crown instead of CFBH, be sure to note that in the Metadata information.

Crown classes (Field 12) represent the position in the canopy of the crown of the tree in question and describe how much light is available to the crown of that tree (see Figure 6.5). Six categories describe the crown class:

O—**Open grown** (the tree is not taller than other trees in the stand but still receives light from all directions)

E—**Emergent** (the crown is totally above the canopy of the stand)

D—**Dominant** (the crown receives light from at least three to four directions)

C—**Codominant** (the crown receives light from at least one to two directions)

I—**Intermediate** (the crown only receives light from the top)

S—**Suppressed** (the crown is entirely shaded and underneath the stand canopy)

Crown class may be used for data report stratification, but it is also an important variable in the computation of tree biomass and leaf area.

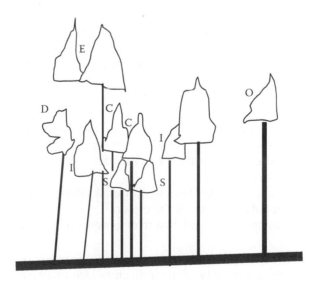

FIGURE 6.5
Use this illustration of crown classes to help you describe the crown class of the tree you are measuring.

TREE CROWN CONDITION INDICATOR: FOREST INVENTORY AND ANALYSIS[*]

Many physical and biological components influence forest trees. Individual sapling and tree growth and vigor are determined by a variety of physiological and external effects such as age, available light, water, and nutrients. Because tree crowns are a component of forest ecosystem structure, they directly affect the composition, processes, and vigor of the understory plant and animal components of the forest. Information about crowns can be used to answer such questions about forest ecosystems as

- What proportion of trees (by species and forest type) in a region have more crown dieback or less crown density than normal?
- Is there a regional change in the proportion of trees (by species and forest type) showing a change in crown condition?

The crown is one component of net primary production, and its dimensions reflect general tree health. Large, dense crowns are associated with potential or previous vigorous growth rates. Small, sparse crowns suggest unfavorable site conditions (such as competition from other trees, moisture stress, or moisture excess) or other influences (such as insect defoliation, foliage diseases, or hail storms). Some changes in crown measurements may be temporary, especially when the stressing condition can be eliminated or reduced. Continued poor conditions however, may be reflected in a tree with increasingly poor vigor. Tree crown information contributes to the investigation of several key forest ecosystem attributes, including biodiversity, productivity, sustainability, aesthetics, forest environment, and wildlife. Much tree crown information can be obtained during crown evaluations. During these tree crown evaluations, seven measurements are made:

- *Uncompacted live crown ratio* estimates the percentage of a tree's height that supports live, green foliage that is contributing to total tree growth.
- *Crown light exposure* estimates the amount of a tree's foliage receiving direct sunlight.
- *Crown position* estimates the position of a tree's crown relative to the stand overstory canopy zone.
- *Crown density* estimates the amount of light blocked by branches, reproductive structures, and foliage within a tree's crown.
- *Crown dieback* estimates recent branch mortality in the upper and outer potions of the live crown.
- *Foliage transparency* estimates the amount of skylight visible through the live, normally foliated portion of the crown.
- *Crown vigor* estimates the overall health or general appearance of saplings.

Three levels of crown indicator data are used to evaluate tree crown condition. As the level increases, the complexity of the indicators also increases.

Level 1. *Absolute crown indicators* (uncompacted live crown ratio, crown light exposure, crown position, crown density, crown dieback, foliage transparency, and crown vigor)—These indicators consist of the crown measurements described above. They are simple and readily identifiable measures of crown condition.

Level 2. *Crown structure indicators* (crown volume and crown surface area)—These indicators are computed from the Level 1 indicators and are measures of crown dimension and fullness. To compute the crown structure indicators at this time, crown diameters are being modeled from previous datasets. The crown deflation indicator is computed from the Level 1 indicators (crown density, crown dieback, and foliage transparency) and is a measure of the thickness and quantity of foliage in the crown.

Level 3. *Composite crown indicators* (composite crown volume, and composite crown surface area)—These indicators are computed from Level 2 indicators by combining crown structure and crown defoliation indicators. *Crown production efficiency* is computed using the composite crown indicators and reflects a tree's potential to capture and utilize solar energy.

---

* Adapted from Schornaker, M., *Tree Crown Condition Indicator*, FIA Fact Sheet Series, U.S. Department of Agriculture, Forest Service, Arlington, VA, 2003.

Measuring Tree Age Characteristics

Age characteristics allow better interpretation of the fire monitoring and tree biomass data by identifying important age and growth classes and using them for results stratification. Growth data also allow direct comparison of the change in growth rate as a result of the fire. Tree age is estimated by extracting a core from the tree at stump height (about 1 ft above ground-line) on the downhill side of the tree using an increment borer. Rings on the increment core can be counted in the field or counted later in the office or laboratory. Coring trees to determine tree age is a time-consuming procedure that often requires more than 50% of the sampling time. There are many reasons for this. First, much time and effort are involved in coring the tree, especially large trees, and often you will need to take more than one core per tree to get a usable core because of rot or because you missed the pith. Next, it is difficult to get comfortable when drilling at stump height because the sampler is stooped over in an unpleasant position. Certainly you do not want to have to core every mature tree on your macroplot, so the question is how many samples are enough? One tree per species per 4-in. (10-cm) diameter class is probably adequate to core if you are sampling at the Level III–Detailed intensity. The project manager can increase the diameter class width to 8 to 12 in. (20 to 30 cm) to further limit coring time on the plat. Enter the tree age in Field 13 of the TD Data Form for each tree that was cored. Growth rate should be determined for trees that have been cored for age; the growth rate core should be taken at DBH because the age core taken at stump height is not a good indicator of growth rate. Remove a core that is deep enough to allow you to measure the last 10 years' growth. Growth rate is the distance between the cambium and the 10th growth ring in from the cambium, measured to the nearest 0.01 in. (0.1 mm). If changes in growth rates are a critical part of the monitoring effort (check sampling objectives), then more trees must be cored for growth rate. If growth rate is an important facet of the fire monitoring project, all tagged trees should be cored to determine the 10-year growth increment. Enter the growth rate in Field 14 on the TD Data Form.

Measuring Damage and Severity

This information may not be essential for determining the effects of prescribed and wild-land fires, but it can be useful for describing, interpreting, and stratifying monitoring results. These fields should be completed for most fire and biomass monitoring applications. Biotic/Abiotic Damage and associated Severity codes (see Table 6.6) are entered in Fields 17 to 20 in the TD Data Form. These codes describe and then quantify the degree of damage by biotic (insect, disease, browsing) and abiotic (wind, snow, fire) agents for the mature trees. Additionally, you will be able to use the abiotic codes for snags; again, these are listed in Table 6.6. When trees exhibit multiple forms of damage, record the damage with the greatest severity first, then the damage with the second greatest severity. Keep in mind that it takes a great deal of experience to identify each of the damaging agents present in a tree. If the crew does not have experience or training in the identification of biotic and abiotic damage sources, then do not fill out these fields.

DID YOU KNOW?

Changes over time in the structure of forest resources are largely driven by stand dynamics (rates of regeneration, growth, and mortality), as well as timber removals and changes in land use. Tree growth is an important facet of stand dynamics. Information about growth can be used to identify any unusual spatial or temporal patterns in growth rates or to determine if the balance between growth and mortality is adequate to sustain a forest ecosystem. Tree growth data contribute to the investigation of several key forest ecosystem attributes such as sustainability, productivity, and aesthetics (USFS, 2003b).

TABLE 6.6

Tree and Snag Damage Codes

Damage Code	Description	Severity Code
00000	No damage	**No damage**
10000	General insects	**101**: *Minor*—Bottle brush or shortened leaders, <20% of branches affected, 0 to 2 forks on stem, or <50% of the bole with larval galleries
		102: *Severe*—3 or more forks on bole, 20% or more branches affected, terminal leader dead, or 50% or more on the bole with visible larval galleries
19000	General diseases	**191**: *Minor*—Short-term tree vigor probably not affected
		192: *Severe*—Tree vigor negatively impacted in the short term
25000	Foliage diseases	**251**: *Minor*—<20% of the foliage affected or <20% of crown in brooms
		252: *Severe*—>20% of the foliage affected or >20% of the crown in brooms
50000	Abiotic damage	**501**: *Minor*—<20% of the crown affected; bole damage is <50% circumference
		502: *Severe*—>20% of the crown affected; bole damage is >50% circumference
90000	Unknown	**900**: 0–9% affected
		901: 10–19% affected
		902: 20–29% affected
		903: 30–39% affected
		904: 40–49% affected
		905: 50–59% affected
		906: 60–69% affected
		907: 70–79% affected
		908: 80–89% affected
		909: 90–99% affected

Measuring Fire Severity

Two fire severity measurements are specific to the TD sampling method and apply to the mature trees. The first is bole char height, which is the height of the continuous char measured above the ground, on the downhill side of the tree, or on flat ground the height of the lowest point continuous char. It is used to quantify potential tree mortality, and some mortality prediction equations use bole char height as an independent variable. Measure char height with a logger's tape or cloth tape, holding the tape on the downhill side of the tree; measure to the top of the charred part of the bole, keeping the tape exactly vertical to the tree. Be sure to measure vertical height. Do not measure along the tree bole, which might be tempting if the tree is leaning. Record bole char height to the nearest 0.1 ft (0.3 m) in Field 21 of the TD Data Form. The second fire severity measurement is percent of crown scorched (PCS), which directly relates to tree mortality. It is extremely difficult to estimate because the sampler often does not know if all the charred branches were alive prior to the fire; however, estimates of PCS to the nearest 10% class are usually adequate for use in mortality equations, so the bias introduced by dead branches should not cause major problems. Estimate PCS by trying to rebuild the tree crown in your head, and then estimate the amount of the live crown volume that was damaged or consumed by the fire. The foliage could have been entirely consumed or scorched black or brown with needles or leaves still attached to the branches. Enter percent crown scorched in Field 22 of the TD Data Form using the classes shown in Table 6.5.

Sampling Snags

Two fields on the TD Data Form are specific to snags: snag decay class and primary mortality agent. Snags progress through a series of stages after the tree dies, from standing with red, dead needles still attached to a well-decayed stump. As snags pass through these stages, they function differently in the ecosystem; thus, it is important to record snag decay class so the ecosystem characteristics can be quantified. Enter the appropriate snag decay class code from Table 6.7 into Field 15 of the TD Data Form. Snag characteristics (e.g., wildlife preference and snag persistence) can be greatly influenced by the mortality agent, so it is important to try to identify what killed the tree. This can be difficult for snags that have been standing for a few years, as it is generally accepted that after 5 years it is not possible to determine the cause of mortality with any certainty. Also, typically there is more than one mortality agent. Common practice is to record the primary cause of mortality;

TABLE 6.7

Descriptive Characteristics Used to Determine Snag Decay Class

Snag Code	Limbs	Top of Bole	Bark	Sapwood	Other
1	All present	Pointed	100% remains	Intact	Height intact
2	Few, limbs	May be broken	Some loss, variable	Some decay	Some loss in height
3	Limb stubs only	Usually broken	Start of sloughing	Some sloughing	Broken top
4	Few or no limb stubs	Always broken; some rot	50% or more loss of bark	Sloughing evident	Loss in height always
5	No limbs or limb stubs	Broken and usually rotten	20% bark remaining	Sapwood gone	Decreasing height with rot

TABLE 6.8

Mortality Codes Used to Identify the Primary
(First) Cause That Killed the Tree

Mortality Code	Description
F	Fire caused
I	Insect caused
D	Disease caused
A	Abiotic (flooding, erosion)
H	Harvest caused
U	Unable to determine
X	Did not assess

for example, a fire may injure a tree enough to cause stress that, in turn, will reduce its resistance so it cannot effectively protect itself from beetle attacks, and then the beetles introduce a fungus that eventually kills the tree. In this case, the primary (first) agent of mortality is fire. If you can determine the primary cause of mortality for the tree, record it in Field 16 using the codes shown in Table 6.8. If you cannot determine the primary cause of mortality, use the U code.

Precision Standards

The precision standards listed in Table 6.9 are used for TD sampling.

Fuel Load (FL) Sampling Method*

The Fuel Load (FL) sampling method is another of the modified FIREMON methods. Although these methods are primarily designed for forest fire managers and forest fire planning scenarios, FIREMON and specifically FL methods are also used to estimate biomass content. The FL sampling method is an important tool that can be used to inventory and monitor forest-based biomass. At the very least, FL sampling and other FIREMON sampling methods provide us with a model of how to inventory and monitor our forest resources. In light of this, learning about FL and other forest sampling methods is important to this discussion.

Introduction

The Fuel Load (FL) methods are used to quantify three general components of the fuel complex: dead and down woody debris (DWD), duff and litter, and understory vegetation. Biomass estimates of dead and down woody debris are collected for the size classes that fire scientists and other forest biomass practitioners have found to be important for predicting fire behavior—1-hour, 10-hour, 100-hour, and 1000-hour and greater. DWD

---

* Information contained herein is from or based on USDA, *Fire Service Fuel Load (FL) Sampling Method*, General Technical Report RMRS-GTR-164-CD, U.S. Department of Agriculture, Forest Service, Rocky Mountain Research Station, Fort Collins, CO, 2006.

TABLE 6.9

Precision Guidelines for Tree Data (TD) Sampling

Component	Standard
DBH	±0.1 inch (0.25 cm)
Height	±1 ft (0.3 m)
Live crown percent	±1 class
Live crown base height	±1 ft (0.3 m)
Crown class	±1 class
Age	±10% of total years
Growth rate	±0.01 inch (0.1 mm)
Decay class	±1 class
Mortality code	Best guess
Damage code	Appropriate category
Severity code	±1 severity class
Char height	±1 ft (0.3 m)
Crown scorch	±1 class
Count	±10% of total count
Average height (saplings)	±1 ft (0.3 m)
Average live crown percent	±1 class

measurements are based on the planar intercept methods published by Brown (1974). The sampling area is an imaginary plane extending from the ground, vertically from horizontal (no perpendicular to the slope) to a height 6 ft (2 m) above the ground. Pieces that intercept the sampling plane are measured and recorded. Frequently, the term *line transect sampling* is used when discussing the planar intercept method. As far as the FL methodology is concerned, the two terms can be interchanged as long as samplers recognize that the "line" is really the measuring tape laid on the litter layer, while the "plane" extends above and below the tape, from the top of the litter layer to a height of six feet. Duff and litter are assessed by measuring the depth of the duff/litter profile down to mineral soil and estimating the proportion of the total duff/litter depth that is litter. The biomass of live and dead, woody and nonwoody understory vegetation is estimated using cover and average height estimations. The data collected using the FL methods are used to model fire behavior or to indicate potential fire effects. Forest managers often prescribe fuel treatments, at least partially, on the data collected using the FL methodology. The load of DWD can also be used to estimate the total carbon pool that is stored in the dead material, or DWD data can be used as an indicator of habitat for wildlife. Standing dead trees (snags) are sampled using the FIREMON Tree Data (TD) methods.

The FL methods allow data collection for a wide number of fuel characteristics on each plot; however, field crews are not required to sample every characteristic represented on the Data Form. In fact, FIREMON was developed specifically so that crews only sample the characteristics they are interested in, as determined by the goals and objectives of the project. In most cases, the data collected from plot to plot will be the same, although in some situations a characteristic may be sampled on a subset of the sampling plots.

Dead wood is important in many forest processes. Fire and biomass managers need to have an estimate of down dead fuel because it substantially influences fire behavior and fire effects. Smaller pieces of DWD are generally associated with *fire behavior* because they reach ignition temperature more readily than larger pieces. The time it takes for a flaming front to move across a fuel complex is an example of fire behavior influenced by the

TABLE 6.10

Ecologists, Biomass Practitioners, and Fire Managers Often
Use Different Terms to Define the Same Dead Woody Debris

Dead Woody Debris (DWD) Class		Piece Diameter	
		Inches	Centimeters
FWD	1-hr	0–0.25	0–0.6
	10-hr	0.25–1.0	0.6–2.5
	100-hr	1.0–3.0	2.5–8.0
CWD	≥1000-hr	≥3.0	≥8.0

Note: Typically 1-, 10-, and 100-hr fuels are grouped together by
ecologists and called *fine woody debris* (FWD); they refer to
1000-hr fuels and larger as *coarse woody debris* (CWD).

smaller DWD. Larger pieces of DWD, on the other hand, are usually associated with *fire effects* because, once ignited, these large pieces generally burn longer in both the flaming and smoldering phase of combustion. Soil heating and emissions from combustion are two fire effects closely tied to large DWD. Fire intensity and duration are directly related to fuel load and influence *fire severity* (a general term used to describe the amount of change in the floral and faunal components of a burned site). Logs contribute to forest diversity by providing important nutrient and moisture pools in forest ecosystems. These pools support microfauna and provide sites for the regeneration of understory plants. Logs are frequently used by animals for food storage and cover, as well as feeding and nesting sites. Duff and litter are rich in nutrients and microfauna, both of which are intrinsically related to the overall vigor of herbaceous and wood species. Disturbances that substantially reduce the amount of DWD, duff and litter, and understory vegetation can increase soil movement and cause siltation into steams. Duff and litter also provide a layer of insulation during a fire, reducing heat transfer to the soils below. In the absence of an insulating layer of duff and litter, the high levels of soil heating can reduce soil nutrients and kill microfauna and underground living plant tissues.

A full description of the FL method is provided below; however, to help the sampling crew understand the research behind and the uses for FL sampling, a brief overview is provided first. Two specific components of dead woody fuel are measured using the FL method: fine woody debris (FWD) and coarse woody debris (CWD). Ecologists often refer to FWD and CWD independently because they function differently in forest ecosystems. FWD are pieces less than 3 in. (8 cm) in diameter and include 1-hour, 10-hour, and 100-hour fuels. CWD includes pieces that are 3 inches (8 cm) or greater in diameter and at least 3 feet (1 m) in length, referred to as 1000-hour and greater fire fuels (see Table 6.10).

Pieces of DWD are sampled if they pass through the 6-ft (2-m) high sampling plane. Fine woody pieces are recorded as simple counts. Diameter and decay class are recorded for each piece of CWD. DWD biomass estimation is made using equations published in Brown (1974). FIREMON provides six optional fields for CWD: (1) diameter of the large end of the log, (2) log length, (3) distance along the tape where the piece intercepts the plane, (4) proportion of diameter lost to decay in hollow logs, (5) percent of log length lost to decay in hollow logs, and (6) proportion of the surface of CWD that is charred.

At two points along the base of each sampling plane, measurements are made of the duff/litter depth and estimation of the proportion of the duff/litter profile that is litter. At these same locations, the sampling crew will also estimate the cover of live and dead herbs and shrubs, as well as the average height of herbs and shrubs.

The planar intercept sampling methodology used in the FL protocol was originally developed by Warren and Olson (1964). Brown (1974) revised the original sampling theory to allow for more rapid fuel measurement while still capturing the intrinsic variability of forest fuels. Brown's method was developed strictly to provide estimates of fuel load in the size classes important to fire behavior. He determined the length of the sampling plane needed for each size class and, for FWD, determined the quadratic mean diameter of several species. Planar sampling has been reduced to its most fundamental and efficient level while still providing good estimates of DWD.

The planar intercept technique assumes that DWD is randomly oriented directionally on the forest floor, although typically this assumption does not hold true; for example, in areas of high wind, trees tend to fall with the prevailing winds. FIREMON uses a sampling scheme that reduces bias introduced from nonrandomly oriented pieces by orienting the DWD sampling planes in different directions. This sampling design greatly reduces or eliminates the bias introduced by nonrandomly oriented DWD (Howard and Ward, 1972; Van Wagner, 1982). The planar intercept technique also assumes that pieces are lying horizontal on the forest floor. Brown (1974) developed a non-horizontal correction for FWD and noted that a correction for CWD would not substantially improve biomass estimates; therefore, samplers do not record piece angle as part of the FL methodology. DWD is notoriously variable in its distribution within and between forest stands. Frequently, the standard deviation of DWD samples exceeds the mean. This variability requires large numbers of samples for statistical tests.

Sampling Procedure

Preliminary Sampling Tasks

Before using the FL methods in the field, it is prudent to find a place where the field crew can lay out at least one plot of three transects. This will give the field crew an opportunity to practice and learn the FL method in a controlled environment where they are not battling steep slopes and tall vegetation. Even if the spot chosen does not have DWD, some branches can be laid on the ground to simulate the sampling environment. Be sure to pick a spot where estimates of vegetation cover, vegetation height, and depth of the duff/litter profile can be made. Use the FL Equipment List to determine the materials needed. A number of preparations must be made before proceeding into the field for FL sampling. First, all equipment and supplies on the FL Equipment List must be purchased and packed for transport into the field. Because travel to plots is usually by foot, it is important that supplies and equipment be placed in a comfortable daypack or backpack. It is also important to have spares of each piece of equipment so an entire day of sampling is not lost if something breaks. Spare equipment can be stored in the vehicle rather than the backpack. Be sure all equipment is well maintained and that you have plenty of extra supplies, such as Data Forms, map cases, and pencils.

All FL Data Forms should be copied onto waterproof paper because inclement weather can easily destroy valuable data recorded on standard copier paper. Data Forms should be transported into the field using a plastic, waterproof map protector or plastic bag. The day's sample forms should always be stored in a dry place (e.g., office or vehicle) and not be taken back into the field for the next day's sampling.

If the sampling project is to resample previously installed plots, then it is recommended that plot sheets from the first measurement be copied and brought to the field for reference. These data can be valuable for help in relocating the plot. One person on the field crew, preferably the crew boss, should have a waterproof, lined field notebook for recording logistic and procedural problems encountered during sampling. This helps with future remeasurements and future field campaigns. All comments and details not documented in the sampling methods should be written in this notebook.

Plot locations or directions should be readily available and provided to the crews in a timely fashion. It is beneficial to have plot locations for several days of work in advance in case something happens, such as if the road to one set of plots is washed out by flooding. Plots should be referenced on maps and aerial plots using pin-pricks or dots to make navigation easy for the crew and to provide a check of the referenced coordinates. If possible, the spatial coordinates should be provided if plots were randomly located.

A field crew of three people is probably the most efficient for implementation of the TD sampling method. For safety reasons there should never be a one-person crew. A crew of two people will require excessive walking and trampling on the plot, and more than three people will probably result in some people waiting for critical tasks to be done and unnecessary physical damage to the plot. Assign one person as data recorder and the other two as samplers. Samplers count FWD and measure CWD pieces that intercept the sampling plane, make duff/litter measurements, and make cover and height estimates along each sampling plane. One sampler should count the 1-hour, 10-hour, and 100-hour size classes while the other measures the CWD. The remainder of the sampling tasks—duff/litter measurements and vegetation cover and height estimates—can be divided between the samplers after they have completed their first tasks.

The crew boss is responsible for all sampling logistics, including the vehicle, plot directions, equipment, supplies, and safety. The initial sampling tasks of the field crew should be assigned based on field experience, physical capacity, and sampling efficiently, but sampling tasks should be modified as the field crew gains experience and should be shared to limit monotony.

Determining Piece Size

An important task when sampling fuels (and, to a lesser extent, biomass in general) is to properly determine whether each piece is in the 1-hour, 10-hour, 100-hour, or 1000-hour and greater size class. Often it will be clear simply by looking at what size class the pieces belong in. This is especially true as field crews gain experience sampling fuels; however, while samplers are calibrating their eyes or when pieces are clearly on the boundary between two size classes, samplers need to take the extra effort to measure pieces and assign them to the proper class. Each sampling crew should have at least one set of sampling dowels for this task. The set is made up of two dowels. One dowel is 0.25 in. (0.6 cm) in diameter and 3 in. (8 cm) long; use this dowel to determine whether pieces are in the 1-hour or 10-hour class.

The second dowel is 1 in. (2.5 cm) in diameter and 3 in. (8 cm) long; use this dowel to separate the 10-hour from the 100-hour fuels. Cutting the dowels into 3-in. (8-cm) lengths makes them useful for discerning 100-hour and 1000-hour fuels. The go/no-go gauge is a tool that can speed up the sampling process (see Figure 6.6). The gaps in the tool correspond to the 1-hour and 10-hour fuel sizes, allowing for quick assessment of fuel size. One can be made out of sheet aluminum (about 0.06 in. thick) so it is lightweight and durable. Or, make one out of an old plastic card, such as a grocery store card. It will not be as durable as an aluminum one, but it is easier to make because you can cut the openings using a scissors.

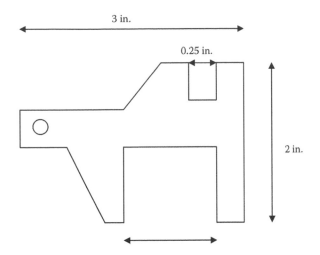

FIGURE 6.6
A go/no-go gauge can help samplers tally 1-hr, 10-hr, and 100-hr fuels more quickly and accurately than wooden dowels.

Fuel Load sampling requires 12 specific tasks for each sampling plane:

1. Lay out the measuring tape, which defines the sampling plane.
2. Measure the slope of the sampling plane.
3. Count FWD.
4. Measure CWD.
5. Measure the depth of the duff/litter profile.
6. Estimate the proportion of the profile that is litter.
7. Estimate the cover of live woody species.
8. Estimate the cover of dead woody species.
9. Estimate the average height of live nonwoody species.
10. Estimate the cover of live nonwoody species.
11. Estimate the cover of dead nonwoody species.
12. Estimate the average height of live and dead nonwoody species.

Tasks 5 through 12 are made at two points along each line. Data are recorded on the FL Data Forms after completing each of steps 2 through 12. You will learn that sampling in order 1 through 12 is not the fastest way to sample a plot. Instead, use the task list provide in Table 6.11 as a general guide for sampling and modify it as necessary to make for the most efficient sampling.

Modifying FL Sampling

In the FL method, sampling is recommended over a 60-ft (20-m) distance with an additional 15 ft (5 m) of buffer provided to keep from distributing fuels around the plot center. The 60-ft (20-m) plane is the shortest recommended for sampling CWD; however, there

TABLE 6.11

General Task Number List for Sampling with the Fuel Load (FL) Method

Task	Recorder	Sampler 1	Sampler 2
Organize materials	1	—	—
Lay out tape	—	1 (guider)	1 (guidee)
Measure slope	2 (record data)	2	2
Count FWD	3 (record data)	3	—
Measure duff/litter and vegetation at 75-ft mark	4 (record data)	—	3
Measure CWD	5 (record data)	—	4
Measure duff/liter and vegetation at 45-ft mark	6 (record data)	4	—
Check for completer forms	7	—	—
Collect equipment	—	5	5

are instances of high fuel loads (e.g., in slash) where shorter planes for DWD may be justified. If the architect wants to use shorter (or longer) sampling planes based on research or expert knowledge, the database can accommodate that data. This discussion assumes that the crew is using the suggested FL method. Additionally, the field crew does not have to use the suggested locations for sampling duff and litter or vegetation. As long as they are thoughtfully placed (e.g., do not sample duff/litter in an area where you will be sampling FWD), these measurements can be made elsewhere along the sampling plane.

Laying Out the Measuring Tape

A measuring tape laid close to the soil surface defines the sampling plane. The sampling plane extends from the top of the litter layer, duff layer, or mineral soil, whichever is visible, to a height of 6 ft (2 m). When laying out the tape, crew members need to step so that they minimize trampling and compacting fuels—DWD, duff/litter, and vegetation. While the data recorder is arranging field forms and so forth, the other two crew members can lay out the measuring tape for the first sampling plane. Have one crew member stand at plot center holding the zero end of the tape, then, using a compass, that person will guide the second crew member on an azimuth of 090 degrees true north. The second sampler will move away from plot center, following the directions of the first crew member, until he or she reaches the 75-ft (15-m) mark on the tape. The process of laying out the tape is typically more difficult than it sounds because the tape must be straight, not zigzagging around vegetation and trees (see Figure 6.7). It pays to sight carefully with the compass and identify potential obstructions before rolling out the tape.

The second crew member must follow the directions given by the first in order to stay on line, but that can take him or her under the low branches of trees and shrubs, through thick brush, ... or worse. The smallest crew member generally has the greatest success at this task, but be sure everyone gets an opportunity. Once the second crew member is at the appropriate location, the first crew member will hold the zero end of the tape over plot center while the second crew member pulls the tape tight. Together, move the tape down as close to the ground as possible without struggling to get it so close to the ground that the debris to be measured is disturbed. In most cases, the tape will end up resting on some of the DWD and low vegetation but below the crowns, of shrubs, seedlings, etc. It is not unusual to get to this point and realize that a large tree, rock, or other obstruction will not

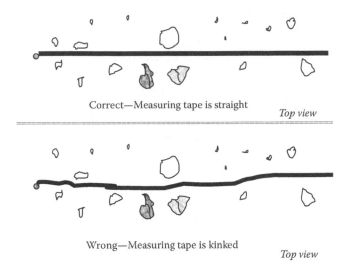

Correct—Measuring tape is straight

Top view

Wrong—Measuring tape is kinked

Top view

FIGURE 6.7
The measuring tape, which represents the lower portion of the sampling plane, should be as straight as possible. If the tape is not straight it needs to be offset left or right until it can be established without kinks or bends.

allow the tape to be laid straight; instead, there is a kink where it hits an obstruction. DWD shouldn't be sampled over a tape that isn't straight, so crew members must lift the tape above the vegetation, move both ends of the line left or right (i.e., keep it oriented at the same azimuth) until the tape will not be influenced by any obstructions, and then place it back down and straight on the soil surface. Usually this offset won't have to be more than a few feet left or right, but on sites with even moderate amounts of tall vegetation offsetting the tape can mean considerable work.

Once established, anchor the tape and do not move its position until all sampling is finished for the sampling plan. Most tapes have a loop on the zero end that a spike can be placed through to keep it anchored, and a spike or stick through the handle on the other end of the tape will hold it in place. Roll-up tapes usually have a winding crank that can be flipped so the knob points toward the reel. In this position, the knob will lock the reel so the tape will not unwind when it is pulled tight. Mark the 0-ft and 75-ft (25-m) marks along the tape so the plane can be easily reestablished. This is especially important when sampling will be done both before and after treatment. Bridge spikes, 8 to 10 in. long, work well because they are relatively permanent when driven completely into the ground and can be relocated with a good metal detector, if needed. Animals such as deer and elk tend to pull survey flags out of the ground so they should not be used as the only indicator of tape position. If spikes and flags are used together do not wrap the survey flag wire around the spike.

Determining the Slope of the Measuring Tape

When the tape has been secured, use a clinometer to measure the percent slope of the line. Aim the clinometer at the eye level of the sampler at the other end of the line. If there is a height difference of the samplers, adjust the height where you are aiming so the slope reading is accurate. Carefully read the percent slope from the proper scale in the instrument and report to the data recorder who will enter it on the FL Data Form.

What Is Woody, Dead, or Down Debris?

Before sampling any DWD, the terms *woody*, *dead*, and *down* should be understood so data gathered with the FL methods is consistent between field crews. *Woody* refers to a plant with stems, branches, or twigs that persist from year to year. The structural parts support leaves, needles, cones, and so forth, and these are the structural components that are tallied along the sampling plane. *Dead* refers to having no live foliage. Sampling deciduous species in the dormant season can be a challenge and should only be done by crews with the expertise to identify dormant vs. dead trees and shrubs.

CWD

Coarse woody debris at an angle of greater than 45 degrees above horizontal where it passes through the sampling plane should only be considered down if it is the broken bole of a dead tree where at least one end of the bole is touching the ground (not supported by its own branches or other live or dead vegetation). If CWD is at an angle of 45 degrees or less above horizontal where it passes through the sampling plane, then it is down regardless of whether or not it is broken, uprooted, or supported in that position. Do not sample a piece of CWD if you believe the central axis of the piece is lying in or below the duff layer where it passes through (actually, under) the sampling plane. These pieces burn more like duff, and the duff/litter methodology will allow field crews to collect a representative sample of this material.

FWD

Pieces of fine woody debris that are woody, dead, and down fall into three general categories: (1) pieces that are not attached to the plant stems or tree boles where they grew and have fallen to the ground, (2) pieces that are not attached to the plant stems or tree boles where they grew but are supported above the ground by live or dead material, and (3) pieces attached to stems or boles of shrubs or trees that are themselves considered dead and down. Note that it is possible for FWD to be considered dead even though it has green foliage attached, because the rules consider any piece severed from the plant where it grew to be both dead and down. Fresh slash and broken branches are examples of green material considered dead. Do not sample dead pieces that are still attached to dead (unless down) and live shrubs and trees, even if those pieces are broken and hanging from the plant where they grew. Piece angle of FWD is not critical in determining whether or not it is down. Do not tally needles, grass blades, pine cones, cone scales, bark pieces, etc., as they are not woody in nature. This material is considered litter and is measured as part of the duff/litter profile.

DWD Sampling Distances

Down woody debris is sampled along a certain portion of the sampling plane based on the size of the piece. The 1-hour and 10-hour fuels are sampled from the 15- to 21-ft (5- to 7-m) marks along the plane, the 100-hr fuels are sampled from the 15- to 30-ft (5- to 10-m) marks, and pieces 3 in. (8 cm) and larger are sampled from the 15- to 75-ft (5- to 25-m) marks along the plane. The distances for sampling FWD are shorter than for CWD because pieces of FWD are more numerous, so a representative sample can be obtained with a shorter sampling distance. DWD is not measured along the first 15 ft (5 m) of the tape because fuels are usually disturbed around plot center by the activity of the sampling crew as they get organized to lay out the tape. The FIREMON Analysis Tools program will accept sampling plane lengths different from the ones suggested here. If you use different lengths, record the reason for changing them in the Metadata table. Enter the sampling plane length for 1-, 10-, 100- and 1000-hr fuels on the FL Data Form.

Sampling FWD

The crew member at the zero end of the tape should sample FWD to maximize sampling efficiency. Count the 1-hour and 10-hour fuels that pass through the sampling plane from the 15- to 21-ft (5- to 7-m) marks on the measuring tape. Remember that the plane extends from the top of the litter layer vertically to a height of 6 ft (2 m). The best way to identify the pieces intercepting the plane is to lean over the tape so your eye is positioned vertically a few feet over the measuring tape at the 15-ft (5-m) mark. Then, while looking at one edge of the tape, maintain your head in that same vertical position over the line and move ahead to the 21-ft mark while making separate counts for the 1-hour and 10-hour fuels that cross under or above the edge of the tape. Each piece needs to be classified as 1-hour or 10-hour fuel by the diameter where it intercepts the sampling plane, defined by one edge of the measuring tape. Samplers should use the dowels or the go/no-go gauge discussed earlier to classify fuels that are close to the size class bounds. Often pieces above the ground will cover pieces below. It is important to locate all the pieces that intercept the plane in order to get accurate fuel load data. When finished tallying the 1-hour and 10-hour fuels, report the counts to the data recorder, who will enter them in Fields 8 and 9 on the data sheet. Use the same basic procedure to count the 100-hour fuels that pass through the sampling plane from the 15- to 30-ft (5- to 10-m) marks on the tape. Report the information to the data recorder, who will enter the count on the data sheet.

Sampling CWD

The CWD sampling plane is 6 ft (2 m) high and extends from the 15- to 75-ft (5- to 25-m) marks along the measuring tape. Sample CWD that intercepts the sampling plane and meets the dead, down, and woody requirements discussed above. In general, at least two fields are recorded for each piece of CWD: diameter and decay class. Proportion of char, log length, diameter of the large end, point of intersect, and estimations of volume lost to decay are additional data fields that may be collected for each piece of CWD. CWD sampling should be done by the crew member who is standing at the 75-ft (25-m) end of the tape while moving toward the zero end. This will keep him or her out of the way of the other sampler and will reduce the chances of the FWD being inadvertently disturbed before being sampled.

Diameter measurement and decay class are determined on each piece of CWD where it passes through the sampling plane. Measure diameter perpendicular to the central axis of each piece to the nearest 0.5 in. (1 cm). If a piece crosses through the sampling plane more than once, measure it at each intersection. A diameter tape or caliper works best for diameter measurements, but a ruler can give good results if it is used so that parallax error does not introduce bias.

Use the descriptions in Table 6.12 to determine the decay class for CWD at the same point where the diameter measurement was made. Decay class can change dramatically from one end of a piece of CWD to the other, and often the decay class at the point where the diameter measurement was taken does not reflect the overall decay class of the piece. However, by recording the decay class at the point where the diameter was measured, the

TABLE 6.12

Descriptions Used to Determine Decay Class Where Log Crosses the Sampling Plane

Decay Class	Description
1	All bark is intact; all but the smallest twigs are present; old needles are probably still present; hard when kicked.
2	Some bark is missing, as are many of the smaller branches; no old needles still on branches; hard when kicked.
3	Most of the bark is missing; most of the branches less than 1 inch in diameter are also missing; still hard when kicked.
4	Looks like a class 3 log but the sapwood is rotten; sounds hollow when kicked, and you can probably remove wood from the outside with your boot; pronounced sagging if suspended for even moderate distances.
5	Entire log is in contact with the ground; easy to kick apart but most of the piece is above the general level of the adjacent ground. If the central axis of the piece lies in or below the duff layer, then it should not be included in the CWD sampling, as these pieces act more like duff than wood when burned.

field crew will collect a representative sample of decay class along each sampling plane. The transect number, sequential piece number (log number), diameter, and decay class for each piece are entered in Fields 16 through 19.

What Are Duff, Litter, and the Duff/Litter Profile?

Duff and litter are two components of the fuel/biomass complex made up of small, woody, and nonwoody pieces of debris that have fallen to the forest floor. Technically, packing ratio, moisture content, and mineral content are used to discriminate the litter and duff layers. Samplers will find it easier to identify each layer by using the following, more general, criteria. *Litter* is the loose layer made up of twigs, dead grasses, recently fallen leaves, needles, and so forth, where the individual pieces are still identifiable and little altered by decomposition. The *duff* layer lies below the litter layer and above the mineral soil. It is made up of litter material that has decomposed to the point that the individual pieces are no longer identifiable. The duff layer is generally darker than the litter layer and is more aggregated because of the fine plant roots growing in the duff material. The *duff/litter profile* is the cross-sectional view of the litter and duff layers. It extends vertically from the top of the mineral soil to the top of the litter layer. The FL methods use the depth of the duff/litter profile and estimation of the proportion of the total duff/litter depth that is litter to estimate the load of each component. Litter usually burns in the flaming phase of consumption because it is less densely packed and has a lower moisture and mineral content than duff, which is typically consumed in the smoldering phase. Litter is usually associated with *fire behavior* and duff with *fire effects*.

Sampling Duff and Litter

The duff and litter layers lie below the sampling plane so they are not sampled using the planar intercept method; instead, duff/litter measurements are made using a duff/litter profile at two points along each sampling plane. The goal is to develop a vertical cross-section of the litter and duff layers without compressing or disturbing the profile. As samplers finish collecting DWD data they can start making the duff/litter measurements. Duff/litter measurements are made at a point within 3 ft (1 m) of the 45-ft (15-m) and 75-ft

(25-m) marks along the tape. Follow the same instruction for each duff/litter measurement. Select a sampling point within a 3-ft (1-m) radius circle that best represents the duff/litter characteristics inside the entire circle. Samplers can make the profile using a trowel or boot heel. Using a boot heel in deep duff and litter generally results in poor profiles, which, in turn, makes measurement difficult. Use the blade of the trowel to lightly scrape just the litter layer to one side, then return the blade to the point where the litter scrape was started, push the trowel straight down as far as possible through the duff layer, and move the material away from the profile. Mineral soil is usually lighter in color than the duff and coarser in composition, often sandy or gravelly. If a boot is used, drive the heel down and drag it toward you. As with the towel, continue working through the duff until mineral soil is noted. It is important not to disturb the profile by compacting it on successive scrapes. The profile that is exposed should allow an accurate measurement of duff/liter depth.

Use a plastic ruler to measure the total depth of the duff/litter profile to the nearest 0.1 in. (0.2 cm). Place the zero end at the point where the mineral soil meets the duff layer, then move either your index finger or thumb down the ruler until it is level with or just touches the top of the litter. While keeping your finger in the same position on the ruler, lift the ruler out of the profile and note the duff/litter depth indicated by your finger. If your ruler is not long enough to measure the duff/litter depth, use the rule to make marks on a stick and measure the profile with the stick. If you use the stick measurement method often, get a longer ruler. Next, examine the duff/litter profile and estimate the proportion of the total depth that is made up of litter, to the nearest 10%. Finally, report the duff/litter depth measurement and litter proportion estimate to the data recorder, who will enter the data on the FL Data Form. Duff and litter measurements are most easily and accurately made on the vertical portion of the profile as long as that portion of the profile is representative of the true duff/litter depth (it wasn't negatively impacted when the profile was developed). Sometimes the most vertical part is where the back of the trowel blade or boot heel went in, and sometimes it is along one side of the profile.

What Is Woody and Nonwoody Vegetation?

The last fuel characteristics that field crews will sample along each sampling plane are the covers of trees, shrubs, and herbs. These can be divided into woody and nonwoody species. Both trees and shrubs are woody species. They are easily identified because their stems persist and growth does not have to start at ground level each growing season. Trees generally have a single, unbranched stem near ground level, and shrubs generally have multiple stems near ground level. Woody species can be evergreen or deciduous. Deciduous species lose their foliage at the end of the growing season, but the aerial woody portions of the plant remain. Herbs are nonwoody plants whose aerial portions die back at the end of the growing season. Most field samplers will have an intuitive idea of which vegetation is woody and which is not. One way to help identify nonwoody plants is to remember that, in general, weather factors, such as wind, rain, snow, and so forth, collapse herb foliage and stems back to or near the ground between growing seasons. Small trees, shrubs, and herbs influence fire behavior because their branches and foliage are suspended above the ground, allowing more efficient heating and burning of the parts. Dense, suspended fuels can lead to fires that are difficult or impossible to control. The fires in chaparral vegetation in the Western United States are an example. By estimating the cover and heights of woody and nonwoody vegetation, fire managers can estimate the volume, density, and biomass of vegetation. All three of the characteristics are strongly associated with fire behavior.

Sampling Vegetation Cover and Height

Estimate vegetation cover and height at the 45-ft (15-m) and 75-ft (25-m) marks on the measuring tape. Field crews will estimate the vertically projected cover of vegetation within an imaginary sampling cylinder 6-ft (2-m) tall by 6-ft (2-m) in diameter. Use the marks on the measuring tape to help visualize the 6-ft (2-m) diameter; for example, when standing at the 45-ft (15-m) mark, the 42-ft (14-m) and 48-ft (16-m) marks will identify the boundary of the cylinder along the tape. Use that measurement to get a good idea of the distance required on each side, perpendicular to the tape, to form the imaginary base of the cylinder. Most people have an arm's width spread that is about 6 ft (2 m). Samplers should measure their arm span and use that measurement to help them visualize the sampling cylinder.

The extent of plant cover, especially of the foliage, is a function of the phenological stage of the plant. Early in the season, many plants may not have completely leafed out; in midseason, plant cover reaches its maximum. Later season plant material, especially herbaceous vegetation, moves from the live to dead class. Clearly, where vegetation is concerned, it is important to sample at the same phenological stage when monitoring, and this should be taken into account when starting a monitoring program. In the Western United States, fires typically burn late in the season, after maximum biomass has been reached, so the best assessment of surface fuels, in terms of potential fire fuels, probably would be made later in the season. The biomass equations in FIREMON are based on oven-dry weight, so when cover is equal there is no difference in biomass between live and dead plants. Fire behavior, however, may be significantly different between mid- and late season because of the lower plant moisture.

Six attributes are measured at each vegetation sampling point. These include four cover estimations for vegetation—(1) live wood species (trees and shrubs), (2) dead woody species, (3) live nonwoody species (herbs), and (4) dead nonwoody species. Also, two height estimations are made for the (5) woody component and (6) nonwoody component. *Cover* is the vertically projected cover contributed by each of the four categories within the sampling cylinder. It includes plant parts from plants rooted in the sampling cylinder and plant parts that project into the sampling cylinder from plants rooted outside (e.g., live and dead branches). Estimate cover by imagining all the vegetation in the class being sampled (say, live shrub cover) compressed straight down to the ground. The percent of the ground covered by the compressed vegetation inside the 6-ft (2-m) diameter sampling area is what is being sampled. The cover of dead branches on a live plant should be included in the dead cover estimate. On the other hand, the cover of the cross-sectional area of vertically oriented single-stemmed trees in the live or dead wood cover estimates are not generally included. The stems do not really count as surface fuel because they do not contribute much to fire behavior or fire effects. Also, if the sampling cylinder was located on an area with an unusually high number of tree stems, the vertical projection of foliage would probably overlap the area of the stems, thus the actual cover would be the same with or without the stems.

Two conditions make cover estimations difficult and, frequently, inaccurate. First, the equations used to estimate biomass assume that all of the plant parts for each species are included in the cover and height estimation. In other words, if looking at the cover of a woody shrub species, samplers need to estimate the cover of all the parts, even things such as the foliage, which are not woody. Second, estimating cover is not something people do very often; it is only with practice and experience that good estimations of plant cover can be made. Fortunately, the cover classes used in FIREMON are typically 10%, so the precision of cover estimates is secondary to accuracy (see Table 6.13).

In addition to the cover estimates, samplers will make two height estimates at each vegetation sampling location, one for the average height of the live and dead woody species and one for the average height of the live and dead nonwoody species. Make your height estimate by noting the maximum height of all of the plants in the class and then recording the typical or average of all the maximum heights. Some people like to envision a piece of plastic covering just the plants in one class and then estimate the average height of the plastic above the ground. Either method will work and give answers that are of adequate precision. Estimate height to the nearest 0.5 ft (0.2 m). Remember, for both the cover and height estimation, to include only the vegetation that is within the sampling cylinder. A fast way to make accurate height assessments is for samplers to measure their ankle, knee, and waist heights and then estimate vegetation height based on those points. Record the vegetation cover classes and height data on the FL Data Form.

Finishing Tasks

The most critical task before moving to the next sampling plane or plot is to make certain that all of the necessary data have been collected. This task is the responsibility of the data recorder. Also, the recorder should write down any comments that might be useful; for example, you might comment on some unique characteristic on or near the plot that will help samplers relocate the plot. Include notes about other plot characteristics, such as "evidence of deer browse" or "deep litter and duff around trees." Finally, collect the sampling equipment and move ahead to start sampling the next plane.

Successive Sampling Planes

On each FL plot, the field crew will collect data for at least three sampling planes. Follow the FL plot design in Figure 6.8. The first sampling plan is always oriented at an azimuth of 090 degrees true north, the second is oriented 330 degrees, and the third at 270 degrees. Planes are oriented in multiple directions to avoid bias that could be introduced by DWD

TABLE 6.13

Record the Cover of Each of the Four Vegetation Categories on the Field Form in One of These Classes

Class	Cover
0	No cover
0.5	>0 to 1% cover
3	>1–5% cover
10	>5–15% cover
20	>15–25% cover
30	>25–35% cover
40	>35–45% cover
50	>45–55% cover
60	>55–65% cover
70	>65–75% cover
80	>75–85% cover
90	>85–95% cover
98	>95–100% cover

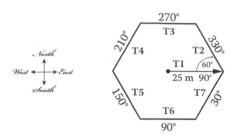

FIGURE 6.8
The FL plot design allows a representative sample of DWD to be obtained while reducing or eliminating the bias introduced by nonrandomly oriented pieces. Data are collected on and along three to seven sampling planes.

pieces that are not randomly oriented on the forest floor. The DWD biomass estimate with the FL methods is an average across all of the sampling planes. It is not necessary for one sampling plane to start at the exact 75-ft (25-m) mark of the previous one. In fact, it is better if the start of the new line is 5 ft or so away from the end of the last so that the activity around the new start does not adversely impact the fuel characteristics at the end of the last one. The duff/litter layer and woody and nonwoody vegetation at the 75-ft (25-m) mark, in particular, can be disturbed by field crew traffic, and that can bias the data when the plot is resampled. Make sure that no portion of the new sampling plane will be crossing fuels that were sampled on the previous plane. When the start of the new sampling plane has been determined, collect the data as you did on the first plane. Look ahead and see which starting point will guarantee a straight line before you start laying out the next sampling plane. If the sample sampling planes are going to be remeasured, be sure to carefully mark the 0-ft and 75-ft end of each sampling plane.

Determining the Number of Sampling Planes

After the crew has finished sampling three planes, the data recorder will sum up the counts of all the DWD pieces (1-hour, 10-hour, 10-hour, and 1000-hour pieces), and if that number is greater than 100 then the crew is finished sampling DWD for the plot. If the count is less than 100 then the crew needs to sample another line. If another line is needed, refer to the FL plot design (Figure 6.8), lay out the next sampling plane, and collect the FL data. When finished with that plane, recalculate the DWD piece count. Again, if the count is greater than 100 the sampling is finished; if not, another plane needs to be sample. Continue sampling until either total piece count is greater than 100 or seven planes have been sample. When sampling has begun on a sampling plane, data must be collected for the *entire* plane. When the sampling is completed, record the number of planes that you sampled on the FL Data Form.

Down woody debris is only one part of the surface fuels complex. The 100-piece rule is also meant to help guide the sample size for the duff, litter, and vegetation components of the complex; thus, even if there is little DWD, the duff, litter, and vegetation should be sampled sufficiently. Conversely, if you are sampling numerous pieces of DWD, in principle that material should be carrying most of the fire so a reduced number of litter, duff, and vegetation assessments should not be an issue. If this is not the case, modify your plot level sampling so you get dependable estimates of all of the FL components. Record any sampling modifications in the data table for the project.

Two potential shortcomings can be encountered when using the 100-piece rule. First, the greater the clumping or aggregation of fuels on the forest floor, the greater the opportunity of having a high number of piece counts on one or more sampling planes. These clumps can lead to an overestimation of DWD biomass. Let's say that as an experiment a field crew wants to compare biomass values using the 100-piece rule vs. sampling with five sampling plans. The first three sampling planes are in the exact same location for the comparison. In the experiment, it just happens that the third plane crosses over a spot where there is an accumulation of FWD and when the third sampling plane was finished the crew had sampled 112 pieces—the end of sampling for the 100-piece rule data. They continued to sample two more planes for their five-plane comparison but recorded no more data because the planes crossed a small grassy area. Back at the office they ran their data through the database and noted that they sampled 5.3 tons/acre of material using the 100-piece rule but only 3.2 tons/acre when five sampling planes were used even though in the field they sampled exactly the same pieces. This is because the tons/acre value that comes from the planar intercept calculation is the average across all of the planes sampled. In the first case the denominator was 3 and in the second it was 5. The example presents an extreme case, but recognize that any aggregation of fuels can lead to overestimation—and always an overestimation—and the earlier in the sampling plane sequence that the aggregation is encountered the greater the opportunity for overestimation. The second shortcoming of the 100-piece rule is that, for comparison, when plots are resampled the number of sampling planes has to be the same as the first time the plot was sampled. It can be time consuming (and presents an opportunity for errors) to look up all of the original plots in the database and note the number of planes sampled at each. Despite these shortcomings, the 100-piece rule works well most of the time. Finally, the 100-piece rule is especially useful in inventory sampling where plots are sampled only once.

If the 100-piece rule is not used for the DWD sampling, then the architect must determine the number of sampling planes that will be used throughout the project. The task is to sample with sufficient intensity to capture the variation while not wasting time sampling too intensively. This is made more difficult when fuels vary greatly across the project area. Assuming that the project funding is not limiting the sampling intensity, we suggest determining the number of sampling planes per plot using a pilot study. Install pilot plots in the study area or a similar ecosystem and sample using the 100-piece rule (you don't need to measure any attributes; just count pieces of DWD). Be sure to put plots in areas representing the range of DWD piece densities you will be sampling in your study area. Depending on the variability of the fuels, after sampling 10 or 20 plots you will be able to identify a good number of sampling planes to use in your project. You should pick the number that lets you meet the 100-piece limit on at least 80% of your plots; for example, say that you had 20 plots in your pilot study and the number of sampling planes needed to count 100 pieces of each plot was

3 sampling planes, 2 plots

4 sampling planes, 5 plots

5 sampling planes, 10 plots

6 sampling planes, 3 plots

For your project, then, you could use five sampling planes per plot and be getting sufficient estimates of DWD. (Be sure to enter this information on each Data Form and make a note of the methods used to determine the number of sampling planes for the project in

TABLE 6.14

Precision Guidelines for Fuel Load (FL) Sampling

Component	Standard
Slope	±5%
FWD	±3%
CWD diameter	±0.5 inch/1 cm
CWD decay class	±1 class
Duff/litter depth	±0.1 inch/0.2 cm
Percent litter estimation	±10%
Vegetation cover estimation	±1 class
Vegetation height estimation	±0.5 ft/0.2 m

the Metadata table.) An absolute minimum of three planes per plot should be sampled for the DWD. Generally, DWD is the most variable of the FL attributes, so the duff, litter, and vegetation sampling intensity should be adequate when sampled at the DWD intensity.

Resampling FL Plots

The FL methods are unique in FIREMON in that they allow a variable number of sample planes on each FL plot, based on piece count. When resampling a FL plot always sample the same number of planes as were sampled when the plot was sampled the *first* time. Never use the 100-piece rule when resampling. Instead, look through the database and record the number of sampling planes that were used when each plot was first sampled, and then sample only that number in subsequent sampling.

What If ...

... no matter where I start my next line, it runs off a cliff? There is no way to foresee every problem samplers will encounter in the field. The best way for a crew to deal with unique situations is to apply the FL methods as well as they can to sample the appropriate characteristics based on the project objectives, then make a record in the comment section of what was encountered and how it was handled. For instance, if a crew *were* going to lay out a line that was headed off a cliff, then the crew would use the next azimuth from the FL plot design and lay out the sampling plane in that direction.

Precision Standards

Use the standards shown in Table 6.14 when collecting data by the FL methods.

Sampling Design Customization: Optional Fields

- *Percent of log that is charred*—Measured to assess extent and severity of fire. Record the percent of the surface of each individual piece of CWD passing through the sampling plane that has been charred by fire using the classes in Table 6.15.
- *Diameter at large end of log*—Measured for wildlife concerns. Record the diameter of the large end of the log to the nearest inch (2 cm). If a piece is broken but the sections are touching consider that one log. If the broken sections are not touching then consider them to be two logs and record the diameter of the large end of the piece that is passing through the sampling plane.

TABLE 6.15

Classes Used to Record the Amount of Surface Charred by Fire for Each Piece of Coarse Woody Debris (CWD)

Class	Char
0	No char
0.5	>0–1%
3	>1–5%
10	>5–15%
20	>15–25%
30	>25–35%
40	>35–45%
50	>45–55%
60	>55–65%
70	>65–75%
80	>75–85%
90	>85–95%
98	>95–100%

TABLE 6.16

Classes Used to Record the Percent of Diameter and Length Lost to Rot in Coarse Woody Debris (CWD)

Class	Percent Lost to Decay
0	No loss
0.5	>0–1%
3	>1–5%
10	>5–15%
20	>15–25%
30	>25–35%
40	>35–45%
50	>45–55%
60	>55–65%
70	>65–75%
80	>75–85%
90	>85–95%
98	>95–100%

- *Log length*—Important for wildlife concerns and useful for rough determination of piece density. Record the length of CWD to the nearest 0.5 ft (0.1 m). If a piece is broken but the two parts are still touching, then record the length end-to-end or sum the lengths for broken pieces not lying in a straight line. If piece is broken and the two parts are not touching, then only measure the length of the piece that intercepts the sample plane.

- *Distance from beginning of line to log*—This measurement makes relocation of specific logs easier, which is especially important when calculating fuel consumption on a log-by-log basis. Frequently, logs that were included in prefire sampling roll away from the sampling plane during a fire, and other logs not originally sampled roll into the plane. Recording the distance from the start of the line, in addition to permanently marking the logs with tags, will make postfire sampling easier. Record the distance from the start of the measuring tape to the point where the diameter was measured.

- *Proportion of log that is hollow*—This characteristic is important for wildlife concerns but also allows more accurate estimates of carbon. Estimate the percent diameter and percent length that have been lost to decay. Record the data using the classes shown in Table 6.16.

Line Intercept (LI) Sampling Method*

The Line Intercept (LI) sampling method is designed to sample within-plot variation and quantify changes in plant species cover and height over time using transects located within the macroplot. Transects have random starting points and are oriented

---

* Material in this section is based on Caratti, J.F., *Line Intercept (LI) Sampling Method*, General Technical Report RMRS-GTR-164-CD, U.S. Department of Agriculture, Forest Service, Rocky Mountain Research Station, Fort Collins, CO, 2006.

perpendicular to the baseline. First, samplers record the transect length and number of transects, then, along each transect, they record cover intercept and average height for each plant species. This method is primarily used when the fire manager wants to monitor changes in plant species cover and height. This method is primarily designed to sample plant species with dense crowns or large basal areas. The LI method works best in open grown woody vegetation (e.g., Western U.S. shrub communities), especially shrubs greater than 3 ft (1 m) in height. The Cover/Frequency (CF) method is generally preferred for sampling herbaceous plant communities with vegetation less than 3 ft (1 m) in height; however, the LI method can be used in junction with the CF method if shrubs greater than 3 ft (1 m) exist on the plot (e.g., CF quadrats can be used to sample herbaceous vegetation, then the transect used to locate the quadrats can be used to sample shrubs using the LI methods).

The CF method is probably the best approach to sampling cover in mixed plant communities with grasses, shrubs, and trees. This method is not well suited for sampling single stemmed plants or dense grasslands. The PO method is better suited for sampling fine-textured herbaceous communities such as dense grasslands and wet meadows. Cover measured with the LI method is less prone to observer bias than ocular estimates of cover in quadrats (CF method); however, if rare plant species are of interest, then the CF methods are preferred since rare species are easier to sample with quadrats than with points of lines.

Tansley and Chipp (1926) introduced the Line Intercept method. A line transect, typically a measuring tape stretched taut on the ground or at a height that contacts the vegetation canopy, is used to make observations of plant cover. The method consists of measuring the length of intercept for each plant that occurs over or under the tape. If basal cover is of interest, then the tape is placed at ground level. Percent cover is sampled by recording the length of intercept for each plant species measured along a tape by noting the point on the tape where the plant canopy or basal portion begins and where the plant canopy or basal portion ends. When these intercept lengths are summed and divided by the total tape length, the result is a percent cover of the plant species along the transect.

The line transect can be any length and, if modified, is usually done so based on the type of vegetation being sampled (Bonham, 1989). In general, cover in herbaceous communities can be estimated with short lines (typically less than 50 m), while longer lines (50 m or greater) should be used in some shrub and tree communities. Canfield (1941) recommended using a 15-m transect for areas with 5 to 15% cover and using a 30-m line when cover is less than 5%. The amount of time needed to measure a transect can also be used to determine the length of the transect (Bonham, 1989). Canfield recommended a transect length in which canopy intercepts can be measured by two people in approximately 15 minutes.

The Line Intercept method is most efficient for plant species that have a dense crown cover (e.g., shrubs or matted plants) or have a relatively large basal area (e.g., bunchgrasses) and is best suited where the boundaries of individual plants are easily determined. Line intercept is not an effective method for estimating the cover of single-stemmed plant species or dense grasslands (e.g., rhizomatous species).

Most plant species have some gaps in their canopies, such as bunchgrasses with dead centers or shrubs with large spaces between branches. Rules for dealing with gaps must be clearly defined, as observers treat gaps differently. One solution is for the observer to assume that a plant has a closed canopy unless a gap is greater than some predetermined width. Gaps less than 2 in. (5 cm) should be considered part of the canopy.

Sampling Procedure

The sampling procedure is described in the order of the fields that need to be completed on the LI Data Form. The sampling procedure described here is the recommended procedure for this method. Later sections will describe how the FIREMON three-tier sampling design can be used to modify the recommended procedure to match resources, funding, and time constraints. This method assumes that the sampling strategy has already been selected and the macroplot has already been located.

Preliminary Sampling Tasks

Before setting out to conduct field sampling, lay out a practice area with easy access. Try to locate an area with the same species or vegetation life form you plan on sampling. Get familiar with the plot layout and the data that will be collected. This will give you a chance to assess the method and will help you think about problems that might be encountered in the field; for example, how will you take into account gaps in the foliage of the same plant? It is better to answer these questions before the sampling begins so you are not wasting time in the field. This will also let you determine if any pieces of equipment must be ordered.

A number of preparations must be made before proceeding into the field for LI sampling. First, all equipment and supplies must be purchased and packed for transport into the field. Because travel to plots is usually by foot, it is important that supplies and equipment be placed in a comfortable daypack or backpack. It is also important to have spares of each piece of equipment so an entire day of sampling is not lost if something breaks. Spare equipment can be stored in the vehicle rather than the backpack. Be sure all equipment is well maintained and that you have plenty of extra supplies, such as Data Forms, map cases, and pencils.

All LI Data Forms should be copied onto waterproof paper because inclement weather can easily destroy valuable data recoded on standard copier paper. Data Forms should be transported into the field using a plastic, waterproof map protector or plastic bag. The day's sample forms should always be stored in a dry place (e.g., office or vehicle) and not be taken back into the field for the next day's sampling.

One person on the field crew, preferably the crew boss, should have a waterproof, lined field notebook for recording logistic and procedural problems encountered during sampling. This helps with future remeasurements and future field campaigns. All comments and details not documented in the sampling methods should be written in this notebook; for example, snow on the plot might be described in the notebook, which would be helpful in plot remeasurement.

It is beneficial to have plot locations for several days of work in advance in case something happens, such as the road to one set of plots being closed. Plots should be referenced on maps and aerial photos using pin-pricks or dots to make navigation easy for the crew and to provide a check of the georeferenced coordinates. It is very easy to transpose Universal Transverse Mercator (UTM) coordinate digits when recording georeferenced positions on the plot sheet; marked maps can help identify any erroneous plot positions. If possible, the spatial coordinates should be provided if plots were randomly located.

A field crew of two people is probably the most efficient for implementation of the LI sampling method. For safety reasons, there should never be a one-person field crew, and any more than two people will probably result in some people waiting for critical tasks to be done and unnecessary trampling on the plot. The crew boss is responsible for all sampling logistics, including the vehicle, plot directions, equipment, supplies, and safety. The crew boss should be the note taker, and the technician should perform most point intercept measurements. The initial sampling tasks of the field crew should be assigned based on field experience, physical capacity, and sampling efficiency, but sampling tasks can be modified as the field crew gains experience and should be shared to limit monotony.

DID YOU KNOW?

The Universal Transverse Mercator (UTM) grid system provides a simple and accurate method for recording the geographic location of a historic site. Its greatest advantage over the Geographic Coordinate System (latitude/longitude) is its reliability, because its measurements are cited in linear, decimal units, rather than in angular, non-decimal ones (Cole, 1977).

Designing the LI Sampling Method

General criteria recorded on the LI Data Form allow for user-specified design of the LI sampling method. Each general LI field must be designed so that the sampling captures the information required to successfully complete the management objective within the time, money, and personnel constraints. These general fields should be decided before the crews go into the field and should reflect a thoughtful analysis of the expected problems and challenges in the fire/biomass monitoring project. However, some of these fields, particularly the number and length of transects, might be adjusted after a pilot study is conducted in the field to determine a sufficient sample size.

Determining the Sample Size

The size of the macroplot ultimately determines the length of the transects and the length of the baseline along which the transects are placed. The amount of variation in plant species composition and distribution determines the number and length of transects required for sampling. The typical macroplot sampled in the LI method is a 0.10 ac (0.04 ha) square, measuring 66 × 66 ft (20 × 20 m), which is sufficient for most monitoring applications. Shrub-dominated ecosystems will generally require larger macroplots when sampling with the LI method. If you are not sure of the plot size to use, contact someone who has sampled the same vegetation that you will be sampling. The size of the macroplot should be adjusted to accommodate the size of the vegetation; however, it is more efficient if you use the same plot size for all sampling methods on the plot. The recommended transect length is 66 ft (20 m) for a macroplot that is 66 × 66 ft (20 × 20 m), although the macroplot size may be adjusted to accommodate longer or shorter transects based on variability in plant species composition and distribution. Five transects within the macroplot are recommended unless a situation arises where more transects should be sampled.

Conducting LI Sampling Tasks

Locating the Baseline for Transects

Once the plot has been monumented, a permanent baseline is set up as a reference from which you will orient all transects. The baseline should be established so that the sampling plots for all of the methods overlap as much as possible. The recommended baseline is

66 ft (20 m) long and is oriented upslope with the 0-ft (0-m) mark at the lower permanent marker and the 66-ft (20-m) mark at the upper marker. On flat areas, the baseline runs from south to north with the 0-ft (0-m) mark on the south end and the 66-ft (20-m) mark on the north end.

Locating the Transects

Transects are located within the macroplot, perpendicular to the baseline and across the slope. For permanent plots, determine the compass bearing of the transects and record it on the plot layout map. All transects should be at the same azimuth. Starting locations for each transect are determined using the FIREMON Random Transect Locator program or from supplied tables. If the LI method is used in conjunction with other replicated sampling methods (CF, PO, RS, or DE), use the same transects for all methods. Remember that in successive remeasurement years it is essential that transects be placed in the same location. Carefully stretch a measuring tape, which represents the transect, from the starting point on the baseline out 66 ft (20 m) at an azimuth perpendicular to the baseline.

The measuring tape will be stretched taut and straight on the ground or at a height above the vegetation canopy if measuring crown cover. If basal cover is recorded, then the tape is always placed at ground level. For two reasons, the tape must be as straight as possible and not zigzagging around the vegetation. First, a tape stretched straight between the permanently marked transect ends will ensure that the same vegetation sampled during the initial visit will be resampled on subsequent visits. Second, the crown cover estimate could be biased if the tape is bent around the stems because more of the tape lies under the plant canopy. If the tape is to be stretched above vegetation, the crew will need some way to hold it tight and above the canopy. One method is to drive a rebar at each of the transects and then slip a piece of metal electrical conduit over the bar and attach the tape ends to the conduit with wire, hooks, or tape. Rebar is an excellent way to permanently mark the ends of the transects, but it should not be left in place if horses, other livestock, or people frequent the study site because the rebar can injure feet and legs. Also, in areas where people are recreating, any visible rebar may be objectionable because it is incongruous with natural surroundings.

Line Intercept Sampling

First, enter the transect number on the LI Data Form. Next, enter the plant species or item code for each item recorded. FIREMON provides plant species codes from the Natural Resources Conservation Service (NRCS) Plants Database; however, local or customized plant species codes are also allowed in FIREMON. Next enter the plant species status on the LI Data Form. Status describes the general health of the plant species as live or dead using the following codes:

L—Live (plant with living tissue)

D—Dead (plant with no living tissue visible)

NA—Not applicable

Plant status is purely qualitative, but it does provide an adequate characteristic for stratification of plant health.

TABLE 6.17

Tree Size Class Codes

Class Code		Tree Size	
		English	Metric
TO	Total cover	—	—
SE	Seedling	<1.0 in. DBH or <4.5 ft height	<2.5 cm DBH or <1.5 m height
SA	Sapling	1.0 to < 5.0 in. DBH	2.5 to <12.5 cm DBH
PT	Pole tree	5.0 to <9.0 in. DBH	12.5 to <25 cm DBH
MT	Medium tree	9.0 to <21.0 in. DBH	25 to <50 cm DBH
LT	Large tree	21.0 to <33.0 in. DBH	50 to <80 cm DBH
VT	Very large tree	33.0+ in. DBH	80+ cm DBH
NA	Not applicable	—	—

Size Class

Plant species size classes represent different layers in the canopy. For example, the upper canopy layer could be defined by large trees, while pole-size trees and large shrubs might dominate the middle layer of the canopy, and the lower canopy layer could include seedlings, saplings, grasses, and forbs. Size class data provide important structural information such as the vertical distribution of plant cover. Size classes for trees are typically defined by height for trees less than 4.5 ft (1.37 m) tall and diameter at breast height (DBH) for larger trees. Size classes for shrubs, grasses, and forbs are typically defined by height. If the vegetation being sampled has a layered canopy structure, then cover can be recorded by plant species and by size class. Total size class cover for a plant species can equal more than 100% for each plant species due to overlap between different size classes. FIREMON uses a size class stratification based on the ECODATA sampling methods (Jensen et al., 1994). Field crews can group individual plants by species into one or more tree size classes (Table 6.17) or shrub, grass, and forb size classes (Table 6.18). There can be multiple size classes for each species.

If cover is being recorded by size class, enter the size class code for each plant species in Field 5 on the LI Data Form. If size class data are not recorded, then record only the total cover for each plant species. When recording total cover for a species, enter the TO code to indicate that the cover estimate is for all of the size classes. Enter the transect length on the LI Data Form. Transect length is entered by item and size class, allowing the transect length to vary by species and size class.

TABLE 6.18

Shrub, Grass, and Herb Size Class Codes

Class Code		Shrub/Grass/Herb Height	
		English	Metric
TO	Total cover	—	—
SM	Small	<0.5 ft	<0.15 m
LW	Low	0.5 to <1.5 ft	0.15 to <0.5 m
MD	Medium	1.5 to <4.5 ft	0.5 to <1.5 m
TL	Tall	4.5 to <8 ft	1.5 to <2.5 m
VT	Very tall	8+ ft	2.5+ m
NA	Not applicable	—	—

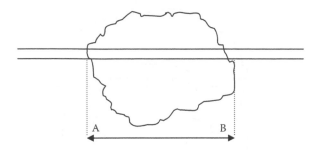

FIGURE 6.9
Measure canopy intercept in feet (m) along the measuring tape. Because canopy intercept can vary on each side of the measuring tape, measure intercept on one side of the measuring tape only. Use the right side as you move along the tape. Record the start of the plant intercept (A) in the Start field and the end intercept (B) in the Stop field.

Estimating Cover (Intercept)

The procedure for measuring the live crown intercept bisected by the transect line is illustrated in Figure 6.9. Proceed from the baseline toward the opposite end of the tape and measure the horizontal linear length of each plant that intercepts the licn. The start and stop points for each intercept are recorded in feet (meters). When measuring intercepts in feet, use a tape marked in 10ths and 100ths of feet. Measure the intercept of grasses and grass-like plants, along with rosette-forming plants, at ground level. For forbs, shrubs, and trees measure the vertical project of the vegetation intercepting one side of the tape. Be sure not to inadvertently move the tape, and carefully look under tall dense crowns to be sure you are sampling all species and size classes. The measurements are recorded by plant species or item in the Start and Stop fields on the LI Data Form to the nearest 0.1 ft (0.03 m). The FIREMON data entry screens populate the intercept field automatically when the start and top points are entered. Canopy overlap within a species is not distinguished, but canopy overlap between different species is recorded (Figure 6.10).

Percent cover is calculated by totaling the intercept measurements for all individuals of that species (in the Intercept field) along the transect and dividing by the total length of the transect. Most plant species have some gaps in their canopies, such as bunchgrasses with

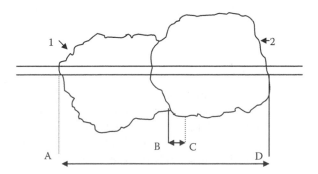

FIGURE 6.10
Canopy overlap (points B to C) is not measured if the canopies of two or more plants of the same species overlap; for example, if shrubs 1 and 2 are the same species, then the canopy intercept is measured from points A to D. If shrubs 1 and 2 are different species, then canopy intercept is measured from points A to C for shrub species 1 and from points B to D for shrub species 2.

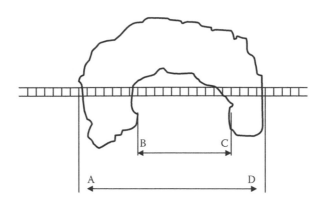

FIGURE 6.11
Gaps in the canopy (points B to C) greater than 2 inches (5 cm) are not measured. The canopy intercept for this shrub is measured from point A to D if the distance from B to C is less than or equal to 2 inches (95 cm) or from points A to B and points C to D if the gap is greater than 2 inches (5 cm).

dead centers or shrubs with large spaces between branches. Examiners must determine how to deal with gaps in the canopy. One solution is for the observer to assume a closed canopy unless the gap is greater than some predetermined length. We recommend that gaps less than 2 in. (5 cm) be considered part of the canopy (Figure 6.11).

Estimating Average Height

Estimate the average height in feet (meters), within ±10%, for each plant species. The estimation should only be for the part of the plant that is intercepted by the tape, not the entire plant. Enter plant height in the Height field on the LI Data Form for each item or species intercept. If plant species are recorded by size class, measure the average height for the plant species by each size class recorded. Plant height may be recorded at one intercept representing an average for the entire transect, at a few intercepts, or at every intercept. Be sure to record this information in the Metadata table.

Precision Standards

Use the precision standards listed in Table 6.19 for the LI sampling.

Sampling Design Customization

User-Specific LI Sampling Design

There are many ways the user can adjust the LI sample fields to make sampling more efficient and meaningful for non-fire-related forest biomass inventory/monitoring and local situations. Adjust the number and length of transects based on plant species size and distribution. Longer transects capture more variability in plant species cover, reducing the number of transects required to accurately estimate cover (Elzinga et al., 1998). The LI method is generally used for sampling shrub communities with vegetation greater than 3 ft (1 m) in height; however, this method can be used to sample taller vegetation if sighting devices are used to record the start and stop points for each intercept along a transect.

TABLE 6.19

Precision Guidelines for Line Intercept (LI) Sampling

Component	Standard
Size class	±1 class
Start	±0.1 ft (0.03 m)
Stop	±0.1 ft (0.03 m)
Height	±10% average height

Sighting devices may be mounted to a tripod and pointed downward to sample shorter vegetation (e.g., grasses, forbs, and small shrubs) and pointed upward to sample taller vegetation (e.g., tall shrubs and trees).

Sampling Hints and Techniques

Examiners must be able to identify plant species and know how to collect cover data by measuring canopy intercepts along a measuring tape. Line intercepts should be recorded only along one edge of the measuring tape—the right side as you proceed to the end of the tape. It is important to prevent the tape from moving such that certain plants are inadvertently included or excluded; for example, it can be very difficult to sample using the LI method when it is windy because the tape will not be stationary.

The accuracy of this method depends on how well the crew can estimate the vertical projection of vegetation along the tape. Observer bias occurs because the sighting line used to determine canopy starts and stops is not perpendicular to the tape. This bias can be minimized by using two measuring tapes (one above and one below) and sighting along the right side of the top tape to the right side of the bottom tape. Another solution is to suspend the measuring tape above the vegetation and use a plumb bob to record intercepts. For overhead vegetation, a pole with a level can be used. When measuring low and high vegetation, the most accurate method is to use a type of optical sighting device.

Measuring tapes come in a variety of lengths, increments, and materials. Examiners should choose English tapes for this method and select a tape that is at least as long as, or a little longer than, the transect length being sampled. When measuring plant species intercepts in feet, use a tape marked in 10ths and 100ths of feet. Steel tapes do not stretch and are the most accurate over the life of the tape. Steel is probably the best choice for permanent transects where remeasurement in exactly the same place each time is important. Cloth and fiberglass tapes will stretch over the life of the tape but are easier to use then steel tape because they are lighter and do not tend to kink.

When entering data on the LI Data Form, examiners will most likely run out of space on the first page. The form was designed to print one copy of the first page and several copies of the second page. The second page can be used to record more plant species intercepts for the plant species or items recorded on the first page or for additional plant species and items. The second page of the Data Form allows the examiner to write the intercept number on the form. This allows the examiner to design the form to accommodate the number of intercepts sampled. Print out enough pages to record all species on all transects for the required number of intercepts. The FIREMON data entry screens and database allow a maximum of 25 intercepts for each plant species along a transect.

Rare Species (RS) Sampling Method*

The Rare Species (RS) sampling method is used to assess changes in uncommon, perennial plant species when other monitoring methods are not effective. This method monitors individual plants and statistically quantifies changes in plant survivorship, growth, and reproduction over time. Plants are spatially located using distance along and from a permanent baseline, and individual plants are marked using a permanent tag. Data are collected for status (living or dead), stage (seedling, nonreproductive, or reproductive), size (height and diameter), and reproductive effort (number of flowers and fruits). This method is primarily used for threatened and endangered species and uncommon grass, forb, shrub, and tree species of special interest. When plants become rare, most sampling methods are not effective because the species will not likely occur in the sampling unit; therefore, it becomes necessary to follow individual plants in order to determine fire effects for that species. The RS method was designed to quantify temporal changes in plant survivorship, growth, and reproduction for uncommon perennial plant species; this method is not effective for rare annual species. A permanent baseline is established, and baseline length and characteristics about the general sample design are recorded. For each plant, a unique ID number, distance along the baseline, and perpendicular distance from the baseline are recorded. This method is primarily used when the fire/biomass manager wants to monitor changes in threatened and endangered perennial plant species and uncommon species of special interest. This method is suited for rare grass, forb, shrub, and tree species that are not effectively monitored by other methods. This sampling method uses attributes of individual plants to assess changes in survivorship, growth, and production over time.

Sampling Procedure

Preliminary Sampling Tasks

As mentioned with the other sampling procedures described in this text, before setting out for your rare species sampling, it is import to lay out a practice area with easy access. Try to locate an area with the same species or vegetation life form you plan on sampling. Get familiar with the plot layout and the data that will be collected. This will give you a chance to assess the method and will help you think about problems that might be encountered in the field. For example, will you really be able to identify the species of interest once you are in the field? It is better to answer these questions before the sampling begins so that you are not wasting time in the field. This will also let you determine if any pieces of equipment must be ordered.

Many preparations must be made before proceeding into the field for RS sampling. First, all equipment and supplies must be purchased and packed for transport into the field. Travel to plots is usually by foot, so it is important that supplies and equipment be placed in a comfortable daypack or backpack. It is also important to have spares of each piece of equipment so an entire day of sampling is not lost if something breaks. Spare equipment can be stored in the vehicle rather than the backpack. Be sure all equipment is well maintained and that you have plenty of extra supplies, such as Data Forms, map cases, and pencils.

---

* Material in this section is based on Sutherland, S., *FIREMON: Fire Effects Monitoring and Inventory System*, General Technical Report RMRS-GTR-164-CD, U.S. Department of Agriculture, Forest Service, Rocky Mountain Research Station, Fort Collins, CO, 2006.

All RS Data Forms should be copied onto waterproof paper because inclement weather can easily destroy valuable data recorded on standard paper. Data Forms should be transported into the field using plastic, waterproof map protectors or plastic bags. The day's sample forms should always be stored in a dry place (e.g., office or vehicle) and not be taken back into the field for the next day's sampling.

One person on the field crew, preferably the crew boss, should have a waterproof, lined field notebook for recording logistic and procedural problems encountered during sampling. This helps with future remeasurements and future field campaigns. All comments and details not documented in the sampling methods should be written in this notebook; for example, snow on the plot or a rock or mudslide might be described in the notebooks, which would be helpful in plot remeasurement.

Again, as with the other sampling protocols, it is beneficial to have plot locations for several days of work in advance in case something happens, such as if the road to one set of plots is washed out by flooding. Plots should be referenced on maps and aerial photos using pin-pricks or dots to make navigation easy for the crew and to provide a check of the georeferenced coordinates. It is easy to transpose UTM coordinate digits when recording georeferenced positions on the plot sheet, so marked maps can help identify any erroneous plot positions. If possible, the spatial coordinate should be provided if FIREMON plots were randomly located.

As with the other sampling techniques covered in this text, a field crew of two people is probably the most efficient for implementation of the RS sampling method. There should never be a one-person field crew for safety reasons, and any more than two people will probably result in some people waiting for critical tasks to be done. The crew boss is responsible for all sampling logistics, including the vehicle, plot directions, equipment, supplies, and safety. The crew boss should be the note taker, and the technician should perform most quadrat measurements. The initial sampling tasks of the field crew should be assigned based on field experience, physical capacity, and sampling efficiency, but sampling tasks should be modified as the field crew gains experience. Tasks should also be shared to limit monotony.

Designing the RS Sampling Method

A set of general criteria recorded on the RS Data Form allows the user to customize the design of the RS sampling method so that the sampling captures the information needed to successfully complete the management objective within the time, money, and personnel constraints. These general fields should be determined before the crews go into the field and should reflect a thoughtful analysis of the expected problems and challenges in the fire monitoring project.

Plot ID Construction

A unique plot identifier must be entered on the RS sampling form. This is the same plot identifier used to describe general plot characteristics in the Plot Description (PD) sampling method.

Determining the Sample Size

The size of the rare species population ultimately determines the length of the baseline from which the individual plants are located. The baseline length is recorded on the RS Data Form. If the population is divided along several patches, several baselines can be established. The size of the rare species population also determines the number of individuals sampled. We recommend sampling 25 individuals within the population; this should be sufficient for most studies. In some situations, however, more individuals should be sampled, and in these cases the project objectives should identify the sampling intensity. The RS Data Form and data entry screen allow an unlimited number of individuals to be measured per baseline. If the population is smaller than 25 individuals, measure all individuals. This will then become a census rather than a sample, and a statistical analysis will not be necessary.

> **DID YOU KNOW?**
>
> A census of the population counts or measures every individual. The main advantage of this approach is that the measure is a count and not an estimate based on sampling. No statistics are required. The changes measured from year to year are real, and the only significance of concern is biological.

Conducting RS Sampling Tasks

Establish the Baseline for Locating Individual Plants

The baseline serves as a georeferenced starting point for relocating individual plants. If other methods are implemented at the same sample site, then, as much as possible, the plots should be set up so they correspond with one another. In most cases, this will not be advisable when monitoring rare plants, as the additional sampling will negatively impact the individuals being monitored (e.g., by trampling). When the rare species population is located, a permanent baseline is set up as a reference from which you will locate individual plants. On flat areas, the baseline runs from south to north with the 0-ft (0-m) mark on the south end. On slopes, the baseline runs upslope with the 0-ft (0-m) mark on the bottom (down slope) end. The length of the baseline will be determined by the size of the rare species population.

Locating Individual Plants

Locate individual plants within the rare species population by running a tape perpendicular to the baseline to the plant. If the rare species population being sampled is to the right of the baseline when looking uphill (or north on flat ground), then distances to plants are recorded as positive (+) numbers and distances to plants to the left are recorded as negative (−) numbers. Because the distance along the baseline and the distance from the baseline will be used to relocate plants in successive years, it is essential that these measurements be exact and that the lines are perpendicular (see Figure 6.12). Measure distances to the nearest 0.1 ft (0.02 m).

Tagging Individual Plants

Individual plants need to be permanently tagged with a unique plant number. If the plant is woody (shrub or tree), the tag can be attached to the plant. If the plant is herbaceous (grass or forb), the tag should be anchored in the soil adjacent to the plant in a standard location (e.g., directly downslope of the plant). Because these tags will uniquely identify the individual plant, it is essential that they be located in a manner that will eliminate any confusion between individuals.

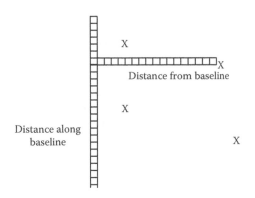

X = Rare plants

FIGURE 6.12
When measuring the location of rare plants, it is critical that the distance locating each plant—along and from the baseline—be accurately measured and that the measuring tapes are perpendicular.

Rare Species Sampling

Plant Identity

First enter the species code on the RS Data Form. FIREMON uses plant species codes from the NRCS Plants Database; however, you may use your own species codes. To uniquely identify each individual of the rare species, enter the plant number, distance along baseline, and distance from the baseline on the RS Data Form.

Status

Next enter the plant status on the RS Data Form. Status describes the individual plant as live or dead using the following codes:

L—Live (plant with living tissue)

D—Dead (plant with no living tissue visible)

NA—Not applicable

Plant stage is purely qualitative but determines postburn survivorship and population health. Use care in determining plant status during the dormant season.

Stage

Enter the plant species stage on the RS Data Form. Stage describes the individual plant as a seedling, nonreproductive adult, or reproductive adult.

S—Seedling: (plant less than 1 year old)

NR—Nonreproductive adult (plant 1 year old or older without flowers or fruits)

R—Reproductive adult (plant 1 year old or older with flowers or fruits)

Plant stage is also qualitative but provides information on plant growth and reproduction and population health. Again, use care in determining plant reproductive status during the dormant season.

Size

Size information provides data on growth rates and population vigor. Four measures of plant size are entered on the RS Data Form: two diameter measures, maximum height, and number of stems. For grasses, forbs, shrubs, and trees less than 1 in. (2 cm) in DBH, measure the diameter of the plant canopy at two places. First, measure the widest part and record the diameter. Make a second measurement at a right angle to the first and record it. Make both measurements in inches (centimeters) to the nearest 0.1 in. (0.2 cm). For trees at least 1 in. (2 cm) in DBH, record the DBH. Measure the maximum height of the plant in feet (meters) and record it to the nearest 0.1 ft (0.03 m). Regardless of the plant size, record the number of plant stems for the individual you are measuring. Use care in determining individual plants when measuring size.

Reproduction

Flower and fruit counts provide data on reproductive rates and population viability. Count the number of flowers and fruits on the plant and record that figure on the RS Data Form.

Precision Standards

Use the precision standards in Table 6.20 for the RS sampling.

Sampling Design Customization

User-Specific RS Sampling Design for Forest Biomass Applications

The user can modify the RS sample fields in a number of ways to make sampling more efficient and meaningful for non-fire-related forest biomass inventory/monitoring and for local situations. This will usually mean adjusting the number of individuals sampled as needed for the specific task. As mentioned earlier, estimating tree biomass (weight) based on parameters that are easily measured in the field using RS and other modified FIREMON sampling methods described in this text is becoming a fundamental task in forestry and forest-based biomass technology. Traditionally, the cubic-foot or board-foot volume of merchantable products, such as sawlogs or pulpwood, adequately described forest stands; however, the intensity of forest utilization has increased in recent years due to whole-tree harvesting and the use of wood for energy. All aboveground branches, leaves, bark, roots, small trees, and trees of poor form or vigor are now commonly included in the harvested product and are listed as biomass of whole trees (WT) or individual components. With this increasing

TABLE 6.20

Precision Guidelines for Rare Species (RS) Sampling

Component	Standard
Distance along baseline	±0.1 foot (0.03 m)
Distance from baseline	±0.1 foot (0.3 m)
Maximum diameter	±0.1 inch (0.2 cm)
Diameter 2	±0.1 inch (0.2 cm)
Height	±0.1 foot (0.03 m)
Stems	±3% total count
Flowers	±3% total count
Fruits	±3% total count

emphasis on complete tree utilization and use of wood as a source of energy, tables and equations have been developed to show the whole tree biomass as weights of total trees and their components. Again, in order to use biomass equations and tables to measure total tree biomass content, various parameters on the target tree population must first be collected in the field. When these data are collected, use the Metadata form to record any changes in sampling methods that are modified from the standard or to remark on any other RS matter that needs to be explained or defined for subsequent sampling and data use.

Sampling Hints and Techniques

Examiners must be able to identify the target plant species and identify individual plants. It can be difficult to distinguish individual plants for some species such as sod-forming grasses. If individual plants are difficult to identify, guidelines should be determined before sampling as to what constitutes the individual count unit. Some examples include counting individual stems in aspen communities, culm (from Latin *culmus* for "stalk") groups in rhizomatous grasses, and flowering stems for mat-forming forbs; however, the counting unit chosen to be monitored should reflect a real change in the plant community.

Measuring tapes come in a variety of lengths, increments, and materials. Examiners should choose English tapes for this method and select a tape that is at least as long as, or a little longer than, the baseline length being sample. Steel tapes do not stretch and are the most accurate over the life of the tape. Steel is probably the best choice where remeasurement in exactly the same place each time is important. Cloth and fiberglass tapes will stretch over the life of the tape but are easier to use than steel tapes because they are light and do not tend to kink.

Because the purpose of resampling is to determine change over time, it is essential that the plots are resampled when plants are in the same phonologic condition as when they were originally sampled. Often this means resampling on the same date as when you originally sampled. If you do not resample when the plants are in the same phonologic condition, then you may be documenting annual growth cycles rather than biomass content.

Point Intercept (PO) Sampling Method*

The modified FIREMON Point Intercept (PO) sampling method is used to assess changes in plant species cover or ground cover for a macroplot. This method uses a narrow-diameter sampling pole or sampling pins placed at systematic intervals along line transects to sample within-plot variation and to quantify statistically valid changes in plant species cover and height over time. Plant species or ground cover classes that touch the pin are recorded as hits along a transect.

This method uses transects located within the macroplot. First, a baseline is established from which to orient the transects, then transects are placed randomly along the baseline. Characteristics, such as transect length, number of transects, and number of points per transect, are recorded about the general sample design. A sampling pole or sampling pins

---

* Material in this section is based on Caratti, J.F., *Point Intercept (PO) Sampling Method*, General Technical Report RMRS-GTR-164-CD, U.S. Department of Agriculture, Forest Service, Rocky Mountain Research Station, Fort Collins, CO, 2006.

are systematically lowered along each transect and hits are tallied when contact is made with a plant species or ground cover class. Percent cover is calculated by dividing the number of hits for each plant species or ground cover class by the total number of points along the transect. This method is primarily suited for vegetation types less that 3 ft (1 m) in height and is particularly useful for recording ground cover.

This method is primarily used when managers want to monitor changes in plant species cover and height or ground cover and is best suited for sampling ground cover and grasses, forbs, and shrubs less than 3 ft (1 m) in height. The Point Intercept method works well for fine textured herbaceous communities, which can be more difficult to estimate with the Line Intercept method, and it provides a more objective estimate of cover than the ocular estimates used in the Cover/Frequency (CF) sampling method. It can be difficult to detect rare plants with the PO method unless many points are used for sampling. Point Intercept sampling requires many points to sample rare species (200 points to sample at 0.5% cover). Quadrats sample more area and have a greater chance of detecting rare species. If rare plant species are of interest, the CF or RS methods are preferred because it is more effective to sample rare species using quadrats or by marking individual plants than with points or lines. Use the PO method if you are primarily interested in monitoring changes in ground cover. The PO method may be used in conjunction with the CF method (discussed later) to sample ground cover by using the CF sampling quadrat as a point frame.

The Point Intercept method is considered one of the most objective ways to sample cover (Bonham, 1989). The observer needs to decide only whether a point intercepts a plant species or the ground. No cover estimates are required. Points offer quick and efficient data collection and can be used to estimate cover values with minimal bias and error; however, errors can be caused by plants moving in the wind or sampling poles being lowered incorrectly. The points themselves have dimension and can be considered small quadrats. In theory, if you sampled an infinite number of plants in an area, you could measure the exact cover for each plant species. Points are either the end of the sampling pole or the intersection of cross-hairs in a sampling frame.

Cover or ground cover is estimated using individual points or collections of points. Collections of points are sampled either with sampling pins grouped into a pin frame (typically 10 pins) or by cross-hairs grouped into a rectangular sighting frame. When using pin frames, the sampling pole is replaced with a pin. Pins are generally smaller in diameter than a sampling pole so they are less prone to sampling error (see below). The pin frame itself helps protect sampling pins from damage. Pin spacing should be determined according to plant species and vegetation patterns; for example, pins in a collection should not be placed so close together that all pins hit bare ground between clumps of grasses or all fall on one clump of grass. The number of points used determines the percent-cover values that can be estimated; for example, if 50 points are sampled along a transect, then cover can be estimated in 2% intervals (1/50, 2/50, and so forth) for the transect. Cover is estimated by dividing the number of hits per species or ground cover category by the total number of points measured. More than one species may be tallied for each pin location depending on project objectives.

Sampling pole and pin diameter can influence the accuracy of cover estimates. This is mostly an issue with large-diameter sampling piles, which overestimate cover, especially for narrow or small-leaved species. Pins less than 0.1 in. (2.5 mm) are impractical in the field because they move in the wind and are easily damaged (Bonham, 1989). Overestimation of cover is not a problem if the monitoring objective is to note relative cover changes rather than absolute changes in cover. Because of the effect of sampling pole and pin diameter on cover estimations, it is necessary to always use the same diameter poles or pins when remeasuring cover.

Point Sampling Techniques

Single Points

Each sample point is defined by a sampling pole guided vertically to the ground. A sturdy 0.25-in. (0.635-cm) diameter sampling pole is recommended when sampling with the PO method. Smaller diameter poles (0.125 in., 0.3175 cm) may be used for more precise measurements and fewer observer decisions; however, thin poles are more flexible, require more finesse to place in a straight line, and are easily bent in the field. A fiberglass tent pole, wooden dowel, or aluminum rod could all be used as a sampling pole. It should be longer than the vegetation that will be sampled is tall and long enough that field crews can sample without leaning over (40 in., 100 cm), and it should be sharpened on one end and have a loop or bend on the other. Individual points placed at systematic intervals along a transect can give a more precise cover estimate than points grouped into point frames or grid frames, given that the same number of points are sampled (Blackman, 1935; Goodall, 1952; Greig-Smith, 1983). Using individual points requires approximately one-third the number of points compared to using points in groups (Bonham, 1989). The distance between systematically located pins along a transect depends on plant size, plant distribution, and the distance between plants. The recommended FIREMON PO sampling method uses the single-point sampling approach.

Point Frames

Point frames are more practical and more commonly used than grid frames. Point frames are built of wood or metal and have two legs and two cross arms, typically containing 10 pins (see Figure 6.13). The pins can be made of any material as long as they are relatively small diameter (0.25 in., 0.635 cm), rigid enough not to bend or break, and long enough to touch the ground. When sampling, the pins are lowered to the ground cover through the holes and the interceptions are tallied. The size of the frame needs to be designed to suit local vegetation conditions because plant height and distribution patterns affect the spacing of pins and height of the frame. In vegetation types with large plants or clumped distributions, groups of points may intercept plants more frequently with more hits per frame, resulting in an overestimation of cover. Some point frames are built to allow the

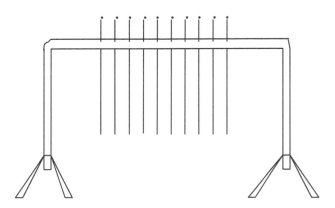

FIGURE 6.13
Example of a point frame with 10 pins.

pins to be slid at an angle into the vegetation, and this can have some sampling benefits in certain types of vegetation. FIREMON provides an alternative data entry form and data entry software to accommodate data gathered with the point frame method.

Grid Quadrat Frame

A quadrat is typically a square frame constructed of plastic (i.e., PVC), metal rod, or wood that is placed directly on top of the vegetation. Quadrats are also commonly called plots. Typically, a grid frame is made from metal or wood. Rows of thin wire or light string are attached to the inside vertical and horizontal pieces of the frame, resulting in a number of intersections or "cross-hairs." Cross-hairs of grid quadrats are considered point quadrats, and the vertical interceptions of cross-hairs with plant parts are considered hits. A double grid of cross-hairs prevents error due to observers viewing the cross-hairs at different angles. Stanton (1960) designed a grid frame for estimating shrub cover in communities. This grid of cross-hairs consisted of 25 points spaced 2.95 inches (7.5 cm) apart and was supported with metal legs. This type of frame is good for measuring cover up to 4.5 ft (1.5 m) tall in sparse vegetation. FIREMON provides a data entry form and data entry software to accommodate data gathered with the grid quadrat frame method.

Sampling Procedure

As with other FIREMON sampling protocols or methods discussed in this text, it is good practice to lay out a practice area with easy access. Try to locate an area with the same species or vegetation life form you plan on sampling. Get familiar with the plot layout and the data that will be collected. This will give you a chance to assess the method and will help you think about problems that might be encountered in the field. For example, will you be recording the plant status—dead or alive—for the part of the plant hit by the sampling pin or the entire plant? It is better to answer these questions before the sampling begins so you do not waste time in the field. This will also let you determine if any pieces of equipment must be ordered. Before heading into the field for PO sampling, many preparations must be made. First, all equipment and supplies must be purchased and packed for transport into the field. Travel to plots is usually by foot, so it is important that supplies and equipment be placed in a comfortable daypack or backpack. It is also important to have spares of each piece of equipment so an entire day of sampling is not lost if something breaks. Spare equipment can be stored in the vehicle rather than the backpack. Be sure all equipment is well maintained and that you have plenty of extra supplies, such as Data Forms, map cases, and pencils.

All PO Data Forms should be copied onto waterproof paper because inclement weather can easily destroy valuable data recorded on standard paper. Data Forms should be transported into the field using a plastic, waterproof map protector or plastic bag. The day's sample forms should always be stored in a dry place (e.g., office or vehicle) and not be taken back into the field for the next day's sampling.

One person on the field crew, preferably the crew boss, should have a waterproof, lined field notebook for recording logistic and procedural problems encountered during sampling. This helps with future remeasurements and future field campaigns. All comments and details not documented in the sampling methods should be written in this notebook.

It is beneficial to have plot locations for several days of work in advance in case something happens, such as if the road to one set of plots is washed out by flooding. Plots should be referenced on maps and aerial photos using pin-pricks or dots to make navigation easy for the crew and to provide a check of the georeferenced coordinates. It is easy to transpose UTM coordinate digits when recording georeferenced positions on the plot sheet, and marked maps can help identify any erroneous plot positions. If possible, the spatial coordinates should be provided if plots were randomly located.

A field crew of two people is probably the most efficient for implementation of the PO sampling method. For safety reasons there should never be a one-person crew, and any more than two people will probably result in part of the crew waiting for tasks to be completed and unnecessary trampling on the macroplot. The crew boss is responsible for all sampling logistics including the vehicle, plot directions, equipment supplies, and safety. The crew boss should be the note taker, and the technician should perform most quadrat measurements. The initial sampling tasks of the field crew should be assigned based on field experience, physical capacity, and sampling efficiency. As the field crew gains experience, switch tasks so that the entire crew is familiar with the different sampling responsibilities and to limit monotony.

Designing the PO Sampling Method

A set of general criteria recoded on the PO Data Form allows the user to customize the design of the PO sampling method so the sampling captures the information needed to successfully complete the management objective within time, money, and personnel constraints. These general fields should be decided before the crews go into the field and should reflect a thoughtful analysis of the expected problems and challenges in the fire monitoring project; however, some of these fields, particularly the number of points per transect and number of transects, might be adjusted after preliminary sampling is conducted in the field to determine a sufficient sample size.

Plot ID Construction

A unique plot identifier must be entered on the PO sample form. This is the same plot identifier used to describe general plot characteristics in the Plot Description (PD) sampling method.

Determining Sample Size

The size of the macroplot ultimately determines the length of the transects and the length of the baseline along which the transects are placed. The amount of variation in plant species composition and distribution determines the number and length of transects and the number of quadrats required for sampling. The typical macroplot sampled in the PO method is a 0.10-ac (0.04 ha) square measuring 66 × 66 ft (20 × 20 m), which is sufficient for most monitoring applications. If you are not sure of the plot size to use, contact someone who has sampled the same vegetation that you will be sampling. The size of the macroplot may be adjusted to accommodate different numbers and lengths of transects and number of points per transect. It is more efficient if you use the same plot size for all sampling methods on the plot.

The sampling unit for the PO method is the transect. Based on personal experience, it is probably best to sample five transects within the macroplot; however, in some situations more transects should be sampled. Enter the number of transects on the PO Transect Data Form. The recommended transect length is 6 ft (20 m) for a 66 × 66 ft (20 × 20 m) macroplot, although the macroplot size may be adjusted to accommodate longer or shorter transects based on the variability in plant species composition and distribution. For example, transects may be lengthened to accommodate more points per transect or more widely spaced points. Enter the transect length on the PO Transect Data Form. The FIREMON PO Data Form and data entry screen allow a maximum of 10 transects.

When sampling with a metric tape, sampling at every 0.25 m for a total of 80 points per transect is recommended. When sampling with a tape in inch and foot increments, 66 points should be placed 1 ft (0.3 m) apart along each 66-ft (20-m) transect. The number of points and spacing should be adjusted based on plant species size and spacing; for example, points should not be placed so close together that all sample points hit bare ground between clumps of grasses or all sample points fall on grass clumps. The number of points along a transect determines the resolution of cover recorded. If 50 points are recorded along a transect, cover values can be recorded in increments of 2% (1/50, 2/50, etc.). At a minimum, you want enough points to sample at least some of the species of interest along each transect. Enter the number of points per transect on the PO Data Form.

Conducting PO Sampling Tasks

Establish the Baseline for Transects

Once the plot has been monumented, a permanent baseline is set up as a reference from which you will orient all transects. The baseline should be established so the sampling plots for all of the method overlap as much as possible. The recommended baseline is 66 ft (20 m) long and is oriented upslope with the 0-ft (0-m) mark at the lower permanent marker and the 66-ft (20-m) mark at the upper marker. On flat areas, the baseline runs from south to north with the 0-ft (0-m) mark on the south end and the 66-ft (20-m) mark on the north end.

Locating the Transects

Transects are placed perpendicular to the baseline and are sampled starting at the baseline. On flat areas, transects are laid out east starting at the baseline. For permanent plots, determine the compass bearing of each transect, record these on the plot layout map, and permanently mark each end of the transect. Starting locations for each transect are determined by selecting a sampling scheme using the FIREMON random transect locator or from supplied tables. If the PO method is used in conjunction with other replicated sampling methods (CF, LI, RS, or DE), use the same transects for all methods. In successive measurement years, it is essential that transects be placed in the same location.

Sampling Points

Points are sampled at equal intervals along the length of a transect by lowering the sampling pole vertically to the ground, not perpendicular to the slope. If 66 points are sampled along a 66-ft transect, the first point is recorded at 1 ft, then every foot to the end of the tape. If 80 points are sampled at 0.25 m along a 20-m transect, the first point is sampled at 0.25 m and the last point at 20 m.

Point Intercept Sampling

Recording Hits

The FIREMON PO method may be used to sample just species cover, just ground cover, or species and ground cover together. If the sampling crew is collecting species cover data, record only the plant species that are hit by the sampling pole. FIREMON provides plant species codes from the NRCS Plants Database; however, local or customized plant species codes are allowed in FIREMON. When using the PO method to sample species cover in

TABLE 6.21

FIREMON Ground Cover Codes

Code	Description	Code	Description
ASH	Ash (inorganic, from fire)	LICH	Lichen
BAFO	Basal forb	LITT	Litter and duff
BAGR	Basal graminoid	MEGR	Medium gravel (5–20 mm)
BARE	Bare soil (soil particles <2 mm)	MOSS	Moss
BARR	Barren	PAVE	Pavement
BASH	Basal shrub	PEIC	Permanent ice
BATR	Basal tree	PEIS	Permanent ice and snow
BAVE	Basal vegetation	PESN	Permanent snow
BEDR	Bedrock	ROAD	Road
BOUL	Boulders (round and flat)	ROBO	Round boulder (>600 mm)
CHAN	Channels (2–150 mm long)	ROCK	Rock
CHAR	Char	ROST	Round stone (250–600 mm)
CML	Cryptograms, mosses and lichens	STON	Stones (round and flat)
COBB	Cobbies (75–250 mm)	TEPH	Tephra volcanic
COGR	Coarse gravel (20–75 mm)	TRIC	Transient ice
CRYP	Cryptogamic crust	TRIS	Transient ice and snow
DEVP	Developed land	TRSN	Transient snow
FIGR	Fine gravel (2–5 mm)	UNKN	Unknown
FLAG	Flag stones (150–380 mm long)	WATE	Water
FLBO	Flat boulders (>600 mm long)	WOOD	Wood
FLST	Flat stone (380–600 mm long)	X	Did not assess
GRAV	Gravel (2–75 mm)		

clumped or sparse vegetation, you may find that data are only recorded at a subset of the sampling locations because there may not be vegetation at every point that is sampled. If ground cover is being sampled, use the cover codes in Table 6.21 to record sampling hits. Unlike species sampling, every point should have a ground cover code recorded. If species and ground cover are being sampled at the same time, record the species name for the plant that the pole hits and the appropriate ground cover code from Table 6.21. If, for example, you are lowering a sampling pole and it first contacts a blade of blue grama grass and then, as you continue to lower it, it hits the basal portion of the plant, record the NRCS Plants Database species code (BOGR2) and the ground cover code (BAGR or BAVE). Enter the plant species and ground cover code on the PO Transect Data Form.

The number of hits that are recorded for each sampling point is dependent on the project objectives. If the objective is simply to monitor ground cover, samplers need to record only the ground cover hits. To develop a complete species list, samplers should record all unique species hits at each sampling point. Multiple hits for each species can be recorded if measuring biomass, volume, or species composition. Again, this is important information to be recorded in the Metadata table.

The angle of the sampling pole has an effect on cover estimates. Vertically lowered sampling poles hit flat-bladed species (forbs) more often than grasses. A pole lowered at an angle tends to favor gasses (Winkworth, 1955). Most cover measurements use vertical placement of poles but will underestimate narrow-leafed species (such as grasses). Other angles are used to increase the number of hits; however, angled sampling eliminates the intuitive visualization of vegetation on the ground. In this modified form of FIREMON, we recommend using the vertical orientation of the sampling pole. The angle used in sampling should be entered in the Metadata table so data collected in subsequent visits is compatible—especially if some orientation other than vertical pole placement was used for the point sampling.

At each interval, lower the sampling pin to the ground and record one hit or each plant species that touches the pole. Record only one hit for each plant species, even if the pole touches the same plant or plant species more than once. When measuring ground cover you will generally record only the first or uppermost hit; for example, if the pin passes through an ash layer to bare soil, record only the ash layer.

Tally the hits for each species using a dot tally in the workspace column for each transect on the PO Transect Data Form. Enter the total number of hits for each species or ground cover class, by transect, on the PO Transect Data Form.

Recording Plant Status

Next enter the plant status on the PO Transect Data Form. Status describes the general health of the plant species as live or dead using the following codes:

L—Live (plant with living tissue)

D—Dead (plant with no living tissue visible)

NA—Not applicable

This may be an evaluation of the entire plant or just the part of the plant that comes in contact with the sampling pole depending on the project objectives. In modified FIREMON, we recommend recording the status of the plant part that touches the sampling pole. Recognize that an accurate assessment of plant status may be difficult during the dormant season. Plant status is purely qualitative, but it provides an adequate characteristic for stratification of plant health.

Estimating Average Height

At the end of each transect, estimate the average height of each plant species you tallied, in feet (meters) within ± 10% of the mean plant height. This is not the height above the ground where the sampling pole touches the vegetation, but rather the average total height of the plants that are tallied.

Precision Standards

Use the precision standards listed in Table 6.22 for the PO sampling.

TABLE 6.22

Precision Guidelines for Point Intercept (PO) Sampling

Component	Standard
Hits	±3% total hits
Height	±10% average height

Sampling Design Customization

Users can adjust the PO sample fields in a number of ways to make sampling more efficient and meaningful for local situations. Examiners may adjust the number of transects and points per transect based on plant species size and distribution. Points could be sampled in quadrats or frames placed along a transect rather than using individual points. Collections of points are pins grouped into a pin frame (usually 10 pins) or cross-hairs grouped into a sighting frame. If 10 pins are used in a frame, percent cover can be estimated within 10% intervals (1/10, 2/10, etc.) for each frame. If the point frames or grid frames are placed far enough apart (they are independent samples), the frames can be the sample units rather than the transects.

The PO method may be used in conjunction with the CF method (discussed later) to sample ground cover by using the CF sampling quadrat as a point frame. A pencil or pen is used to record ground cover hits at the four corners and the four midpoints on each side of the quadrat. A total of eight points are recorded for each quadrat.

If grid frames or point frames are being sampled along transects rather than individual points, then determine the number of quadrats or point frames sampled per transect. Record the number of frames per transect, the number of points per frame, and the transect number on the PO Point Frame Data Form.

The PO method is typically used for grasses, forbs, and small shrubs less than 3 ft (1 m) in height; however, this method can be modified to sample large shrubs and trees as well. Instead of using pins that drop to the ground or a grid frame that the observer looks down on, you could use a sighting device, such as sighting tube or "moosehorn," which allows you to look down for small plants and look up into the canopy for larger species.

The type of cover typically estimated by points is total cover. Cover can also be measured within defined vegetation layers or by different species. If measuring cover, record only the first hit of each pin. Multiple hits for each species can be recorded if measuring biomass, volume, or species composition.

Sampling Hints and Techniques

Examiners must be able to identify plant species, be familiar with ground cover category codes, and know how to collect cover data using a pin. If only ground cover is being sampled, plant identification skills are not required. Measuring tapes are made from a variety of materials and are available in varying lengths and increments. Examiners should choose metric tapes for this method, and the tapes should be at least as long as, or a little longer than, the transect length being sampled. Steel tapes do not stretch and are the most accurate over the life of the tape. Steel is probably the best choice for permanent transects where remeasurement in exactly the same place each time is important. Cloth and fiberglass tapes will stretch over the life of the tape but are easier to use than steel tapes because they are lighter and do not tend to kink.

Point intercept cover is easily calculated if 100 points are sampled per transect. Cover values are equal to the number of hits for each item on the transect. Metric tapes are easily divided into 100 intervals, and sampling a 20-m tape at 20-cm intervals is relatively simple; however, if sampling with English tapes marked in inches and feet, sampling 100 points is impractical unless the transect length is a multiple of 100 (50, 100, 200). On a 66-ft transect, 100 points must be placed at 8-in. intervals. It is time consuming to place the point at the right mark on the tape. If a high-resolution of cover is desired (e.g., 1%), one solution is to sample along a 66-ft transect at every 0.5 ft for 132 points. Another solution is to increase the transect length to 100 ft and place a point at very foot. When deciding how many points to sample per transect, it is better to sample more transects than place more points per transect. Sampling with fewer points and more transects will often sample more variability within the plot at a slightly lower resolution of cover (e.g., 1/66 = 1.5% versus 1/100 = 1%).

When entering data on the PO Transect Data Form, examiners will most likely run out of space on the first page. The form was designed to print one copy of the first page and several copies of the second page. The second page can be used to record more plant species on the first three transects or to record data for additional transects. The second page of the Data Form allows the examiner to write the transect number on the form. This allows the examiner to design the form to accommodate the number of transects sampled. Print out enough pages to record all species on all transects or the required number of intercepts. The FIREMON data entry screens and database allow a maximum of 10 transects.

When entering data on the PO Point Frames Data Forms, examiners will most likely run out of space on the first page. As with the PO Transect form, the Frames form was designed to print one copy of the first page and several copies of the second page. The second page can be used to record more plant species on the first five point frames or to record data for additional frames. The second page of the Data Form allows the examiner to write the frame number on the form. This allows the examiner to design the form to accommodate the number of frames sampled per transect. Print out enough pages to record all species on all transects for the required number of intercepts. The FIREMON PO data entry screens and database allow a maximum of 20 transects per plot.

Plot Description (PD) Sampling Method*

The Plot Description (PD) form is used to describe general characteristics of the FIREMON macroplot to provide ecological context for data analyses. The PD data characterize the topographical setting, geographic reference point, general plant composition and cover, ground cover, fuels, and soils information. The method provides the general ecological data that can be used to stratify or aggregate fire/biomass monitoring results. The PD method also has comment fields that allow for documentation of plot conditions and location using photos and notes. The PD method assumes that the sampling strategy has already been selected and the macroplot has already been located. The PD sampling methods described here are the recommended procedures for this method. This section, formatted differently from the preceding modified FIREMON protocol examples for illustrative purposes, describes the sampling procedure in the order of the fields that must be completed on the PD Data Form. In addition, the PD Data Form includes a Comments field where data regarding an unplanned or unexpected equipment failure or other unforeseen circumstances can be recorded. The first four fields are standard, key information entries for the FIREMON database: Registration Code, Project Code, Plot Number, and Sampling Date. Again, these entries are important and necessary and

---

* Material in this section is based on Keane, R.E., *Plot Description (PD) Sampling Method*, General Technical Report RMRS-GTR-164-CD, U.S. Department of Agriculture, Forest Service, Rocky Mountain Research Station, Fort Collins, CO, 2006.

must be entered; however, for the purposes of this text, the fields of most importance are Fields 5 through 80, which include plot information fields, biophysical setting fields, vegetation fields, ground cover fields, fire behavior and effects fields, common fields, and comments fields.

Plot Information Fields

Field 5: Examiner Name

The name of the crew boss or lead examiner should be entered, up to eight characters. This is a nonstandarized field so anything can be entered here, but convention calls for the first letter in the first name followed by a dot followed by the entire last name. So, Smokey Bear would be s.bear and Rusty Nail would be r.nail. There should be no blanks in the text; for example, don't enter Rusty Nail as r. nail.

Field 6: Units

Enter "E" if you will be collected data using English units or "M" if you are using metric units. These units are used for all measurements in the sampling. The only exception is the Error Unit field associated with the GPS location. GPS error may be in English or metric units regardless of what is entered in Field 6.

Plot Size

The macroplot is the area where you will be applying the sampling methods. The size of the macroplot ultimately dictates the representative area to be sampled (see Table 6.23). If vegetation is dense, large plot sizes usually take longer to sample because it is difficult to transverse the plot; however, some ecosystems have large trees scattered over large areas so large plot sizes are required to obtain realistic estimates. Studies have attempted to identify the optimum plot size for different ecosystems but have provided only mixed results. Table 6.23 is offered to help determine the plot size that matches the fire/biomass monitoring application. Plot size and shape selection should be determined by the project leader prior to entering the field.

Usually, the 0.1-acre circular plot will be sufficient for most ecosystems, and this size should be used if no other information is available. A general rule of thumb is that the plot should be big enough to capture at least 20 trees above 4 in. in diameter at breast height (DBH) on average (across all plots in your project). It is important that the plot size stay constant across all plots in a sampling project; for example, if a project contains shrublands,

TABLE 6.23

Suggested FIREMON Macroplot Plot Sizes

Average Plant Height (ft)	Plant Cover (%)	Suggested Plot Size (acres)	Plot Radius (ft)	Suggested Plot Size (m²)	Plot Radius (m)
$X < 15$	<50	0.10	37.2	400	11.3
	>50	0.05	26.3	200	8.0
$15 < X < 100$	<50	0.10	37.2	400	11.3
	>50	0.08	33.3	300	9.8
$X > 100$	<50	0.40	74.5	1000	17.8
	>50	0.13	42.5	500	12.6

grasslands, and forests, do not change the plot size when you sample each one. Select the largest plot size (forest, in this example) and use it for all ecosystems. In general, using a circular PD macroplot is recommended.

Two fields in the PD method are used to describe plot shape and size. If the plot shape is circular, then enter the plot radius/length in Field 7 and enter 0 (zero) in Field 8. If a rectangular plot shape is required, the length of the macroplot is entered in Field 7 and the width is entered in Field 8. No other plot shapes are used in FIREMON.

Field 7: Plot Radius (ft/m)

If the macroplot is circular, enter the radius of the macroplot. Enter the length of the macroplot if is rectangular.

Field 8: Plot Width (ft/m)

Enter the width of the plot if it is rectangular, or enter zero (0) or leave the field blank if the macroplot shape is circular.

Sampling Information

FIREMON data can be collected on *monitoring plots* or *control plots*. Monitoring plots are located inside the treatment area so you can compare the effects of different treatments on the sampled attributes. Control plots are placed outside the treatment area and are used to check that any changes in the sampled attributes were actually due to the treatments and not some unrelated factor.

Field 9: Plot Type

Enter "M" if you are sampling a monitoring plot or "C" if you are sampling a control plot.

Field 10: Sampling Event

Monitoring requires that sampling be stratified by space and time. Because monitoring is a temporal sampling of repeated measures, it is essential that you record the reason for sampling to provide a context for analysis. The Sampling Event field is used to document why the plot is being measured at this particular time (as recorded by date). The Sampling Event will help you track changes at the plot level more easily than if you used only the sampling date. The codes used for this field are as follows: (1) P is the pretreatment measurement of the plot, (2) R is the post-treatment remeasurement of the plot, and (3) IV indicates an inventory plot that is not permanently monumented and won't be resampled (see Table 6.24). The codes P and R are followed by a numeric value that indicates the sampling visit of the current sampling; for example, if you sampled a plot once before a prescribed fire then the code would be P1, and when you sample after the fire the code would be R1 for the first sampling, R2 for the second sampling, and so on. When you change event codes from P to R, you should start the sequential sample number over at 1. The FIREMON database

TABLE 6.24

Sampling Event Codes

Code	Event
Pn	Pretreatment measurement, sequential sample number
Rn	Posttreatment remeasurement of a plot, sequential sample number
IV	Inventory plot, not a monitoring plot

will accept data for up to three pretreatment measurements. When you are sampling a plot that has been sampled once or more before, you will have to consult previously collected FIREMON data so you use the appropriate sequential sample number. For simplicity, we have only provided standardized codes for pre- and post-treatment measurements. This may be a problem if, for example, you plan on three measurements: one preharvest, one postharvest/preburn, and one postburn. In practice, P is used before any treatments are applied, then R codes are used after the first treatment.

Linking Fields

Field 11: Fire ID

Enter a Fire ID of up to 15 characters. The ID number or name relates the fire that burned this plot to the same fire described in the Fire Behavior (FB) table. This field links these plot-scale data with the fire-scale data in the FB method. There may be many FIREMON plots referencing one fire. This field will be empty until after the burn has been completed.

Field 12: Metadata ID

Enter a code of up to 15 characters that links the plot data to the Metadata table. The table is used to store information on the sampling intensity and methods used in the monitoring project. This field is highly recommended so important information will be recorded for future reference.

Georeferenced Plot Positions

The next set of fields is important for relocating sample plots and for using plot data in mapping and map validation of remote sensing projects. These fields fix the geographic location of the plot center. Geographic coordinates are nearly always obtained from a Geographic Positioning System (GPS). GPS technology uses data from at least four orbiting satellites to triangulate your position in three dimensions (X, Y, and Z, or North, East, and Elevation) to within 3 to 50 meters of accuracy. GPS receivers are available from many sources, and there is a wide range of GPS models to choose from depending on various sampling criteria. GPS selection and training are not part of the FIREMON sampling methods; however, a number of resources provide advice on purchasing the right GPS for your sampling needs. A wide variety of public and private agencies also provide excellent training. Georeferenced coordinates for FIREMON plots should be taken from a GPS receiver and not from paper maps such as U.S. Geological Survey (USGS) quadrangle maps because of the high degree of error. Average the plot location over at least 200 readings to reduce the location error.

Many map projections are available to record plot georeferenced coordinates. Users can use either latitude–longitude (lat–long) or the Universal Transverse Mercator (UTM) coordinate system. If you are using UTM coordinates, record easting and northing to the nearest whole meter. If you are using lat–long coordinates, record latitude and longitude to the sixth decimal place using decimal degrees (this corresponds to about 1 meter of ground distance at 45 degrees latitude). The downside of lat–long coordinates is that it is difficult to visualize the measurements on the ground (e.g., how far is 0.05 degrees latitude?). Be especially alert because units of degrees, minutes, and seconds look similar to decimal degrees. If using lat–long coordinates, enter the data in Fields 14, 15, 19, 20, and 21. If using UTM coordinates enter data in Fields 16 to 21.

Field 13: Coordinate System

This field is automatically filled based on the data entered in Fields 14 to 21. The user does not see this field.

Field 14: Latitude

If using the lat–long system, enter the latitude in decimal degrees to six decimal places.

Field 15: Longitude

If using the lat–long system, enter the longitude in decimal degrees to six decimal places.

Field 16: Northing

If using the UTM system, enter the UTM northing to the nearest whole meter.

Field 17: Easting

If using the UTM system, enter the UTM easting to the nearest whole meter.

Field 18: Zone

If using the UTM system, enter the UTM zone of the plot center.

Field 19: Datum

If using the UTM system, enter the datum used in conjunction with the UTM coordinates.

Field 20: Position Error

Enter the position error value provided by the GPS unit. This should be entered regardless of whether you are using lat–long or UTM coordinates.

Field 21: Error Units (E/M)

Enter the units associated with the GPS error; these may differ from the units listed in Field 6.

Fields 5 though 21 make up the information that is critical to have for every macroplot, regardless of the sampling intensity or methods you will be using to collect data. The following sections describe the measurement or estimation of various ecosystem characteristics that are important to fire effects/biomass monitoring.

Biophysical Setting Fields

The biophysical setting describes the physical environment of the plot relative to the organisms that grow there. Many site characteristics can be included in a description of biophysical setting, but only topography, geology, soils, and landform fields are implemented in modified FIREMON.

Topography

Field 22: Elevation (ft/m)

Enter the elevation above mean sea level (MSL) of the plot in feet (meters) to the nearest 100 feet (30 m). Elevation can be estimated from three sources. Most GPS readings include an estimate of elevation, and these estimates are usually fairly accurate. Elevation can also be estimated from an altimeter. There are many types of altimeters, but most are barometric, estimating elevation from atmospheric pressure. Altimeters are notoriously fickle and require calibration nearly every day. When there are frequent weather systems passing the area, altimeters should be calibrated every 4 hours. Finally, elevation can be taken from USGS topographic maps.

Field 23: Plot Aspect

Enter the aspect of the plot in degrees true north to the nearest 5 degrees. Aspect is the direction the plot is facing. For example, a slope that faces exactly west would have an aspect of 270 degrees true north. Be sure to record the aspect that best represents the mac-roplot as a whole and not just the point where you are standing. Also, be sure you check your compass reading with your knowledge of the area to be sure that the aspect indicated is really correct. Often, metal on sampling equipment, or iron rebar at plot center, can influence the estimation of aspect.

Field 24: Slope

Record the plot slope using the percent scale to the nearest 5%. The slope is measured as an average of uphill and downhill slope from plot center. Be sure the recorded slope reflects the slope of the entire plot and not just the line where you are standing. Slope values should always be positive.

Field 25: Landform

Enter up to a four-character code that best describes the landform containing the macro-plot from Table 6.25.

Field 26: Vertical Slope Shape

Enter up to a two-character code using the classes in Table 6.26 that best describes the general contour of the terrain upslope and downslope from plot center. As you look up and down the slope, estimate a shape class that best describes the horizontal contour of the land (see Figure 6.14).

Field 27: Horizontal Slope Shape

Enter up to a two-character code using the classes in Table 6.26 that best describes the general contour of the terrain upslope and downslope from plot center. This is an estimate of the general shape of the slope parallel to the contour of the slope. As you look across the slope along the contour, estimate a shape class that best describes the horizontal contour of the land (Figure 6.14).

TABLE 6.25

Landform Codes

Code	Landform
GMF	Glaciated mountains foothills
UMF	Unglaciated mountains and foothills
BRK	Breaklands, river breaks, and badlands
PLA	Plains, rolling planes, and plains with breaks
VAL	Valleys, swales, and draws
HIL	Hills, low ridges, and benches
X	Did not assess

TABLE 6.26

Slope Shape Codes

Code	Slope Shape
LI	Linear or planar
CC	Depression or concave
PA	Patterned
CV	Rounded or convex
FL	Flat
BR	Broken
UN	Undulating
OO	Other shape
X	Did not assess

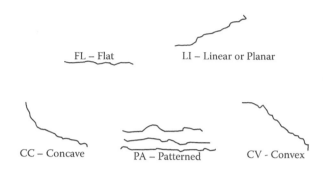

FIGURE 6.14

These illustrations depict the different types of vertical slope shapes. Horizontal slope shapes use the same classification but are determined by examining the across-slope profile, rather than up and down the slope.

Geology and Soils Fields

Field 28: Primary Surficial Geology

This is the first of five fields used to describe geology and soils. Determine the geological rock type composing the parent material at the plot and enter the appropriate code from Table 6.27 into that field. Generally, identification of surficial geology requires someone with specialized training and experience.

Field 29: Secondary Surficial Geology

Use this field only if you have coded a primary surficial geology type. Determine the secondary geological rock type composing the parent material at the plot and enter the appropriate code from Table 6.28 into the field. Generally, identification of surficial geology requires someone with specialized training and experience. Table 6.28 provides an abridged list of common surficial type.

Field 30: Soil Texture Class

The description of soil on the plot is limited to a general description because fire effects are not influenced by fine-scale soil characteristics. Generally, identification of soil texture requires someone with specialized training and experience. Many fire effects can be described by general soil characteristics, and soil texture is one of those general characteristics. Enter the code that best describes the texture of the soil on the macroplot (see Table 6.29). These soil textures are described in many soils textbooks. If you are unsure of how to evaluate soil texture or have no confidence in your estimates, then used the X code or leave the field blank.

TABLE 6.27

Common Primary Surficial Geology Codes

Primary Code	Rock Type 1
IGEX	Igneous extrusive
IGIN	Igneous extrusive
META	Metamorphic
SEDI	Sedimentary
UNDI	Undifferentiated
X	Did not assess

TABLE 6.28

Common Secondary Surficial Geology Codes

Secondary Code	Rock Type 2	Secondary Code	Rock Type 2
ANDE	Andesite	CONG	Conglomerate
BASA	Basalt	DOLO	Dolomite
LATI	Latite	LIME	Limestone
RHYO	Rhyolite	SANS	Sandstone
SCOR	Scoria	SHAL	Shale
TRAC	Trachyte	SILS	Siltstone
DIOR	Diorite	TUFA	Tufa
GABB	Gabbro	MIEXME	Mixed extrusive and metamorphic
GRAN	Granite	MIEXSE	Mixed extrusive and sedimentary
QUMO	Quartz monzonite	MIIG	Mixed igneous (extrusive and intrusive)
SYEN	Syenite	MIIGME	Mixed igneous and metamorphic
GNEI	Gneiss	MIIGSE	Mixed igneous and sedimentary
PHYL	Phyllite	MINME	Mixed intrusive and metamorphic
QUAR	Quartzite	MINSE	Mixed intrusive and sedimentary
SCHI	Schist	MIMESE	Mixed metamorphic and sedimentary
SLAT	Slate	X	Did not assess
ARGI	Argillite		

Field 31: Erosion Type

Erosion is an important second-order fire effect that must be documented. We have based the Erosion Type on the classification used by the Natural Resources Conservation Service *National Soil Survey Handbook* (Table 6.30). If your macroplot is on a site that has moved in its entirety through landslip, include that information in the Comments field of the PD form, then include the code with the code that identifies the erosion conditions you are seeing on the surface. Be sure to record erosion on preburn plots in order to provide the reference conditions. The types of erosion are listed along with their codes in Table 6.30. Enter the code that best describes the erosion occurring on the plot.

TABLE 6.29

Soil Texture Codes

Code	Description	Code	Description
C	Clay	S	Sand
CL	Clay loam	SC	Sandy clay
COS	Coarse sand	SCL	Sandy clay loam
COSL	Coarse sandy loam	SI	Silt
FS	Fine sand	SIC	Silty clay
FSL	Fine sandy loam	SICL	Silty clay loam
L	Loam	SIL	Silt loam
LCOS	Loamy coarse sand	SL	Sandy loam
LFS	Loamy fine sand	VFS	Very fine sand
LS	Loamy sand	VFSL	Very fine sandy loam
LVFS	Loamy very fine sand	X	Did not assess

Field 32: Erosion Severity

The severity of the erosion event is extremely difficult to assess and is best estimated by those who have some experience with erosion processes. We have based Erosion Severity on the classification used by the USDA Natural Resources Conservation Service *National Soil Survey Handbook* (Table 6.31). The severity codes use the depth and extent of erosion to quantify severity. Enter the code that best fits the severity of the erosion on the plot in this field. Severity codes do not apply to tunnel erosion. If you have tunnel erosion on your plot, enter –1 in this field.

TABLE 6.30

Erosion Type Codes

Code	Erosion Type
S	Stable, no erosion evident
R	Water erosion, rill
H	Water erosion, sheet
G	Water erosion, gully
T	Water erosion, tunnel
W	Wind erosion
O	Other type of erosion
X	Did not assess

TABLE 6.31

Erosion Severity Codes

Code	Erosion Severity
0	Stable, no erosion is evident.
1	Low erosion severity: Small amounts of material are lost from the plot. On average less than 25% of the upper 8 inches (20 cm) of soil surface have been lost across the macroplot. Throughout most of the area the thickness of the soil surface layer is within the normal range of variability of the uneroded soil.
2	Moderate erosion severity: Moderate amounts of material are lost from the plot. On average, between 25 and 75% of the upper 8 inches (20 cm) of soil surface have been lost across the macroplot. Erosion patterns may range from small, uneroded areas to small areas of severely eroded sites.
3	High erosion severity: Large amounts of material are lost from the plot. On average, 75% or more of the upper 8 inches (20 cm) of soil surface have been lost across the macroplot. Material from deeper horizons in the soil profile is visible.
4	Very high erosion severity: Very large amounts of material are lost from the plot. All of the upper 8 inches (20 cm) of soil surface have been lost across the macroplot. Erosion has removed material from deeper horizons of the soil profile throughout most of the area.
–1	Unable to assess.

Vegetation Fields

These PD fields describe general aspects of the vegetation using percent canopy cover as the measurement unit. All vegetation fields require an estimate of percent vertically projected canopy cover recorded by class (Table 6.32). The seasonal timing of cover estimates can lead to substantially different cover estimations, especially for the shrub and herbaceous components. Critically consider how and when cover should be estimated based on project objects, resources, and the sampling experience of the crew. One option may be to attempt to estimate what the cover would be at the peak of the growing season. Doing so can remove some of the seasonal variation in vegetation sampling; however, it can also lead to error in the cover estimates. Cover of herbaceous plants often appears greater when they are dormant because they fall over and lie flat on the ground. To get accurate values for these species,

TABLE 6.32

Canopy Cover Codes Used to Record Vegetation Cover in Fields That Require Cover Estimation

Code	Cover Class
0	0
0.5	>0–1%
3	>1–5%
10	>5–15%
20	>15–25%
30	>25–35%
40	>35–45%
50	>45–55%
60	>55–65%
70	>65–75%
80	>75–85%
90	>85–90%
98	>95–100%

estimate cover as if they were erect. Vegetation cover in these PD fields is stratified by life-form and size class. This makes determining canopy cover difficult because estimations require quite a bit of experience to arrive at consistent assessments of lifeform and size class cover when lifeforms and classes are unevenly distributed in all three dimensions. If you are unable to make an estimation for any reason, leave the field blank and note the reason in the comments section. Always enter the code 0 (zero) when there is no cover for that ground element. Vegetation cover does not need to sum to 100% by lifeform because there will probably be overlapping cover across all lifeforms; however, the total cover for each lifeform must always be greater than any of the covers estimated for the size classes within that lifeform.

Vegetation—Trees

The following fields provide an estimate of tree cover by size class.

Field 33: Total Tree Cover

Enter the percent canopy cover of all trees using the canopy cover codes presented in Table 6.32. This estimate includes cover of *all* tree species from the smallest of seedlings to the tallest of old growth stems. It includes all layers of canopy vertically projected to the ground.

Field 34: Seedling Tree Cover

Enter the percent canopy cover of all trees that are less than 4.5 ft (1.4 m) tall using the codes in Table 6.32. This cover estimate includes only the small seedlings.

Field 35: Sapling Tree Cover

Enter the percent canopy cover of all trees that are greater than 4.5 ft (1.4 m) tall and less than 5.0 inches (13 cm) DBH using the codes in Table 6.32.

Field 36: Pole Tree Cover

Enter the percent canopy cover of all trees that are greater than 5 in. (13 cm) DBH and less than 9 in. (23 cm) DBH using the codes in Table 6.32.

Field 37: Medium Tree Cover

Enter the percent canopy cover of all trees that are greater than 9 in. (23 cm) DBH up to 21 in. (53 cm) DBH using the codes in Table 6.32.

Field 38: Large Tree Cover

Enter the percent canopy cover of all trees that are greater than 21 in. (53 cm) DBH up to 33 in. (83 cm) DBH using the codes in Table 6.32.

Field 39: Very Large Tree Cover

Enter the percent canopy cover of all trees that are greater than 33 in. (83 cm) DBH using the codes in Table 6.32.

Vegetation—Shrubs

The next set of fields allows the sampler to estimate shrub cover in three height size classes.

Field 40: Total Shrub Cover

Enter the percent canopy cover of all shrubs on the plot into using the canopy cover in Table 6.32. This cover estimate includes vertically projected cover of all shrub species of all heights.

Field 41: Low Shrub Cover

Enter the percent canopy cover of all shrubs that are less than 3 ft (1 m) tall on the plot using the codes in Table 6.32.

Field 42: Medium Shrub Cover

Enter the percent canopy cover of all shrubs that are greater than 3 ft (1 m) tall and less than 6.5 ft (2 m) tall on the plot using the codes in Table 6.32.

Field 43: Tall Shrub Cover

Enter the percent canopy cover of all shrubs that are greater than 6.5 ft (2 m) tall on the plot using the codes in Table 6.32.

Vegetation—Herbaceous

Cover of grasses, forbs, ferns, mosses, and lichens are entered in the next set of vegetation fields. If you feel uncomfortable distinguishing between species within and across lifeforms, try to get some additional training from the ecologist, forester, or other resource specialist at your local office. Phenological adjustments must be made for many herbaceous species because most cure during the dry season, making overestimation difficult.

Field 44: Graminoid Cover

Enter the percent canopy cover of all graminoid species on the plot using the codes in Table 6.32. Graminoid cover includes all grasses, sedges, and rushes in all stages of phenology. This cover is for all sizes and species of graminoids.

Field 45: Forb Cover

Enter the percent canopy cover of all forbs on the plot using the cover codes in Table 6.32.

Field 46: Fern Cover

Enter the percent canopy cover of all ferns on the plot using the cover codes in Table 6.32.

Field 47: Moss and Lichen Cover

Enter the percent canopy cover of all mosses and lichens on the plot using the codes in Table 6.32. These mosses and lichen can be on the ground or suspended from plants in the air (arboreal).

DID YOU KNOW?

The composition of an epiphytic lichen community is one of the best biological indicators of nitrogen and sulfur-based air pollution in forests. Their sensitivity results from their total reliance on atmospheric sources of nutrition. Because lichens are so sensitive to these pollutants, they are useful as an early indicator of improving or deteriorating air quality (USFS, 2005).

Vegetation—Composition

The following fields document the dominant plant species in each of three layers or strata on the plot. These fields are used to describe the existing vegetation community based on dominance in cover. These descriptions are especially useful in satellite classification for mapping vegetation, developing existing vegetation community classifications, and stratifying FIREMON fire effects results. For a species to be dominant it has to have at least 10% canopy cover in that stratum, and the species must have higher cover than any other species in that stratum. In the PD method, two species per stratum are used to describe dominance. The first species (Species 1) is the most dominant in terms of canopy cover, and the second species (Species 2) is the second most dominant. There are three strata for stratifying dominant existing vegetation. The first stratum is called the Lower Stratum and is the cover of all plants less than 3 ft (1 m) tall, the Mid Stratum is for plants 3 to 10 ft (1 to 3 m) tall, and the Upper Stratum is for plants taller than 10 ft tall (3 m). Only species cover within the stratum is used to assess dominance. Many shade-tolerant tree species can be dominant in all three strata. If there are no species above 10% cover in a stratum, enter the code N to indicate that there are no species that qualify for dominance. The same applies if there is no secondary species for dominance.

Field 48: Upper Dominant Species 1

Enter the species code of the most dominant species in the upper-level stratum of the plot. This is the stratum that is greater than 10 ft (3 m) above ground level.

Field 49: Upper Dominant Species 2

Enter the species code of the second most dominant species in the upper-level stratum of the plot. This is the stratum that is greater than 10 ft (3 m) above ground level.

Field 50: Mid Dominant Species 1

Enter the species code of the most dominant species in the mid-level stratum of the plot. This is the stratum that is greater than 3 ft and less than 10 ft (1 to 3 m) above ground level.

Field 51: Mid Dominant Species 2

Enter the species code of the second most dominant species in the mid-level stratum of the plot. This is the stratum that is greater than 3 ft and less than 10 ft (1 to 3 m) above ground level.

Field 52: Lower Dominant Species 1

Enter the species code of the most dominant species in the lowest level stratum of the plot. This is the stratum that is less than 3 ft (1 m) above ground level.

Field 53: Lower Dominant Species 2

Enter the species code of the second most dominant species in the lowest level stratum of the plot. This is the stratum that is less than 3 ft (1 m) above ground level.

Potential Vegetation

An important characteristic for describing biotic plant communities, especially in the Western United States, is the potential vegetation type. Potential vegetation generally describes the capacity of a site by describing the vegetation that would eventually occupy a site in the absence of disturbance over a long time. For example, an alpine site can only

TABLE 6.33

Potential Lifeform Codes

Code	Potential Lifeform
AQ	Aquatic—lake, pond, bog, river
NV	Nonvegetated—bare soil, rock, dunes, scree, talus
CF	Coniferous upland forest—pine, spruce, hemlock
CW	Coniferous wetland or riparian forest—spruce, larch
BF	Broadleaf upland forest—oak, beech, birch
BW	Broadleaf wetland or riparian forest—tupelo, cypress
SA	Shrub-dominated alpine—willow
SU	Shrub-dominated upland—sagebrush, bitterbrush
HA	Herbaceous-dominated alpine—dryas
HU	Herbaceous-dominated upland—grasslands, bunchgrass
HW	Herbaceous-dominated wetland or riparian—ferns
ML	Moss or lichen-dominated upland or wetland
OT	Other potential vegetation lifeform
X	Did not assess

support herbaceous communities because these sites are too cold for shrubs or trees, whereas a clear-cut cedar–hemlock site has the potential to support coniferous forest ecosystems. Potential vegetation classifications are highly ecosystem specific and are locally developed for certain regions, so a standardized potential vegetation classification for the entire United States does not currently exist.

Field 54: Potential Vegetation Type ID

Potential vegetation types are the foundation of many management decisions. Many forest plans and project designs stratify treatments by potential vegetation type to achieve better results. Unfortunately, there is no national standard list of potential vegetation types in the United States; instead, a generic field is provided for the user to enter his or her own PVT code to stratify results. This field is not standardized, and any combination of alpha or numeric characters can be used. Do not use spaces in the text (e.g., enter ABLA/VASC). Be sure you document your codes in the FIREMON Metadata table. There are 16 characters available in this field.

Field 55: Potential Lifeform

Enter the potential lifeform code that best describes the community lifeform that would eventually inhabit the plot in the absence of disturbance (Table 6.33).

Ground Cover Fields

The next set of PD fields describes the fuels complex on the FIREMON plot. The first group of fuels fields characterizes ground cover by various characteristics important for evaluating fire effects. The standard FIREMON percent cover class codes (Table 6.32) are used to quantify ground cover. Ground cover is critical for describing fuel continuity and cover, but it is also used for evaluation of erosion potential and for classification of satellite imagery. A group of generalized fuel attributes are used to describe biomass characteristics for the entire plot. The first fields describe surface fuel characteristics through standardized fuel models, while the last fields describe crown fuel characteristics important for fire modeling.

Ground Cover

Ground cover attempts to describe important attributes of the forest floor or soil surface. Ground cover is estimated in 10 categories, each category being important for calculating subsequent or potential fire effects. Ground cover is another difficult sampling element. Cover within a category is evaluated as the vertically projected cover of that category that occupies the ground. *Only elements that are in direct contact with the ground are considered in the estimation of ground cover.* Ecosystem components suspended above the ground, such as branches, leaves, and moss, are not considered in the estimation of ground cover. Ground cover is described by a set of 10 fields where the sum *must add to 100%* (unlike the PD vegetation cover fields), plus or minus 10%. The following procedure is recommended for making these cover estimates. First, estimate ground cover for those categories with the least ground cover. These categories are the easiest to estimate with high accuracy. Scan the entire plot to check for mineral soil, moss/lichen, and rock ground cover. Next, estimate the basal vegetation field to the cover codes 0.5, 3, or 10 (basal vegetation rarely exceeds 15% ground cover). Finally, use the ground cover fields with the most cover (this is often only one or two fields, such as duff/litter) to make your estimate add to 100%. If you are unable to make an estimation for any reason, leave the field blank and note the reason in the Comments section (Field 81). Always enter the code 0 (zero) when there is no cover for that ground element.

Field 56: Bare Soil Ground Cover

Estimate the percent ground cover of bare soil using the codes in Table 6.32. Bare soil is considered to be all those mineral soil particles less than 1/16 in. (2 mm) in diameter. Bare soil does not include any organic matter. The bare soil can be charred or blackened by the fire.

DID YOU KNOW?

Soils represent the basic support system for terrestrial ecosystems because of their role in providing nutrients, water, oxygen, heat, and mechanical support to vegetation. Any environmental stressor that alters the natural function of the soil has the potential to influence the productivity, species composition, and hydrology of forest systems (USFS, 2003c).

Field 57: Gravel Ground Cover

Estimate the percent ground cover of gravel using the codes in Table 6.32. Gravel is those mineral soil particles greater than 1/16 in. (2 mm) in diameter to 3 in. (80 mm) in diameter. Again, gravel does not include any organic soil colloids. The gravel can be charred or blackened by the fire.

Field 58: Rock Ground Cover

Estimate the percent ground cover of rock using the codes in Table 6.32. Rock ground cover is considered to be all those mineral soil particles greater than 3 in. (8 cm) in diameter, including boulders. Rocks can be blackened by the fire.

Field 59: Litter and Duff Ground Cover

Estimate the percent ground cover of all *uncharred* litter and duff on the soil surface using the codes in Table 6.32. Litter and duff cover is mostly organic material, such as partially decomposed needles, bark, and leaves, deposited on the ground. Do not include any woody material in this ground cover category unless it is highly decomposed twigs or logs that appear to be part of the duff. Sometimes after a fire the litter and duff will be charred, but the cover of this charred litter/duff is estimated into the Charred Ground Cover field and not here. Other ground cover elements that are included in this category include plant fruits, buds, seeds, animal scat, and bones. If human litter appears on the plot, pick it up, throw it away, and do not include it in the ground cover estimate.

Field 60: Woody Ground Cover

Estimate the percent ground cover of all uncharred woody material using the codes in Table 6.32. Woody ground cover is only those wood particles that are recognizable as twigs, branches, or logs. Do not include cover of suspended woody material, such as dead branches connected on shrub or tree stems, into this field.

Field 61: Moss and Lichen Cover

Enter the percent canopy cover of all mosses and lichens on the plot using the codes in Table 6.32. These mosses and lichens can be on the ground or suspended from plants in the air (arboreal). This is the same estimate as in Field 43. The duplication is because some people consider moss and lichens ground cover and some consider them to be vegetation.

Field 62: Charred Ground Cover

Estimate the percent ground cover of all *charred organic* material using the codes in Table 6.32. Char is the blackened charcoal left from incomplete combustion of organic material. Char can occur on any piece of organic matter, such as duff, litter, logs, and twigs, and cover of all char is lumped into this category. Do not include ash with the charred ground cover. If it is difficult to distinguish char and black lichen, try to scrape the black area with your fingernail and then rub your nail onto your plot sheet. Char will usually leave a mark.

Field 63: Ash Ground Cover

Estimate the percent ground cover of all ash material using the codes in Table 6.32. Ash can sometimes look like mineral soil, but mineral surface feels sandy or gritty when touched, while ash will often feel like a powder. Ash can occur in a variety of colors (red, gray, white), but light gray is often the primary shade.

DID YOU KNOW?

The *down woody material (DWM) indicator* is a set of variables collected on Forest Inventory and Analysis (FIA) forest health plots. The DMW indicator is designed to estimate the biomass of forest ecosystem components not sampled during the FIA inventory. These biomass components include coarse woody debris, fine woody debris, duff, litter, shrubs/herbs, slash piles, and fuelbed depths. The DMW indicator provides the only nationally consistent and extensive inventory of down woody biomass. DMW data can be used to explore important fire, wildlife, and carbon questions. The DMW indicator serves not only as a data source for estimation and monitoring of down woody material biomass but also as a broad indicator of forest health (USFS, 2003d).

> **DID YOU KNOW?**
>
> The *vegetation indicator* is a set of variables collected on FIA forest health plots. It is designed to assess the type, abundance, and spatial arrangement of all trees, shrubs, herbs, grasses, ferns, and fern allies (horsetails and club mosses) occurring on the plots. Measuring vegetation allows us to report on the relative diversity of both native and introduced species. Information about the abundance and arrangement of species (structure) allows us to classify plots into community types. By remeasuring plots over time, we can monitor for change outside expected rates (USFS, 2003e).

Field 64: Basal Vegetation Ground Cover

Estimate the percent ground cover of basal vegetation using the codes in Table 6.32. Basal vegetation is the area of the cross-section of the stem where it enters the ground surface expressed as a percent of plot cover. This category is extremely difficult to estimate, but fortunately it has some repeatable characteristics. First, basal vegetation rarely exceeds 15% cover, so it will only get four valid FIREMON cover codes: 0, 0.5, 3, or 10. Next, it is highly ecosystem specific. Usually only forested ecosystems have high basal vegetation ground covers. This field is only used for vascular plant species. All nonvascular species are estimated in the Moss/Lichen Ground Cover field.

Field 65: Water Ground Cover

Estimate the percent ground cover of standing water using the codes in Table 6.32. Water ground cover includes rainwater puddles, ponding, runoff snow, ice, and hail. Do not include wet surfaces of other ground cover categories in this estimate. Although water is often only ephemeral, its cover must be recorded to make cover estimates sum to 100.

The remaining fields used in the Plot Description (PD) sampling method deal primarily with fire-related data such as general fuel characteristics and fire behavior and fire effects. Although these fields are important for U.S. Forest Service fire sampling and monitoring purposes, they are not applicable to forest biomass sampling/monitoring for energy production and are therefore not included in this text.

> **DID YOU KNOW?**
>
> Reserved forest land has tripled since 1953 and now stands at 10% of all forest land in the United States. This reserved forest land includes state and federal parks and wilderness areas but does not include conservation easements, areas protected by nongovernmental organizations, many wildlife management areas, and most urban and community parks and reserves. Significant additions to federal forest reserves occurred after passage of the Wilderness Act in 1964 (USFS, 2009).

Species Composition (SC) Sampling Method*

The Species Composition (SC) sampling method is designed to provide plant species composition and cover and height estimates to describe the plant community found on the FIREMON plot. This method uses a circular macroplot to record plant species characteristics. Cover, height, and optional user-specific attributes are recorded for each plant species or ground cover within the macroplot. Plant height is measured in feet or meters. This method is primarily used when the user wants to acquire inventory data over a large area using few examiners. The SC method is useful for documenting important changes in plant species and composition over time; however, this method is not designed to monitor statistically significant changes in vegetation over time due to the subjective nature of the cover estimations. The SC sampling method primarily addresses individual plant species cover and height for vascular and nonvascular plants, by size class.

Cover is an important vegetation attribute that is used to determine the relative influence of each species on a plant community. Cover is a commonly measured attribute of plant community composition because small, abundant species and large, rare species have comparable cover values. In this method, we record foliar cover as the vertical projection of the foliage and supporting parts on the ground; therefore, total cover on a plot can exceed 100% due to overlapping layers in the canopy. When cover is summed by size class, the total cover can equal more than 100% for a plant species due to overlap in the canopy between the different size classes.

Ocular estimates of cover are usually based on cover classes. The two most common are Daubenmire (1959) and Braun-Blanquet (1965). The range of cover values, 0 to 100%, is divided into classes, and each class is assigned a rating or number. In broadly defined cover classes, there is little chance for consistent human error in assigning the cover class (Daubenmire, 1959). The lowest cover classes are sometimes split into finer units (Bailey and Poulton, 1968; Jensen et al., 1994), because many species fall into the lowest cover classes. These systems are more sensitive to species with low cover. A finer breakdown of scale toward the lower scale values allows better estimation of less abundant species. In this method, we use a cover class system that splits the lowest classes into finer units. The midpoint of each class can be used for numerical computations. The use of midpoints for actual values is based on the assumption that actual cover values are distributed symmetrically about the midpoint. Plant height measurements are used to estimate the average height of individual species or species by size class. Plant heights give detailed information about the vertical distribution of plant species cover on the plot. In addition, height measurements allow the examiner to calculate plant species volume (cover × height) and to estimate biomass using the appropriate biomass equations based on cover and height. Plant height is measured with a yard stick (or meter stick) for small plants (<10 ft or 3 m) and with a clinometer for larger plants (>10 ft or 3 m).

Sampling Procedure

This method requires that the same preliminary sampling tasks, planning requirements, equipment outfitting, data logging recordings, and safety precautions have been completed, as was required with the sampling methods described earlier. However, there is a set of general criteria particular to the SC method that must be recorded on the SC Data

---

* Material in this section is based on Caratti, J.F., *Species Composition (SC) Sampling Method*, General Technical Report RMRS-GTR-164-CD, U.S. Department of Agriculture, Forest Service, Rocky Mountain Research Station, Fort Collins, CO, 2006.

Form. Moreover, each general SC field must be designed so that the sampling captures the information required to successfully complete the management objective within the time, money, and personnel constraints.

Designing the SC Sampling Method

Macroplot Size

The typical macroplot sampled in the SC method is a 0.10-ac (0.04-ha) circular plot having a radius of 37.2 ft. (1128 m). This plot size will be sufficient for most forest ecosystems and should be used if no other information is available. It is more efficient to use the same macroplot shape and size for all of the FIREMON sampling methods on the plot; for example, if you are using other FIREMON sampling methods (adapted for biomass sampling or personal purposes) that require a baseline and transects, then a rectangular plot of 66 ft × 66 ft (20 m × 20 m) would be used. Also, if the FIREMON DE method (described later) is being used to estimate density for some species or if ocular cover is being calibrated using the FIREMON CF method (described later) or LI method, then use a rectangular macroplot.

Plant Species ID Level

This field is used to determine the sampling-level intensity for the SC method. Enter the percent cover about which all plants are identified on the SC Data Form; for example, if you were interested in only sampling plants with at least 5% cover, then the number 5 would be entered. Entering a 0 (zero) in this field indicates that all plant species on the plot will be identified (i.e., a full species will be recorded). Because most changes in plant species cover occur in species that are already present on the plot, full species lists should be collected when feasible. Full species lists are especially useful if your data will be analyzed for biodiversity calculations, community classification, or species inventory. For example, because biodiversity calculations use the number of individual species as part of the calculation, it is important that each species be recorded.

Conducting the SC Sampling Tasks

Initial Plot Survey

Once the plot boundary is delineated, walk around the plot and become familiar with the plant species and vegetation layers. As you go, record species and ground cover codes you identify. Only record the items that you are interested in sampling. If you are only interested in monitoring the cover of noxious weeds, you would only record those species. After examining the macroplot, return to the center and start to record cover and height for all appropriate plant species as described below.

Sampling Cover

Starting with the first species on your list, enter the plant species status on the SC Data Form. Status describes the general health of the plant species as live or dead using the following codes:

L—Live (plant with living tissue)

D—Dead (plant with no living tissue visible)

NA—Not applicable

Plant status is purely qualitative but it does provide an adequate characteristic for stratification of plant health.

Size Class

Plant species size classes represent different layers in the canopy; for example, the upper canopy layer could be defined by large trees, while pole size trees and large shrubs might dominate the middle layer of the canopy, and the lower canopy layer could include seedlings, saplings, grasses, and forbs. Size class data provide important structural information, such as the vertical distribution of plant cover. Size classes for trees are typically defined by height for seedlings and diameter at breast height (DBH) for larger trees. Size classes for shrubs, grasses, and forbs are typically defined by height. If the vegetation being sampled has a layered canopy structure, then cover can be recorded by plant species and by size class. Total size class cover for a plant species can equal more than 100% due to overlap between different size classes. Group individual plants by species into one or more tree size classes (Table 6.17) or shrub, grass, and forb size classes (Table 6.18). If you are recording cover by size class, enter the size class code for each plant species on the SC Data Form. If size class data are not recorded, then record only the total cover for each plant species. When recording total cover for a species, use the code TO, for total cover.

Estimating Cover

Cover is the vertical projection of the foliage and supporting parts onto the ground. When estimating total cover for a plant species, do not include overlap between canopy layers of the same plant species. When estimating cover by size class for a plant species, the cover for each size class is recorded and includes canopy overlap between different size classes. Select one of the class codes from Table 6.32 to describe the cover estimate for the species. Enter the cover class code on the SC Data Form.

Measuring Average Height

Measure the average height for each plant species in feet (meters) within ±10% of the mean plant height. If plant species are recorded by size class, measure the average height for a plant species by each size class. Enter plant height on the SC Data Form.

Using the Optional Fields

There are two optional fields for user-defined codes or measurements; for example, codes can be entered to record plant species phenology or wildlife utilization of plant species. If standing dead trees (snags) were recorded on the SD Data Form, one could enter a decay class code in one of the optional fields. Enter the user-defined codes or measurements on the SC Data Form.

Precision Standards

Use the precision standards listed in Table 6.34 for the SC sampling.

TABLE 6.34

Precision Guidelines for Species Composition (SC) Sampling

Component	Standard
Size class	±1 class
Cover	±1 class
Height	±10% average height

Cover/Frequency (CF) Sampling Protocol*

Introduction

The Cover/Frequency (CF) method is designed to sample within-plot variation and quantify changes in plant species cover, height, and frequency over time. This method uses quadrats that are systematically placed along transects located within the macroplot. First, a baseline is established along the width of the plot, and transects are oriented perpendicular to the baseline and placed at random starting points along the baseline. Quadrats are then placed systematically along each transect. Characteristics are recorded about the general CF sample design and for individual plant species within each quadrat. First the transect length, number of transects, quadrat size, and number of quadrats per transect are recorded. Within each quadrat, depending on the project objectives, any combination of cover, frequency, and height is recorded for each plant species. This method is primarily used when the manager wants to monitor statistically significant changes in plant species cover, height, and frequency.

The CF sampling method is most appropriate for vascular and nonvascular plants less than 3 ft (1 m) in height. The Line Intercept (LI) method is better suited for estimating cover of shrubs greater than 3 ft (1 m) in height (e.g., Western U.S. shrub communities; mixed plant communities of grasses, trees, and shrubs; and open grown woody vegetation). The CF methods can also be used to estimate ground cover; however, the Point Intercept (PO) methods are better suited for estimating ground cover. It is probably best to use the PO method if you are primarily interested in monitoring changes in ground cover. The PO method may be used in conjunction with the CF method to sample ground cover by using the CF sampling quadrat as a point frame. The PO method is also better suited for sampling fine-textured herbaceous communities (e.g., dense grasslands and wet meadows); however, if rare plant species are of interest, the CF methods are preferred since it is easier to sample rare species with quadrats than with points or lines.

Estimating Cover and Height

Cover is an important vegetation attribute used to determine the relative influence of each species on a plant community. Vegetation can give important indicators of ecological processes occurring on a site and can also be a valuable management indicator for monitoring (Muir and McClaran, 1997). Cover is a commonly measured attribute of plant community composition because small, abundant species and large, rare species have comparable cover values. There are four types of cover:

- *Basal cover* is the proportion of the plant that extends into the soil. Generally, basal cover is more stable from year to year and less sensitive to changes due to climatic fluctuation. Moreover, it is not affected by utilization by grazing animals. Basal cover is usually used for trend comparisons or for calculation of species composition; however, it can be difficult to measure for forbs with a single, small stem.

---

* Material in this section is based on Caratti, J.F., *Cover/Frequency (CF) Sampling Method*, General Technical Report RMRS-GTR-164-CD, U.S. Department of Agriculture, Forest Service, Rocky Mountain Research Station, Fort Collins, CO, 2006.

- *Canopy cover* is an estimate of the area of influence of the plant. It includes the influence of the roots below the leaf/stem canopy and ignores gaps in the canopy. Canopy cover is a vertical projection of the outmost perimeter of the natural spread of foliage of plants. For any given area, the total canopy cover can exceed 100% because plants can overlap.
- *Ground cover* is the cover of the soil surface with plants, litter, biotic soils crust (i.e., lichens or mosses, or both), rocks, or gravel. Ground cover is most often used to determine the watershed stability of the site.
- *Foliar cover* is a vertical projection of exposed leaf area. The cover would equal the shadow cast if the sun was directly overhead. Small openings in the canopy or overlap within the plant are excluded. Can be difficult to measure except for some growth forms. Easiest to measure for forbs, succulents with large leaves, or shrubs.

For our purposes, we record foliar cover as the vertical projection of the foliage and supporting parts onto the ground; therefore, total cover on a plot can exceed 100% due to overlapping layers in the canopy. Estimating cover in quadrats is more accurate that estimating cover on a macroplot because samplers record cover in small quadrats more consistently than in large areas. Sampling with quadrats is also more effective than the Point Intercept method at locating and recording rare species. Point Intercept sampling requires many points to sample rare species (e.g., 200 points to sample at 0.5% cover). Quadrats sample more area and have a greater chance of detecting rare species.

Cover is typically based on a visual estimate of cover classes that range from 0 to 100%. These classes are broadly defined, lowering the chance for consistent human error in assigning the cover class. The lowest cover classes are sometimes split into finer units, because many species fall into the lowest cover classes. These systems are more sensitive to species with low cover. A finer breakdown of scale toward the lower scale values allows better estimation of less abundant species. In this text, we use a cover class system that splits the lowest classes into finer units. The midpoint of each class can be used for numerical computations. The use of midpoints for actual values is based on the assumption that actual cover values are distributed symmetrically about the midpoint.

Plant height measurements are used to estimate the average height of individual plant species. Plant heights give detailed information about the vertical distribution of plant species cover on the plot. In addition, height measurements allow the examiner to calculate plant species volume (cover × height) and to estimate biomass using the appropriate bulk density equations. Plant height is measured with a yardstick (meter stick) for plants less that 10 ft tall (3 m); taller plants are measured with a clinometer and tape measure.

Estimating Frequency

Frequency is a simple vegetation attribute used to describe the abundance and distribution of species and can be used to detect changes in vegetation over time. It is typically defined as the number of times a species occurs in the total number of quadrats sampled, usually expressed as a percent; for example, if a species has a frequency of 75%, we expect it to occur in three out of every four quadrats examined. Frequency is one of the fastest and easiest methods for monitoring vegetation because it is objective and repeatable and requires just one decision: whether or not a species is rooted within the quadrat frame. Frequency is a useful tool for comparing two plant communities or to detect changes within one plant community over time.

Frequency is most commonly measured with square quadrats. The size and shape of the frequency quadrat influence the results of the frequency recorded. If a plot is too small, rare plants may not be recorded. If you use a large quadrat, you will have individual species in all quadrats and frequency values of 100%, which will not allow you to track increases in frequency. If you have small quadrats, you will record low frequency values that are not very sensitive to declining frequency values for a species. A reasonable sensitivity to change results from frequency values between 20 to 80%. Frequencies less than 5% or greater than 95% typically result in heavily skewed distributions.

For this reason, nested plots, or subplots, are usually used to sample frequency. Plot sizes are nested in a configuration that gives frequencies between 20 and 80% for the majority of species. Nested subplots allow frequency data to be collected in different size subplots of the main quadrat. Because frequency of occurrence can be analyzed for different sized plots, this eliminates the problems of comparing data collected from different size quadrats. In this text, we use a nested plot design of four subplots within one quadrat and record the smallest subplot number in which the plant is rooted. This frequency measurement is typically referred to as nested rooted frequency (NRF).

Because plant species frequency is highly sensitive to the size and shape of quadrats, changes in frequency can be difficult to interpret, possibly resulting from changes in cover, density, or pattern of distribution. For this reason, if money and time are available it is best to collect cover data along with frequency data; however, if the main concern is to document any change in vegetation that may have occurred, then frequency is the most rapid method.

Sampling Procedure

As with the other sampling protocols discussed to this point, the Cover/Frequency sampling method entails the use of a sampling strategy and the performance of preliminary sampling tasks before the actual sampling is conducted. If the standard FIREMON sampling strategy is used, then the sampling procedure is described in the order of the fields that need to be completed on the CF form. If a modified Cover/Frequency sampling method is used to determine the extent of biomass in the sampled area, the procedure will be modified to fit the need and the goal of the method used. When using either procedure, however, basically the same preliminary sampling tasks and equipment and safety precautions must be planned for or considered.

Designing the CF Sampling Method

As with any vegetation sampling activity, the first step is to design a protocol that can accomplish the assessment or monitoring goal with available time and resources. Once this is set, the CF sampling method can be conducted and general data can be recorded on the CF Data Form.

Determining the Sample Size

The size of the macroplot ultimately determines the length of the transects and the length of the baseline along which the transects are placed. The amount of variability in plant species composition and distribution determines the number and length of transects and the number of quadrats required for sampling. The typical macroplot sampled in the CF method is a 0.10-ac (0.04 ha) square measuring 66 × 66 ft (20 × 20 m), which is sufficient for most forest understory and grassland monitoring applications. Shrub-dominated ecosystems will generally require larger macroplots when sampling with the CF method. If you are not sure of the plot size to use, contact someone who has sampled the same vegetation that you will be sampling. The size of the macroplot may be adjusted to accommodate different numbers and lengths of transects. In general, it is more efficient if you use the same sample plot size for all of the sampling methods on the plot; however, this is not always feasible. Sampling five transects within the macroplots should be sufficient for most studies, although in some situations more transects should be sampled. Enter the number of transects on the CF Data Form. Again, the recommended transect length is 66 ft (20 m) for a 66 × 66 ft (20 × 20 m) macroplot, although the macroplot size may be adjusted to accommodate longer or shorter transects based on variability in plant species composition and distribution. Transects may be lengthened to accommodate more quadrats per transect or to allow more distance between quadrats. Enter the transect length on the CF Data Form.

DID YOU KNOW?

When selecting the appropriate quadrat size, we need to ensure that the quadrats are large enough to contain at least one plant of interest and should include enough plants to get a good estimate of density. Conversely, the quadrat must be small enough that the count is conducted in a reasonable amount of time. In other words, you don't want to measure hundreds of individuals per quadrat. A standard rule of thumb on quadrat size: If more than 5% of sampling units have none of the plants of interest, increase the plot size.

Determining the Quadrat Size

Quadrats are required for estimating several vegetation attributes, including cover and frequency. With regard to frequency, it is typically recorded in square quadrats. The standard quadrat for measuring nested rooted frequency is a 20 × 20 in. (50 × 50 cm) square with four nested subplot sizes. A nested frame allows frequency data to be collected in different size subplots of the main quadrat. Measuring frequency this way is commonly referred to as nested rooted frequency (NRF). Plot sizes are nested in a configuration that gives frequencies between 20 to 80% for the majority of species. Table 6.35 and Table 6.36 list common quadrat and subplot sizes in English and metric dimensions for recording nested rooted frequency on the CF Data Form. Cover can be estimated using the same quadrat frames and recorded at the same time frequency is recorded. Enter the quadrat length and quadrat width in inches (centimeters) on the CF Data Form.

Recording the Subplot Size Ratio and NRF Numbers

If nested rooted frequency is being recorded, then enter the percent area of the quadrat contained by each subplot on the CF Data Form. Start with the smallest subplot and end with the largest subplot. For example, the subplot ratio for the standard 20 × 20 inch (50 × 50 cm) quadrat would be 1:25:50:100. Subplot 1 is 2 × 2 in. (5 × 5 cm) and is 1% of the quadrat. Subplot 2 is 10 × 10 (25 × 25 cm) and is 25% of the quadrat. Subplots 3 and 4 are 10 × 20 in. (25 × 50 cm) and 20 × 20 in. (50 × 50 cm), which correspond to 50 and 100% of the quadrat, respectively. If nested rooted frequency is being recorded, then enter the corresponding frequency numbers of each subplot on the CF Data Form. Start numbering with the smallest subplot and end with the largest subplot; for example, 1:2:3:4 would correspond to the 1:25:50:10 percentages of total plot when using the standard 20 × 20 inch (50 × 50 cm) quadrat.

TABLE 6.35

Commonly Used Quadrat Sizes for Recording Nested Rooted Frequency (English)

NRF Numbers	Standard Communities	Grassland Communities	Sagebrush	Pinyon–Juniper
Subplot 1	2 × 2 in.	—	2 × 2 in.	—
Subplot 2	10 × 10 in.	2 × 2 in.	4 × 4 in.	8 × 8 in.
Subplot 3	10 × 20 in.	4 × 4 in.	8 × 8 in.	20 × 20 in.
Subplot 4	0 × 20 in.	8 × 8 in.	20 × 20 in.	40 × 40 in.

TABLE 6.36

Commonly Used Quadrat Sizes for Recording Nested Rooted Frequency (Metric)

NRF Numbers	Standard Communities	Grassland Communities	Sagebrush	Pinyon–Juniper
Subplot 1	5 × 5 cm	—	5 × 5 cm	—
Subplot 2	25 × 25 cm	5 × 5 cm	10 × 10 cm	20 × 20 cm
Subplot 3	25 × 50 cm	10 × 10 cm	20 × 20 cm	50 × 50 cm
Subplot 4	50 × 50 cm	20 × 20 cm	50 × 50 cm	100 × 100 cm

Conducting CF Sampling Tasks

Establishing the Baseline for Transects

Once the plot has been monumented, a permanent baseline is set up as a reference from which you will orient all transects. The baseline should be established so that the sampling plots for all of the methods overlap as much as possible. The recommended baseline is 66 ft (20 m) long and is oriented upslope with the 0-ft (0-m) mark at the lower permanent marker and the 66-ft (20-m) mark at the upper marker. On flat areas, the baseline runs from south to north with the 0-ft (0-m) mark on the south end and the 66-ft (20-m) mark on the north end. Transects are placed perpendicular to the baseline and are sampled starting at the baseline. On flat areas, transects are located from the baseline to the east.

Locating the Transects

Locate transects within the macroplot perpendicular to the baseline and parallel with the slope. For permanent plots, determine the compass bearing of each transect and record these on the plot layout map or the comment section of the CF Data Form. Permanently mark the beginning and ending of each transect (e.g., using concrete reinforcing bar). Starting locations for each transect can be determined randomly on every plot or systematically with the same start locations used on every plot in the project. In successive remeasurement years, it is essential that transects be placed in the same locations as in previous visits. If the CF method is used in conjunction with other replicated sampling methods (LI, PO, RS, or DE), use the same transects for all methods, whenever possible.

Locating the Quadrats

Sample five quadrats located at 12-ft (4-m) intervals along a transect, with the first quadrat placed 12 ft (4 m) from the baseline. If macroplots are being sampled for permanent remeasurement, quadrats must be placed at specified intervals along a measuring tape placed along each transect. When sampling macroplots that are not scheduled for permanent remeasurement, the distance between quadrats maybe estimated by pacing after the examiner measures the distance between quadrats. Each quadrat is placed on the uphill side of the transect line with the quadrat frame placed parallel to the transect. The lower left corner of the quadrat frame will be placed at the foot (meter) mark for the quadrat location. Figure 6.15 displays the proper placement of a quadrat frame.

Quadrat Sampling

First, enter the transect number being sampled on the CF Data Form. Next, enter the plant species or item code. We use the NRCS Plants Database species codes; however, you may use your own species codes. If ground cover is being sampled, use the ground cover codes in Table 6.21. Next enter the plant species status on the CF Data Form. Status describes the general health of the plant species as live or dead using the following codes:

L—Live (plant with living tissue)

D—Dead (plant with no living tissue visible)

NA—Not applicable

Plant status is purely qualitative. Be careful when making this assessment on plants in their dormant season.

Cover

As mentioned earlier, cover is the vertical projection of the vegetation foliage and supporting parts onto the ground. Estimating cover within quadrats is made easier by using subplot sizes and the percent of quadrat area they represent. Subplots are used to estimate cover for a plant species by mentally grouping cover for all individuals of a plant species into one of the subplots. The percent size of that subplot, in relation to the size of the quadrat being sampled, is used to make a cover class estimate for the species. For each plant species or ground cover class in the quadrat, estimate its percent cover within the quadrat and enter a cover class code (Table 6.32) to denote that value. Enter the class for each quadrat.

Nested Rooted Frequency

The standard 20×20 in. (50×50 cm) quadrat is partitioned into four subplots for recording nested rooted frequency (see Figure 6.16 and Table 6.37). Species located in the smallest subplot are given the frequency value of 1. Plants in successively larger subplots have frequency values of 2, 3, and 4. Decisions about counting boundary plants, plants that have a

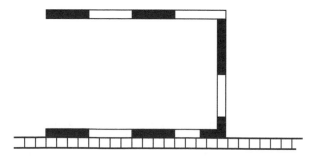

FIGURE 6.15
An example of quadrat placement along a transect.

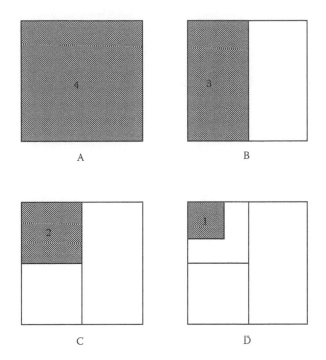

FIGURE 6.16

The numbers inside the plot frame denote the value recorded if a plant is present in that area of frame. The number 4 corresponds to the entire quadrat (A). The sampling area for number 3 is the entire top half of the quadrat (B). The sampling areas for the numbers 2 and 1 are the upper left quarter and the upper left corner (1%) of the quadrat, respectively (C and D). Each larger subplot contains all smaller subplots. Subplots aid the sample in estimating cover by mentally grouping cover for all individuals of a plant species into one of the subplots.

portion of basal vegetation intersecting the quadrat, need to be applied systematically to each quadrat. Record the smallest size subplot in which each plant species is rooted. Begin with subplot 1, the smallest subplot. If the basal portion of a plant species is rooted in that subplot, record 1 for the species. Next find all plant species rooted in subplot 2 that were not previously recorded for subplot 1, and record a 2 for these plant species. Then identify all plant species rooted in subplot 3 that were not previously recorded for subplots 2 and 1, and record 3 for these species. Finally, record a 4 for each species rooted in subplot 4, the remaining half of the quadrat, that were not previously recoded in subplots 3, 2, and 1. Enter the subplot number in the NRF field for each species on the CF Data Form.

TABLE 6.37

Percent of Quadrat Represented by the Four Subplots Used to Record Nested Rooted Frequency within Standard 20 × 20 in. (50 × 50 cm) Quadrat

Subplot Number for Rooted Frequency	Size of Subplot	Percent Area of a 20 × 20 in. (50 × 50 cm) Quadrat
1	2 × 2 in. (5 × 5 cm)	1%
2	10 × 10 in. (25 × 25 cm)	25%
3	10 × 20 in (25 × 50 cm)	50%
4	20 × 20 in. (50 × 50 cm)	100%

Estimating Height

Measure the average height for each plant species in feet (meters) within ±10% of the mean plant height. Enter plant height in the Height field for each quadrat.

Precision Standards

Use the precision standards listed in Table 6.38 for CF sampling.

TABLE 6.38

Precision Guidelines for Cover/Frequency (CF) Sampling

Component	Standard
Cover	±1 class
NRF	No error
Height	±10%

Density (DE) Sampling Method*

Introduction

The Density (DE) sampling method was designed to sample within-plot variation and quantify changes in plant species density and height over time. This method uses multiple quadrats to sample herbaceous plant density and belt transects to sample shrub and tree density. First, a baseline is established from which to run the transects. Transects are placed randomly along the baseline. Quadrats for sampling herbaceous plants are then placed systematically along each transect. Belt transects for sampling shrub and tree density are placed along the length of each transect. Characteristics are recorded about the general sample design and for individual plant species. First the transect length, number of transects, and number of quadrats per transect are recorded. Within each quadrat or belt transect, density and average height are recorded for each plant species. The quadrat length and width or belt transect width are also recorded for each species. Different size quadrats and belts can be used for different plant species.

This method is used primarily when the biomass production manager wants to monitor changes in plant species numbers. This method is best suited for grasses, forbs, shrubs, and small trees that are easily separated into individual plants or counting units, such as stems; however, the author recommends using the modified FIREMON Tree Data sampling method, described earlier. Based on experience, it seems to be the best methodology for measuring tree density. This method uses density to assess changes in plant species numbers over time. The quadrat size and belt width vary with plant size or item and size class, allowing different size sampling units for different size plants or items. Quadrat size and belt width should be adjusted according to plant size and distribution.

Though the term *abundance* is often used synonymously with *density*, density is unique because it is specifically defined as the number of items per unit area; that is, it is related to an amount of spaces (e.g., trees per acre or plants per m²). When sampling density, the examiner must be able to recognize and define individual plants. This may be relatively easy for single stemmed plants, but it is more difficult for plants that reproduce vegetatively (not sexually) such as rhizomatous plants (e.g., western wheatgrass) or clonal species (e.g., aspen). It is critical to define exactly what item will be counted before sampling. The item counted may be individual plants for single-stem plants or individual stems for a clonal species such as quaking aspen.

---

* Material in this section is based on Caratti, J.F., *Density (DE) Sampling Method*, General Technical Report RMRS-GTR-164-CD, U.S. Department of Agriculture, Forest Service, Rocky Mountain Research Station, Fort Collins, CO, 2006.

Density is used for monitoring an increase or decrease in the number of individuals or counting units. Density is more effective for detecting changes in recruitment or mortality than changes in vigor. It is not a practical monitoring method when plants respond to management treatments or disturbance with decreased cover rather than mortality. In such cases, density may not change much, although cover and biomass may change considerably.

The accuracy of the density estimate depends largely on the size and shape of the quadrat or belt transect. Note that a belt transect is essentially a long, narrow quadrat. Pound and Clements (1898) considered plant dispersion, quadrat size and shape, and number of observations required as being important characteristics of sample design. Van Dyne et al. (1963) reviewed results from studies on quadrat sizes and shapes used to sample grassland communities of the Western United States. Long, narrow quadrats tend to include more species, as vegetation tends to occur in clumps rather than be randomly distributed. The desired size and shape for a quadrat depends largely on the distribution of plant species being sampled. In general, sampling in sparse vegetation requires the use of larger quadrats; however, quadrats should not be too large, since counting large numbers of plants in a quadrat can be overwhelming and can lead to errors. In order to increase the sampling area, it is better to sample more quadrats than to use overly large quadrats.

With small quadrats, there is a greater chance of boundary error because of the greater perimeter-to-area ratio. Boundary problems are caused by erroneously including or excluding plants near the quadrat perimeter. Some portion of basal vegetation must intersect the quadrat boundary for a plant to be considered a boundary plant. Boundary rules must be established before sampling. Good boundary rules for forest-based biomass production involve counting boundary plants as in on two adjacent sides of a quadrat and as out on the other adjacent sides.

Sampling Procedure

Again, as with the other sampling protocols discussed to this point, the Density sampling method entails the use of a sampling strategy and the performance of preliminary sampling tasks before the actual sampling is conducted. If the standard FIREMON sampling strategy is used, then the sampling procedure is described in the order of the fields that need to be completed on the DE form. If a modified FIREMON Density sampling method is used to determine extent of biomass in the sampled area, the procedure will be modified to fit the need and the goal of the method used. Whichever procedure is used, basically the same preliminary sampling tasks, equipment, and safety precautions must be planned for or considered.

Designing the DE Sampling Design

A set of general criteria recorded on the DE Data Form provides the basis for user-specified design of the DE sampling method. Each general DE field must be designed so the sampling captures the information required to successfully complete the management objectives within the time, money, and personnel constraints. These general fields should be determined before the crews go into the field and should reflect a thoughtful analysis of the expected problems and challenges in the forest-based biomass monitoring project.

Plot ID Construction

A unique plot identifier must be entered on the DE sampling form. This is the same plot identifier used to describe general plot characteristics in the Plot Description (PD) sampling method. Enter the plot identifier at the top of your DE Data Form.

Determining Sample Size

The size of the macroplot ultimately determines the length of the transects and the length of the baseline along which the transects are placed. The amount of variability in plant species composition and distribution determines the number and length of transects and the number of quadrats required for sampling. The typical macroplot sampled in the DE method is a 0.10-ac (0.04-ha) square measuring 66 × 66 ft (20 × 20 m), which is sufficient for most monitoring applications. The size of the macroplot may be adjusted to accommodate different numbers and lengths of transects; however, it is more efficient if you use the same plot size for all modified FIREMON sampling methods on the plot. When sampling shrubs and trees, practice has shown that it is best to sample five belt transects within the macroplot. This should be sufficient for most studies, although in some situations more transects should be sampled. Enter the number of transects on the DE Belt Transect Data Form. The recommended transect length is 66 ft (20 m) for a macroplot that is 66 × 66 ft (20 × 20 m); however, the macroplot size may be adjusted to accommodate longer or shorter transects based on the variability in plant species composition and distribution.

Determining the Belt Transect Size and Quadrat Size

Density is recorded in quadrats for herbaceous species and in belt transects for shrubs. We recommend using a 66 × 3 ft (20 × 1 m) belt width for sampling smaller shrubs (<3 ft or 1 m average diameter) and a 66 × 6 ft (20 × 2 m) belt width for larger shrubs (>3 ft or 1 m average diameter). Belt length and width may be adjusted to accommodate different sizes and densities of shrubs and trees. Longer, wider transects can be sampled for larger or sparsely distributed shrubs and trees, and shorter, narrower transects can be sampled for smaller, more dense shrubs and trees. Enter the belt transect length and width on the DE Transect Data Form. You may vary the belt length and width for different plant species you encounter on the macroplot, but enter the appropriate length and width for each species on the Data Form. Experience demonstrates that using 3 × 3 ft (1 × 1 m) quadrats for sampling herbaceous vegetation is best; however, quadrat size and transect length may be adjusted to accommodate the size and spacing of the plants being sampled. Larger plans can be counted in larger quadrats on longer transects and smaller plants in smaller quadrats on shorter transects. Enter the quadrat length and wide in feet (meters) on the DE Quadrat Data Form.

Conducting DE Sampling Tasks

As with preceding sampling protocol methodologies, once the plot DE has been monumented, a permanent baseline is set up as a reference from which you will orient all transects. The baseline should be established so the sampling plots for all of the methods overlap as much as possible. The recommended baseline is 66 ft (20 m) long and is oriented upslope with the 0-ft (0-m) mark at the lower permanent marker and the 66-ft (20-m) mark at the upper marker. On flat areas, the baseline runs from south to north with the 0-ft (0-m) mark on the south end and the 66-ft (20-m) mark on the north end.

Locating the Transects

Transects are placed perpendicular to the baseline and are sampled starting at the baseline. On flat areas, transects are located from the baseline to the east. Starting locations for each transect are determined using the FIREMON random transect locator or from supplied tables. If the CF method is used in conjunction with other replicated sampling methods (LI, PO, RS, or CF), use the same transects for all methods, if possible. In successive remeasurement years, it is essential that transects be placed in the same location.

Locating the Quadrats

There will typically be five 3 × 3 ft (1 × 1 m) quadrats located at 12-ft (4-m) intervals along a transect, with the first quadrat placed 12 ft (4 m) from the baseline; however, the spacing of the quadrats will depend on the size of the quadrat and the length of the transect. If macroplots are being sampled for permanent remeasurement, quadrats must be placed in the same location in successive sampling. When sampling macroplots that are not scheduled for permanent remeasurement, the distance between quadrats may be estimated by pacing.

Density Sampling

When sampling herbaceous species, enter the transect number on the DE Quadrat Data Form. The transect number is not entered on the DE Transect Data Form for sampling shrub or trees species using belt transects. Enter the species code or item code (e.g., moose pellets) on the DE Quadrat Option Data Form on the Transect Option Data Form. Modified FIREMON uses the NRCS Plants Database species codes; however, you may use your own species codes. Next enter the plant species status on the DE Quadrat Data Form and on the DE Transect Data Form. Status describes the general health of the plant species as live or dead using the following codes:

L—Live (plant with living tissue)

D—Dead (plant with no living tissue visible)

NA—Not applicable

Plant status is purely qualitative, but it does provide an adequate characteristic for stratification of plant health. Use care in determining plant status during the dormant season.

Size Class

Plant species size classes represent different layers in the canopy. The upper canopy layer, for example, could be defined by large trees, while pole-size trees and large shrubs might dominate the middle layer of the canopy, and the lower canopy layer could include seedlings, saplings, grasses, and forbs. Size class data provide important structural information, such as the vertical distribution of plant cover. Size classes or trees are typically defined by height for seedlings and diameter at breast height (DBH) for larger trees. Size classes for shrubs, grasses, and forbs are typically defined by height. If the vegetation being sampled has a layered canopy structure, the density can be recorded by plant species and by size class. FIREMON uses a size class stratification based on the ECODATA sampling methods (Jensen et al., 1994). Group individual plants by species into one or more trees size classes (Table 6.17) or shrub, grass, and forb size classes (Table 6.18). Each species can have multiple size classes. If recording density by size class, enter the size class code for each plant species on the DE Quadrat Data Form and on the DE Transect Data Form. If size class data are not recorded, indicate that by entering the code use the code TO, for total cover.

Density

Record the number of individual plants for each plant species or individual plants for each plant species by size class within the quadrat. Decisions about counting boundary plants (plants for which a portion of their basal vegetation intersects the quadrat) need to be applied systematically to each quadrat. Enter the count for each plant species or plant species by size class in the count field by quadrat on the DE Quadrat Data Form or by transect on the DE Transect Data Form. Use the workspace below each Count field for a dot tally.

TABLE 6.39

Precision Guidelines for Density (DE) Sampling

Component	Standard
Count	±10% total count
Average height	±10% average height
Size class	±1 class

Estimating Average Height

Measure the average height for each plant species in feet (meters) within ±10% of the mean plant height. If plant species are recorded by size class, measure the average height for individual species by each size class recorded. Enter the plant height in the Height field for each transect on the DE Transect Data Form or for each quadrat on the DE Quadrat Data Form.

Precision Standards

Use the precision standards listed in Table 6.39 for the DE sampling.

Landscape Assessment (LA) Sampling Method*

Because Landscape Assessment (LA) sampling method addresses the need to identify and quantify fire effects over large areas, at times involving many burns, the methodology is not applicable to forest-based biomass energy production and is therefore not discussed in detail in this text. As pointed out, the previously presented modified FIREMON sampling and analysis methods can be applied to forest-product monitoring, surveying, and inventorying and as such were explained in some detail. Landscape Assessment is used to obtain a landscape perspective on fire effects—spatial data on burn severity throughout a whole burn. Landscape assessment methods show the results of fire in the context of regional biophysical characteristics, such as topography, climate, vegetation, hydrography, fuels, and soil. These results can be used to isolate burned from unburned surroundings, measure the amount burned at various levels of effect, and gauge the spatial heterogeneity of the burn.

Chapter Review Questions

6.1   A forested tract is stratified into three forest cover types, with the following percent of total area in each class:

Class I: 26%

Class II: 32%

Class III: 42%

---

* Material in this section is based on Key, C.H. and Benson, N.C., *Landscape Assessment (LA) Sampling and Analysis Methods*, General Technical Report RMRS-GTR-164-CD, U.S. Department of Agriculture, Forest Service, Rocky Mountain Research Station, Fort Collins, CO, 2006.

   a. How will a sample of 50 plots be allocated if using stratified random sampling?

   b. Simple random sampling?

6.2 The total _____ area of all trees or specified classes of trees per unit area is a useful characteristic of a forest stand.

6.3 How well forest sampling is done depends on what?

6.4 When sample selection is allowed with replacement, each unit is allowed to appear in the sample as _____ as it is selected.

6.5 Small plots will be almost always be more _____ than large ones.

6.6 The _____ indicates the average change in volume per unit change in area between the sampling units in the sample and the population.

6.7 _____ is commonly used when the population mean or total of the supplementary variable is unknown.

References and Recommended Reading

Avery, T.E. and Burkhart, H.E. (2002). *Forest Measurements*, 5th ed., McGraw-Hill, New York.

Bailey, A.W. and Poulton, C.E. (1968). Plant communities and environmental relationships in a portion of the Tillamook Burn, northwest Oregon, *Ecology*, 49:114.

Barbour, M.G., Burk, J.H., and Pitts, W.D. (1987). *Terrestrial Plant Ecology*, Benjamin/Cummings, San Francisco, CA, pp. 182–208.

Blackman, G.E. (1935). A study of statistical methods of the distribution of species in grassland associations, *Annals of Botany*, 49:749–777.

Bonham, C.D. (1989). *Measurements for Terrestrial Vegetation*, John Wiley & Sons, New York.

Braun-Blanquet, J. (1965). *Plant Sociology: The Study of Plant Communities*, Hafner Publishing, London.

Brown, J.K. (1974). *Handbook for Inventorying Downed Woody Material*, General Technical Report INT-16, U.S. Department of Agriculture, Forest Service, Intermountain Forest and Range Experiment Station, Ogden, UT, 24 pp.

Busing, R., Rimar, K., Solte, K.W., and Stohlgren, T.J. (1999). *Forest Health Monitoring: Vegetation Pilot Field Methods Guide: Vegetation Diversity and Structure Down Woody Debris Fuel Loading*, National Forest Health Monitoring Program, Research Triangle Park, NC.

Canfield, R.H. (1941). Application of the line interception method in sampling range vegetation, *Journal of Forestry*, 39:388–394.

Cochran, W.G. (1953). *Sampling Techniques*, Wiley, New York.

Cole, W.P. (1977). *Using the UTM Grid System to Record Historic Sites*, U.S. Department of Interior, Washington, D.C.

Cook, C.W. and Stubbendieck, J. (1986). *Range Research: Basic Problems and Techniques*, Society for Range Management, Denver, CO.

Coulloudon, B., Podborny, P., Eshelman, K., Rasmussen, A., Gianola, J., Robels, B., Habich, N., Shaver, P., Hughes, L., Spehar, J., Johnson, C., Willoughby, J., and Pellant, M. (1999). *Sampling Vegetation Attributes*, Technical Reference 1734-4, U.S. Department of Interior, Bureau of Land Management, Denver, CO, 164 pp.

Daubenmire, R.R. (1959). A canopy-coverage method, *Northwest Science*, 33:43–64.

Despain, D.W., Ogden, P.R., and Smith, E.L. (1991). Plant frequency sampling for monitoring rangelands, in Ruyle, G.B., Ed., *Some Methods of Monitoring Rangelands and Other Natural Area Vegetation*, Extension Report 9043, University of Arizona, College of Agriculture, Tucson.

Diggle, P.J. (1975). Robust density estimation using distance methods, *Biometrika*, 62:43–64.

Dilworth, J.R. (1989). *Log Scaling and Timber Cruising*, Oregon State University, Corvallis.

Dilworth, J.R. and Bell, J.F. (1973). *Variable Probability Sampling: Variable Plot and 3P*, Oregon State University, Corvallis.

Elzinga, C.I., Salzer, D.W., and Willoughby, J.W. (1998). *Measuring and Monitoring Plant Populations*, Technical Reference 1730-1, U.S. Department of Interior, Bureau of Land Management, Denver, CO.

Evans, R.A. and Love, R.M. (1957). The step-point method of sampling—a practical tool in range research, *Journal of Range Management*, 10:208–212.

Floyd, D.A. and Anderson, J.E. (1987). A comparison of three methods for estimating plant cover, *Journal of Ecology*, 75:229–245.

Freese, F. (1976). *Elementary Forest Sampling*, Agriculture Handbook No. 232, U.S. Department of Agriculture, Washington, D.C.

Goebel, C., DeBano, L.F., and Lloyd, R.D. (1958). A new method of determining forage cover and production on desert shrub vegetation, *Journal of Range Management*, 11:244–246.

Goldsmith, F.B., Harrison, C.M, and Morton, A.J. (1986). Description and analysis of vegetation in methods, in Moore, P.D. and Chapman, S.B., Eds., *Plant Ecology*, Blackwell Scientific, New York, pp. 437–424.

Goodall, D.W. (1952). Some considerations in the use of pint quadrats for the analysis of vegetation, *Australian Journal of Scientific Research*, 5:1–41.

Greig-Smith, P. (1983). *Quantitative Plant Ecology*, University of California Press, Berkeley, pp. 1–53.

Hanley, T.A. (1978). A comparison of the line interception and quadrat estimation methods of determining shrub canopy coverage, *Journal of Range Management*, 31:60–62.

Hansen, H.C. (1962). *Dictionary of Ecology*, Crown, New York, 382 pp.

Hironaka, M. (1985). Frequency approaches to monitor rangeland vegetation, in Krueger, W.C., Chairman, *Proceedings of the 38th Annual Meeting*, Society for Range Management, February 11–14, Salt Lake City, UT, pp. 84–86.

Howard, J.O. and Ward, F.R. (1972). *Measurement of Logging Residue—Alternative Applications of the Line Transect Method*, Research Note PNW-183, U.S. Department of Agriculture, Forest Service, Pacific Northwest Research Station, Portland, OR, 6 pp.

Husch, B., Miller, C.I., and Beers, T.W. (1982). *Forest Mensuration*, John Wiley & Sons, New York.

Hyder, D.N., Bement, R.E., Remmenga, E.E., and Terwilliger, Jr., C. (1965). Frequency sampling of blue gramma range, *Journal of Range Management*, 18:90–94.

Jasmer, G.E. and Holechek, J. (1984). Determining grazing intensity on rangelands, *Journal of Soil and Water Conservation*, 39(1):32–35.

Jensen, M.E., Hann, W.H., Keane, R.E., Caratti, J., and Bourgeron, P.S. (1994). ECODATA—A multire-source database and analysis system for ecosystem description and analysis, in Jensen, M.E. and Bourgeron, P.S., Eds., *Eastside Forest Ecosystem Health Assessment*. Vol. II. *Ecosystem Management: Principles and Applications*, General Technical Report GTR-PNW-318, U.S. Department of Agriculture, Forest Service, Portland, OR, pp. 203–216.

Kennedy, K.A. and Addison, P.A. (1987). Some considerations for the use of visual estimates of plant cover in biomonitoring, *Journal of Ecology*, 75:151–157.

Krebs, C.J. (1989). *Ecological Methodology*, Harper & Row, New York.

Lapin, L.L. (1993). *Statistics for Modern Business Decisions*, 6th ed., Dryden Press, Orlando, FL.

Laycock, W.A. (1987). *Setting Objectives and Picking Appropriate Methods for Monitoring Vegetation on Rangelands: Rangeland Monitoring Workshop Proceedings*, U.S. Department of Interior, Bureau of Land Management, Golden, CO.

Lucas, H.A. and Seber, G.A.F. (1977). Estimating coverage and particle density using the line intercept method, *Biometricka*, 64:618–622.

Lutes, D.C. et al. (2006). *FIREMON: Fire Effects Monitoring and Inventory System*, General Technical Report RMRS-GTR-164-CD, U.S. Department of Agriculture, Forest Service, Rocky Mountain Research Station, Fort Collins, CO.

Muir, S. and McClaran, M.P. (1997). *Rangeland Inventory, Monitoring, and Evaluation*, University of Arizona, Tucson (http://rangelandswest.arid.arizona.edu/rangelandswest/jsp/module/az/inventorymonitoring/frequency.jsp).

Nudds, T.D. (1977). Quantifying the vegetative structure of wildlife cover, *Wildlife Society Bulletin*, 5:113–117.

Pechanec, J.F. and Pickford, G.D. (1937). A weight-estimate method for the determination of range or pasture protection, *Journal of the American Society of Agronomy*, 29:894–904.

Pound, P. and Clements, F.E. (1989). A method of determining the abundance of secondary species, *Minnesota Botanical Studies*, 2:19–24.

Riser, P.G. (1984). *Methods for Inventory and Monitoring of Vegetation, Litter, and Soil Surface Condition: Developing Strategies for Rangeland Monitoring*, National Research Council, National Academy of Sciences, Washington, D.C.

Robel, R.J., Briggs, J.N., Dayton, A.D., and Hulbert, L.C. (1970). Relationships between visual obstruction measurements and weight of grassland vegetation, *Journal of Range Management*, 23:295.

Salzer, D. (1994). *An Introduction to Sampling and Sampling Design for Vegetation Monitoring*, unpublished paper prepared for Training Course 173-5, U.S. Department of Interior, Bureau of Land Management Phoenix, AZ.

Schornaker, M. (2003). *Tree Crown Condition Indicator*, FIA Fact Sheet Series, U.S. Department of Agriculture, Forest Service, Arlington, VA.

Schreuder, H.T. Gregoire, T.G., and Wood, G.B. (1993). *Sampling Methods for Multiresource Forest Inventory*, John Wiley & Sons, New York.

Shivers, B.D and Borders, B.E. (1996). *Sampling Techniques for Forest Resource Inventory*, John Wiley & Sons, New York.

Spalinger, D.E. (1980). *Vegetation Changes on Eight Selected Deer Ranges in Nevada Over a 15-Year Period*, Bureau of Land Management, Nevada State Office, Reno.

Stanton, F.W. (1960). Ocular point frame, *Journal of Range Management*, 13:153.

Tansley, A.G. and Chipp, T.F., Eds. (1926). *Aims and Methods in Study of Vegetation*, British Empire Vegetation Committee, London, 383 pp.

USDA. (1976). *National Range Handbook*, U.S. Department of Agriculture, Soil Conservation Service, Washington, D.C.

USFS. (2003a). *Fact Sheet: Tree Mortality*, U.S. Department of Agriculture, Forestry Service, Arlington, VA (www.fia.fs.fed.us).

USFS. (2003b). *Fact Sheet: Tree Growth*, U.S. Department of Agriculture, Forestry Service, Arlington, VA (www.fia.fs.fed.us).

USFS. (2003c). *Fact Sheet: Soil Quality Indicator*, U.S. Department of Agriculture, Forestry Service, Arlington, VA (www.fia.fs.fed.us).

USFS. (2003d). *Fact Sheet: Down Woody Materials Indicator*, U.S. Department of Agriculture, Forestry Service, Arlington, VA (www.fia.fs.fed.us).

USFS. (2003e). *Fact Sheet: Vegetation Indicator*, U.S. Department of Agriculture, Forestry Service, Arlington, VA (www.fia.fs.fed.us).

USFS. (2005). *Fact Sheet: Lichen Communities Indicator*, U.S. Department of Agriculture, Forestry Service, Arlington, VA (www.fia.fs.fed.us).

USFS. (2009). *U.S. Forest Resource Facts and Historical Trends*, U.S. Department of Agriculture, Forestry Service, Arlington, VA (http://fia.fs/fed/us/).

Van Dyne, G.M., Vogel, W.G., and Fisser, H.G. (1963). Influence of small plot size and shape on range herbage production estimates, *Ecology*, 44:746–759.

Van Wagner, C.E. (1982). *Practical Aspects of the Line Intersect Method*, Information Report PI-X-12, Canadian Forest Service, Ottawa, Ontario.

Warren, W.G. and Olsen, P.F. (1964). A live intersect technique for assessing logging wastes, *Forest Service*, 10:267–276.

West, N.E. (1985). Shortcomings of plant frequency-based methods for range condition and trend, in Krueger, W.C., Chairman, *Proceedings of the 38th Annual Meeting*, Society for Range Management, February 11–14, Salt Lake City, UT, pp. 87–90.

Whysong, G.L. and Brady, W.W. (1987). Frequency sampling and type II errors, *Journal of Range Management*, 40:172–174.

Winkworth, R.E. (1955). The use of point quadrats for the analysis of heathland, *Australian Journal of Botany*, 3, 68–81.

Yates, D.S., Moore, D.S., and Starnes, D.S. (2008). *The Practice of Statistics*, 3rd, ed., Freeman, New York.

7

Timber Scaling and Log Rules*

With regard to scaling and log rules associated with timber, the measurement of timber to be harvested (the cruise), timber cut and removed from the forest (scaling), timber not recovered in the harvesting process (waste processing), and the use of formulas or tables to estimate net yield for logs (log rule) are the basis of forest-based biomass harvesting operations.

Theory of Scaling

Scaling is the determination (measuring) of the gross and net volume of logs by the customary commercial units for the product involved; volume may be expressed in terms of board feet, cords, cubic feet, cubic meter, linear feet, or number of pieces. The *cubic foot* is an amount of wood equivalent to a solid cube that measures 12 × 12 × 12 inches and contains 1728 cubic inches. The *cubic meter,* used in countries that have adopted the metric system, contains 35.3 cubic feet. The *board foot* is a plank 1 inch thick and 12 inches square; it contains 144 cubic inches of wood. Scaling is not guessing; it is an art founded on applying specific rules in a consistent manner based on experienced judgment as to how serious certain external indicators of defect are in a specific locality.

> **DID YOU KNOW?**
> Tree cross-sections rarely form true circles, but they are normally presumed to be circular for purposes of computing cross-sectional areas.

The measuring standard used in scaling logs, called a *log rule,* is a table intended to show amounts of lumber that may be sawed from logs of different sizes under assumed conditions. At best, a log rule can only approximate salable manufactured volume because of constant changes in markets, machinery, manufacturing practices, and even the varying skills of individual sawyers. Thus, a log rule is an imaginary measure. Its application must not be varied according to the mill in which logs are sawed. The scaled volume of logs must be independent of variations in manufacture. The difference between the volume of log scale and the actual volume of lumber sawed from the same logs is called *overrun* if the volume tally exceeds log scale or *underrun* if it is less.

---

* This chapter is based on USDA, *National Forest Log Scaling Handbook,* FSH 2409.11, U.S. Department of Agriculture, Forest Service, Washington, D.C., 1941; Freese, F., *A Collection of Log Rules,* General Technical Report FPL-01, U.S. Department of Agriculture, Forest Service, Forest Products Laboratory, Madison, WI, 1973; USDA, *National Forest Log Scaling Handbook,* FSH 2409.11, U.S. Department of Agriculture, Forest Service, Washington, D.C., 2006.

DID YOU KNOW?

When measuring the cross-sectional area of a log, diameter instead of the radius of the log is measured; thus, this area in square inches may be determined by

$$\text{Area in square inches} = \pi D^2/4$$

There will generally be an overrun or an underrun when logs are scaled by a particular rule in a given locality and sawed by a mill. Basic assumptions in the log rules and assumptions in utilization practices cause overrun to vary with the size of the average log. Experience proves that this is true even for the International 1/4-Inch rule, although not to the same degree as for the Scribner Decimal C rule (both rules are discussed further below). This fact does not change scaling practices. Overrun (or underrun) is estimated in the process of appraising National Forest timber for sale and presumably by purchasers in determining what prices they will bid. Overrun or underrun is not considered in log scaling, even though it is very important to any mill.

As a general rule, timber is appraised, sold, and measured by customary commercial units for the products involved. Standard practice is to scale saw timber by a board-foot log rule scale, mining timbers by the piece or linear foot, telephone poles by the linear foot or the piece of stated length, piling by the linear foot, pulpwood by the solid cubic foot or cord, and fuelwood, shingle bolts, and similar material by the cord. Other units may be used when better adapted to local trade customs or local situations.

Authorized Log Rules

The Scribner Decimal C rule, the International 1/4-Inch rule, and the Smalian Cubic Volume rule are used by the U.S. Forest Service for uniform scaling of sawtimber. With the exception of the Smalian cubic volume rule, all specific rules are board-foot rules. Each board-foot rule is represented by a table showing an arbitrary estimate of the amount of lumber a log of given length and diameter can produce. Inasmuch as the tables for each rule have a different base, the scale of identical logs will differ according to the rule used.

Scribner Decimal C Rule

The Scribner Decimal C rule, developed by J.M. Scribner around 1846, is a standard rule for U.S. Forest service saw log scaling. This rule was derived from diagrams of 1-in. boards drawn to scale within cylinders of various sizes (see Figure 7.1). The rule rounds contents to the nearer 10 board feet; for example, logs that, according to the Scribner rule, have volumes between 136 and 145 board feet are rounded to 140 board feet and shown as 14.

DID YOU KNOW?

The Scribner rule gives a relatively high overrun (up to 30%) for logs under 14 in. Above 14 in., the overrun gradually decreases and flattens out around 28 in. to about 3 to 5%.

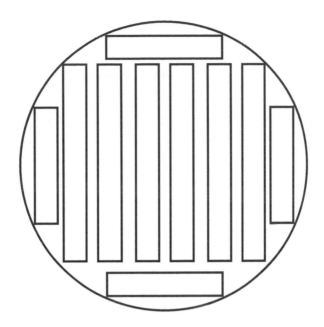

FIGURE 7.1
Diagram showing the number of 1-inch boards that can be cut from a specific log.

International 1/4-Inch Rule

Developed in 1906, the International 1/4-Inch rule is based on a reasonably accurate mathematical formula that probably gives a closer lumber-volume estimate than other log rules in common use. This rule measures logs to the nearest 5 board feet. As the name implies, it allows for a saw kerf (width of the saw cut) of 1/4 inch. It is a rule based on a formula applied to each 4-foot section of the log. For practical purposes, the scaling cylinder becomes a part of a cone (a frustum) with a taper of 2 inches over 16 feet. This rule generally results in a log scale relatively close to lumber tally when logs are sawed in a reasonably efficient mill.

Smalian Cubic Volume Rule

The Smalian Cubic Volume rule requires measurement of the two inside bark diameters and the length. It can be shown generally in the form:

$$V = \frac{A + a}{2} \times L \tag{7.1}$$

where:

  V = Volume in cubic feet (ft³)

  A = Large-end cross-section area (ft²)

  a = Small-end cross-section area (ft²)

  L = Log length (ft)

Log Rules

A *log rule* is a table or formula showing the estimated net yield for logs of a given diameter and length. Ordinarily, the yield is expressed in terms of board feet of finished lumber, although a few rules give the cubic volume of the log or some fraction of it. As noted earlier, the board foot is equivalent to a plank 1 inch thick by 12 inches (1 foot) square; it contains 144 cubic inches of wood. Although the board foot has been a useful and fairly definitive standard for the measure of sawed lumber, it is an ambiguous and inconsistent unit for log scaling. Built into each log rule are allowances for losses due to such things as slabs, saw kerf, edgings, and shrinkage. The volume commonly used for determining the board-foot content of saw lumber is

$$\text{Board feet} = \frac{\text{Thickness (in.)} \times \text{Width (in.)} \times \text{Length (ft)}}{12} \qquad (7.2)$$

At first glance and with the use of a simple equation (Equation 7.2), it would seem to be a relatively simple matter to devise such a rule and having done so that should be the end of the problem. But, it would seem so only to those who are unfamiliar with the great variations in the dimensions of lumber that may be produced from a log, with variations in the equipment used in producing this lumber and the skills of various operators, and, finally, with variations in the logs. All of these have an effect on the portion of the total log volume that ends up as usable lumber and the portion that becomes milling residue.

Because no industrial lumber organization or government agency has control over the measurement of logs, districts and individual buyers have devised their own rules to fit a particular set of operating conditions. Thus, in the United States and Canada there are over 95 recognized rules bearing about 185 names. In addition, there are numerous location variations in the application of any given rule.

Three methods are employed to develop a new log rule. The most obvious is to record the volume of lumber produced from straight, defect-free logs of given diameters and lengths and accumulate such data until all sizes of logs have been covered. These "mill scale" or "mill tally" rules have the virtue of requiring no assumptions and of being perfectly adapted to all the conditions prevailing when the data were obtained. Their disadvantage, aside from the amount of recordkeeping required, is that they may have been produced in such a restricted set of conditions that the values are not applicable anywhere else.

The second method is to prescribe all of the pertinent conditions (e.g., allowance for saw kerf and shrinkage, thickness and minimum width and length of boards, taper assumptions) and then to draw diagrams in circles of various sizes, representing the sawing pattern on the small end of a log. These "diagram rules," of which the Scribner is an example, will be good or bad, depending on how well the sawmilling situation fits the assumptions used in producing the diagrams.

DID YOU KNOW?

By keeping careful lumber tallies of boards cut from various sized logs, any sawmill may construct its own empirical log rule. Such rules may provide excellent indicators of log volume at the particular sawmills where they are compiled.

The third basic procedure is to start with the formula for some assumed geometric solid and then make adjustments to allow for losses to saw kerf, edgings, and so forth. These are referred to as "formula rules," and, as is the case for any type of rule, their applicability will depend on how well the facts fit all of the assumptions.

The development of a rule may involve more than one of these procedures; thus, the step-like progression of values in a mill tally or diagram rule may be smoothed out by fitting a regression equation. Or, the allowance to be used for slabs and edgings in a formula rule may be estimated from mill tally data. Finally, there are the "combination" rules such as the Doyle–Scribner, which uses values from the Doyle rule for small logs and from the Scribner rule for large logs. The aim, of course, is to take advantage of either the best or the worst features of the different rules.

Basal Area Measurements

As mentioned, the cross-sectional area at tree stem breast height is called *basal area*. This is important because tree-stem measurements are often converted to cross-sectional areas. To compute tree basal area, one commonly assumes that the tree stem is circular in cross-section at breast height. Thus, the formula for calculating basal area in square feet (where the DBH is measured in inches) is

$$\text{Basal Area (ft}^2) = \frac{\pi(\text{DBH})^2}{4 \times 144} \tag{7.3}$$
$$= 0.005454 \times (\text{DBH})^2$$

If metric units are used, basal area should be expressed in square meters (m²), and DBH is measured in centimeters:

$$\text{Basal Area (m}^2) = \frac{\pi(\text{DBH})^2}{4 \times 10,000} \tag{7.4}$$
$$= 0.00007854 \times (\text{DBH})^2$$

■ *Example 7.1*

Problem: Calculate the basal area (in ft²) for a tree measuring 6 in. DBH.
Solution:

$$\text{Basal Area (ft}^2) = \frac{\pi(\text{DBH})^2}{4 \times 144} = 0.005454 \times (\text{DBH})^2$$
$$= \frac{3.14 \times 6^2}{4 \times 144}$$
$$= \frac{113.04}{576} = 0.19625 \text{ ft}^2$$

Wood Density and Weight Ratios

The weight (lb) per cubic foot of any tree species may be computed by using the moisture content and specific gravity (based on oven-dry weight and green volume):

$$\text{Density} = \text{Specific Gravity} \times 62.4 \left(1 + \frac{\% \text{ moisture content}}{100} \right) \tag{7.5}$$

■ *Example 7.2*

Problem: Determine the volume (ft³) contained in the following wood weight, given that the specific gravity and moisture content of the wood, respectively, are 15,530 lb and 0.53, 100%.
Solution:

$$\text{Density} = \text{Specific Gravity} \times 62.4 \left(1 + \frac{\% \text{ moisture content}}{100} \right)$$

$$= 0.53 \times 62.4 \left(1 + \frac{100}{100} \right)$$

$$= 33.07 \times 2 = 66.1 \text{ lb/ft}^3$$

$$\text{Volume} = \frac{15,530 \text{ lb}}{66.1 \text{ lb/ft}^3} = 234.9 \text{ ft}^3$$

Chapter Review Questions

7.1   Scaling is not guessing; it is an _____.
7.2   A _____ is a table intended to show amounts of lumber that may be sawed from logs of different sizes.
7.3   _____ requires measurement of the two inside bark diameters and the length.
7.4   The cross-sectional area at tree stem breast height is called _____.
7.5   Calculate the basal area (in ft²) for a tree measuring 12-in. DBH.

References and Recommended Reading

Bonhan, C.D. (1989). *Measurements for Terrestrial Vegetation*, John Wiley & Sons, New York.
Cain, S.A. and Castro, G.M. de O. (1959). *Manual of Vegetation Analysis*, Harper & Brothers, New York.
Daubenmire, R.E. (1968). *Plant Communities: A Textbook of Plant Synecology*, HarperCollins, New York.
Elzinga, C.L., Salzer, D.W., and Willoughby, J.W. (1998). *Measuring and Monitoring Plant Populations*, Technical Reference 1730-1, U.S. Department of Interior, Bureau of Land Management, Denver, CO.

Section II

Forest-Based Biomass

In discussing forest-based biomass we must focus our discussion, of course, on the Tree. In seeking a suitable description of what a tree is we could seek out (and we did) any of the many explanations in authoritative volumes that have been written about trees. However, in providing a description suitable for this text it is probably appropriate to follow the advice of Einstein when he said, "I never stopped thinking like a child." Thus, like a child, we simply state the obvious: "A tree is a big plant with a stick up the middle."

—Colin Tudge (*The Tree*, Three Rivers Press, 2005)

Any fool can destroy trees. They cannot run away; and if they could, they would still be destroyed, chased, and hunted down as long as fun or a dollar could be got out of their bark hides, branching horns, or magnificent bole backbones.

—John Muir (*Atlantic Monthly*, 1897)

8

Bioenergy

The wretched and the poor look for water and find none,
   their tongues are parched with thirst;
   but I the Lord will give them an answer,
   I, the God of Israel, will not forsake them.
I will open rivers among the sand-dunes and wells in the valleys;
   I will turn the desert into pools.
   and the dry land into springs of water;
I will plant cedars in the wastes,
   and acacia and myrtle and wild olive;
   the pine shall grow on the barren heath
   side by side with fir and box ...

—Isaiah **41:17–20**

Biomass

Biomass is a scientific term for living matter, but the word biomass is also used to denote products derived from living organisms—wood from trees, harvested grasses, plant parts, and residues such as twigs, stems, and leaves, as well as aquatic plants and animal wastes. Stated differently, biomass consists of plant and plant-derived materials, including animal manure, starch, sugar, and oil crops that are already used for food and energy. All of the Earth's biomass exists in the thin layer we call the *biosphere*. This represents only a tiny fraction of the total mass of the Earth, but in human terms it is an enormous store of energy—as fuel and as food. More importantly, it is a store that is being replenished continually. The source that supplies the energy is, of course, the sun, and although only a tiny fraction of the solar energy reaching the Earth each year is converted into biomass it is nevertheless equivalent to over five times the total world energy consumption. *Bioenergy* (stored energy from the sun) has been used since early times when people began burning wood to cook food and to keep warm.

Biomass has great potential to provide renewable energy for America's future; for example, biomass can be converted to energy-producing ethanol and wood pellets. Biomass recently surpassed hydropower as the largest domestic source of renewable energy and currently provides over 3% of the total energy consumption in the United States. In addition to the many benefits common to renewable energy, biomass is particularly attractive because it is the only current renewable source of liquid transportation fuel. This, of course, makes it invaluable in reducing oil imports—one of our most pressing economic, political, and energy needs. An obvious question, however, is how large a role biomass could play in responding to the nation's energy demands. This depends, of course, on how economic and financial policies and advances in conversion technologies make biomass fuels and products more economically viable. The need for energy contributions is real (and soon to be pressing); however, the major question is would the biomass potential be sufficiently large to justify the necessary capital replacements in the fuels and automobiles sectors?

The short answer to the question of whether biomass fuels and products can be made more economically feasible and in quantities to augment our energy needs is "yes." Looking at just the two largest potential biomass sources, forestland and agricultural land, Schreuder et al. (2004) estimated that more than 1.3 billion dry tons of biomass (368 dry tons of forestland biomass and 998 million dry tons of agricultural biomass) can be economically produced per year—more than enough to produce biofuels to meet more than one-third of the current demand for transportation fuels.

Of the 368 dry tons of forest-based biomass produced annually in the contiguous United States, this amount includes 52 million dry tons of fuelwood harvested from forests, 145 million dry tons of residues from wood-processing mills and pulp and paper mills, 47 million dry tons of urban wood residues (including construction and demolition debris), 64 million dry tons of residues from logging and site-clearing operations, and 60 million dry tons of biomass from fuel treatment operations to reduce fire hazards. All of these forest resources are sustainably available on an annual basis (Schreuder et al., 2004). It excludes organic material that has been transformed by geological processes into substances such as coal or petroleum. The biomass industry is one of the fastest growing industries in the United States.

DID YOU KNOW?

Renewable energy is any energy source that can be replenished either continuously or within a moderate time frame as a result of natural energy flows. The so-called "renewables" include solar energy (heat and electricity), bioenergy, wind power, hydropower, and geothermal power.

Bioenergy vs. Biomass

In discussions concerning renewable energy, it is not uncommon to read or to hear the terms *biomass* and *bioenergy* used interchangeably. *Bioenergy* or *biomass energy* includes any solid, liquid, or gaseous fuel, or any electrical power or useful chemical products derived from organic matter, whether directly from plants or indirectly from plant-derived industrial, commercial, or urban wastes, or agricultural and forestry residues. This bioenergy can be derived from a wide range of raw materials and produced in a variety of ways. Because of the wide range of potential feedstocks and the variety of technologies available to produce them and process them, bioenergy is usually considered as a series of many different feedstocks/technology combinations. In practice, we tend to use different terms for different end uses (e.g., electric power or transportation). The term *biopower* describes biomass power systems that use biomass feedstock instead of the usual fossil fuels (natural gas or coal) to produce electricity, and the term *biofuel* is used primarily for liquid transportation fuels that substitute for petroleum products such as gasoline or diesel. *Biofuel* is short for *biomass fuel*.

> **DID YOU KNOW?**
> It takes less energy to grow forest biomass and convert it to ethanol than it takes to grow corn and convert it to ethanol. In addition, the entire process emits fewer greenhouse gases when using forest biomass.

Biomass Feedstocks

A variety of biomass feedstocks can be used to produce transportation fuels, biobased products, and power. Currently, a majority of the ethanol produced in the United States is made from corn or other starch-based crops; however, the current trend in research is to develop biomass fuels from non-grain foodstocks. For example, the focus is on the development of cellulosic feedstocks—non-food-based feedstocks such as switchgrass, corn stover, and woody material—and on technologies to convert cellulosic material into transportation fuels and other products. Using cellulosic bioenergy feedstocks not only can alleviate the potential concern of diverting food crops to produce fuel but also offers a variety of environmental benefits. *Bioenergy crops*, also called *energy crops*, are fast-growing crops that are grown for the specific purpose of producing energy (electricity or liquid fuels) from all or part of the resulting plant. The plants that have been selected by the U.S. Department of Energy for further development as energy crops are mostly perennials, such as forest-based willow and poplar. They were selected for their advantageous environmental qualities, such as erosion control, soil organic matter build-up, and reduced fertilizer and pesticide requirements. Many other perennial plant species could be used for energy crops; in addition, some parts of traditional agricultural crops such as the stems or stalks of alfalfa, corn, or sorghum may be used for energy production.

Environmental benefits of using biomass energy include greatly reducing greenhouse gas emissions. Burning biomass releases about the same amount of carbon dioxide as burning fossil fuels, but fossil fuels release carbon dioxide captured by photosynthesis millions of years ago—an essentially "new" greenhouse gas. The burning of conventional fossil fuels such as gasoline, oil, coal, or natural gas results in an increase in carbon dioxide in the atmosphere; carbon dioxide is the major greenhouse gas thought to be responsible for global climate change. Some nitrogen oxides inevitably result from biomass burning (as with all combustion process), but these are comparable to emissions from natural wildfires and are generally lower than those from burning fossil fuels. Other greenhouse gas emissions are associated with the use of fossil fuels by farm equipment and with the application of inorganic fertilizers to the bioenergy crop; however, these may be offset by the increase in carbon storage in soil organic matter compared with conventional crops. Utilization of biomass residues that would have otherwise been dumped in landfills (e.g., urban and industrial residues) greatly reduces greenhouse gas emissions by preventing the formation of methane. Moreover, biomass releases carbon dioxide that is largely balanced by the carbon dioxide captured in its own growth (depending on how much energy was used to grow, harvest, and process the fuel). Another huge benefit of using biomass for fuel is that it can reduce dependence on foreign oil, as biofuels are the only renewable liquid transportation fuels available.

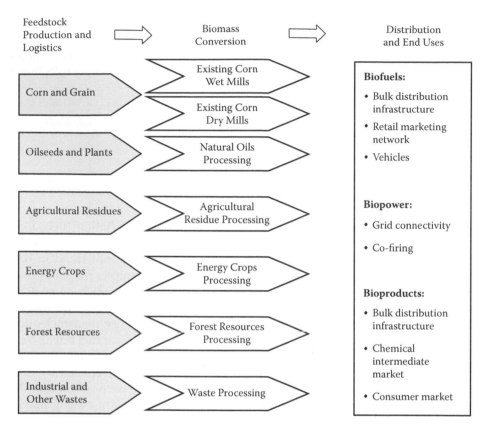

FIGURE 8.1
Resource-based biorefinery pathways.

Biomass energy supports U.S. agricultural and forest-products industries. The main biomass feedstocks for power are paper-mill residues, lumber-mill scrap, and municipal waste. For biomass fuels, the feedstocks are corn (for ethanol) and soybeans (for biodiesel), both surplus crops. In the near future—and with developed technology—agricultural residues such as corn stover (the stalks, leaves, and husks of the plant) and wheat straw will also be used. Long-term plans include growing and using dedicated energy crops (purpose-grown biomass), such as fast-growing trees and grasses that can grow sustainably on land that will not support intensive food crops.

Because such a wide variety of cellulosic feedstocks can be used for energy production, potential feedstocks are grouped into categories, or pathways. Figure 8.1 shows some of the specific feedstocks in each of these areas.

Biomass Resources

The concept of the forest as a source of energy is not new. Since humans first harnessed fire, wood, and other lignocellulosic biomass, it has been a source of heat, light, and power. Before petroleum became commonplace, wood and biomass were the primary source of such chemicals as acetic acid, turpentine, pine tar, resin, and rubber. With industrial advances, larger quantities and more concentrated sources of energy were required, thus petroleum and coal largely displaced wood. Although firewood is still important in some less-developed countries, in most developed countries it is used only for residential wood stoves and, to some extent, for commercial electrical power generation. Today, the primary use for wood is as a raw material for furniture and building products (Winandy et al., 2008).

Well, as Dylan put it, "The Times They Are A-Changin'," which is apropos to our current energy status: The days of easily accessible and cheap petroleum-based fuels are ending—the times are indeed a-changin'. We should not panic, though, because the day of the tree is returning; moreover, the conversion of standing-crop, tree-based biomass to energy is growing. Worldwide, total standing-crop biomass (99% on land, and 80% in trees) is a huge resource, equivalent to about 60 years of world energy use in the year 2000 (1250 metric tons of dry plant matter, containing 560 billion metric tons of carbon). For the United States alone, standing vegetation has been variously estimated at between 65 and 90 billion metric tons of dry matter (30 to 40 billion metric tons of carbon), equivalent to 14 to 19 years of current U.S. primary energy use. Every year, the Earth grows about 130 billion metric tons of biomass on land (60 billion metric tons of carbon) and a further 100 billion metric tons in the rivers, lakes, and oceans (46 billion metric tons carbon). The energy content of this annual biomass production is estimated to be more than 6 times world energy use of 2640 exajoules (2500 Quads—short for quadrillion or 10^{15} Btu) on land, with an additional 2024 exajoules (1920 Quads) in the waters.

DID YOU KNOW?

The Joule (pronounced "jewel") is a derived unit of energy or work in the International System of Units. It is equal to the energy expended in applying a force of 1 newton through a distance of 1 meter, or in passing an electrical current of one ampere though a resistance of 1 ohm for 1 second. An exajoule is equivalent to 10^{18} J.

Estimates of how much of the Earth's land-based production of biomass is used by the human population worldwide range from a low figure of about 5% to a high of over 30% (this includes food, animal fodder, timber, and other products, as well as bioenergy). The higher estimates include a lot of wasted material and inefficient activities such as forest clearance, as well as losses of productivity due to human activity.

Worldwide, biomass is the fourth largest energy resource after coal, oil, and natural gas—estimated at about 14% of global primary energy (and much higher in many developing countries). In the United States, biomass today provides about 3.4% of primary energy (depending on the method of calculation). Biomass is used for heating (such as wood stoves in homes and for process heat in bioprocessing industries), cooking (especially in many parts of the developing world), transportation (fuels such as ethanol), and, increasingly, for electric power production. Installed capacity of biomass power generation worldwide is about 35,000 megawatts (MW), with about 7000 MW in the United States derived from residues from the forest-product industry and agriculture (plus an additional 2500 MW of municipal solid-waste-fired capacity, which is often not counted as part of biomass power, and 500 MW of landfill gas-fired and other capacity). Much of this 7000-MW capacity is currently found in the pulp and paper industry, in combined heat and power (cogeneration) systems.

Nearly every part of the world has a biomass resource that can be tapped to make biofuels and generate electrical power. From coconut or rice husks to fast-growing trees, from perennial grasses to scrap wood, there is a (probably underutilized) form of biomass almost anywhere you go on Earth. With regard to electrical power generation, wood and wood-processing residues and byproducts are the most widely used biomass fuels in the United States today (where more than 500 electrical power plants operate on biomass). These conventional steam-cycle power systems have the longest operating experience and are well understood.

Biomass Technology and Biorefineries

Chemical companies are exploring the use of low-cost biomass processes to make chemicals and plastics that are now made from more expensive petrochemical processes. New innovative processes such as biorefineries may become the foundation of the new bioindustry. A biorefinery is analogous to a petroleum refinery, except that it is based on the conversion of biomass feedstocks rather than crude oil. Biorefineries, in theory, would use multiple forms of biomass to produce a flexible mix of products, including fuels, power, heat, chemicals, and materials. By producing multiple products, it would be able to take advantage of the differences in lignocellulosic components and intermediates and maximize the value derived from the lignocellulosic feedstock. In a biorefinery, biomass would be converted into high-value chemical products and fuels (both gas and liquid). Byproducts and residues, as well as some portion of the fuels produced, would be used to fuel onsite power generation or cogeneration facilities producing heat and power. The biorefinery concept has already proven successful in the U.S. agricultural and forest-products industries, where such facilities now produce food, feed fiber, or chemicals, as well as heat and electricity to run plant operations. Biorefineries offer the most potential for realizing the ultimate opportunities of the bioenergy industry.

The most common liquid fuel from biomass in the United States is ethanol, which is produced by fermentation. Typically, sugars are extracted from the biomass feedstock by crushing and washing or, in the case of starchy feedstocks such as corn (maize), by the breakdown of starch to sugars. The sugar syrup is then mixed with yeast and kept warm, so the yeast breaks down the sugars into ethanol; however, the fermented product is only about 10% ethanol, so a further stage of distillation is required to concentrate the ethanol to 95%. If the ethanol is intended for blending with gasoline, a "dehydration" phase may be required to make 100% ethanol. In the near future, ethanol may be made from cellulose, again by breakdown into sugars for fermentation. Cellulose is widely and cheaply available from many other biomass feedstocks, energy crops, agricultural, and forestry residues.

Another form of liquid fuel from biomass is biodiesel, which is derived from vegetable oils extracted by crushing oilseeds, although waste cooking oil or animal fats (tallow) can also be used. The oil is strained and usually esterified by combining the fatty acid

DID YOU KNOW?

Based on a rough rule of thumb, it takes a little under 1000 acres (400 hectares) of poplar (grown as a short-rotation crop at a usable yield of 5 dry U.S. tons/acre, or 1 metric ton/hectare) to supply an electric power plant with a capacity of 1 megawatt.

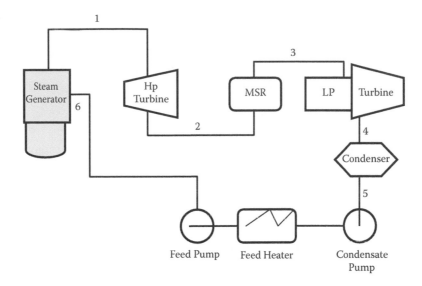

FIGURE 8.2
Simple steam cycle and the processes that comprise the cycle. (1–2) Saturated steam from the steam generator is expanded in the high-pressure (HP) turbine to provide shaft work output at a constant entropy (i.e., the
amount of thermal energy not available to do work). (2–3) The moist steam from the HP turbine is dried and
superheated in the moisture separator reheater (MSR). (3–4) Superheated steam from the MSR is expanded in
the low-pressure (LP) turbine to provide shaft work output at a constant entropy. (4–5) Steam exhaust from the
turbine is condensed in the condenser, where heat is transferred to the cooling water under a constant vacuum
condition. (5–6) The feedwater is compressed as a liquid by the condensate and feedwater pump, and the feedwater is preheated by the feedwater heaters. (6–1) Heat is added to the working fluid in the steam generator
under a constant pressure condition.

molecules in the oil with methanol or ethanol. Vegetable oil esters have been shown to
make good-quality, clean-burning diesel fuel; in 2008, there were 100 biodiesel processing
plants worldwide, mostly in Europe.

Simple Steam Cycle

When biomass is burned, it produces heat, as in any simple fire pit, fireplace, or furnace
in the cave or in the house. In most power plants (*steam-cycle* or *steam-turbine* systems; see
Figure 8.2), this heat is captured by boiling water to generate steam, which turns turbines
and drives generators that convert the energy into electricity. New technologies now being
evaluated include several types of biomass gasifiers in which biomass is heated to convert
it into a gas. This gas is used directly in a gas turbine, which drives a generator (a simple
gas turbine system). In some cases, the waste heat from the gas turbine may be used to
drive a secondary steam turbine, thus converting more of the fuel energy into electricity
(a combined-cycle system).

Co-Firing

Co-firing refers to the blending of biomass with coal in the furnace of a conventional
coal-fired steam-cycle electric power plant. This is currently one of the simplest ways of
utilizing biomass to displace fossil fuels, requiring no new investment or specialized technology. Between 5% and 15% biomass (by heat content) may be used in such facilities

at an additional cost estimated at less than 0.5¢/kWh (compared with coal-firing alone). Co-firing is known to reduce carbon dioxide emissions, sulfur dioxide (SO_2) emissions, and potentially some emissions of nitrogen oxides (NO_x), as well. Many electric utilities around the United States have experimented successfully with co-firing, using wood chips, urban waste wood, and forestry residues.

Gasification

The main requirement for further development of biomass–electric technology is improving the efficiency of energy conversion to lower emissions and to reduce costs. Gasification offers greater flexibility, both in the range of possible biomass feedstocks and in the end use of energy; for example, in addition to driving a gas turbine, the gas from a gasifier can power a fuel cell to generate electricity, or it can be used to generate steam in a gas boiler, sometimes in combination with natural gas. Large steam-turbine systems in power plants 200 MW or larger (such as most coal-fired power plants) are relatively efficient at energy conversion. Smaller biomass-fired steam-turbine systems (20 to 100 MW) require further research to improve their cost competitiveness with fossil fuels, but biomass gasification systems may be able to combine high efficiency with cost competitiveness in this size range. Other research is continuing to develop small modular biomass conversion systems (100 kW to 5 MW) to provide electricity cost-effectively to communities and industries. A variety of biomass gasifier types have been developed. They can be grouped into four major classifications: fixed-bed updraft, fixed-bed downdraft, bubbling fluidized-bed, and circulating fluidized-bed. Differentiation is based on the means of supporting the biomass in the reactor vessel, the direction of flow of both the biomass and oxidant, and the way heat is supplied to the reactor.

Updraft Gasification

Also known as counter-flow gasification, the updraft configuration is the oldest and simplest form of gasifier; it is still used for coal gasification (Ciferno and Marano, 2002). Biomass is introduced at the top of the reactor, and a grate at the bottom of the reactor supports the reacting bed. Air or oxygen and/or steam are introduced below the grate and diffuse up through the bed of biomass and char. Complete combustion of char takes place at the bottom of the bed, liberating CO_2 and H_2O. These hot gases (~1000°C) pass through the bed above, where they are reduced to H_2 and CO and cooled to 750°C. Continuing up the reactor, the reducing gases (H_2 and CO) pyrolyze the descending dry biomass and finally dry the incoming wet biomass, leaving the reactor at a low temperature (~500°C) (Bridgwater and Evans, 1993; Reed and Siddhartha, 2001; Stultz and Kitto, 1992). Examples are the PUROX and the Sofresid/Caliqua technologies. The advantages of updraft gasification are

- The process is simple and low cost.
- It is able to handle biomass with a high moisture and high inorganic content (e.g., municipal solid waste).
- It is a proven technology.

The primary disadvantage of updraft gasification is

- Syngas contains 10 to 20% tar by weight, requiring extensive syngas cleanup before engine, turbine, or synthesis applications.

Downdraft Gasification

Also known as cocurrent-flow gasification, the downdraft gasifier has the same mechanical configuration as the updraft gasifier except that the oxidant and product gases flow down the reactor, in the same direction as the biomass (Ciferno and Marano, 2002). A major difference is that this process can combust up to 99.9% of the tars formed. Low-moisture biomass (<20%) and air or oxygen are ignited in the reaction zone at the top of the reactor. The flame generates pyrolysis gas/vapor, which burns intensely, leaving 5 to 15% char and hot combustion gas. These gases flow downward and react with the char at 800 to 1200°C, generating more CO and H_2 while being cooled to below 800°C. Finally, unconverted char and ash pass through the bottom of the grate and are sent to disposal (Paisley et al., 2000; Reed and Siddhartha, 2001; Stultz and Kitto 1992). The advantages of downdraft gasification are

- Up to 99.9% of the tar formed is consumed, requiring minimal or no tar cleanup.
- Minerals remain with the char/ash, reducing the need for a cyclone.
- It is a proven, simple, low-cost process.

The disadvantages of downdraft gasification are

- It requires feed drying to a low moisture content (<20%).
- Syngas exiting the reactor is at a high temperature, requiring a secondary heat recovery system.
- About 4 to 7% of the carbon remains unconverted.

Bubbling Fluidized Bed

Most biomass gasifiers under development employ one of two types of fluidized bed configurations: bubbling fluidized bed and circulating fluidized bed. A bubbling fluidized bed consists of fine, inert particles of sand or alumina, which have been selected for size, density, and thermal characteristics. As gas (oxygen, air, or steam) is forced through the inert particles, a point is reached when the frictional force between the particles and the gas counterbalances the weight of the solids. At this gas velocity (minimum fluidization), bubbling and channeling of gas through the media occur, such that the particles remain in the reactor and appear to be in a "boiling state" (Craig et al., 1996). The fluidized particles tend to break up the biomass fed to the bed and ensure good heat transfer throughout the reactor (Bridgwater and Evans, 1993; Paisley et al., 2000). The advantages of bubbling fluidized-bed gasification are

- The process yields a uniform product gas.
- It exhibits a nearly uniform temperature distribution throughout the reactor.
- It is able to accept a wide range of fuel particle sizes, including fines.
- It provides high rates of heat transfer among the inert material, fuel, and gas.
- High conversion is possible with low tar and unconverted carbon.

The main disadvantage of bubbling fluidized-bed gasification is

- Large bubble size may result in gas bypass through the bed.

Circulating Fluidized Bed

Circulating fluidized bed gasifiers operate at gas velocity higher than the minimum fluidization point, resulting in entrainment of the particles in the gas stream. The entrained particles in the gas exit the top of the reactor, are separated in a cyclone, and then return to the reactor (Bridgwater and Evans, 1993; Paisley et al., 2000). The advantages of circulating fluidized-bed gasification are

- The process is suitable for rapid reactions.
- High heat transport rates are possible due to the high heat capacity of the bed material.
- High conversion rates are possible with low tar and unconverted carbon.

The disadvantages of circulating fluidized-bed gasification are

- Temperature gradients occur in the direction of solid flow.
- Fuel particle size determines the minimum transport velocity; high velocities may result in equipment erosion.
- Heat exchange is less efficient than the bubbling fluidized bed.

Chapter Review Questions

8.1 Define biomass: _____.

8.2 What are the two largest potential biomass sources?

8.3 _____ describes a system that uses biomass feedstock instead of the usual fossil fuels to produce electricity.

8.4 An exajoule is equivalent to _____ J.

8.5 What process is the blending of biomass with coal in the furnace of a conventional coal-fired steam-cycle electrical power plant?

References and Recommended Reading

Bridgwater, A.V. and Evans G.D. (1993). *An Assessment of Thermochemical Conversion Systems for Processing Biomass and Refuse*, Energy Technology Support Unit Report B/T1/00207/REP, Department of Trade and Industry, London.

Ciferno, J.P. and Marano, J.J. (2002). *Benchmarking Biomass Gasification Technologies for Fuels, Chemicals, and Hydrogen Production*, U.S. Department of Energy, National Energy Technology Laboratory, Philadelphia, PA.

Craig, K.R., Mann, M.K., and Bain, R.L. (1996). *Cost and Performance Analysis of Biomass-Based Integrated Gasification Combined-Cycle (BIGCC) Power Systems*, Task No. BP611717, National Renewable Energy Laboratory, Golden, CO.

Paisley, M.A., Farris, M.C., Black, J., Irving, J.M., and Overend, R.P. (2000). Commercial demonstration of the Battelle/FERCO biomass gasification process: startup and initial operating experience, in *A Growth Opportunity in Green Energy and Value-Added Products: Proceedings of the Fourth Biomass Conference of the Americas*, Elsevier, London, pp. 1061–1066.

Reed, T.B. and Siddhartha, G. (2001). *A Survey of Biomass Gasification*, 2nd ed., Biomass Energy Foundation, Franktown, CO.

Schreuder, H.T., Ernst, R., and Ramirez-Maldonado, H. (2004). *Statistical Techniques for Sampling and Monitoring Natural Resources*, U.S. Department of Agriculture, Washington, D.C.

Stultz, S.C. and Kitto, J.B. (1993). *Steam: Its Generation and Use*, Babcock & Wilcox, Barberton, OH.

Winandy, J.E., Williams, R.S., Rudie, A.W., and Ross, R.J. (2008). Opportunities for using wood and biofibers for energy, chemical feedstocks, and structural applications, in Pickering, K., Ed., *Properties and Performance of Natural-Fibre Composites*, Woodhead Publishing, Cambridge, U.K., pp. 330–355.

9

Biomass: Plant Basics*

The thing I love most about America is that there's always somebody here who doesn't get the word—and they go out and do the right thing or invent the new thing, no matter what's going on politically or economically.

—Thomas Friedman (*The Virginian-Pilot*, December 18, 2010)

Nature's peace will flow into you as sunshine flows into trees.

—John Muir (*Our National Parks*, 1901)

Plant Basics

To optimize woody biomass, including limbs, tops, needles, leaves, and other woody parts, grown in a forest, woodland, or rangeland environment and to achieve more efficient processing we must understand plants and plant cell-wall structure and function. The plant kingdom ranks second in importance only to the animal kingdom (at least from the human point of view). The importance of plants and plant communities to humans, bioenergy production, and their environment cannot be overstated. Some of the important things plants provide are listed below:

- *Aesthetics*—Plants add to the beauty of the places where we live.
- *Medicine*—80% of all medicinal drugs originate in wild plants.
- *Food*—90% of the world's food comes from only 20 plant species.
- *Industrial products*—Plants are very important for the goods they provide (e.g., fibers for clothing, wood to build homes).
- *Recreation*—Plants form the basis for many important recreational activities, including fishing, nature observation, hiking, and hunting.
- *Air quality*—The oxygen in the air we breathe comes from the photosynthesis of plants.
- *Water quality*—Plants aid in maintaining healthy watersheds, streams, and lakes by holding soil in place, controlling stream flows, and filtering sediments from water.
- *Erosion control*—Plant cover helps to prevent wind or water erosion of the top layer of soil that we depend on.

---

* Material in this chapter is adapted from Spellman, F.R., *Biology for the Non-Biologist*, Government Institutes Press, Lanham, MD, 2009.

TABLE 9.1

Plants vs. Animals

Plants	Animals
Plants contain chlorophyll and can make their own food.	Animals cannot make their own food and are dependent on plants and other animals for food.
Plants give off oxygen and take in carbon dioxide given off by animals.	Animals give off the carbon dioxide that plants need to make food and they take in the oxygen given off by plants that they need to breathe.
Plants generally are rooted in one place and do not move on their own.	Most animals have the ability to move fairly freely.
Plants have either no or a very basic ability to sense.	Animals have a much more highly developed sensory and nervous system.

- *Climate*—Regional climates are impacted by the amount and type of plant cover.
- *Fish and wildlife habitat*—Plants provide the necessary habitat for wildlife and fish populations.
- *Ecosystem*—Every plant species serves an important role or purpose in its community.
- *Feedstock for bioenergy production*—Some important fuel chemicals come from plants, such as ethanol from corn and soy diesel from soybeans.

Though both are important kingdoms of living things, plants and animals differ in many important aspects. Some of these differences are summarized in Table 9.1. Before discussing the basics of plants, it is important to first define a few key plant terms.

Plant Terminology

Apical meristem consists of meristematic cells located at the tip (apex) of a root or shoot.

Cambium is the lateral meristem in plants.

Chloroplasts are disk-like organelles with a double membrane that are found in eukaryotic plant cells.

Companion cells are specialized cells in the phloem that load sugars into the sieve elements.

Cotyledon is a leaf-like structure (sometimes referred to as *seed leaf*) present in the seeds of flowering plants.

Dicot is one of the two main types of flowering plants; they are characterized by having two cotyledons.

Diploid refers to having two of each kind of chromosome.

Guard cells are specialized epidermal cells that flank stomata and whose opening and closing regulate gas exchange and water loss.

Haploid refers to having only a single set of chromosomes.

Meristem is a group of plant cells that can divide indefinitely to provide new cells for the plant.

Monocot is one of the two main types of flowering plants; monocots are characterized by having a single cotyledon.

Periderm is a layer of plant tissue derived from the cork cambium; it replaces the epidermis and acts as a protective covering.

Phloem is complex vascular tissue that transports carbohydrates throughout the plant.

Sieve cells are conducting cells in the phloem of vascular plants.

Stomata are pores on the underside of leaves that can be opened or closed to control gas exchange and water loss.

Thallus is the main plant body, not differentiated into a stem or leaves.

Tropism is plant behavior that controls the direction of plant growth.

Vascular tissue, composed of the xylem and phloem, transports water, nutrients, and carbohydrates in vascular plants.

Xylem is a vascular tissue of plants that transports water and dissolved minerals from the roots upward to other parts of the plant; xylem often also provides mechanical support against gravity.

Tree Terminology

Bark is the productive outside covering of a wood stem or root.

Cambium is a thin layer of living cells that produce a new layer of wood each year, visible as tree rings that can be counted to tell the age of a tree. The cambium lies between the xylem and phloem layers.

Cellular respiration is the chemical breakdown of glucose to produce energy; this process is the opposite of photosynthesis.

Chlorophyll is the green substance found in leaves and needles that captures the sun's energy.

Heartwood (old xylem) is the hard, inactive wood at the center of the tree.

Phloem is a layer of inner bark cells that transport food made in the leaves by photosynthesis to the rest of the plant.

Roots are the network below ground that anchors the tree in the soil; root hairs push their way through the soil and absorb moisture and minerals from the soil.

Sapwood (living wood) is comprised of the newly formed wood cells that lie just inside the cambium. It acts as a major conductor of water and minerals for the tree and is also known as *xylem*.

Although not typically acknowledged, plants are as intricate and complicated as animals. Plants evolved from photosynthetic protists and are characterized by photosynthetic nutrition, cell walls made from cellulose and other polysaccharides, lack of mobility, and a characteristic life cycle involving an alternation of generations. The phyla/division of plants and examples are listed in Table 9.2.

The Plant Cell

A brief summary of plant cells is provided here (see Figure 9.1):

TABLE 9.2

The Main Phyla/Division of Plants

Phylum/Division	Examples
Bryophyta	Mosses, liverworts, and hornworts
Coniferophyta	Conifers such as redwoods, pines, and firs
Cycadophyta	Cycads and sago palms
Gnetophyta	Shrub trees and vines
Ginkophyta	*Ginkgo* (only genus)
Lycophyta	Lycopods (look like mosses)
Pterophyta	Ferns and tree-ferns
Anthophyta	Flowering plants, including oak, corn, maize, and herbs

- *Plants have all the organelles that animal cells have* (e.g., nucleus, ribosomes, mitochondria, endoplasmic reticulum, Golgi apparatus).
- *Plants have chloroplasts.* Chloroplasts are special organelles that contain chlorophyll and allow plants to carry out photosynthesis.
- *Plant cells can sometimes have large vacuoles for storage.*
- *Plant cells are surrounded by a rigid cell wall made of cellulose,* in addition to the cell membrane that surrounds animal cells. Those walls provide support.

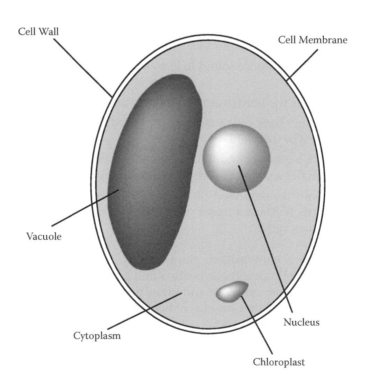

FIGURE 9.1
Plant cell.

Vascular Plants

Vascular plants, also called *tracheophytes*, have special vascular tissue for the transport of necessary liquids and minerals over long distances. Vascular tissues are composed of specialized cells that create tubes that allow materials to flow throughout the body of the plant. These vessels are continuous throughout the plant, allowing for the efficient and controlled distribution of water and nutrients. In addition to this transport function, vascular tissues also support the plant. The two types of vascular tissue are xylem and phloem:

- *Xylem* consists of a tube or a tunnel (pipeline) in which water and minerals are transported throughout the plant to leaves for photosynthesis. In addition to distributing nutrients, xylem (wood) provides structural support. After a time, the xylem at the center of older trees ceases to function in transport and takes on a supportive role only.
- *Phloem* tissue consists of cells called *sieve tubes* and *companion cells*. Phloem tissue moves dissolved sugars (carbohydrates), amino acids, and other products of photosynthesis from the leaves to other regions of the plant.

The two most important tracheophytes are gymnosperms (*gymno*, "naked"; *sperma*, "seed") and angiosperms (*angio*, "vessel"):

- *Gymnosperms* are characterized by the generation of sporophytes, which occurs during the spore-producing phase in the lifecycle of a plant that exhibits alternation of generations. Gymnosperms were the first tracheophytes to use seeds for reproduction. The seeds develop in protective structures called *cones*. A gymnosperm contains some cones that are female and some that are male. Female cones produce spores that, after fertilization, become eggs enclosed in seeds that fall to the ground. Male cones produce pollen, which is taken by the wind and fertilizes female eggs by that means. Unlike flowering plants, the gymnosperm does not form true flowers or fruits. Coniferous trees such as firs and pines are good examples of gymnosperms.
- *Angiosperms*, the flowering plants, are the most highly evolved plants and the most dominant today. They have stems, roots, and leaves. Unlike gymnosperms such as conifers and cycads, the seeds of angiosperms are found in a flower. Angiosperm eggs are fertilized and develop into a seed in an ovary that is usually in a flower.

The two types of angiosperms are monocots and dicots:

- *Monocots* start with a single seed leaf (cotyledon) during embryonic development, thus their name. Monocots include grasses, grains, and other narrow-leaved angiosperms. The main veins of their leaves are usually parallel and unbranched, the flower parts occur in multiples of three, and a fibrous root system is present. Monocots include orchids, lilies, irises, palms, grasses, and wheat, corn, and oats.
- *Dicots* grow two seed leaves (cotyledons). Most plants are dicots, including maples, oaks, elms, sunflowers, and roses. Their leaves usually have a single main vein or three or more branched veins that spread out from the base of the leaf.

Leaves

The principal function of leaves is to absorb the sunlight necessary for the manufacturing of plant sugars in photosynthesis. The broad, flattened surfaces of leaves gather energy from sunlight, while apertures on their undersides bring in carbon dioxide and release oxygen. Leaves develop as a flattened surface to present a large area for efficient absorption of light energy. The exterior part of the leaf has layers of epidermal cells that secrete a waxy, nearly impermeable cuticle (chitin) to protect against water loss (dehydration) and fungal or bacterial attack. Gases diffuse in or out of the leaf through *stomata*, small openings on the underside of the leaf. The opening or closing of the stomata occurs through the swelling or relaxing of *guard cells*. If the plant wants to limit the diffusion of gases and the transpiration of water, the guard cells swell together and close the stomata. Leaf thickness is kept to a minimum so gases that enter the leaf can diffuse easily throughout the leaf cells. From a tree's perspective, the leaf is the factory where raw materials (carbon dioxide and sunlight) are converted into food (energy) that is used by the entire tree for growth and survival.

Chlorophyll and Chloroplast

The green pigment in leaves is *chlorophyll*. Chlorophyll absorbs red and blue light from the sunlight that falls on leaves, so the light reflected by the leaves is diminished in red and blue and appears green. The molecules of chlorophyll are large. They are not soluble in the aqueous solution that fills plant cells; instead, they are attached to the membranes of disc-like structures, called *chloroplasts*, inside the cells. Chloroplasts are the site of photosynthesis, the process in which light energy is converted to chemical energy. In chloroplasts, the light absorbed by chlorophyll supplies the energy used by plants to transform carbon dioxide and water into oxygen and carbohydrates. Chlorophyll is not a very stable compound; bright sunlight causes it to decompose. To maintain the amount of chlorophyll in their leaves, plants continuously synthesize it. The synthesis of chlorophyll in plants requires sunlight and warm temperatures; in the summer, chlorophyll is continuously broken down and regenerated in the leaves of trees.

Photosynthesis

Because our quality of life, and indeed our very existence, depends on photosynthesis, it is essential to understand it. In photosynthesis, plants (and other photosynthetic autotrophs) use the energy from sunlight to create the carbohydrates necessary for cell respiration. More specifically, plants take water and carbon dioxide and transform them into glucose and oxygen:

$$6CO_2 + 6H_2O + \text{Light Energy} \rightarrow C_6H_{12}O_6 + 6O_2$$

This general equation of photosynthesis represents the combined effects of two different stages. The first stage is called the *light reaction* and the second stage is called the *dark reaction*. The light reaction is the photosynthesis process in which solar energy is harvested and transferred into the chemical bonds of adenosine triphosphate (ATP); it can only occur in light. The dark reaction is the process in which food (sugar) molecules are formed from carbon dioxide from the atmosphere with the use of ATP; it can occur in the dark as long as ATP is present.

> **DID YOU KNOW?**
> Charles Darwin was the first to discuss how plants respond to light. He found that new shoots of grasses bend toward the light because the cells on the dark side grow faster than those on the lighted side.

Roots

Roots absorb nutrients and water, anchor the plant in the soil, provide support for the stem, and store food. They are usually below ground and lack nodes, shoots, and leaves. There are two major types of root systems in plants. Taproot systems have a stout main root with a limited number of side-branching roots. Examples of taproot system plants include nut trees, carrots, radishes, parsnips, and dandelions. Taproots make transplanting difficult. The second type of root system, fibrous, has many branched roots. Examples of fibrous root plants are most grasses, marigolds, and beans. Radiating from the roots is a system of root hairs, which vastly increase the absorptive surface area of the roots. Roots also anchor the plant in the soil.

Growth in Vascular Plants

Vascular plants undergo two kinds of growth (growth is primarily restricted to meristems). *Primary growth* occurs relatively close to the tips of roots and stems. It is initiated by apical meristems and is primarily involved in extension of the plant body. The tissues that arise during primary growth are called *primary tissues*, and the plant body composed of these tissues is called the *primary plant body*. Most primitive vascular plants are entirely made up of primary tissues. *Secondary growth* thickens the stems and roots of some plants. Secondary growth results from the activity of lateral meristems called *cambia*:

- *Vascular cambium* gives rise to secondary vascular tissues—secondary xylem on the inside and secondary phloem on the outside.
- *Cork cambium* forms the *periderm* (bark), which replaces the epidermis in woody plants.

Plant Hormones

Plant growth is controlled by plant hormones, which influence cell differentiation, elongation, and division. Some plant hormones also affect the timing of reproduction and germination:

- *Auxins* affect cell elongation (tropism), apical dominance, and fruit drop or retention. Auxins are also responsible for root development, secondary growth in the vascular cambium, inhibition of lateral branching, and fruit development. Auxin is involved in absorption of vital minerals and fall color. As a leaf reaches its maximum growth, auxin production declines. In deciduous plants, this triggers a series of metabolic steps that lead to the reabsorption of valuable materials (such as chlorophyll) and their transport into the branch or stem for storage during the winter months. When the chlorophyll is gone, the other pigments typical of fall color become visible.

- *Kinins* promote cell division and tissue growth in the leaf, stem, and root. Kinins are also involved in the development of chloroplasts, fruits, and flowers. In addition, they have been shown to delay senescence (aging), especially in leaves, which is one reason why florists use cytokinins on freshly cut flowers. When treated with cytokinins, the leaves stay green, protein synthesis continues, and carbohydrates do not break down.

- *Gibberellins* are produced in the root growing tips and act as messengers to stimulate growth, especially elongation of the stem, and can also end the dormancy period of seeds and buds by encouraging germination. Additionally, gibberellins play a role in root growth and differentiation.

- *Ethylene* controls the ripening of fruits. Ethylene may ensure that flowers are carpelate (female), whereas gibberellin confers maleness on flowers. Ethylene also contributes to the senescence of plants by promoting leaf loss and other changes.

- *Inhibitors* restrain growth and maintain the period of dormancy in seeds and buds.

Tropisms: Plant Behavior

Tropism is the growth or movement of an organism in response to an external stimulus (e.g., growth in plants). Tropisms are controlled by hormones and are a unique characteristic of sessile organisms such as plants that allows them to adapt to different features of their environment—gravity, light, water, touch—so they can flourish. The three main tropisms are

- *Phototropism*, which is the tendency of plants to grow or bend in response to light. Phototropism results from rapid elongation of cells on the dark side of the plant which causes the plant to bend in the opposite direction; for example, the stems and leaves of a geranium plant growing on the windowsill always turn toward the light.

- *Gravitropism*, which refers to a plant's tendency to grow toward or against gravity. Positive gravitropism can be seen in plant roots, which grow downward, toward the center of Earth; gravity causes the roots of plants to grow down so the plant is anchored in the ground and has enough water to grow and thrive. Negative gravitropism can be seen in plant stems, which grow upward, away from the Earth. Gravitropism is affected by the hormone auxin. In a horizontal root or stem, auxin is concentrated in the lower half, pulled by gravity. In a positively gravitropic plant, this auxin concentration will inhibit cell growth on the lower side, causing the stem to bend downward. In a negatively gravitropic plant, this auxin concentration will inspire cell growth on that lower side, causing the stem to bend upward.

- *Thigmotropism*, which is growing or bending in response to touch. Some people notice that their houseplants grow better when they touch them and pay attention to them. Touch causes parts of the plant to thicken or coil as they touch or are touched by environmental entities; for example, tree trunks grow thicker when exposed to strong winds, and vines tend to grow straight until they encounter a substrate to wrap around.

Photoperiodism

Photoperiodism is the response of an organism, such as a plant, to naturally occurring changes in light over a 24-hour period. The photoperiodic conditions are perceived by the leaves of plants. Sunflowers are particularly known for their photoperiodism, as they open and close in response to the changing position of the sun throughout the day. All flowering plants have been placed in one of three categories with respect to photoperiodism:

- *Short-day plants*—Flowering is promoted by day lengths shorter than a certain critical day length; this category includes poinsettias, chrysanthemums, goldenrod, and asters.
- *Long-day plants*—Flowering is promoted by day lengths longer than a certain critical day length; this category includes spinach, lettuce, and most grains.
- *Day-neutral plants*—Flowering response is insensitive to day length; this category includes tomatoes, sunflowers, dandelions, rice, and corn.

Plant Reproduction

Plants reproduce either sexually or asexually. Each type of reproduction has its benefits and disadvantages.

Sexual Reproduction

- Sexual reproduction occurs when a sperm nucleus from the pollen grain fuses with an egg cell from the ovary of the pistil (the female reproductive structure in flowers, consisting of the stigma, style, and ovary).
- Each brings a complete set of genes and produces genetically unique organisms.
- The resulting plant embryo develops inside the seed and grows when the seed is germinated.

Asexual Reproduction

- Asexual reproduction occurs when a vegetative part of a plant, root, stem, or leaf gives rise to a new offspring plant whose genetic content is identical to the parent plant. An example would be a plant reproducing by root suckers, shoots that come from the root system (e.g., breadfruit tree).
- Asexual reproduction is also called *vegetative propagation*. It is an important way for plant growers to obtain many identical plants from one very quickly.
- Asexual reproduction allows plants (e.g., crabgrass) to spread and colonize an area quickly.

Biomass Plant Cell Walls

Figure 9.1 shows the basic organelles within a standard plant cell, including the cell wall. It should be pointed out, however, that plants can have two types of cell walls, primary and secondary (see Figure 9.2). Primary cell walls contain cellulose, consisting of hydrogen-bonded chains of thousands of glucose molecules, in addition to hemicellulose and other materials all woven into a network. Certain types of cells, such as those in vascular tissues, develop secondary walls inside the primary wall after the cell has stopped growing. These cell-wall structures also contain lignin, which provides rigidity and resistance to compression. The area formed by two adjacent plant cells, the middle lamella, typically is enriched

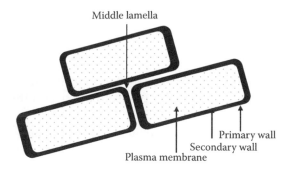

FIGURE 9.2
Plant cell walls.

with pectin. Cellulose in higher plants is organized into microfibrils, each measuring about 3 to 6 nm in diameter and containing up to 36 glucan chains having thousands of glucose residues. Like steel girders stabilizing a skyscraper's structure, the mechanical strength of a primary cell wall is due mainly to the microfibril scaffold (i.e., crystalline cellulose core) (USDOE, 2005).

> **DID YOU KNOW?**
> Cellulose microfibrils are composed of linear chains of glucose molecules bound by hydrogen.

Composition of Biomass

The ease with which biomass can be converted to useful products or intermediates is determined by the composition of the biomass feedstock. Biomass contains a variety of components, some of which are readily accessible and others that are much more difficult and costly to extract. The composition and subsequent conversion issues for current and potential biomass feedstock compounds are listed and described below.

Starch (Glucose)

Starch is readily recovered and converted from grain (e.g., corn, wheat, rice) for use in a variety of products. Starch is the main energy reserve of superior vegetal plants. Starch from corn grain provides the primary feedstock for today's existing and emerging sugar-based bioproducts, such as polylactide, as well as the entire fuel ethanol industry. Corn grain serves as the primary feedstock for starch used to manufacture today's biobased products. Corn wet mills use a multistep process to separate starch from the germ, gluten (protein), and fiber components of corn grain. The starch streams generated by wet milling are highly pure, and acid or enzymatic hydrolysis is used to break the glycosidic linkages of starch to yield glucose. Glucose is then converted into a multitude of useful products (Paster et al., 2003).

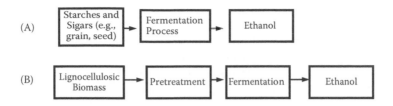

FIGURE 9.3

(A) Starch and sugars directly fermented to ethanol, and (B) pretreatment of lignocellulosic biomass before fermentation into ethanol.

Lignocellulosic Biomass

Lignocellulosic biomass is the non-grain portion of biomass (e.g., cobs, stalks), often referred to as *agricultural stover* or *residues*; energy crops such as switchgrass also contain valuable components, but they are not as readily accessible as starch. These lignocellulosic biomass resources (also called *cellulosic*) are comprised of cellulose, hemicellulose, and lignin and are available in various forms, including agricultural residues, forestry residues, wood, energy crops, and wastes from the food, paper, and pulp industries. Generally, lignocellulosic material contains 30 to 50% cellulose, 20 to 30% hemicellulose, and 20 to 30% lignin. Some exceptions to this are cotton (98% cellulose) and flax (80% cellulose). Lignocellulosic biomass is perceived as a valuable and largely untapped resource for the future bioindustry. Demirbas (2006) pointed out that pyrolysis and gasification are promising technologies for the production of fuels and chemicals from lignocellulosic biomass; however, recovering the components in a cost-effective way represents a significant technical challenge. In addition to forest and agriculture biomass residues, dedicated energy crops are another source of lignocellulosic biomass for biofuel production. Among a rather large number of herbaceous and woody crops that can be used for biofuel production, poplar (*Populus* spp., 18.4% hemicellulose) and willow (*Salix* spp., 21.3% hemicellulose) have been the focus of most research efforts (Drapcho et al., 2008; Esteghlalian et al., 1997; Sassner et al., 2006). It is important to point out that the process of producing ethanol from various feedstocks can vary according to the feedstock used; for example, feedstocks such as grain or seeds that contain starches and sugars can be directly fermented with minimal pretreatment, as illustrated in Figure 9.3A. Lignocellulosic biomass, however, is comprised of lignin, cellulose, and hemicellulose and requires more pretreatment before it can be fermented. In the process of lignocellulosic biomass fermentation, the feedstock must first be pretreated with hydrolysis or one of several other treatments (see Figure 9.3B). Pretreatment breaks the feedstock into component sugars that can then be fermented, as shown in Figure 9.3A.

Cellulose

Cellulose is one of nature's polymers and is composed of glucose, a six-carbon sugar. It is synthesized within living cells from a glucose-based sugar nucleotide. The glucose molecules are joined by glycosidic linkages, which allow the glucose chains to assume an extended ribbon conformation that favors its organization in oriented packs of about 50 to 100 molecules. Hydrogen bonding between chains leads to the formation of the flat sheets that lay on top of one another in a staggered fashion, similar to the way staggered bricks add strength and stability to a wall. As a result, cellulose is very chemically stable and insoluble and serves as a structural component in plant walls.

Hemicellulose

Hemicellulose is a polymer containing primarily 5-carbon sugars such as xylose and arabinose, with some glucose and mannose dispersed throughout. Hemicellulose is a key component of plant cell walls, comprising up to 25 to 30% of woody plant tissues; it forms a short-chain polymer that interacts with cellulose and lignin to form a matrix in the plant wall, strengthening it. Hemicellulose is more easily hydrolyzed than cellulose. Much of the hemicellulose in lignocellulosic material is solubilized and hydrolyzed to pentose and hexose sugars. Hemicellulose can be hydrolyzed into its component sugars and used as a fermentation feedstock for the production of ethanol.

Lignin

Lignin binds the cellulosic/hemicellulose matrix in plant cell walls and shields them from enzymic and chemical degradation while adding flexibility to the mix. It is the third most abundant structural polymeric material found in plant cells. The molecular structure of lignin polymers is very random and disorganized and consists primarily of carbon ring structures (benzene rings with methoxyl, hydroxyl, and propyl groups) interconnected by polysaccharides (sugar polymers). The ring structures of lignin have great potential as valuable chemical intermediates; however, separation and recovery of the lignin are difficult.

Oils and Protein

The seeds of certain plants offer two families of compounds with great potential for bioproducts: oils and protein. Oils and protein are found in the seeds of certain plants (soybeans, castor beans) and can be extracted in a variety of ways. Plants raised for this purpose include soy, corn, sunflower, safflower, rapeseed, and others. A large portion of the oil and protein recovered from oilseeds and corn is processed for human or animal consumption, but they can also serve as raw materials for lubricants, hydraulic fluids, polymers, and a host of other products.

Vegetable Oils

Vegetable oils are composed primarily of triglycerides, also referred to as triacylglycerols. Triglycerides contain a glycerol molecule as the backbone with three fatty acids attached to glycerol's hydroxyl groups.

Proteins

Proteins are natural polymers with amino acids as the monomer unit. They are incredibly complex materials, and their functional properties depend on molecular structure. The 20 amino acids are each differentiated by their side chain or R-group, and they can be classified as nonpolar and hydrophobic, polar uncharged, or ionizable. The interactions among the side chains, the amide protons, and the carbonyl oxygen help create the three-dimensional shape of the protein.

Chapter Review Questions

9.1 _____ consists of meristematic cells located at the tip of a root or shoot.

9.2 _____ is characterized by having two cotyledons.

9.3 _____ is a group of plant cells that can divide indefinitely, providing new cells for the plant.

9.4 _____ is characterized by having a single cotyledon.

9.5 _____ is the vascular tissue of plants that transports water and dissolved minerals from the roots upward to other parts of the plant.

9.6 _____ is living wood.

9.7 _____ tissues are composed of specialized cells that create tubes through which material can flow throughout the plant body.

9.8 _____ absorb the sunlight necessary to manufacture plant sugars.

9.9 _____ are the site of photosynthesis.

9.10 Lateral meristems are called _____.

9.11 Root tip messengers are called _____.

References and Recommended Reading

Demirbas, M.F. (2006). Hydrogen from various biomass species via pyrolysis and steam gasification processes, *Energy Sources, Part A, Recovery, Utilization, and Environmental Effects*, 28:245–252.

Drapcho, C.M., Nhuan M.P., and Walker, T.H. (2008). *Biofuels Engineering Process Technology*, McGraw-Hill, New York.

Esteghlalian, A. et al. (1997). Modeling and optimization of the dilute sulfuric-acid pretreatment of corn stover, poplar and switchgrass, *Bioresource Technology*, 59:129–136.

Paster, M., Pellgrino, J.L., and Carole, T.M. (2003). *Industrial Products: Today and Tomorrow*, Energetics, Inc., for the U.S. Department of Energy, Office of Energy Efficiency and Renewable Energy, Office of the Biomass Program, Washington, D.C.

Sassner, P., Galbe, M., and Zacchi, G. (2006). Bioethanol production based on simultaneous saccharification and fermentation of steam-pretreated *Salix* at high dry-matter content, *Enzyme and Microbial Technology*, 39:756–762.

USDOE. (2005). *Breaking the Biological Barriers to Cellulosic Ethanol: A Joint Research Agenda*, Workshop sponsored by the U.S. Department of Energy, December 7–9, 2005, Rockville, MD.

Vaca-Garcia, C. (2008). Biomaterials, in Clark, J. and Deswarte, F., Eds., *Introduction to Chemicals from Biomass*, John Wiley & Sons, New York.

10

Forest-Based Biomass Feedstock

> The sustainable development of the chemical and energy industry is an indispensable component of our sustainable society. However, the traditional chemical and energy industry depends heavily on such nonrenewable fossil resources as oil, coal, and natural gas. Its feedstock shortage and the resultant environmental and climatic problems pose a great threat for any type of sustainable development. Lignocellulosic materials are the most abundant renewable resources in the world, and their efficient utilization provides a practical route to address these challenges.
>
> —Cheng and Zhu (2009)

Introduction

Interest in using biomass feedstocks to produce power, liquid fuels, and chemicals in the United States is increasing. Perhaps this increased interest has been stoked by a recent statement by John Hofmeister, who said that, "Americans could be paying $5 for a gallon of gasoline by 2012" (Segall, 2010). In that same article, Tom Kloza, chief oil analyst with Oil Price Information Service, stated that the "wolf is out there and it's going to be at the door. … I agree with [Hofmeister] that we'll see those numbers at some point this decade but not yet" (Segall, 2010). Paul Thompson (2008) warned that "for years, the experts have been warning of the dangers of oil depletion. They have been accused of crying wolf. This time, the wolf really *is* at the door." Statements such as these catch our attention and make us aware of a pending dilemma—the realization that cheap, readily accessible oil and oil products are a thing of the past with a future best characterized as uncertain at best.

What it really boils down to is awareness. This growing awareness of our uncertain current and future energy supply is quite apparent to motorists at the gas pump who pay an increasing amount each time they fill their gas tanks. That $3, $4, or $5+ price per gallon of gasoline is a real eye-opener and bank account drainer. NRDC (2001) reported that Americans spend more than $200,000 per minute on foreign oil—$13 million each hour. More than $25 billion a year goes for Persian Gulf imports alone. This NRDC analysis considered oil demand and supply projections and how our current policy of oil dependence affects our economy and security. Our dependence on oil is a threat to our national security and our economy. Growing demand and shrinking domestic production mean that we are importing more and more oil each year—much of it from the world's most unfriendly

---

DID YOU KNOW?

From May 2010 until December 2010, the price of oil increased 34%, and the average price of gas was $3.07.

or unstable regions. These are important considerations, and when Thompson warns that the wolf is at the door it is difficult to argue against such a view, because if he is not at the door then he is certainly in our front yard.

Not all is lost, however. Because we possess that one magical attribute, freedom, we have the opportunity to think freely, to invent, and to innovate. It is innovation that will solve our energy crisis. Just like the cure for cancer and the cure for all other horrendous diseases, solutions to our energy problems are out there—we simply have to find them. We must remember that every problem has a solution. Whether we produce biofuels from used vegetable oil or by growing, harvesting, and processing switchgrass, corn, wheat, algae, gold-of-pleasure (*Camelina sativa*), elephant yeast, sugar, starch, myco-diesel, *Miscanthus*, or lignocellulosic feedstocks, the solution is out there. In a rather comic sidenote, while I was writing this text one of my learned associates asked me: "Do you really think ... I mean, do you really believe that we will be able to chop down a tree and use it to produce energy in the future?" My reply: "Absolutely!" One part of my reply I kept to myself, though: "Heck, even cavemen knew the value of an energy-producing tree."

Before presenting a detailed discussion of forest-derived biofuel resources, it is important to set the stage for our biomass discussion by providing a basic overview of biofuels classified by generational levels. Following is a description of the potential availability of these biofuel feedstocks in the United States.

First-Generation Feedstocks*

Simply, first-generation biofuels are biofuels made from sugar, starch, and vegetable oils. The type of major end-use product is easily categorized for first-generation feedstocks (see Table 10.1). For ethanol production, corn and sugarcane are the most commonly used feedstocks. Soybean and other vegetable oils and animal fats are used for biodiesel production (and bioproducts). Manure and landfill organic waste are used for methane production and the generation of electricity. Corn is used for ethanol production and currently is the leading feedstock used in United States.

Several factors favor a positive outlook for further near-term growth in corn-ethanol production. Continued high oil prices will provide economic support for the expansion of all alternative fuel programs, including corn ethanol. Technology improvements that increase feedstock productivity and fuel conversion yields and positive spillovers from second-generation technologies (biomass gasification in ethanol refiners) will also help to lower production costs for corn ethanol.

Among the factors likely to limit future growth of corn-ethanol production are increased feedstock and other production costs; increased competition from unconventional liquid fossil fuels (from oil sands, coal, heavy oil, and shale); the emergence of cellulosic ethanol as a low-cost competitor; and new policies to reduce greenhouse gas emissions (GHGs) that could favor advanced biofuels over corn ethanol.

Biodiesel is another biofuel experiencing expansion. Although its production costs are higher than ethanol, biodiesel has some environmental advantages, including biodegradability and lower sulfur and carbon dioxide emissions when burned. Biodiesel production

---

* This section is based on USDA, *The Economics of Biomass Feedstocks in the United States: A Review of the Literature*, Biomass Research and Development Board, U.S. Department of Agriculture, Washington, D.C., 2008.

TABLE 10.1

Biomass Feedstocks and Major Bioenergy End-Use Applications

		First-Generation Fuels			Second-Generation Fuels	
		Ethanol	Biodiesel	Methane	Cellulosic Ethanol	Thermoch Fuels (e.g., Ethanol, Diesel, Butanol)
First-Generation Feedstocks						
Starch and sugar-feedstock for ethanol	Corn	X				
	Sugarcane	X				
	Molasses	X				
	Sorghum	X				
Vegetable oil and fats for biodiesel	Vegetable oils		X			
	Recycled fats and grease		X			
	Beef tallow		X			
Second-Generation Feedstocks (Short-Term)						
Agricultural residues and livestock byproducts	Corn stover			X		
	Wheat straw					
	Rice straw					
	Bagasse					
	Manure					
Forest biomass	Logging residues				X	X
	Fuel treatments				X	X
	Conventional wood				X	X
Urban woody waste and landfills	Primary wood products				X	X
	Secondary mill residues				X	X
	Municipal solid waste			X	X	X
	Landfills			X	X	
Second-Generation Feedstocks (Long-Term)						
Herbaceous energy crops	Switchgrass				X	X
	Miscanthus				X	X
	Reed canary grass				X	X
	Alfalfa				X	X
Short-rotation woody crops	Willow				X	X
	Hybrid popular				X	X
	Cottonwood pines				X	X
	Sycamore pines				X	X
	Eucalyptus				X	X

in the United States increased rapidly from less than 2 million gallons in 2000 to about 500 million gallons in 2007. Policy incentives in the Energy Independence and Security Act of 2007 are expected to sustain demand for 1 billion gallons per year of this fuel after 2011.

A variety of oil-based feedstocks are converted to biodiesel using a process known as *transesterification*. This is the process of exchanging the organic group R″ of an ester with the organic group R′ of an alcohol. These reactions are often catalyzed by the addition of an acid or base as shown below:

$$\text{Alcohol} + \text{Ester} \rightarrow \text{Different Alcohol} + \text{Different Ester}$$

The oil-based feedstocks include vegetable oils (mostly soy oil), recycled oils and yellow grease, and animal fats such as beef tallow. It takes 3.4 kg of oil/fat to produce 1 gallon of biodiesel (Baize, 2006). Biodiesel production costs are high compared to ethanol, with feedstocks accounting for 80% or more of total costs.

The biodiesel industry consists of many small plants that are highly dispersed geographically. Decisions about plant location are primarily determined by local availability and access to the feedstock. Recent expansion in biodiesel production is affecting the soybean market. Achieving a nationwide target of 2% biodiesel blend in diesel transportation fuel, for example, would require 2.8 million metric tons (MT) of vegetable oil, or about 30% of current U.S. soybean oil production (WAOB, 2008).

One of the key biodiesel byproducts is glycerin. At the present time, there is concern about producing glycerin mainly because there are no existing markets for the product; however, recent technological developments include an alternative chemical process that would produce biodiesel without glycerin. In addition, new processes are being tested that further transform glycerin into propylene glycol, which is used in the manufacture of antifreeze (Anon., 2007).

Second-Generation Feedstocks: Short-Term Availability

Agricultural residues, a second-generation biomass feedstock (considered an advanced biofuel), are a potentially large and readily available biomass resource, but sustainability and conservation constraints could place much of it out of reach (see Table 10.1). Given current U.S. cropland use, corn and wheat offer the most potentially recoverable residues; however, these residues play an important role in recycling nutrients in the soil and maintaining long-term fertility and land productivity. Removing too much residue could aggravate soil erosion and deplete the soil of essential nutrients and organic matter. Methodologies have been developed to estimate the safe removal rates for biomass (based on soil erosion). Methodologies to determine safe removal rates while safeguarding soil fertility and meeting conservation objectives still need to be developed. Studies show that under current tillage practices the national average safe removal rate based on soil erosion for corn stover is less than 30%. Actual rates vary widely depending on local conditions. In other words, much of the generated crop residues may be out of reach for biomass use if soils conservation goals are to be achieved (Graham et al., 2007).

The estimated delivery costs for agricultural residues vary widely depending on crop type, load resource density, storage and handling requirements, and distance and transportation costs. Moreover, existing estimates are largely derived from engineering models, which may not account for economic conditions. Agricultural residue feedstocks (such as corn stover) have a significant advantage in that they can be readily integrated into the expanding corn-ethanol industry; however, dedicated energy crops (such as switchgrass) may have more benign environmental impacts.

Another significant second-generation biomass source (the focus of this text) is *forest-derived biomass*, which can be immediately available should the bioenergy market develop. Logging residues are associated with timber industry activities and constitute significant biomass resources in many states, particularly in the Northeast, North Central, Pacific Northwest, and Southeast. In the Western United States, the predominance of public lands and environmental pressure reduces the supply potential for logging residues, but there is a vast potential for biomass from thinning undertaken to reduce the risk of forest fires. The few analyses that have examined recoverability of logging residues cite the need to account for factors such as the scale and location of biorefineries and biopower plants, as well as regional resource density.

The potential for forest residues may be large but the actual quantities available for biomass conversion may be low due to the economics of harvesting, handling, and transporting the residues from forest areas to locations where they could be used. It is not clear how these residues compete with fossil fuels in the biopower and co-firing industries. In addition, there are competing uses for these products in the pulp and paper industry, as well as different bioenergy end-uses. Economic studies of logging residues suggest a current lack of competitiveness with fossil fuels (coal, gas), but logging residues could become more cost competitive with further improvements in harvesting and transportation technologies and with policies that require a more full accounting of the social and environmental benefits from converting forest residues to biopower or biofuels. Another source of forest residues that could be recovered in significant quantities is biomass from fuel treatments and thinnings. Fuel treatment residues are the byproduct of efforts to reduce the risk of loss from fire, insects, and disease; therefore, they present substantially different challenges than logging residues.

The overall value of forest health benefits such as clean air and water is generally believed to exceed the cost of treatment; however, treated forests are often distant from end-use markets, resulting in high transportation costs to make use of the harvested material. Road or trail access, steep terrain, and other factors commonly limit thinning operations in Western forests.

Transportation costs can be a significant factor in the cost of recovering biomass. As much as half the cost of the material delivered to a manufacturing facility may be attributed to transportation. An alternative to the high cost of transporting forest thinnings is onsite densification of the biomass. This could entail pelletization, fast pyrolysis (to produce bio-oil), or baling. The economics of transporting thinned woody residues vs. onsite densification depends on the distance to end-use markets. Densification may be more economical if power generation facilities are far away. In addition to co-firing or co-generation

DID YOU KNOW?

The recent discovery of the fungus *Gliocladium roseum* points toward the production of myco-diesel from cellulose. This organism was discovered in the rainforests of northern Patagonia and has the unique capability of converting cellulose into medium-length hydrocarbons typically found in diesel fuel (Chadwick, 2008).

facilities, improvements in thermochemical conversion efficiency and establishment of small-scale conversion facilities using gasification or pyrolysis may favor the use of forest residues for biofuel production (Polagye et al., 2007).

A third major category of immediately available second-generation biomass is wood residues from *secondary mill products* and *urban wood waste*. Urban wood waste provides a relatively inexpensive feedstock to supplement other biomass resources (Wiltsee, 1998). Urban wood waste encompasses the biomass portion of commercial, industrial, and municipal solid waste (MSW), while secondary mill residues include sawdust, shavings, wood trims, and other byproducts generated from processing lumber, engineered wood products, or wood particles. Both urban wood waste and secondary mill residues have several primary uses and disposal methods. Urban wood waste not used in captive markets (such as the pulpwood industry) could be used as biomass either to generate electricity or to produce cellulosic ethanol when it becomes commercially viable.

The amount of urban wood wastes produced in the United States is significant, and their use as biomass could be economically viable, particularly in large urban centers (Wiltsee, 1998). Several national availability estimates exist for various types of urban wood wastes, but estimates vary depending on methodology, product coverage, and assumptions about alternative uses (McKeever, 2004; Wiltsee, 1998). One of the challenges facing the potential availability of urban woody waste is to sort out the portion that is available (not currently used) and determine alternative uses, including those used by captive markets (not likely to be diverted to bioenergy). One assessment of urban wood waste found that 36% of total biomass generated is currently sold to noncaptive markets, and 50% of the unused residues are not available due to contamination, quality, or recoverability.

One source of recurring and potentially available carbon feedstock is municipal solid waste. In 2005, the EPA estimated that 245.7 million tons of MSW were generated in the United States, of which 79 million tons were recycled, 33.4 millions tons were diverted to energy recovery, and 133.3 million tons were disposed of in landfills. As such, landfilled material represents a potentially significant source of renewable carbon that could be used for fuel/energy production or in support of biofuel production.

Second-Generation Feedstocks: Long-Term Availability

Large-scale biofuels production, in the long run, will require other resources, including dedicated energy crops. Dedicated feedstocks include perennial grasses (e.g., switchgrass) and trees grown as crops specifically to provide the required raw materials to bioenergy producers. A steady supply of low-cost, uniform, and consistent-quality biomass feedstock will be critical for the economic viability of cellulosic ethanol production. In the late 1980s, the Department of Energy sponsored research on perennial herbaceous (grassy) biomass crops, particularly switchgrass, which is considered a model energy crop because of its many perceived advantages: native to North America, high biomass yield per acre, wide regional coverage, and adaptability to marginal land conditions. An extensive research program on switchgrass in the 1990s generated a wealth of information on high-yielding varieties, regional adaptability, and management practices. Preliminary field trials showed that the economic viability of switchgrass cultivation depends critically on the initial establishment success. During this phase, seed dormancy and seedling sensitivity to soil and weed conditions require that recommended practices be closely followed by growers. Viable yields require fertilization rates at about half the average for corn.

ATTRIBUTES OF THE PERFECT ENERGY CROP

- High biomass
- Improved composition and structure
- Disease and pest resistance
- Optimized architecture
- Salt, pH, and aluminum tolerance
- Rapid and cost-effective propagation
- Stand establishment (cold germination and cold growth)
- Perennial
- Deep roots

Switchgrass (*Panicum virgatum*), a hardy, deep-rooted, perennial rhizomatous warm-season grass native to North America, is believed to be the most suitable for cultivation in marginal lands, low-moisture lands, and lands with lower opportunity costs such as pastures, including lands under the Conservation Reserve Program (CRP) where the federal government pays landowners annual rent for keeping land out of production (McLaughlin and Kzos, 2005). Additionally, a large amount of highly erodible land in the Corn Belt is unsuitable for straw or stover removal but potentially viable for dedicated energy crops such as switchgrass. Factors favoring adoption of switchgrass include selection of suitable lands, environmental benefits (carbon balances, improved soil nutrients and quality), and use of existing hay production techniques to grow the crop. Where switchgrass is grown on CRP lands, payments help to offset production costs. Factors discouraging switchgrass adoption include no possibility for crop rotation; farmers' risk aversion for producing a new crop because of lack of information, skills, and know-how; potential conflict with on-farm and off-farm scheduling activities; and a lack of compatibility with long-term land tenure. Overall, production budget and delivery cost assessments suggest that switchgrass is a high-cost crop (under current technology and price conditions) and may not compete with established crops, except in areas with low opportunity costs (e.g., pasture land, marginal lands).

Substantial variability is apparent in the economics of switchgrass production and assessments of production budgets and delivered costs. Factors at play include methods of storage and handling, transport distances, yields, and types of land used (cropland vs. grassland). When delivered costs of switchgrass are translated into breakeven prices (compared with conventional crops), it becomes apparent that cellulosic ethanol or bio-power plants would have to offer relatively high prices for switchgrass to induce farmers to grow it (Rinehart, 2006). The economics of switchgrass could be improved if growers benefited from CRP payments and other payments tied to environmental services (such as carbon credits). In the long run, the viability of an energy crop such as switchgrass hinges on continued reductions in cellulosic ethanol conversion costs and sustained improvements in switchgrass yield and productivity through breeding, biotechnology, and agronomic research.

Although an important biofuels crop, switchgrass does have limitations. Switchgrass is not optimally grown everywhere; for example, in the upper Midwest under wet soils, reed canary grass (*Phalaris arundinacea*) is more suitable, and semitropical grass species are better adapted to the Gulf Coast region. State and local efforts are underway to test alternatives to switchgrass, such as reed canary grass, which is a tall, perennial grass that forms extensive single-species stands along the margins of lakes and streams, and *Miscanthus* (often confused with "elephant grass"), which offers rapid growth, low mineral content, and high biomass yield.

With regard to the perspective of long-run sustainability, the ecology of perennial grassy crops favors a multiplicity of crops or even a mix of species within the same area. Both ecological and economic sustainability favor the development of a range of herbaceous species for optimal use of local soil and climatic condition. A mix of several energy crops in the same region would help reduce the risk of epidemic pests and disease outbreak and optimize the supply of biomass to an ethanol or biopower plant, as different grasses mature and can be harvested at different times. Moreover, development of future energy crops must be evaluated from the standpoint of their water use efficiency, impact on soil nutrient cycling, effect on crop rotations, and environmental benefits (improved energy use efficiency and reduced greenhouse gas emissions, nutrient runoff, pesticide runoff, and land-use impacts). In the long run, developing a broad range of grassy crops for energy use is compatible with both sustainability and economic viability criteria.

Short-Rotation Woody Crops

Short-rotation woody crops (SRWC) represent another important category of future dedicated energy crops. Energy crops include perennial grasses and trees that are grown through traditional agricultural practices primarily to be used as feedstocks for energy generation. The U.S. Department of Energy estimated that about 190 million acres of land in the United States could be used to produce energy crops (Antares Group, 2003). Among the SRWC, hybrid poplar, willow, American sycamore, sweetgum, and loblolly pine have been extensively researched for their very high biomass yield potential. Breeding programs and management practices continue to be developed for these species. SRWC are based on a high-density plantation system and more frequent harvesting (every 3 to 4 years for willow and 7 years for hybrid poplar).

- *Hybrid poplar*—This species is very site specific and has a limited growing niche. It requires an abundant and continuous supply of moisture during the growing season. Soils should be moist, but not continually saturated, and should have good internal drainage. Poplar prefers damp, well-drained, fine sandy-loam soils located near streams, where coarse sand is first deposited as flooding occurs. Hybrid poplars are among the fastest-growing trees in North America; in just six growing seasons, hybrid poplars can reach 60 feet or more in height. They are well suited for the production of bioenergy, fiber, and other biobased products.

- *Willow (Salix) species*—In folklore and myth, a willow tree is believed to be quite sinister, capable of uprooting itself and stalking travelers. The reality is that willow is used for biofiltration, constructed wetlands, ecological wastewater treatment systems, hedges, land reclamation, landscaping, phytoremediation, streambank stabilization (bioengineering), slope stabilization, soil erosion control, shelterbelt and windbreak, soil building, soil reclamation, tree bog composting toilet, and wildlife habitat. Willow is grown for biomass or biofuel in energy forestry systems as a consequence of its high energy in/energy out ratio, large carbon mitigation potential, and fast growth (Aylott, 2008). Willow is also grown and harvested for making charcoal.

- *American sycamore*—The sycamore prefers alluvial soils along streams in bottom-lands. Sycamore growth and yield are less than for cottonwood and willow.

- *Sweetgum (Liquidambar styraciflua)*—Sweetgum is a species tolerant of a variety of soils, but it grows best on the rich, moist, alluvial clay and loam soils of river bottoms.
- *Loblolly pine (Pinus taeda)*—Loblolly pine is quite adaptable to a variety of sites. It performs well on both poorly drained bottomland flats and modestly arid uplands. Biomass for energy is currently being obtained from precommercial thinnings and from logging residues in loblolly pine stands. Utilization will undoubtedly increase, and loblolly pine energy plantations may become a reality.

In many parts of the country, plantations of willow, poplar, pines, and cottonwood have been established and are being commercially harvested. Willow plants are being planted in New York, particularly following enactment of the Renewable Portfolio Standard (RPS) and other state incentives. Over 30,000 hectares of poplar are grown in Minnesota, and several thousand hectares are also grown as part of a DOE-funded project to provide biomass for a power utility company in southern Minnesota. The Pacific Northwest has large plantations of hybrid poplars, estimated at 60,000 hectares in 2007; however, the production of most of these plantations is currently used for pulp wood, with little volume being used for bioenergy. Because SRWC can be used either for biomass or as feedstock for pulp and other products, pulp demand will influence the cost of using it for bioenergy production.

An important consideration for energy crops (e.g., switchgrass, poplar, willow) is the potential for increasing yields and developing other desirable characteristics. Most energy crops are unimproved or have been bred only recently for biomass yield, whereas corn and other commercial food crops have undergone substantial improvements in yield, disease resistance, and other agronomic traits. A more complex understanding of biological systems and application of the latest biotechnological advances would accelerate the development of new biomass crops with desirable attributes. These attributes include increased yields and processability, optimal growth in specific microclimates, better pest resistance, efficient nutrient use, and greater tolerance to moisture deficits and other sources of stress. Agronomic and breeding improvements of these new crops could provide a significant boost to future energy crop production. Table 10.2 presents the energy content and typical costs for common forest-derived (or agriculturally grown) energy crops.

Energy crops could represent a significant source of additional farm income. The advantages of using crops specifically grown for energy production include consistency in moisture content, heat content, and processing characteristics. Disadvantages include relatively higher overall costs than many fossil fuels, higher value alternative land uses that further drive up costs, added expenses associated with harvesting and processing, and farmers' and power plant owners' unfamiliarity with energy crops.

DID YOU KNOW?

In the 1980s, the U.S. Department of Energy (DOE) gradually shifted its focus to technologies that could have large-scale impacts on national consumption of fossil energy. Many of the DOE's publications from this period reflect a philosophy of energy research that might, somewhat pejoratively, be called the "quads mentality." A quad is a short-hand name for quadrillion Btu, the unit of energy often used by the DOE to describe the amounts of energy that a given technology might be able to displace. A quad represents 10^{15} (1,000,000,000,000,000) Btu of energy. This perspective led the DOE to focus on the concept of immense algae farms.

TABLE 10.2

Energy and Cost Characteristics of Forest-Derived Energy Crops

Energy Crop	Energy Content Wet (Btu/lb)	Energy Content Dry (Btu/lb)	Cost Range (per ton)	Cost/MMBtu
Hybrid poplar	4100	8200	$39 to $60	$4.78 to $7.32
Hybrid willow	6060	8670	$39 to $60	$4.76 to $7.32

Source: Adapted from ODOE, *Bioenergy in Oregon: Biomass Technologies*, Oregon Department of Energy, Salem, 2010; Walsh, M. et al., *Biomass Feedstock Availability in the United States: 1999 State Level Analysis*, Oak Ridge National Laboratory, Knoxville, TN, 1999.

Third-Generation Biofuels Produced from Algae*

Third-generation biofuels are produced from algae. Algae are low-input, high-yield feedstocks for the production of biofuels. Based on laboratory experiments, algae can produce up to 30 times more energy per acre than land crops such as soybeans (Hartman, 2008). The higher prices of fossil fuels have led to increased interest in algae farming (algaculture). Actually, the algae-to-biodiesel alternative renewable fuel source program got its start earlier, during the Carter Administration, in response to the 1970s Arab fuel embargo. At that time, the Carter Administration consolidated all federal energy activities under the auspices of the newly established U.S. Department of Energy (DOE). Among the various programs established to develop all forms of alternative energy, the DOE initiated research on the use of plant life as a source of transportation fuels (today known as the Biofuels Program). The DOE estimated that if algae fuel replaced all of the petroleum fuel in the United States, producing it would require only 15,000 square miles (38,849 square kilometers), which is roughly the size of Maryland (Hartman, 2008). Before discussing the algae-to-biofuels process in detail, it is important to present a fundamental explanation of the algae-to-biofuels process; it is important for the reader to be grounded in basic algal concepts.

The protists that perform photosynthesis are called *algae*. Algae can be both a nuisance and an ally. Many ponds, lakes, rivers, streams, and bays (e.g., Chesapeake Bay) in the United States (and elsewhere) are undergoing *eutrophication*, the enrichment of an environment with inorganic substances (phosphorus and nitrogen). When eutrophication occurs, filamentous algae such as *Caldophora* break loose in a pond, lake, stream, or river and wash ashore, thus making their stinking, noxious presence known. Algae are allies in many wastewater treatment operations. They can be valuable in long-term oxidation ponds, where they aid in the purification process by producing oxygen. Before discussing the specifics and different types of algae, it is important to be familiar with algal terminology.

Algae Terminology

Algae—Large and diverse assemblages of eukaryotic organisms that lack roots, stems, and leaves but have chlorophyll and other pigments for carrying out oxygen-producing photosynthesis.

---

* This section is based on Spellman, F.R., *The Science of Renewable Energy*, CRC Press, Boca Raton, FL, 2011.

Algology or *phycology*—The study of algae.

Antheridium—Special male reproductive structures where sperm are produced.

Aplanospore—Nonmotile spores produced by sporangia.

Benthic—Algae attached and living on the bottom of a body of water.

Binary fission—Nuclear division followed by division of the cytoplasm.

Chloroplasts—Packets that contain chlorophyll *a* and other pigments.

Chrysolaminarin—Carbohydrate reserve in organisms of the division Chrysophyta.

Diatoms—Photosynthetic, circular, or oblong chrysophyte cells.

Dinoflagellates—Unicellular, photosynthetic protistan algae.

Epitheca—Larger part of the frustule (diatoms).

Euglenoids—Contain chlorophylls *a* and *b* in their chloroplasts; representative genus is *Euglena*.

Fragmentation—A type of asexual algal reproduction in which the thallus breaks up and each fragmented part grows to form a new thallus.

Frustule—The distinctive two-piece wall of silica in diatoms.

Hypotheca—The small part of the frustule (diatoms).

Neustonic—Algae that live at the water–atmosphere interface.

Oogonia—Vegetative cells that function as female sexual structures in the algal reproductive system.

Pellicle—A *Euglena* structure that allows for turning and flexing of the cell.

Phytoplankton—Made up of algae and small plants.

Plankton—Free-floating, mostly microscopic aquatic organisms.

Planktonic—Algae suspended in water as opposed to attached and living on the bottom (benthic).

Protothecosis—A disease in humans and animals caused by the green algae *Prototheca moriformis*.

Thallus—The vegetative body of algae.

Algae are autotrophic, contain the green pigment chlorophyll, and are a form of aquatic plant. Algae differ from bacteria and fungi in their ability to carry out photosynthesis, the biochemical process requiring sunlight, carbon dioxide, and raw mineral nutrients. Photosynthesis takes place in the chloroplasts. The chloroplasts are usually distinct and visible. They vary in size, shape, distribution, and numbers. In some algal types, the chloroplast may occupy most of the cell space. They usually grow near the surface of water because light cannot penetrate very far through water. Although in mass (multicellular forms such as marine kelp) they can easily be seen by the unaided eye, many of them are microscopic. Algal cells may be nonmotile or motile by one or more flagella, or they may exhibit gliding motility, as diatoms do. They occur most commonly in water, be it freshwater, saltwater, or polluted. They may be suspended in the water, as in the case of plankton, or attached and living on the bottom (referred to as *benthic*). A few algae that live at the water–atmosphere interface are termed *neustonic*. Within these freshwater and saltwater environments, they are important primary producers (the start of the food chain for other organism). During their growth phase, they are important oxygen-generating organisms and constitute a significant portion of the plankton in water.

According to the five kingdom system of Whittaker, the algae belong to seven divisions distributed between two different kingdoms. Although seven divisions of algae occur, only five divisions are discussed in this text:

- *Chlorophyta*—Green algae
- *Euglenophyta*—Euglenoids
- *Chrysophyta*—Golden-brown algae, diatoms
- *Phaeophyta*—Brown algae
- *Pyrrophyta*—Dinoflagellates

The primary classification of algae is based on cellular properties. Several characteristics are used to classify algae, including: (1) cellular organization and cell-wall structure; (2) the nature of chlorophylls present; (3) the type of motility, if any; (4) the carbon polymers that are produced and stored; and (5) the reproductive structures and methods. Table 10.3 summarizes the properties of the five divisions discussed in this text.

Algae show considerable diversity in the chemistry and structure of their cells. Some algal cell walls are thin, rigid structures usually composed of cellulose modified by the addition of other polysaccharides. In other algae, the cell wall is strengthened by the deposition of calcium carbonate, and still other forms have chitin present in the cell wall. Complicating the classification of algal organisms are the euglenoids, which lack cell walls. In diatoms, the cell wall is composed of silica. The frustules (shells) of diatoms have extreme resistance to decay and remain intact for long periods of time, as the fossil records show.

The principal feature used to distinguish algae form other microorganisms (e.g., fungi) is the presence of chlorophyll and other photosynthetic pigments in algae. All algae contain chlorophyll *a*. Some, however, contain other types of chlorophylls. The presence of

TABLE 10.3

Comparative Summary of Algal Characteristics

Algal Group	Common Name	Structure	Pigments	Carbon Reserve	Motility	Reproduction
Chlorophyta	Green algae	Unicellular to multicellular	Chlorophylls *a* and *b*, carotenes, xanthophylls	Starch, oils	Most are nonmotile	Asexual and sexual
Euglenophyta	Euglenoids	Unicellular	Chlorophylls *a* and *b*, carotenes, xanthophylls	Fats	Motile	Asexual
Chrysophyta	Golden-brown algae	Multicellular	Chlorophylls *a* and *b*, special carotenoids, xanthophylls	Oils	Gliding by diatoms; others by flagella	Asexual and sexual
Phaeophyta	Brown algae	Unicellular	Chlorophylls *a* and *b*, carotenoids, xanthophylls	Fats	Motile	Asexual and sexual
Pyrrophyta	Dinoflagellated	Unicellular	Chlorophylls *a* and *b*, carotenes, xanthophylls	Starch	Motile	Asexual; sexual rare

these additional chlorophylls is characteristic of a particular algal group. In addition to chlorophyll, other pigments encountered in algae include fucoxanthin (brown), xanthophylls (yellow), carotenes (orange), phycocyanin (blue), and phycoerythrin (red).

Many algae have flagella (threadlike appendages). As mentioned, the flagella are locomotor organelles that may be of the single polar or multiple polar types. The *Euglena* is a simple flagellate form with a single polar flagellum. Chlorophyta have either two or four polar flagella. Dinoflagellates have two flagella of different lengths. In some cases, algae are nonmotile until they form motile gametes (a haploid cell or nucleus) during sexual reproduction. Diatoms do not have flagella but do have gliding motility.

Algae can be either autotrophic or heterotrophic. Most are photoautotrophic; they require only carbon dioxide and light as their principal source of energy and carbon. In the presence of light, algae carry out oxygen-producing photosynthesis; in the absence of light, algae use oxygen. Chlorophyll and other pigments are used to absorb light energy for photosynthetic cell maintenance and reproduction. One of the key characteristics used in the classification of algal groups is the nature of the reserve polymer synthesized as a result of utilizing carbon dioxide present in water.

Algae may reproduce either asexually or sexually. Three types of asexual reproduction occur: binary fission, spores, and fragmentation. In some unicellular algae, binary fission occurs where the division of the cytoplasm forms new individuals like the parent cell following nuclear division. Some algae reproduce through spores. These spores are unicellular and germinate without fusing with other cells. In fragmentation, the thallus breaks up and each fragment grows to form a new thallus. Sexual reproduction can involve the union of cells where eggs are formed within vegetative cells called *oogonia*, which function as female structures, and sperm are produced in a male reproductive organ called the *antheridia*. Algal reproduction can also occur through a reduction of chromosome number or the union of nuclei.

Characteristics of Algal Divisions

Algae are photosynthetic organisms that are far from monolithic. Biologists have categorized microalgae in a variety of classes, mainly distinguished by the pigmentation, lifecycle, and basic cellular structure:

- *Chlorophyta* (green algae)—The majority of algae found in ponds belong to this group; they also can be found in saltwater and soil. Several thousand species of green algae are known today. Many are unicellular; others are multicellular filaments or aggregated colonies. The green algae have chlorophylls *a* and *b*, along with specific carotenoids, and they store carbohydrates as starch. Few green algae are found at depths greater than 7 to 10 meters, largely because sunlight does not penetrate to that depth. Some species have a holdfast structure that anchors them to the bottom of the pond and to other submerged inanimate objects. Green algae reproduce by both sexual and asexual means. Multicellular green algae have some division of labor, producing various reproductive cells and structures.

- *Euglenophyta* (euglenoids)—These are a small group of unicellular microorganisms that have a combination of animal and plant properties. Euglenoids lack a cell wall, possess a gullet, have the ability to ingest food, have the ability to assimilate organic substances, and, in some species, are absent of chloroplasts. They occur in fresh, brackish, and salt waters, and on moist soils. A typical *Euglena* cell is elongated and bound by a plasma membrane; the absence of a cell wall makes them very flexible in movement. Inside the plasma membrane is a structure called the *pellicle* that gives the organisms a definite form and allows the cell to turn and flex. Euglenoids that are photosynthetic contain chlorophylls *a* and *b*, and they always have a red eyespot (*stigma*) that is sensitive to light (photoreceptive). Some euglenoids move about by means of flagella; others move about by means of contracting and expanding motions. The characteristic food supply for euglenoids is a lipopolysaccharide. Reproduction in euglenoids is by simple cell division.

Key Point: Some autotrophic species of *Euglena* become heterotrophic when light levels are low.

- *Chrysophyta* (golden-brown algae)—The Chrysophyta division is quite large, having several thousand diversified members. They differ from green algae and euglenoids in that: (1) chlorophylls *a* and *c* are present; (2) fucoxanthin, a brownish pigment, is present; and (3) they store food in the form of oils and leucosin, a polysaccharide. The combination of yellow pigments, fucoxanthin, and chlorophylls causes most of these algae to appear golden brown in color. The Chrysophyta are also diversified in cell-wall chemistry and flagellation. The division is divided into three major classes: golden-brown algae, yellow-brown algae, and diatoms. Some Chrysophyta lack cell walls; others have intricately patterned coverings external to the plasma membrane, such as walls, plates, and scales. The diatoms are the only group that has hard cell walls of pectin, cellulose, or silicon, constructed in two halves (the epitheca and the hypotheca) called a *frustule*. Two anteriorly attached flagella are common among Chrysophyta; others have no flagella. Most Chrysophyta are unicellular but a few are multicellular. Asexual cell division is the usual method of reproduction in diatoms; other forms of Chrysophyta can reproduce sexually. Diatoms have direct significance for humans. Because they make up most of the phytoplankton of the cooler ocean parts, they are the ultimate source of food for fish. Water and wastewater operators understand the importance of their ability to function as indicators of industrial water pollution. As water quality indicators, their specific tolerances to environmental parameters such as pH, nutrients, nitrogen, concentration of salts, and temperature have been determined.

Key Point: Diatoms secrete a silicon dioxide shell (frustule) that forms the fossil deposits known as diatomaceous earth, which is used in filters and as an abrasive in polishing compounds.

- *Phaeophyta* (brown algae)—With the exception of a few freshwater species, all algal species of this division exist in marine environments as seaweed. They are a highly specialized group, consisting of multicellular organisms that are sessile (attached and not free moving). These algae contain essentially the same pigments seen in the golden-brown algae, but they appear brown because of the predominance of and the masking effect of a greater amount of fucoxanthin. Brown algal cells store food as the carbohydrate laminarin and some lipids. Brown algae reproduce asexually. They are used in foods, animal feeds, and fertilizers and as a source for alginate, a chemical emulsifier added to ice cream, salad dressing, and candy.

- *Pyrrophyta* (dinoflagellates)—The principal members of this division are the dino-flagellates. The dinoflagellates comprise a diverse group of biflagellated and non-flagellated unicellular, eukaryotic organisms. The dinoflagellates occupy a variety of aquatic environments, although the majority live in marine habitats. Most of these organisms have a heavy cell wall composed of cellulose-containing plates. They store food as starch, fats, and oils. These algae have chlorophylls *a* and *c* and several xanthophylls. The most common form of reproduction in dinoflagellates is by cell division, but sexual reproduction has also been observed.

DID YOU KNOW?

Cell division in dinoflagellates differs from most protistans, as chromosomes attach to the nuclear envelope and are pulled apart as the nuclear envelope stretches. During cell division in most other eukaryotes, the nuclear envelope dissolves.

Algal biomass contains three main components:

- Carbohydrates
- Protein
- Natural oils

Biodiesel production applies exclusively to the natural oil fraction, the main product of interest to us in this section. The bulk of natural oil made by oilseed crops is in the form of triacylglycerols (TAGs). TAGs consist of three long chains of fatty acids attached to a glycerol backbone. Algae species can produce up to 60% of their body weight in the form of TAGs. Thus, algae represent an alternative source of biodiesel, one that does not compete with the exiting oilseed market.

Algae-to-Biodiesel Production

Algae can produce up to 300 times more oil per acre than conventional crops, such as rape-seed, palms, soybean, or jatropha (Christi, 2007). Algae can grow 20 to 30 times faster than food crops (McDill, 2009). The harvesting cycle of algae is from 1 to 10 days, permitting several harvests within a very short time frame, which differs considerably from yearly crops. Algae can also be grown on land that is not suitable for other established crops, such as arid land, land with excessively saline soil, and drought-stricken land. This minimizes the issue of taking away pieces of land from the cultivation of food crops (Schenk et al., 2008). Algae can be produced and harvested for biofuel using various technologies. These include photobioreactors (plastic tubes full of nutrients exposed to sunlight), closed-loop systems (not exposed to open air), and open ponds.

Open-Pond Algae Farm

Algae farms consist of open, shallow ponds in which some source of waste carbon dioxide (CO_2) can be bubbled into the ponds and captured by the algae. As shown in Figure 10.1, the ponds in an algae farm are "raceway" designs, in which the algae, water, and nutrients circulate around a racetrack. Paddlewheels provide the flow, and the algae are kept

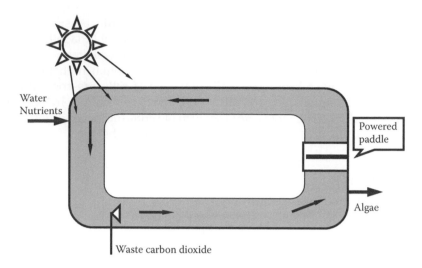

FIGURE 10.1
Raceway-design algae pool.

suspended in the water. They are circulated back up to the surface on a regular frequency. The ponds are kept shallow because of the need to keep the algae exposed to sunlight and the limited depth to which sunlight can penetrate the pond water. The ponds are operated continuously; that is, water and nutrients are constantly fed to the pond, while algae-containing water is removed at the other end. Some kind of harvesting system is required to recover the algae, which contain substantial amounts of natural oil.

Figure 10.2 illustrates the concept of an algae farm. The size of these ponds is measured in terms of surface area (as opposed to volume), because surface area is so critical to capturing sunlight. Their productivity is measured in terms of biomass produced per day per unit of available surface area. Even at levels of productivity that would stretch the limits of an aggressive research and development program, such a system would require acres of land. At such large sizes, it is more appropriate to think of these operations on the scale of a farm.

Waste CO_2 is readily available from a number of sources. Every operation that involves combustion of fuel for energy is a potential source. Generally, coal and other fossil-fuel-fired power plants are targeted as the main sources of CO_2. Typical coal-fired power plants emit flue gas from their stacks containing up to 13% CO_2. This high concentration of CO_2 enhances transfer and uptake of CO_2 in the ponds. The concept of coupling a coal-fired power plant with an algae farm provides a win–win approach to recycling the CO_2 from coal combustion into a usable liquid fuel.

Fourth-Generation Biofuels

Simply, fourth-generation biofuels are created from processes other than first-generation ethanol and biodiesel, second-generation cellulosic ethanol, and third-generation algae biofuel. Some fourth-generation technology pathways include gasification, pyrolysis, upgrading, solar-to-fuel, and genetic manipulation of organisms to secret hydrocarbons (Kagan, 2010).

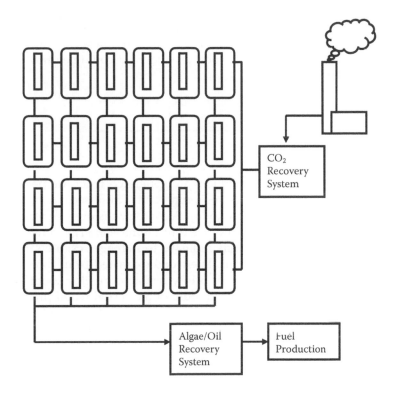

FIGURE 10.2
Algae farm.

Forestland Resource Base*

Approximately 749 million acres in the United States is forestland—about one-third of the nation's total land area. Most of this land is owned by private individuals or by the forest industry. Two-thirds of the forestland (504 million acres) is classified as timberland, which, according to the Forest Service, is land capable of growing more than 20 ft³ per acre of wood annually (Smith et al., 2004). Although timberland is not legally reserved from harvesting, much of it is inaccessible or inoperable by forestry equipment. In addition, 168 million acres of forestland is classified by the Forest Service as "other." This "other" forestland is generally incapable of growing 20 ft³ per acre of wood annually. The lower productivity is due to a variety of factors or site conditions that adversely affect tree growth (e.g., poor soils, lack of moisture, high elevation, rockiness). As a result, this land tends to be used for livestock grazing and extraction of some non-industrial wood products. The remaining 77 million acres of forestland are reserved from harvesting and are intended for a variety of non-timber uses, such as parks and wilderness. The 504 million acres classified as timberland is the source of nearly all current forest-derived bioenergy consumption and

---

* This section is based on Perlack, R.D., Wright, L.L., Turhollow, A.F. et al., *Biomass as Feedstock for a Bioenergy and Bioproducts Industry: The Technical Feasibility of a Billion-Ton Annual Supply*, U.S. Department of Energy, U.S. Department of Agriculture, Washington, D.C., 2009.

the source of most of the potential. The other forestland is included because it has accumulated excess biomass that poses wildland fire risks and hazards. Much of this excess biomass is not suitable for conventional wood products but could be utilized for a variety of bioenergy and biobased product uses.

Forest Resources

Significant mill residues and pulping liquors are generated from the processing of harvested forest products, such as sawlogs and pulpwood. These secondary forest residues constitute the majority of biomass in use today. Secondary residues generated in the process of forest products account for 50% of current biomass energy consumption. These materials are used by the forest products industry to manage residue streams, produce energy, and recover important chemicals. Fuelwood extracted from forestlands for residential and commercial use and electrical utility use accounts for about 35 million dry tons of current consumption. In total, the amount of harvested wood products from timberlands in the United States is less than the annual forest growth and considerably less than the total forest inventory, suggesting substantial scope for expanding the biomass resource base from forestlands. In addition to these existing uses, forestlands have considerable potential to provide biomass from two primary sources:

- Residues associated with the harvesting and management of commercial timberlands for the extraction of sawlogs, pulpwood, veneer logs, and other conventional products
- Currently nonmerchantable biomass associated with the standing forest inventory

This latter source is more difficult to define, but generally would include rough and rotten wood not suitable for conventional forest products and excess quantities of smaller diameter trees in overstocked forests. A large amount of this forest material has been identified by the Forest Service as needing to be removed to improve forest health and to reduce fire hazards risks (Miles, 2004; USDA, 2003).

These two categories of forest resources constitute what is defined as the primary source of forest residue biomass in addition to the fuelwood that is extracted for space heating applications in the residential and commercial sectors and for some feedstocks by electric utilities. Perennial wood crops (also referred to as short-rotation woody crops, or SRWC) are also a potential primary biomass resource. There is also a relatively large tertiary, or residue, source of forest biomass in the form of urban wood residues—a generic category that includes yard trimmings, packaging residues, discarded durable products, and construction and demolition debris.

All of these forest resources can contribute an additional 226 million dry tons to the current forest biomass consumption (approximately 142 million dry tons)—an amount still only a small fraction of the total biomass timberlands inventory of more than 20 billion dry tons. Specifically, these forest resources include the following:

- Recovered residues generated by traditional logging activities and residues generated from forest cultural operations or clearing of timberlands (about 67 tons of residues) (Smith et al., 2004; USDA, 2004)

- Recovered residues generated from fuel treatment operations on timberland and other forestland (about 8 billion tons of biomass) (Miles, 2004)
- Direct conversion of roundwood to energy (fuelwood) in the residential, commercial, and electric utility sectors (35 million dry tons of biomass are currently extracted by the residential and commercial sectors and by the electric power sector)
- Forest products industry residues and urban wood residues
- Forest growth and increase in the demand for forest products

Biomass Feedstock Availability*

In the United States, nearly 16 million cubic feet of roundwood are harvested and processed annually to produce sawlogs, paper, veneers, composites, and other fiber products. For the purpose of this book, biomass feedstocks are classified as forest residues, forest thinnings, mill residues, urban wood wastes, or dedicated energy crops (discussed earlier under SRWC). The extensive forest acreage and roundwood harvest generate logging residues and provide the potential to harvest nonmerchantable wood for energy. Processing of the wood into fiber products creates substantial quantities of mill residues that could potentially be used for energy.

Forest Residues

Forest wood residues are commonly grouped into the following categories: (1) logging residues; (2) rough, rotten, and salvable dead wood; (3) excess saplings; and (4) small pole trees. Specifically, logging residues are the unused portion of the growing of stock trees—commercial species with a diameter at breast height (DBH) greater than 5 inches, excluding cull trees—that are cut or killed by logging and left behind. Rough trees are those that do not contain a sawlog (i.e., 50% or more of live cull volume) or are not a currently merchantable species. Rotten trees are trees that do not contain a sawlog because of rot (i.e., 50% or more of the live cull volume). Salvable dead wood includes downed or standing trees that are considered currently or potentially merchantable. Excess saplings are live trees having a DBH of between 1.0 and 4.9 inches. Small pole trees are trees with a DBH greater than 5 inches but are smaller than saw timber trees.

The primary advantage of using forest residues for power generation is that an existing collection infrastructure is already set up to harvest wood in many areas. Companies that harvest wood already own equipment and transportation options that could be extended to gathering forest residues.

The forest wood residues supplies that could potentially be available for energy use in the United States are estimated using an updated version of a model originally developed by McQuillan et al. (1984). The McQuillan model estimates the total quantities of forest wood residues that can be recovered by first classifying the total forest inventory by the above wood categories (for both softwood and hardwood) and by volume, haul distances,

---

* This section is based on USEPA, *Biomass Combined Heat and Power Catalog of Technologies*, U.S. Environmental Protection Agency, Washington, D.C., 2007.

and equipment operability constraints. This total inventory is then revised downward to reflect the quantities that can be recorded in each class due to constraints on equipment retrieval efficiencies, road access to a site, and impact of site slope on harvest equipment choice.

Forest residues typically have an energy content of 5140 Btu/lb wet and 8570 Btu/lb dry. The costs for obtaining recoverable forest wood residues include collecting harvesting, chipping, loading, transportation, and unloading; a stumpage fee; and a return for profit and risk. The cost of forest residue can be as low as $15 to $25 per ton, or between $1.46 and $3.243/million Btu (MMBtu); however, the average price in most parts of the country is roughly $30/ton, or $2.92/MMBtu (Curtis et al., 2003; Walsh et al., 1999).

> **DID YOU KNOW?**
>
> Retrieval efficiency accounts for the quantity of the inventory that can actually be recovered due to technology or equipment (assumed to be 40%). It is assumed that 50% of the resource is accessible without having to construct roads, except for logging residues for which 100% of the inventory is assumed accessible. Finally, inventory that lies on slopes greater than 20% or where conventional equipment cannot be used is eliminated for cost and environmental reasons.

Forest Thinnings

Forest thinnings are defined as underbrush and saplings smaller than 2 inches in diameter, as well as fallen or dead trees. These substances are sometimes known as *ladder fuels*, because they can accelerate a forest fire's vertical spread (i.e., live or dead vegetation allows a fire to climb up from the forest floor into the tree canopy). Large volumes of forest biomass should be available from implementation of the U.S. Department of the Interior, U.S. Department of Agriculture/U.S. Forest Service, and Bureau of Land Management joint initiatives to reduce fire risk in national forests; however, the actual business of harvesting, collecting, processing, and transporting loose forest thinnings is costly and presents an economic barrier to the recovery and utilization for energy. Typically, the wood waste from forest thinnings is disposed of through controlled burning due to the expense of transporting it to a power generation facility. In areas that are not already used for wood harvesting, there is no existing infrastructure to extract forest thinnings. A study for the Colorado Office of Energy Management and Conservation found that the delivered cost of forest thinnings was nearly $100 per dry ton, making it difficult to compete with other fuels at a cost of $5.83 to $9.73/MMBtu (USEPA, 2007b). Forest thinnings typically have an energy content of 5140 Btu/lb wet and 8570 Btu/lb dry. The use of forest thinnings for power generation is concentrated in the Western United States. Nevada, Arizona, Idaho, and New Mexico have the greatest potential to generate power from forest thinnings.

Primary Mill Residues

Primary mill residues are waste wood from manufacturing operations that would otherwise be sent to a landfill. Manufacturing operations that produce mill residues usually include sawmills, pulp and paper companies, and other millwork companies involved in producing lumber, pulp, veneers, and other composite wood fiber materials. Primary mill residues are usually in the form of bark, chips, sander dust, edgings, sawdust, or slabs. Due to the fact that primary mill residues are relatively clean, homogeneous, and concentrated at one source, nearly 98% of all residues generated in the United States are currently used as fuel or to produce other fiber products. Of the 21.6 million dry tons of bark produced in the United States, 76.6% is used for fuel and 20.6% for other purposes such as mulch, bedding, and charcoal. Overall, the USDA estimates that 2 to 3% of primary mill residues are available as an additional fuel resource because they are not being used for other purposes. Because most primary mill residues are fairly dry after they have been through a manufacturing process, they fall at the upper level of the energy content range for wood (8570 Btu/lb). Producing power from primary mill residues is highly advantageous in the wood products industries because they have a "free" (i.e., no additional cost) source of fuel with no transportation costs and a secure supply that they control. The cost of these residues is actually negative for most wood products industries because if the residues are not used onsite companies have to pay for disposal. When purchasing mill residues, the price can vary considerably from $8 to $50 per oven-dry ton, corresponding to a cost of $0.46 to $2.2/MMBtu (Walsh et al., 1999). This high variability occurs on a site-by-site basis depending on whether the site is already using the residues.

Urban Wood Waste

Urban wood wastes include yard trimmings, wood construction and demolition (C&D) wastes; site-clearing wastes; and pallets, wood packaging, furniture, and other miscellaneous commercial and household wood wastes that are generally disposed at MSW landfills or C&D landfills. Urban wood wastes are available across the United States, but they are mainly concentrated in populous areas. Woody yard trimmings are an abundant source of wood sent to landfills. In 1996, yard trimmings were the second largest component of the MSW stream at 29.3 million tons (McKeever, 1998). Yard trimmings can be generated from residential landscaping and right-of-way trimming near roads, railways, and utility systems such as power lines. Yard trimmings comprise about 14% of the MSW stream; because approximately 36% of yard trimmings are recoverable, roughly 5% of the total MSW stream for each state is available yard trimming residue.

Construction and demolition waste is woody material generated from C&D activity. Wood debris makes up around 26% of the total C&D stream, or approximately 35.1 million tons (Sandler, 2003). Approximately 30% of that debris, about 10.5 million tons/year, is uncontaminated by chemical treatment and available for recovery (Antares Group, 2003).

Other wood wastes include discarded consumer wood products and wood residues from non-primary mill manufacturers, such as discarded wooden furniture, cabinets, pallets and containers, and scrap lumber. Approximately 7% of the entire MSW stream is other wood residues; of this, 44% is generally available.

Wood waste costs can be lower than for other forms of biomass because wood waste that is burned for energy generation purposes is usually offsetting disposal costs from otherwise being landfilled; therefore, some urban wood wastes can actually be collected at a negative cost. Typically, urban wood waste costs range from $3 to $24/ton. The energy content of urban wood waste is 4600 Btu/lb wet and 6150 Btu/lb dry, or between $0.33 and $2.61/MMBtu (Antares Group, 2003; Walsh et al., 1999). One drawback to using urban wood waste for energy generation is that wood used for construction and consumer wooden goods can contain high levels of impurities caused by chemical treatments to extend the useful life of the wood. These impurities can cause emission problems when burned and might require wood waste boilers to have extra filtration and control equipment to curb contaminants or would require effective separation of the contaminated items prior to burning.

Gasoline Gallon Equivalent (GGE)

With regard to forest-based biomass energy sources and energy or heat combustion potential (Btu/oven-dry pound) of stems and branches of >6-inch trees and southern pine trees, the reader should have a fundamental understanding of the difference between conventional gasoline and diesel fuel energy output as compared to nonconventional renewable products. Typically, this comparison is made utilizing a standard engineering parameter known as the gasoline gallon equivalent (GGE), which is the ratio of the number of British thermal units (Btu) available in 1 U.S. gallon of gasoline to the number of British thermal units available in 1 gallon of the alternative substance in question. In 1994, NIST (2007) defined a gasoline gallon equivalent as 5660 pounds of natural gas. The GGE parameter allows consumers to compare the energy content of competing fuels against a commonly known fuel—gasoline. Table 10.4 provides GGE and Btu/unit value comparisons for various fuels. A very useful parameter to know when dealing with biomass is the heat value, or heat of combustion, which is defined as the total amount of heat obtainable from oven-dry material when burned in an enclosure of constant volume, allowing no deductions for heat losses. Table 10.5 lists the heat of combustion of stems and branches of 6-inch trees of hardwood species growing on southern pine sites, and Table 10.6 lists the heat of combustion of various materials from southern pine trees.

TABLE 10.4

Gasoline Gallon Equivalents (GGE)

Fuel	GGE	Btu/Unit
Gasoline (regular)	1 U.S. gallon	114,100 Btu/gal
Gasoline (conventional, summer)	0.996 U.S. gallon	114,500 Btu/gal
Gasoline (conventional, winter)	1.013 U.S. gallon	112,500 Btu/gal
Gasoline (reformulated gasoline, ethanol	1.019 U.S. gallon	111,836 Btu/gal
Gasoline (reformulated gasoline, ETBE)	1.019 U.S. gallon	111,811 Btu/gal
Gasoline (reformulated gasoline, MTBE)	1.020 U.S. gallon	111,745 Btu/gal
Gasoline (10% MBTE)	1.02 U.S. gallon	112,000 Btu/gal
Gasoline (regular unleaded)	1 U.S. gallon	114,100 Btu/gal
Diesel #2	0.88 U.S. gallon	129,500 Btu/gal
Biodiesel (B100)	0.96 U.S. gallon	118,300 Btu/gal
Biodiesel (B20)	0.90 U.S. gallon	127,250 Btu/gal
Liquid natural gas (LNG)	1.52 U.S. gallon	75,000 Btu/gal
Compressed natural gas	1.26 ft^3 (3.587 m^3)	900 Btu/ft^3
Hydrogen at 101.325 kPa	357.37 ft^3	319 Btu/ft^3
Hydrogen by weight	0.997 kg (2.198 lb)	119.9 MJ/kg (51,500 Btu/lb)
Liquefied petroleum gas (LPG)	1.35 U.S. gallon	84,300 Btu/gal
Methanol fuel (M100)	2.01 U.S. gallon	56,800 Btu/gal
Ethanol fuel (E100)	1.500 U.S. gallon	76,100 Btu/gal
Ethanol (E85)	1.39 U.S. gallon	81,800 Btu/gal
Jet fuel (naphtha)	0.97 U.S. gallon	118,700 Btu/gal

Sources: Gable, C. and Gable, S., *Fuel Energy Comparisons: Gasoline Gallon Equivalents (GGE)*, About.com (http://alternativefuels.about.com/od/resources/a/gge.htm); NAFA, *Energy Equivalents of Various Fuels*, NAFA Fleet Management Association, Princeton, NJ, 2011 (http://www.nafa.org/Template.cfm?Section=Energy_Equivalents); USDOE, *Transportation Energy Data Book*, Edition 22, ORNL-6967, U.S. Department of Energy, Oak Ridge National Laboratory, Center for Transportation Analysis, Washington, D.C., 2002; USEPA, *Fuel Economy Impact Analysis of RFG*, U.S. Environmental Protection Agency, Office of Mobile Sources, Ann Arbor, MI, 2007.

TABLE 10.5

Heat of Combustion (Btu/Oven-Dry Pound) of Stems and Branches of 6-Inch Hardwood Trees

Species	Stemwood	Stembark	Branchwood	Branchbark
Ash, green	7695	7472	7727	7606
Ash, white	8033	7695	8013	7816
Elm, American	7770	6840	7857	6904
Elm, winged	7917	7019	7869	6889
Hackberry	7882	7147	7867	7141
Hickory species	8183	7586	7931	7259
Maple, red	7846	7595	7829	7384
Oak, black	7680	7642	7692	7847
Oak, blackjack	7739	7766	7739	7907
Oak, cherrybark	7848	7582	7737	7655
Oak, laurel	7828	7897	7653	7806
Oak, northern red	7791	7879	7776	7926
Oak, post	7889	7191	7845	7728
Oak, scarlet	7798	8041	7673	7894
Oak, Shumard	7789	7970	7745	7913
Oak, southern red	7919	7983	7839	7798
Oak, water	7876	7930	7833	7918
Oak, white	7676	7328	7507	7574
Sweetbay	7736	7822	7802	7886
Sweetgum	7667	7200	7690	7214
Tupelo, black	7867	7788	7814	8176
Yellow poplar	7774	7696	7811	7666
Average	7827	7593	7784	7632

Source: USDA, *Biomass Research*, U.S. Department of Agriculture, Forest Service, Southern Research Station, Auburn, AL, 2010 (http://www.srs.fs.usda.gov/forestops/biomass.htm).

TABLE 10.6

Heat of Combustion (Btu/Oven-Dry Pound) of Selected Materials from Southern Pine Trees

Material	Average	Range
Resin	—	14,625 to 16,250
Charcoal	12,335	11,225 to 12,740
Needles	9030	8935 to 9105
Loblolly pine stemwood	8600	8310 to 9352
Earlywood	8610	8470 to 8760
Latewood	8585	8385 to 8755
Rootwood	8605	8560 to 8680
Tops (bark and wood at 1-inch top)	8395	8015 to 8745
Old cones	8130	8085 to 8190
Kraft black liquor	5965	5820 to 6130

Source: USDA, *Biomass Research*, U.S. Department of Agriculture, Forest Service, Southern Research Station, Auburn, AL, 2010 (http://www.srs.fs.usda.gov/forestops/biomass.htm).

Chapter Review Questions

10.1   First-generation biofuels are biofuels made from _____, _____, and _____.

10.2   What is one of the key biodiesel byproducts?

10.3   _____ is a biofuels crop not optimally grown everywhere.

10.4   What type of tree is among the fastest growing in North American, achieving considerable height in just six growing seasons?

10.5   _____ is short for quadrillion Btu.

10.6   The protists that perform photosynthesis are called _____.

10.7   _____ is the vegetative body of algae.

10.8   What are the three main components of algal biomass?

10.9   _____ are the unused portion of the growing of stock trees.

10.10   _____ is the ratio of the number of Btu available in 1 U.S. gallon of gasoline to the number of Btu available in 1 gallon of the alternative substance in question.

References and Recommended Reading

Allen, H.L., Fox, T.R., and Campbell, R.G. (2005). What's ahead for intensive pine plantation silviculture? *Southern Journal of Applied Forestry*, 29:62–69.

Anon. (2007). BQ-9000 inductee earns accreditation, certification, *Biodiesel Magazine*, January 24.

Antares Group, Inc. (2003). *Assessment of Power Production at Rural Utilities Using Forest Thinnings and Commercially Available Biomass Power Technologies*, prepared for the U.S. Department of Agriculture, U.S. Department of Energy, and National Renewable Energy Laboratory, Washington, D.C.

Arthur D. Little, Inc. (2001). *Aggressive Growth in the Use of Bioderived Energy and Products in the United States by 2010*, Final Report, Arthur D. Little, Inc., Cambridge, MA.

Ashford, R.D. (2001). *Ashford's Dictionary of Industrial Chemicals*, 2nd ed., Wavelength Publications, London.

Aylott, M.J. (2008). Yield and spatial supply of bioenergy poplar and willow short-rotation coppice in the U.K., *New Phytologist*, 178(2):358–370.

Baize, J. (2006). *Bioenergy and Biofuels*, Agricultural Outlook Forum, February 17, Washington, D.C.

Baker, J.B. and Broadfoot, W.M. (1979). *A Practical Field Method of Site Evaluation for Commercially Important Southern Hardwoods*, General Technical Report SO-26, U.S. Department of Agriculture, Forest Service, South Forest Experiment Station, New Orleans, LA.

Bender, M. (1999). Economic feasibility review for community-scale farmer cooperatives for biodiesel, *Bioresource Technology*, 70:81–87.

Chadwick, A. (2008). *Scientist Discovers Fungus That Could Fuel a Car*, http://www.npr.org/templates/story/story.php?storyId=96574076.

Cheng, S. and Zhu, S. (2009). Lignocellulosic biorefinery, *BioResources*, 4(2):456–457.

Christi, Y. (2007). Biodiesel from microalgae, *Biotechnology Advances*, 25:294–306.

Coyle, W. (2007). The future of biofuels: a global perspective, *Amber Waves*, November (http://www.ers.usda.gov/AmberWaves/november07/features/biofuels.htm).

Curtis, W., Ferland, C., McKissick, J., and Barnes, W. (2003). *The Feasibility of Generating Electricity from Biomass Fuel Sources in Georgia*, University of Georgia, Center of Agribusiness and Economic Development, Athens.

Dickman, D. (2006). Silviculture and biology of short-rotation wood crops in temperate regions: then and now, *Biomass and Bioenergy*, 30:696–705.

EERE. (2008). *Biomass Program*, U.S. Department of Energy, Energy Efficiency and Renewable Energy, Washington, D.C. (http://www1.eere.energy.gov/biomass/feedstocks_types.html).

EERE. (2010). Alkali content and slagging potential of various biofuels [table], in *Biomass Energy Data Book*, U.S. Department of Energy, Energy Efficiency and Renewable Energy, Washington, D.C. (cta.ornl.gov/bedb/pdf/BEDB3_Chapter3.pdf).

EIA. (2006). *Annual Energy Outlook 2007 with Projections to 2030*, DOE/EIA-0383(2007), U.S. Department of Energy, Energy Information Administration, Office of Integrated Analysis and Forecasting, Washington, D.C.

Elliott, D.C., Fitzpatrick, S.W., Bozell, J.J. et al. (1999). Production of levulinic acid and use as a platform chemical for derived products, in Overend, R.P. and Chornet, E., Eds., *Biomass: A Growth Opportunity in Green Energy and Value-Added Products, Proceedings of the Fourth Biomass Conference of the Americas*, Elsevier Science, Oxford, U.K., pp. 595–600.

EPRI. (1997). *Renewable Energy Technology Characterizations*, TR-109496, U.S. Department of Energy, Electric Power Research Institute, Washington, D.C.

Fahey, J. (2001). Shucking petroleum, *Forbes Magazine*, 168(13):206–208.

Farrell, A.E. and Gopal, A.R. (2008). Bioenergy research needs for heat, electricity, and liquid fuels, *MRS Bulletin*, 33:373–387.

Gentille, S.B. (1996). *Reinventing Energy: Making the Right Choices*, American Petroleum Institute, Washington, D.C.

Graham, R., Nelson, R., Sheehan J., Perlack, R., and Wright L. (2007). Current and potential U.S. corn stover supplies, *Agronomy Journal*, 99:1–11.

Haas, M., McAloon, A., Yee, W., and Foglia, T. (2006). A process model to estimate biodiesel production costs, *Bioresource Technology*, 97:671–678.

Harrar, E.S. and Harrar, J.G. (1962). *Guide to Southern Trees*, 2nd ed., Dover, New York.

Hartman, E. (2008). A promising oil alternative: algae energy, *The Washington Post*, January 6 (http://www.washingtonpost.com/wp-dyn/content/article/2008/01/03/AR2008010303907.html).

Hileman, B. (2003). Clashes over agbiotech, *Chemical & Engineering News*, 81:25–33.

Hoffman, J. (2001). BDO outlook remains healthy, *Chemical Market Reporter*, 259(14):5.

Hughes, E. (2000). Biomass cofiring: economics, policy and opportunities, *Biomass and Bioenergy*, 19:457–465.

ILSR. (2002). *Accelerating the Shift to a Carbohydrate Economy: The Federal Role*, Executive Summary of the Minority Report of the Biomass Research and Development Technical Advisory Committee, Institute for Local Self-Reliance, Washington, D.C.

Kagan, J. (2010). *Third and Fourth Generation Biofuels: Technologies, Markets, and Economics Through 2015*, Greentech Media, Boston, MA (http://www.greentechmedia.com/research/report/third-and-fourth-generation-biofuels).

Kantor, S.L., Lipton, K., Manchester, A., and Oliveira, V. (1997). Estimating and addressing America's food losses, *Food Review*, 20(1):3–11.

Klass, D.I. (1998). *Biomass for Renewable Energy, Fuels, and Chemicals*, Academic Press, San Diego, CA.

Lee, S. (1996). *Alternative Fuels*, Taylor & Francis, Boca Raton, FL.

Markarian, J. (2003). New additives and basestocks smooth way for lubricants, *Chemical Market Reporter*, April 28.

McDill, S. (2009). Can algae save the world—again? *Reuters*, February 20 (http://www.reuters.com/article/2009/02/10/us-biofuels-algae-idUSTRE5196HB20090210?pageNumber=2&virtualBrandChannel=0).

McGraw, L. (1999). Three new crops for the future, *Agricultural Research Magazine*, 47(2):17.

McKeever, D.B. (1998). Wood residual quantities in the United States, *BioCycle: Journal of Composting and Recycling*, 39(1):65–68; as cited in Antares Group, Inc., *Assessment of Power Production at Rural Utilities Using Forest Thinnings and Commercially Available Biomass Power Technologies*, prepared for the U.S. Department of Agriculture, U.S. Department of Energy, and National Renewable Energy Laboratory, Washington, D.C., 2003.

McKeever, D.B. (2004). Inventories of woody residues and solid wood waste in the United States, 2002, in *Proceedings of the Ninth International Conference on Inorganic-Bonded Composite Materials Conference*, October 10–13, Vancouver, British Columbia.

McKendry, P. (2002). Energy production from biomass. Part I. Overview of biomass, *Bioresource Technology*, 83:37–46.

McLaughlin, S.B. and Kzos, L.A. (2005). Development of switchgrass (*Pancium virgatum*) as a bioenergy feedstock in the United States, *Biomass and Bioenergy*, 28:515–535.

McQuillan, A., Skog, K., Nagle, T., and Loveless, R. (1984). *Marginal Cost Supply Curves for Utilizing Forest Waste Wood in the United States*, unpublished manuscript, University of Montana, Missoula.

Miles, P.D. (2004). *Fuel Treatment Evaluator: Web-Application Version 1.0*, U.S. Department of Agriculture, Forest Service, North Central Research Station, St. Paul, MN (http://ncrs2/4801/fiadb/fueltreatment.fueltreatmentwc.asp).

Miles, T.R., Miles, Jr., T.R., Baxter, L.L., Jenkins, B.M., and Oden, L.L. (1993). Alkali slagging problems with biomass fuels, in *First Biomass Conference of the Americas: Energy, Environment, Agriculture, and Industry*, Vol. 1, National Renewable Energy Laboratory, Burlington, VT, pp. 406–421.

Morey, R.V., Tiffany, D.G., and Hartfield D.L. (2006). Biomass for electricity and process heat at ethanol plants, *Applied Engineering in Agriculture*, 22:723–728.

NIST. (2007). Appendix D: definitions, in *Specifications, Tolerances, and Other Technical Requirements for Weighing and Measuring Devices*, NIST Handbook 44, U.S. Department of Commerce, National Institute of Standards and Technology (ts.nist.gov/WeightsAndMeasures/upload/HB44_07_FullDoc_Rev1_LC.pdf).

NRDC. (2001). *Reducing America's Oil Dependence*, Natural Resources Defense Council, New York (http://www.nrdc.org/air/energy/fensec.asp).

ODOE. (2010). *Bioenergy in Oregon: Biomass Technologies*, Oregon Department of Energy, Salem (http://www.oregon.gov/ENERGY/RENEW/Biomass/biofuels.shtml).

Perlin, J. (2005). *A Forest Journey: The Story of Wood and Civilization*, Countryman Press, Woodstock, VT.

Polagye, B., Hodgson, K., and Malte, P. (2007). An economic analysis of bioenergy options using thinning from overstocked forests, *Biomass and Bioenergy*, 31:105–125.

Rinehart, L. (2006). *Switchgrass as a Bioenergy Crop*, National Center for Appropriate Technology, Butte, MT.

Rossell, J.B. and Pritchard, J.L.R., Eds. (1991). *Analysis of Oilseeds, Fats and Fatty Foods*, Elsevier, London.

Sandler, K. (2003). Analyzing what's recyclable in C&D debris, *Biocycle*, 44(11):51–54.

Sauer, P. (2000). Domestic spearmint oil producers face flat pricing, *Chemical Market Reporter*, 258(18):14.

Schenk, P., Thomas-Hall, S., Stephens, R., Marx U., Mussgnug, J., Posten, C., Kruse, O., and Hankamer, B. (2008). Second generation biofuels: high-efficiency microalgae for industrial production, *BioEnergy Research*, 1(1):20–43.

Sedjo, R. (1997). The economics of forest-based biomass supply, *Energy Policy*, 25(6):559–566.

Segall, L. (2010). Ex-Shell president sees $5 gas in 2012, CNNMoney.com, December 27 (http://money.cnn.com/2010/12/27/markets/oil_commodities/index.htm).

Silva, B. (1998). Meadowfoam as an alternative crop, *AgVentures*, 2(4):28.

Sioru, B. (1999). Process converts trash into oil, *Waste Age*, 30(11):20.

Smith, W.B. et al. (2004). *Forest Resources of the United States, 2002*, General Technical Report NC-241, U.S. Department of Agriculture, Forest Service, North Central Research Station, St. Paul, MN.

Spellman, F.R. (2009). *Handbook of Water and Wastewater Treatment Plant Operations*, 2nd ed., CRC Press, Boca Raton, FL.

Thompson, P. (2008). *Wolf Thoughts* [blog], http://wolfatd.blogspot.com/.

Tillman, D. (2000). Biomass cofiring: the technology, the experience, the combustion consequences, *Biomass and Bioenergy*, 19:365–384.

Uhlig, H. (1998). *Industrial Enzymes and Their Applications*, John Wiley & Sons, New York.

USDA. (2003). *A Strategic Assessment of Forest Biomass and Fuel Reduction Treatments in Western States*, U.S. Department of Agriculture, Forest Service, Research and Development, Washington, D.C. (www.fs.fed.us/research/pdf/Western_final.pdf).

USDA. (2004). *Timber Products Output Mapmaker Version 1.0*, U.S. Department of Agriculture, Forest Service, Washington, D.C.

USDOE. (2003). *Industrial Bioproducts: Today and Tomorrow*, U.S. Department of Energy, Washington, D.C.

USDOE. (2006). *Breaking the Biological Barriers to Cellulosic Ethanol: A Joint Research Agenda*, Report from the December 2005 Workshop, DOE-SC-0095, U.S. Department of Energy, Office of Science, Washington, D.C.

USDOE. (2007). *Understanding Biomass: Plant Cell Walls*, U.S. Department of Energy, Washington, D.C. (http://news.mongabay.com/2006/0806-660.html).

USEPA. (1979). *Process Design Manual: Sludge Treatment and Disposal*, EPA/625/625/1-79-011, U.S. Environmental Protection Agency, Office of Research and Development, Washington, D.C.

USEPA. (2004) Overview of biogas technology, in Roos, K.F., Martin, Jr., J.B., and Moser, M.A., Eds., *AgSTAR Handbook*, U.S. Environmental Protection Agency, Washington, D.C. (http://www.epa.gov/agstar/documents/chapter1.pdf).

USEPA. (2006). *Biosolids Technology Fact Sheet: Multi-Stage Anaerobic Digestion*, EPA/832-F-06-031, U.S. Environmental Protection Agency, Office of Water, Washington, D.C.

USEPA. (2007a). *Fuel Economy Impact Analysis of RFG*, U.S. Environmental Protection Agency, Washington, D.C. (http://www.epa.gov/oms/rfgecon.htm).

USEPA. (2007b). *Biomass Combined Heat and Power Catalog of Technologies*, U.S. Environmental Protection Agency, Washington, D.C.

Valigra, L. (2000). Tough as soybeans, *The Christian Science Monitor*, January 20.

Walsh, M. et al. (1999). *Biomass Feedstock Availability in the United States: 1999 State Level Analysis*, Oak Ridge National Laboratory, Knoxville, TN.

WAOB. (2008). *World Agricultural Supply and Demand Estimates*, U.S. Department of Agriculture, World Agricultural Outlook Board, Washington, D.C.

Wilhelm, W.W., Johnson, J.M.F., Karlen, D.L., and Lightle, D.T. (2002). Corn stover to sustain soil organic carbon further constrains biomass supply, *Agronomy Journal*, 99:1665–1667.

Wiltsee, G. (1998). *Urban Wood Waste Resource Assessment*, NREL/SR-570-25918, National Renewable Energy Laboratory, Golden, CO.

Wood, M. (2002). Desert shrub may help preserve wood, *Agricultural Research Magazine*, 50(4):10–11.

11

Biomass Conversion/Power Generation Technologies

Keeping America competitive requires affordable energy. And here we have a serious problem: America is addicted to oil, which is often imported from unstable parts of the world. The best way to break this addiction is through technology; by applying the talent and technology of America, this country can dramatically improve our environment, move beyond a petroleum-based economy, and make our dependence on Middle Eastern oil a thing of the past.

—**President George W. Bush, January 31, 2006**

There are painters who transform the sun to a yellow spot, but there are others who with their help and their art and their intelligence transform a yellow spot into the sun.

—**Pablo Picasso**

Introduction*

Biomass conversion refers to the process of converting biomass into energy that will in turn be used to generate electricity or heat. The principal categories of biomass conversion technologies for power and heat production are direct-fired and gasification systems. Within the direct-fired category, specific technologies include stoker boilers, fluidized-bed boilers, and co-firing. Within the gasification category, specific technologies include fixed-bed gasifiers and fluidized-bed gasifiers. Anaerobic digesters are also considered a biomass conversion technology; however, in this text only a brief, introductory discussion of anaerobic digestion is provided. (*Note:* Readers desiring an in-depth discussion of aerobic and anaerobic digestion should refer to F.R. Spellman, *Handbook of Water/Wastewater Plant Operations*, 2nd ed., CRC Press, 2008.)

Biomass Conversion: Anaerobic Digestion

Anaerobic digestion is an efficient way to refine biomass. Biomass fermentation transforms sugars into an ethanol/water mixture from which the alcohol has to be recovered by means of an energy-intensive distillation process. In contrast, anaerobic digestion transforms many animal and vegetable substances into methane gas with very little expenditure of energy. Methane gas has low solubility in water, evolves naturally, and, as mentioned, can be collected with a minimal expenditure of energy. Anaerobic digestion is the traditional

---

* Material in this section is based on USEPA, *Biomass Combined Heat and Power Catalog of Technologies*, U.S. Environmental Protection Agency, Washington, D.C., 2007.

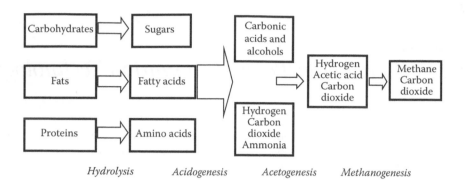

FIGURE 11.1
Key stages of anaerobic digestion.

method for managing waste, sludge stabilization, or releasing energy. The process uses bacteria that thrive in the absence of oxygen. It is slower than aerobic digestion but has the advantage that only a small percentage of the wastes are converted into new bacterial cells; instead, most of the organics are converted into carbon dioxide and methane gas. Anaerobic digestion is a process that is still being actively researched on many levels; its use and operation are well understood in wastewater (sewage) treatment but less is known about potentially important industrial applications.

Note: In an anaerobic digester, the entrance of air should be prevented because air mixed with gas produced in the digester could create an explosive mixture.

Stages of Anaerobic Digestion

Following are the four key biological and chemical stages of anaerobic digestion (see Figure 11.1) (UESPA, 1979, 2006):

1. *Hydrolysis* occurs when proteins, cellulose, lipids, and other complex organics are broken down into smaller molecules and become soluble by utilizing water to split the chemical bonds of the substances. This step solubilizes some normally insoluble substances such as cellulose.

2. *Acidogenesis* occurs when the products of hydrolysis are converted into organic acids (this is where monomers are converted to fatty acids).

3. *Acetogenesis* occurs when the fatty acids are converted into acetic acid, carbon dioxide, and hydrogen.

4. *Methanogenesis* occurs when the organic acids produced during the fermentation step are converted into methane and carbon dioxide.

The efficiency of each phase is influenced by the temperature and the amount of time the process is allowed to react. The organisms that perform hydrolysis and volatile acid fermentation (often called the *acidogenic bacteria*) are fast-growing microorganisms that prefer a slightly acidic environment and higher temperatures than the organisms that perform the methane formation step (*methanogenic bacteria*). A simplified generic chemical equation for the overall processes is

$$C_6H_{12}O_6 \rightarrow 3CO_2 + 3CH_4$$

TABLE 11.1

Typical Contents of Biogas

Matter	Percentage (%)
Methane (CH_4)	50–75
Carbon dioxide (CO_2)	25–50
Nitrogen (N_2)	0–10
Hydrogen (H_2)	0–1
Hydrogen sulfide (H_2S)	0–3
Oxygen (O_2)	0–2

Biogas is the ultimate waste product of the bacteria feeding off the input biodegradable feedstock and is comprised mostly of methane and carbon dioxide, with a small amount of hydrogen and trace hydrogen sulfide (see Table 11.1) (Spellman, 2009). Keep in mind that the ultimate output from a wastewater digester is water; biogas is more of an off-gas that can be used as an energy source.

Biomass Steam Power Systems

Biomass steam power systems are typically below 50 MW in size, compared to coal-fired plants, which are in the range of 100 to 1000 MW. Most of today's biomass power plants are *direct-fired systems*. The biomass fuel is burned in a boiler to produce high-pressure steam that is used to power a steam-turbine-driven power generator. In many applications, steam is extracted from the turbine at medium pressures and temperatures and is used for process heat, space heating, or space cooling. *Co-firing* involves substituting biomass for a portion of the coal in an existing power plant boiler. It is the most economic near-term option for introducing new biomass power generation. Because much of the existing power plant equipment can be used without major modifications, co-firing is far less expensive than building a new biomass power plant. Compared to the coal it replaces, biomass reduces SO_2, NO_x, CO_2, and other air emissions.

DID YOU KNOW?

Biomass power plants burn mostly "slash"—the branches that [timber] companies slash off the tree trunks. The wood costs about half the price of coal.

—Carolyn Shapiro (*The Virginian-Pilot*, April 2, 2011, p. 5)

Biomass *gasification* systems operate by heating biomass in an environment where the solid biomass breaks down to form a flammable gas. The gas produced—synthesis gas, or *syngas*—can be cleaned, filtered, and then burned in a gas turbine in simple or combined-cycle mode, comparable to landfill gas (LFG) or biogas produced from an anaerobic digester. In smaller systems, the syngas can be fired in reciprocating engines, microturbines, Stirling engines, or fuel cells. Gasification technologies using biomass byproducts are popular in the pulp and paper industry, where they improve chemical recovery and

TABLE 11.2

Summary of Biomass Conversion Technologies

Biomass Conversion Technology	Common Fuel Types	Feed Size	Moisture Content	Capacity Range
Stoker grate, underfire stoker boilers	Sawdust, bark, chips, hog fuel, shavings, end cuts, sander dust	0.25–2 in.	10–50%	4 to 300 MW (many in the 20- to 50-MW range)
Fluidized-bed boiler	Wood residue, peat, wide variety of fuels	<2 in.	<60%	Up to 300 MW (many in the 20- to 25-MW range)
Cofiring—pulverized coal boilers	Sawdust, bark, shavings, sander dust	<0.25 in.	<25%	Up to 1000 MW
Cofiring—stoker, fluidized-bed gasifier	Sawdust, bark, shavings, hog fuel	<2 in.	10–50%	Up to 300 MW
Fixed bed gasifier	Chipped wood or hog fuel, rice hulls, shells, sewage sludge	0.25–4 in.	<20%	Up to 50 MW
Fluidized-bed gasifier	Most wood and agriculture residues	0.25–2 in.	15–30%	Up to 25 MW

Source: Based on Wright, L. et al., *Biomass Energy Data Book*, Oak Ridge National Laboratory, Knoxville, TN, 2006.

generate process steam and electricity at higher efficiencies and with lower capital costs than conventional technologies. Pulp and paper industry byproducts that can be gasified include hogged wood (machine-shredded waste wood), bark, and spent black liquor. Table 11.2 provides a summary of biomass conversion technologies for producing heat and power.

Modular systems employ some of the same technologies mentioned above, but on a smaller scale that is more applicable to farms, institutional buildings, and small industry. A number of modular systems are now under development and could be most useful in remote areas where biomass is abundant and electricity is scarce.

Direct-Fired Systems

The most common utilization of solid fuel biomass is direct combustion; the resulting hot flue gases produce steam in a boiler, a technology that goes back to the 19th century. Boilers today burn a variety of fuels and continue to play a major role in industrial process heating, commercial and institutional heating, and electricity generation. Boilers are differentiated by their configuration, size, and quality of the steam or hot water produced. Boiler size is most often measured by the fuel input in 1 million Btu per hour (MMBtu/hr), but it may also be measured by output in pounds of steam per hour. Because large boilers are often

DID YOU KNOW?

When harvested from sustainably managed private forest lands, biomass has the potential to be a key ingredient, along with wind and solar energy, in meeting [future] power needs.

—J.R. Tolbert (*The Virginian-Pilot*, April 2, 2011, p. 5)

used to generate electricity, it can also be useful to relate boiler size to power output in electric generating applications. Using typical boiler and steam turbine generating efficiencies, a heat input of 100 MMBtu/hr provides about 10 MW of electric output. The two most commonly used types of boilers for biomass firing are stoker boilers and fluidized-bed boilers. Either of these can be fueled entirely by biomass fuel or co-fired with a combination of biomass and coal. The efficiency, availability, and operating issues of each of these options are discussed below.

Boilers

Stoker Boilers

Stoker boilers employ direct fire combustion of solid fuels with excess air, producing hot flue gases, which then produce steam in the heat-exchange section of the boiler. The steam is used directly for heating purposes or passed through a steam turbine generator to produce electric power. Stoker-fired boilers were first introduced in the 1920s for coal; in the late 1940s, the Detroit Stoker Company installed the first traveling-grate spreader stoker boiler for wood. Mechanical stokers are the traditional technology that has been used to automatically supply solid fuels to a boiler. All stokers are designed to feed fuel that has been used to automatically supply solid fuels to a boiler. All stokers are designed to feed fuel onto a grate where it burns with air passing up through it. The stoker is located within the furnace section of the boiler and is designed to remove the ash residue after combustion. Stoker units use mechanical means to shift and add fuel to the fire that burns on and above the grate located near the base of the boiler. Heat is transferred from the fire and combustion gases to water tubes on the walls of the boiler.

Modern mechanical stokers consist of four elements: (1) a fuel admission system, (2) a stationary or moving grate assembly that supports the burning fuel and provides a pathway for the primary combustion air, (3) an overfire air system that supplies additional air to complete combustion and minimize atmospheric emissions, and (4) an ash discharge system. A successful stoker installation requires selecting the correct size and type of stoker for the fuel being used and for the load conditions and capacity being served.

Stoker boilers are typically described by their method of adding and distributing fuel. The two general types of systems are *underfeed* and *overfeed*. Underfeed stokers supply both the fuel and air from under the grate, while overfeed stokers supply fuel from above the grate and air from below. Overfeed stokers are further divided into two types—*mass feed* and *spreader*. In the mass feed stoker, fuel is continuously fed onto one end of the grate surface and travels horizontally across the grate as it burns. The residual ash is discharged from the opposite end. Combustion air is introduced from below the grate and moves up through the burning bed of fuel. In the spreader stoker, the most common type of stoker boiler, combustion air is again introduced, primarily from below the grate but the fuel is thrown or spread uniformly across the grate area. The finer particles of fuel combust in suspension as they fall against the upward moving air. The remaining heavier pieces fall and burn on the grate surface, and any residual ash is removed from the discharge end of the grate. Chain-grate, traveling-grate, and water-cooled vibrating-grate stokers are other less common configurations that use various means to maintain an even, thin bed of burning fuel on the grate. Other specialized stoker boilers include balanced-draft, cyclone-fired, fixed-bed, shaker-hearth, tangential-fired, and wall-fired. Practical considerations limit stoker size and, consequently, the maximum steam generation rates. For coal firing, this maximum is about 35,000 pounds per hour (lb/hr); for wood or other biomass firing, it is about 700,000 lb/hr.

DID YOU KNOW?

Boiler efficiency is defined as the percentage of the fuel energy that is converted to steam energy. Major efficiency factors in biomass combustion are moisture content of the fuel, excess air introduced into the boiler, and the percent of uncombusted or partially combusted fuel. According to the Council of Industrial Boiler Owners (CIBO), the general efficiency range of stoker and fluidized-bed boilers is between 65 and 85%.

Fluidized-Bed Boilers

Fluidized-bed boilers are able to burn a wide range of conventional fuels and waste fuels with high moisture, including:

- Wood wastes and bark
- Paper mill sludges
- Recycled paper facility sludges
- Sewage sludge
- Tire-derived fuel
- Oil and natural gas
- Coal, in combination
- Peat
- Biomass
- Sugar cane waste
- Agricultural waste

In this method of combustion, fuel is burned in a bed of hot, inert (incombustible) particles suspended by an upward flow of combustion air that is injected from the bottom of the combustor to keep the bed in a floating or "fluidized" state. The scrubbing action of the bed material on the fuel enhances the combustion process by stripping away the CO_2 and char (solids residue that normally forms around the fuel particles). This process allows oxygen to reach the combustible material more readily and increases the rate and efficiency of the combustion process. One advantage of mixing in the fluidized bed is that it allows a more compact design than in conventional water tube boiler designs. Natural gas or fuel oil can also be used as a start-up fuel to preheat the fluidized bed or as an auxiliary fuel when additional heat is required. The effective mixing of the bed makes fluidized-bed boilers well suited to burn solid refuse, wood waste, waste coals, and other nonstandard fuels.

The fluidized-bed combustion process provides a means for efficiently mixing fuel with air for combustion. When fuel is introduced to the bed, it is quickly heated above its ignition temperature, ignites, and becomes part of the burning mass. The flow of air and fuel to the dense bed is controlled so the desired amount of heat is released to the furnace section on a continuous basis. Typically, biomass is burned with 20% or higher excess air. Only a small fraction of the bed is combustible material; the remainder is comprised of inert material, such as sand. This inert material provides a large inventory of heat in the furnace section, dampening the effect of brief fluctuations in fuel supply or heating value on boiler steam output.

Fuels that contain a high concentration of ash, nitrogen, and sulfur can be burned efficiently in fluidized-bed boilers while meeting stringent emission limitations. Because of long residence time and high intensity of mass transfer, fuel can be efficiently burned in a fluidized-bed combustor at temperatures considerably lower than in conventional combustion processes (1400 to 1600°F compared to 2200°F for a spreader stoker boiler). The lower temperatures produce less NO_x, a significant benefit with high-nitrogen-content wood and biomass fuels. SO_2 emissions from wood waste and biomass are generally insignificant, but where sulfur contamination of the fuel is an issue, limestone can be added to the fluid bed to achieve a high degree of sulfur capture. Fuels that are typically contaminated with sulfur include construction debris and some paper mill sludges.

Fluidized-bed boilers are categorized as either *atmospheric* or *pressurized* units. Atmospheric fluidized-bed boilers are further divided into bubbling-bed and circulating-bed units; the fundamental difference between bubbling-bed and circulating-bed boilers is the fluidization velocity, which is higher for circulating-bed boilers. Circulating fluidized-bed boilers separate and capture fuel solids entrained in the high-velocity exhaust gas and return them to the bed for complete combustion. Atmospheric-pressure bubbling fluidized-bed boilers are most commonly used with biomass fuels. The type of fluid bed selected is a function of the as-specified heating value of the biomass fuel. Bubbling-bed technology is generally selected for fuels with lower heating values. The circulating fluidized-bed boiler is most suitable for fuels of higher heating values.

In a pressurized fluidized-bed boiler, the entire fluidized-bed combustor is encased inside a large pressure vessel. Burning solid fuels in a pressurized fluidized-bed boiler produces a high-pressure stream of combustion gases. After the combustion gases pass through a hot-gas cleanup system, they are fed into a gas turbine to make electricity, and the heat in the hot exhaust gas stream can be recovered to boil water for a steam turbine. For this reason, a pressurized fluidized-bed boiler is more efficient but also more complicated and expensive. Capital costs of pressurized fluidized-bed combustion technology are higher than atmospheric fluidized-beds.

Boiler Efficiency

The efficiency of a biomass stoker and a fluidized-bed boiler is estimated based on the heat-loss method, which is a way of determining boiler efficiency by measuring the individual heat losses (expressed as a percent of the input) and subtracting them from 100%. The largest energy loss in a boiler is the heat that leaves the stack. This loss could amount to as much as 30 to 35% of the fuel input in older, poorly maintained boilers. The primary difference in efficiency between a stoker boiler and a fluidized-bed boiler is the amount of fuel that remains unburned. The efficiency of fluidized-bed boilers compares favorably with stoker boilers due to lower combustion losses. Stoker boilers can have 30 to 40% carbon in the ash and additional volatiles and CO in the flue gases, whereas fluidized-bed boiler systems typically achieve nearly 100% fuel combustion. The turbulence in the

DID YOU KNOW?

Biomass boilers are typically run with a considerable amount of excess air so they can achieve complete combustion, but this has a negative impact on efficiency. A CIBO rule of thumb is that boiler efficiency can be increased 1% for each 15% reduction in excess air (Harrell, 2002).

combustor combined with the thermal inertia of the bed material provide for complete, controlled, and uniform combustion. These factors are key to maximizing the thermal efficiency, minimizing char, and controlling emissions. It should be noted that efficiency is not constant through the entire operating range of a boiler. Peak efficiency generally occurs at a particular boiler output that is determined by design characteristics. Whenever boiler operations deviate from this output, the resulting performance is usually below peak efficiency. Operating continuously at peak efficiency is not practical due to seasonal demands, load variations, and fuel property variations; however, operating at a steady load and avoiding cyclic or on/off operation can improve efficiency.

Boiler Operating Availability

The availability of a power generation system is the percentage of time the system can operate or is available to operate. Both planned maintenance and unplanned outages have a negative effect on system availability; therefore, an availability of 100% would represent a system that never broke down or required maintenance, which is impossible to achieve in real operation. Typically, both stoker and fluidized-bed boilers are designed for continuous operation, and design performance is over 90% availability. Seasonal variability in fuel availability or quality can affect the plant availability, but this is a feedstock issue, not an issue of boiler performance. A well-designed biomass steam system has a reasonable expectation of operating at 92 to 98% availability (Energy Products of Idaho, 2011).

Boiler Operating Advantages and Disadvantages

Depending on the fuel characteristics and site requirements, stoker and fluidized-bed boilers have specific operating advantages and disadvantages with biomass fuels. Biomass fuels are extremely variable in terms of heating value, moisture content, and other factors that affect combustion. Wood and most other biomass fuels are composed primarily of cellulose and moisture. As discussed previously, the high proportion of moisture is significant because it acts as a heat sink during the combustion process. The latent heat of vaporization—the amount of heat necessary to change 1 kg of a liquid into a gas— depresses flame temperature, taking heat energy away from steam production and contributing to the difficulty of efficiently burning biomass fuels. Cellulose, in addition to containing the chemical energy released in combustion, contains fuel-bound oxygen. This oxygen decreases the theoretical air requirements for combustion and, accordingly, the amount of nitrogen included in the products of combustion. A few general guidelines for direct firing of wood and biomass in boilers include the following:

- Stable combustions should be maintained, which can be achieved in most water-cooled boilers with fuel moisture contents as high as 65% by weight, as received.

- Use of preheated combustion air reduces the time required for fuel drying prior to ignition and is essential to spreader stoker combustion systems. Design air temperatures will vary directly with moisture content.

- A high proportion of the combustible content of wood and other biomass fuels burns in the form of volatile compounds. A large proportion of the combustion air requirement, therefore, is added above the fuel in stoker and other conventional combustion boilers as overfire air.

- Solid chars produced in the initial stages of combustion of biomass fuels are of very low density. Conservation selection of furnace section size is used to reduce gas velocity and keep char entrainment into the flue gases and possibly out of the stack at acceptable levels.

To ensure smooth fuel feeding, biomass fuels have to be carefully sized and processed. As discussed above, the moisture content of wood and other biomass waste can vary over a wide range, from 10 to more than 60%. To ensure steady heat input into the boiler using volumetric feeders, efficient homogenization of fuel with different moisture contents at the fuel yard is necessary.

Caution: Biomass-based fuels can increase the risk of slagging and fouling of heat transfer surfaces and, in some cases, the risk of fireside corrosion, as well. Potassium ash content is relatively high in fresh wood, green particles, and fast-growing biomass, which causes the ash to melt at low temperatures and leads to a tendency for fouling and slagging. Additionally, biomass fuels can contain chlorine, which, together with alkalis, can induce aggressive corrosion.

Co-Firing

Co-firing refers to the practice of mixing biomass with a fossil fuel in high-efficiency boilers as a supplementary energy source. It is one of the most cost-effective and easily implemented biomass energy technologies. Biomass can substitute for up to 20% of the coal used in the boiler. Co-firing is typically used when either the supply of biomass is intermittent or the moisture content of the biomass is high. At large plants, biomass is co-fired with coal, and more coal is typically used than biomass. At small plants, biomass is co-fired with natural gas, and more biomass is typically used than natural gas because the natural gas is used to stabilize combustion when biomass with high-moisture content is fed into the boiler. Figure 11.2 shows a process diagram for a standard coal-based co-firing plant. Biomass has been co-fired with coal economically in commercial plants, which is principally viewed as a strategy to reduce fuel costs. In certain situations, co-firing has provided opportunities for utilities to get fuel from wood manufacturing and other businesses at

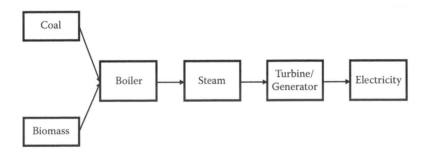

FIGURE 11.2
Biomass cofiring in coal power plant. (Adapted from Antares Group, Inc., *Assessment of Power Production at Rural Utilities Using Forest Thinnings and Commercially Available Biomass Power Technologies*, prepared for the U.S. Department of Agriculture, U.S. Department of Energy, and National Renewable Energy Laboratory, Washington, D.C., 2003.)

zero or negative cost. Overall production cost savings can also be achieved by replacing coal with inexpensive biomass fuel sources such as wood waste and waste paper. Typically, biomass fuel supplies should cost at least 20% less, on a thermal basis, than coal supplies before a co-firing project can be economically attractive.

Biomass co-firing is mainly a retrofit application. A basic principle of co-firing is that significant changes to the boiler are not required beyond some minor burner modifications or additions necessary to introduce and burn the supplemental fuel. To meet this objective, co-firing biomass fuels is usually done on a limited basis, with the amount of biomass ranging from 5 to 15% of the total heat input to the boiler (Fehrs and Donovan, 1999). Biomass fuels that have been successfully co-fired include wood and palletized waste paper. Among electric utilities and other users of coal boilers, interest in co-firing biomass is growing, chiefly due to the need to improve air emissions from coal-burning facilities, as well as to diversify fuel supplies.

Co-Firing Efficiency

Usually, no major changes in boiler efficiency result from co-firing; however, some design and operational changes might be necessary to maximize boiler efficiency while maintaining acceptable opacity, baghouse performance, and other operating requirements. Without these adjustments, boiler efficiency and performance can decrease; for example, at a biomass heat input level of 10%, boiler efficiency losses of 2% were measured during co-firing tests at a facility with a pulverized coal boiler when no adjustments were made (Tillman, 2000). Numerous co-firing projects have demonstrated that efficiency and performance losses can be minimized with proper awareness of operational issues.

Co-Firing Operating Availability

The availability of biomass and coal co-fired boilers is similar to that of regular coal boilers, if proper modifications are made to the system. If some of the potential operating issues mentioned in the next section manifest, then availability might be negatively affected.

Co-Firing Advantages and Disadvantages

Typically, co-firing biomass in an existing coal boiler requires modifications or additions to fuel handling, processing, storage, and feed systems. Slight modifications to existing operation procedures, such as increasing overfire air, might also be necessary, as well as increasing fuel feeder rates to compensate for the lower density and heating value of biomass. Fuel characteristics and processing can greatly affect the ability to use biomass as a fuel in boilers. Wood chips are preferable to mulch-like material for co-firing with coal in stoker boilers because the chips are similar to stoker coal in terms of size and flow characteristics. This similarity minimizes problems with existing coal-handling systems. When using a mulch-like material or a biomass supply with a high fraction of fine particles (sawdust size or smaller), periodic blockage of fuel flow openings in various areas of the conveying, storage, and feed systems can occur. These blockages can cause significant maintenance increases and operations problems; therefore, fuel should be processed to avoid difficulties with existing fuel feeding systems.

Another fuel consideration when dealing with biomass is the potential for problems with slagging, fouling, and corrosion. Some biomass fuels have high alkali (principally potassium) or chlorine content that can lead to unmanageable ash deposition problems on heat-exchange and ash-handling surfaces. Chlorine in combustion gases, particularly at high temperatures, can cause accelerated corrosion of combustion system and flue gas cleanup components. These problems can be minimized or avoided by screening fuel supplies for materials high in chlorine and alkali, limiting the biomass contribution to boiler heat input to 15% or less, using fuel additives, or increasing soot blowing. The most troublesome biomass resource tends to be agricultural residues, including grasses and straws, which have high alkali and chlorine contents. In contrast, most woody materials and waste papers are relatively low in alkali and chlorine and should not present this problem.

Currently, about 25% of the fly ash from coal-fired power plants is used as a feedstock for cement and concrete production, while another 15% is used as a feedstock in other applications (American Coal Ash Association, 2009). According to ASTM C-618, only fly ash from coal combustion qualifies for use in cement/concrete applications. Co-firing biomass in a coal power plant would keep the fly ash from meeting the current standard. Similarly, coal fly ash will sometimes not meet the current standard when certain emissions control techniques are used, such as ammonia injection. Although these restrictions can impact the economics of biomass co-firing, the value of finding a productive use for fly ash and other coal combustion products is primarily the avoidance of a roughly $20/ton landfill fee. For coal with 10% ash content, this value would be worth about $2/ton of the input fuel cost. Even though the current restrictions are a barrier to considering co-firing in some applications, other uses of fly ash are not affected, and researchers are currently studying the impact of using fly ash from biomass and biomass/coal co-firing on concrete characteristics. Early results show that biomass and co-fired fuels do not adversely affect the usefulness of fly ash in cement and concrete, and in fact might have some advantages (Wang and Baxter, 2007). It is likely that this work will eventually lead to a reevaluation of the standard and inclusion of fly ash from co-firing as an acceptable cement/concrete feedstock as has already happened in Europe.

Gasification Technologies

Biomass gasification for power production involves heating solid biomass in an oxygen-starved environment to produce a low- or medium-calorific gas. Depending on the carbon and hydrogen content of the biomass and the gasifier's properties, the heating value of the syngas can range anywhere from 100 to 500 Btu/ft^3 (10 to 50% that of natural gas). The heating value of syngas generally comes from CO and hydrogen produced by the gasification process. The remaining constituents are primarily CO_2 and other incombustible gases. Biomass gasification offers certain advantages over directly burning the biomass because the gas can be cleaned and filtered to remove problem chemical compounds before it is burned. Gasification can also be accomplished using chemicals or biologic action (e.g., anaerobic digestion); however, thermal gasification is currently the only commercial or near commercial option.

The fuel output from the gasification process is generally called *syngas*, though in common usage it might be called *wood gas*, *producer gas*, or *biogas*. Syngas can be produced through direct heating in an oxygen-starved environment, partial oxidation, or indirect heating in the absence of oxygen. Most gasification processes include several steps. The primary conversion process, called *pyrolysis*, is the thermal decomposition of solid biomass (in an oxygen-starved environment) to produce gases, liquids (tar), and char. Pyrolysis releases the volatile components of the biomass feed at around 1100°F through a series of complex reactions. Biomass fuels are an ideal choice for pyrolysis because they have so many volatile components (70 to 85% on a dry basis, compared to 30% for coal). The next step involves a further gasification process that converts the leftover tars and char into CO using steam and partial combustion. In coal gasification, pure oxygen or oxygen-enriched air is preferred as the oxidant because the resulting syngas produced has a higher heating value and the process is more efficient. In biomass gasification, oxygen is generally not used because biomass ash has a lower melting point than coal ash and because the scale of plants is generally smaller. Very-high-temperature processes involving passing the biomass through a plasma arc have been developed and tested primarily for waste remediation, contaminated wastes, and municipal solid waste.

Compared with direct-fired biomass system, gasification is not yet an established commercial technology. There is great interest, however, in the development and demonstration of biomass gasification for a number of reasons:

- A gaseous fuel is more versatile than a solid fuel. It can be used in boilers, process heaters, turbines, engines, and fuel cells; distributed in pipelines; and blended with natural gas or other gaseous fuels.

- Gasification can remove fuel contaminants and reduce emissions compared to direct-fired systems.

- Gasification can be designed to handle a wide range of biomass feedstocks, from woody residues to agricultural residues to dedicated crops, without major changes in the basic process.

- Gasification can be used to process waste fuels, providing safe removal of biohazards and entrainment of heavy metals in nonreactive slag.

A gaseous fuel can be used in a high-efficiency power generation system, such as a gas turbine-combined cycle or fuel cells, provided it is cleaned of contaminants. When equipment is added to recover the heat from the turbine exhaust, system efficiencies can increase to 80%. Like the direct combustion processes described earlier, two principal types of gasifiers have emerged: fixed-bed and fluidized-bed. Fixed-bed gasifiers are typically simpler and less expensive and produce a lower heat content syngas. Fluidized-bed gasifiers are more complicated and more expensive and produce a syngas with a higher heating value.

Gasifiers

Fixed-Bed Gasifiers

Fixed-bed gasifiers typically have a fixed grate inside a refractory-lined shaft. The fresh biomass fuel is typically placed on top of the pile of fuel, char, and ash inside the gasifier. A further distinction is based on the direction of air (or oxygen) flow: downdraft (air flows down through the bed and leaves as biogas under the grate), updraft (air flows up through the grate and biogas is collected above the bed), or crossflow (air flows across the bed, exiting as biogas). Schematics of the primary section of the types of fixed-bed gasifiers are shown in Figure 11.3. Table 11.3 compares fixed-bed gasifier types. Fixed-bed gasifiers are usually limited in capacity and are typically used for generation systems that are able to produce less than 5 MW. The physics of the refractor-line shaft reactor vessel limits the diameter and thus the throughput. Developers have identified a good match between fixed-bed gasifiers and small-scale distributed power generation equipment; however, the variable economics of biomass collection and feeding, coupled with the gasifier's low efficiency, make the economic viability of the technology particularly site specific.

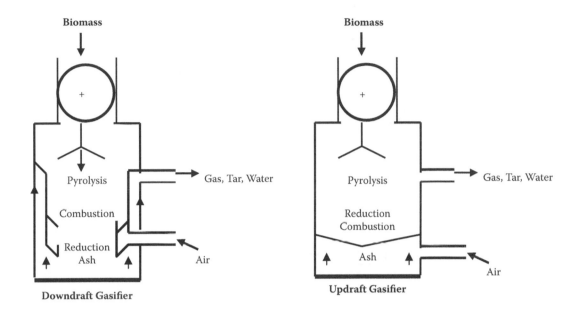

FIGURE 11.3
Fixed bed gasifier types. (Adapted from Bain, R., *Biomass Gasification*, presented at the USDA Thermochemical Conversion Workshop, Pacific Northwest National Laboratory, Richmond, WA, September 6, 2006.)

TABLE 11.3

Comparison of Fixed-Bed Gasification Technologies

	Type of Gasifier		
	Downdraft	**Updraft**	**Crossflow**
Operation	Biomass is introduced from the top and moves downward. Oxidizer (air) is introduced at the top and flows downward. Syngas is extracted at the bottom at grate level.	Biomass is introduced from the top and moves downward. Oxidizer is introduced at the bottom and flows upward. Some drying occurs. Syngas is extracted at the top.	Biomass is introduced from the top and moves downward. Oxidizer is introduced at the bottom and flows across the bed. Syngas is extracted opposite the air nozzle at the grate.
Advantages	Levels of tars and particulates in the syngas are lower, allowing direct use in some engines without cleanup. The grate is not exposed to high temperatures.	Process can handle higher moisture biomass. Higher temperatures can destroy some toxins and slag minerals and metal. Higher tar content adds to heating value.	This is the simplest of designs and provides stronger circulation in the hot zone. Lower temperatures allow the use of less expensive construction materials.
Disadvantages	Biomass must be very dry (<20% moisture content). The syngas is hot and must be cooled if compression or extensive cleanup is required. About 4 to 7% of the carbon is not converted and remains in the ash.	Higher tar content can foul engines or compressors. The grate is exposed to high temperatures and must be cooled or otherwise protected.	This process is more complicated to operate. Issues with slagging have been reported. High levels of carbon (33%) are in the ash.

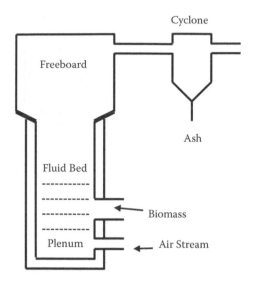

FIGURE 11.4
Fluidized-bed gasifier. (Adapted from Bain, R., *Biomass Gasification*, presented at the USDA Thermochemical Conversion Workshop, Pacific Northwest National Laboratory, Richmond, WA, September 6, 2006.)

Fluidized-Bed Gasifiers

Fluidized-bed gasifiers utilize the same gasification processes and offer higher performance than fixed-bed systems, but with greater complexity and cost. Similar to fluidized-bed boilers, the primary gasification process takes place in a bed of hot inert materials suspended by an upward motion of oxygen-depleted gas (see Figure 11.4). As the amount of gas is augmented to achieve greater throughput, the bed will begin to levitate and become "fluidized." Sand or alumina is often used to further improve the heat transfer. Notable benefits of fluidized-bed devices are their high productivity (per area of bed) and flexibility. Fluidized-bed gasifiers can also handle a wide range of biomass feedstocks with moisture content up to 30% on average.

Three stages of fluidization can occur in the gasifier depending on the design: bubbling, recirculating, or entrained flow. At the lower end of fluidization, the bed expands and begins to act as a fluid. As the velocity is increased, the bed will begin to bubble. With a further increase in airflow, the bed material begins to lift off the bed. This material is typically separated in a cyclone and recirculated to the bed. With still higher velocities, the bed material is entrained (i.e., picked up and carried off by the airflow).

Fluidized-bed gasifiers can be designed to use a portion of the pyrolysis gases to generate the heat to drive the process, or they can be externally fired. Operating the gasifier at higher pressures increases the throughput; however, this also increases the complexity and cost of the gasifier. In these units, the biomass is fully converted after going through the pyrolysis and char conversion processes.

By reducing the quantity of air and process temperature, it is possible to operate fluidized-bed boilers as gasifiers. In this operating mode, the gasifiers produce a gas with a heating value of slightly more than 100 Btu/ft^3. This gas is burned above the bed as additional air supply is injected upstream of the boiler tube section.

Gasification Efficiency

Fixed-bed and fluidized-bed biomass gasification processes use similar types of equipment as direct combustion. The biomass fuel is fed into a combustion/reaction vessel with a fixed, fluidized, or moving bed. The thermodynamics of heat loss are similar, but gasification conditions are different from direct combustion. In direct combustion, 10 to 14 times the weight of the fuel is introduced as air. In gasification, the air entering the reactor, if any, is only one to two times the weight of the fuel. This difference reduces heat losses from the reaction zone. On the other hand, the syngas exits the gasification reactor at very high temperatures (1200 to 1500°F); some of this heat loss can be recovered either directly through the use of heat exchangers in the gas cooling section or indirectly through the use of heat recovery from the combustion of the syngas in the power generation section. To the extent that heat is used to preheat incoming air, to introduce high-temperature steam, or to dry the incoming biomass, the efficiency of biomass to syngas conversion will be increased. Heat that is recovered from the hot-gas cooling section can also be added to the combined heat and power heat recovery. In this case, the intermediate efficiency value of syngas conversion is not increased, but the overall combined heat and power efficiency is. These differences combine to produce biomass to syngas efficiencies (heat value of the syngas divided by the heating value of the biomass) of 60 to 80%. In integrated configurations, however, additional steam can be generated from cooling the hot syngas exiting the reactor prior to cleanup.

Gasification Operating Availability

Because the commercialization of biomass gasification plants is in its early stages, no facility survey information on their availability or reliability is available. Plants are designed for continuous operation, and design performance exceeds 90%. Actual experience with emerging technology tends to result in lower availability than is experienced during broad commercial use, as material handling problems, control issues, and component failures cause more frequent unplanned outages than are seen after accumulating additional operating experience. With a newly established support infrastructure, outages also tend to last longer before being fixed or solved. A well-designed system, however, has a reasonable expectation of operating in the 85 to 95% availability range.

Gasification Operating Issues

As discussed above, moisture content, gas cleanup, and operating pressure can all affect operation of a gasifier. Several operating issues are common to the different types of gasification systems.

Moisture Content

Green biomass, defined as freshly harvested plant material, can contain a significant amount of water by weight (up to 60%). This water does not contribute to the heat content of the syngas but consumes a significant amount of energy in gasification. Even though water cannot be burned (oxidized) at elevated temperatures, it will dissociate into its elemental components—hydrogen and oxygen. The hydrogen will contribute to the calorific value of the syngas. This reaction is very temperature sensitive, and the hydrogen and oxygen will usually recombine into water vapor as the syngas cools; therefore, the moisture content of biomass must be strictly limited. If there is excess moisture, the gasification process cannot sustain itself with an external source of heat. As the moisture content of the biomass increases, the net energy available in the syngas decreases. Fixed-bed gasifiers that use internal combustion of the syngas typically utilize biomass with less than 20% moisture content. Fluidized-bed gasifiers typical require less than 30% moisture content. Green biomass is the most readily available and inexpensive biomass product. The drying process requires a considerable additional capital investment and increases the operating and maintenance (O&M) costs. Unfortunately, the cost of the drying equipment (equipment cost and O&M cost) seldom covers the cost savings of using green biomass.

Gas Cleanup

As syngas leaves the gasifier, it contains several types of contaminants that are harmful to downstream equipment, ash handling, and emissions. The degree of gas cleanup must be appropriately matched to its intended use. For use in reciprocating engines, gas turbines, and especially fuel cells, a very clean gas is required. As shown in Table 11.4, the primary contaminants in syngas are tars, particles, alkali compounds, and ammonia. The types of contaminants that are observed depend on the biomass feedstock and the gasification process used. Because gasification occurs at an elevated temperature, syngas can have as much as a third of its total energy in sensible heat. Cleaning the gas while it is hot would be advantageous from an energy-use perspective, but this task is currently difficult to accomplish. Research is ongoing regarding hot gas filters, which can be applied in coal gasification, as well as other high-temperature processes. Wet scrubbers are currently one of the most reliable and least expensive options for gas cleanup, even though they sacrifice a large portion of the sensible heat of the syngas. Cooling the hot syngas can provide a source of steam for the cleaning process, power generation, or end-use.

Gasification Operating Pressure

Gasifiers can be operated at either atmospheric or elevated pressures. Air-blown, atmospheric gasifiers produce a very low Btu gas—110 to 170 Btu/standard cubic feet (scf). To introduce this gas into a gas turbine in the power generation section of the plant requires considerable compression energy, up to a third of the turbine's output. Therefore, it would

TABLE 11.4

Gas Cleanup Issues

Containment	Description	Treatment
Tar	Tars (creosote) are complex hydrocarbons that persist as condensable vapors.	Wet scrubbers, electrostatic precipitators, barrier filters, catalysts, combustion
Particles	Particles are very small, solid materials that typically include ash and unconverted biomass.	Cyclone separators, fabric filters, electrostatic precipitators, wet scrubbers
Alkali compounds	Potassium, alkali salts, and condensed alkali vapors are part of the chemical composition of biomass.	Cool syngas below 1200°F, causing the alkali vapors to condense. Use cyclone separators, fine fabric filters, electrostatic precipitators, wet scrubbers.
Ammonia	Ammonia is formed from nitrogen (fuel-bound and in air) and hydrogen (in fuel and in moisture content). When syngas is burned, ammonia is converted to NO_x.	Catalysts, hydrocarbon reforming, wet scrubbing

TABLE 11.5

Relative Advantages/Disadvantages of Gasifier Types

Gasifier	Advantages	Disadvantages
Updraft fixed bed	Mature for heat	Feed-size limits
	Small-scale applications	High tar yields
	Can handle high moisture	Scale limitations
	No carbon in ash	Low-Btu gas
		Slagging potential
Downdraft fixed bed	Small-scale applications	Feed-size limits
	Low particulates	Scale limitations
	Low tar	Low-Btu gas
		Moisture sensitive
Bubbling fluid bed	Large-scale applications	Medium tar yield
	Feed characteristics	Higher particle loading
	Direct/indirect heating	
	Can produce higher Btu gas	
Circulating fluid bed	Large-scale applications	Medium tar yield
	Feed characteristics	Higher particle loading
	Can produce syngas	
Entrained flow fluid bed	Can be scaled	Large amount of carrier gas
	Potential for low tar	Higher particle loading
	Potential for low methane	Particle size limits
	Can produce higher Btu gas	

be advantageous to produce the syngas at a high pressure so it can be introduced directly into the combustion section of a gas turbine without additional compression. Pressurized reactors, however, do need to compress any combustion air or oxygen that is introduced into the reactor and maintain a pressure seal on the biomass input and ash removal systems.

Gasification Advantages and Disadvantages

Fixed-bed and fluidized-bed gasifiers have specific operating advantages and disadvantages, with biomass fuels depending on the biomass characteristics and site requirements. Table 11.5 provides a qualitative comparison of gasifier characteristics and operating issues for fixed-bed and fluidized-bed systems.

Modular Systems

Modular biomass-fueled combined heat and power systems are defined as small systems, less than 5 MW, although they are typically smaller. The main operating components come in one or more preengineered and packaged modules for simple installation at the user's site. The systems typically include a fuel processor (combustion or gasification), necessary intermediate fuel cleanup, an electric generator, and heat recovery from both the power generation and energy conversion sections. An automatic fuel storage and delivery system must be added for a complete operating system. Small modular biomass systems can supply electricity to rural areas, farms, businesses, and remote villages. These systems use locally available biomass fuels such as wood, crop waste, animal manure, and landfill gas. Development of biomass-fueled modular power systems is of great interest internationally as a means to bring power to isolated communities in areas lacking power and fuel infrastructure. In the United States, there is interest in small systems to utilize opportunity fuels from a local area, such as crop wastes or fire-control forest thinnings. Modular systems are essentially scaled-down versions of larger systems. Some systems use direct-fired technology with steam power, and others use gasification technology and gaseous fuel burning power technologies, such as internal combustion engines, microturbines, and Stirling engines. There are also direct-fired systems that use Stirling engines for power production, as well as systems that employ gasification, wherein the hot raw gas is combusted to raise steam.

Modular Gasification Systems

In a 75-kW modular biomass basic gasification system (used here for discussion purposes), eight submodules are included; storage and feed submodules are not included. The basic package modules include the following:

- Automatic biomass feed system
- Dryer to reduce the feedstock moisture content
- Chip sorter for sizing
- Heat exchanger that extracts heat from the gasifier for use in the dryer and for onsite thermal applications
- Gasifier feeder
- Downdraft gasifier producing low-Btu gas (heating value of about 110 Btu/scf—higher heating value)
- Filtering stages that remove particulates
- Power module

The power module can be an internal combustion engine designed to run on low-Btu fuel, a microturbine, a Stirling engine, or even a fuel cell. The power module also has heat-recovery equipment to provide additional useable thermal energy for onsite use. Because the gas is of such a low Btu content, propane or natural gas is required on system start-up. After start-up, the system can run on the syngas alone.

Systems such as these will require feedstock storage with an in-place delivery system. An in-ground storage bunker with a moving bed would allow direct delivery of fuel loads into the automated system. This can consist of a permanently installed live-bottom van into which dump trucks can deliver a sized fuel supply.

Modular Combustion Systems

Direct combustion in feed bed combustors is a commercial technology in larger sizes. In these large systems, as characterized previously, power is generated by steam turbines. In modular systems, other power systems are being developed that are more suitable for small-sized applications. The typical power and heat cycles being employed or explored for use are as follows:

- Steam cycle
- Organic Rankine cycle (ORC)
- Brayton cycle, hot-air turbine
- Entrophic cycle, as defined by its developer, similar to organic Rankine cycle but with a higher temperature differential producing higher efficiencies
- Stirling engine, external combustion

Stirling engine technology is used for very small (500 W to 10 kW) modular systems being developed. The generators will convert various biomass fuels (wood, wood pellets, sawdust, chips, or biomass waste) to electricity and useful heat. These systems typically convert 10 to 20% of the fuel energy to electricity; 60 to 70% of fuel energy is then available for heating water and spaces. The burner for the prototype system includes a ceramic fire box and a fuel hopper with a fuel capacity of 24 hours. It accomplishes complete two-stage combustion with comparatively low emissions. The Stirling engine–alternator requires minimal maintenance because its gas bearings eliminate contact, friction, and wear. Its projected life is 40,000 hours.

Modular Hybrid Gasification/Combustion Systems

The modular hybrid gasification/combustion system operates functionally like a direct combustion system. Power is derived by a back-pressure steam turbine that also provides steam for onsite thermal energy requirements. The difference is that the combustion chamber is actually a gasification system that uses a two-chamber gasifier approach. The system is similar to a two-stage combustion boiler design. This approach allows the production of gas in a relatively cool chamber at temperatures from 1000°F to 1400°F and then combustion in a relatively hot chamber (the boiler) at temperatures up to 2300°F. These temperatures allow the complete removal of carbon from the fuel in the gasifier and more complete oxidation of complex organics in the oxidation zone. The combination of these features results in a clean-burning, fuel-efficient system. Combined heat and powers units include small back-pressure steam turbines from 100 kW up to several megawatts. This approach combines the simplicity and low cost of a combustion system with the gasification advantages of more complete carbon conversion and cleaner combustion characteristics.

Modular Gasification/Combustion Process Efficiency

Efficiency is typically fairly low. In applications requiring considerable thermal energy, the overall combined heat and power efficiencies are comparable to gas-fired systems; however, the ratio of electric to thermal for these system is much lower, so more of the total useful energy is delivered in the form of heat rather than in the form of higher value electricity.

Modular Gasification/Combustion Operating Advantages and Disadvantages

The main operating advantages today are in the use of opportunity biomass fuels of low value such as wood chips or forest thinnings. In addition, many of the systems are targeted at remote applications where it would be too costly to connect to grid electricity. The main disadvantage affecting all types of modular systems is the comparatively high capital costs associated with all of the required equipment. This equipment also takes up considerable space compared to conventional gas-fired combined heat and power systems. The engine generator systems occupy only about 5% of the total space required for the modular biomass system. Another disadvantage is the need for maintenance and repairs associated with the many subsystems, particularly the solids-handling components and filters.

Power Generation Technologies

Combined heat and power (CHP) is the sequential or simultaneous generation of multiple forms of useful energy (usually mechanical and thermal) in a single, integrated system. CHP systems consist of a number of individual components—prime mover (heat engine), generator, heat recovery, and electrical interconnection—configured into an integrated whole. The type of equipment that drives the overall system (i.e., the prime mover) typically identifies the CHP system. Prime movers for CHP systems include steam turbines, gas (or combustion) turbines, spark ignition engines, diesel engines, microturbines, and fuel cells. These prime movers are capable of burning a variety of fuels, including biomass/biogas, natural gas, or coal to produce shaft power or mechanical energy. Additional technologies are also used in configuring a complete CHP system, including boilers, absorption chillers, desiccants, engine-driven chillers, and gasifiers.

Although mechanical energy from the prime mover is most often used to drive a generator to produce electricity, it can also be used to drive rotating equipment such as compressor, pumps, and fan. Thermal energy from the system can be used in direct process applications or indirectly to produce steam, hot water, hot air for drying, or chilled water for process cooling.

The industrial sector currently produces both thermal output and electricity from biomass in CHP facilities in the paper, chemical, wood products, and food processing industries. These industries are major users of biomass fuels—utilizing the heat and steam in their processes can improve energy efficiencies by more than 35%. In these applications, the typical CHP system configuration consists of a biomass-fired boiler whose steam is used to propel a steam turbine in addition to the extraction of steam or heat for process use.

The following technologies are discussed in this section, with specific respect to their ability to run on biomass or biogas. A synopsis of key characteristics of each is provided in Table 11.6.

- *Fuel cells* produce an electric current and heat from a chemical reaction between hydrogen and oxygen rather than combustion. They require a clean gas fuel or methanol with various restrictions on contaminants.
- *Gas (combustion) turbines, including microturbines,* use heat to move turbine blades that produce electricity.
- *Reciprocating internal combustion (IC) engines* operate on a wide range of liquid and gaseous fuels but not solid fuels. The reciprocating shaft power can either produce electricity through a generator or drive loads directly.
- *Steam turbines* convert steam energy from a boiler or waste heat into shaft power.
- *Stirling engines* operate on any fuel and can produce either electricity through a generator or drive loads directly.

TABLE 11.6

Comparison of Prime Mover Technologies Applicable to Biomass

Characteristic	Steam Turbine	Gas/Combustion Turbine	Microturbine	Reciprocating IC Engine	Fuel Cell	Stirling Engine
Size	50 kW–150 MW	500 kW–40 MW	30 kW–250 kW	Smaller than 5 MW	Smaller than 1 MW	Smaller than 200 kW
Fuels	Biomass/biogas-fueled boiler for steam	Biogas	Biogas	Biogas	Biogas	Biomass or biogas
Fuel preparation	None	PM filter needed	PM filter needed	PM filter needed	Sulfur, CO, methane can be issues	None
Sensitivity to fuel moisture	N/A	Yes	Yes	Yes	Yes	No
Electric efficiency (electric, HHV)[a]	5–30%	22–36%	22–30%	22–45%	30–63%	5–45%
Turn-down ratio	Fair; responds within minutes	Good; responds within a minute	Good; responds quickly	Wide range; responds within seconds	Wide range; slow to respond (minutes)	Wide range; responds within a minute
Operating issues	High reliability, slow start-up, long life, maintenance infrastructure readily available	High reliability, high-grade heat available, no cooling required, requires gas compressor, maintenance infrastructure readily available	Fast start-up, requires fuel gas compressor	Fast start-up, good load-following, must be cooled when CHP heat is not used, maintenance infrastructure readily available, noisy	Low durability, low noise	Low noise
Field experience	Extensive	Extensive	Extensive	Extensive	Some	Limited
Commercialization status	Numerous models available	Numerous models available	Limited models available	Numerous models available	Commercial introduction and demonstration	Commercial introduction and demonstration
Installed system cost	$350–$750/kW (without boiler)	~$700–$2000/kW	$1100–$2000/kW	$800–$1500/kW	$3000–$5000/kW	Variable; $1000–$10,000/kW
Operation and maintenance (O&M) costs	Less than 0.4¢/kWh	0.6–1.1¢/kWh	0.8–2.0¢/kWh	0.8–2.5¢/kWh	1–4¢/kWh	Around 1¢/kWh

[a] Efficiency calculations are based on higher heating value (HHV) of the fuel, which includes the heat of vaporization of the water in the reaction products.

Source: USEPA, *Biomass Combined Heat and Power Catalog of Technologies*, U.S. Environmental Protection Agency, Washington, D.C., 2007.

Each of the technologies listed in Table 11.6 will require a fuel that has gone through the various preparation steps. For power generation technologies that require steam for fuel (steam turbine), a boiler is used to combust the biomass fuel, converting it to steam. For power generation technologies that require gas to operate (gas turbines, reciprocating engines, fuel cells, Stirling engines), the biomass feedstock will either be gasified or will be collected as biogas from an anaerobic digester or landfill gas (LFG).

Some amount of gas cleaning is required for almost any prime mover to run on biogas, as is standard practice to date. The cleaning would minimally include the removal of solids and liquid water. The prime mover can be damaged when only solids and liquid water are removed, leaving corrosive components and siloxanes (i.e., silicon, oxygen, and alkane), which are found in landfill gas. Some options for further cleanup are beginning to emerge, such as

- Chemical filters to remove sulfur compounds
- Cooling the gas to remove additional moisture
- Filter systems to remove siloxanes
- Chemical systems to remove CO_2

Specific details regarding the use of biomass and biogas fuels in the various prime movers are discussed within each of the following subsections.

Steam Turbine Technologies

A steam turbine is a thermodynamic device that converts the energy in high-pressure, high-temperature steam into shaft power that can in turn be used to turn a generator and produce electric power. Unlike gas turbine and reciprocating engine CHP systems where heat is a byproduct of power generation, steam-turned CHP systems normally generate electricity as a byproduct of the heat (steam) generation. A steam turbine requires a separate heat source and does not directly convert fuel to electric energy. The energy is transferred from the boiler to the turbine through high-pressure steam, which in turn powers the turbine and generator (they are also used to drive pumps, compressors and other mechanical equipment). This separation of functions enables steam turbines to operate with an enormous variety of fuels, from natural gas to solid waste, including all types of coal, wood, wood waste, and agricultural byproducts (sugar cane bagasse, fruit pits, and rice hulls). In CHP applications, steam at lower pressure is extracted from the steam turbine and used directly or is converted to other forms of thermal energy.

In the thermodynamic cycle illustrated in Figure 11.5 (also shown earlier in greater detail in Figure 8.2), called the Rankine cycle, liquid water is converted to high-pressure steam in the boiler and fed into the steam turbine. The steam causes the turbine blades to rotate, creating power that is turned into electricity by a generator. A condenser and pump are used to collect the steam exiting the turbine, feeding it into the boiler and completing the cycle. There are several different types of steam turbines:

- A condensing steam turbine as shown in Figure 11.5 is for power-only applications and expands the pressurized steam to low pressure at which point a steam/liquid water mixture is exhausted to a condenser at vacuum conditions.

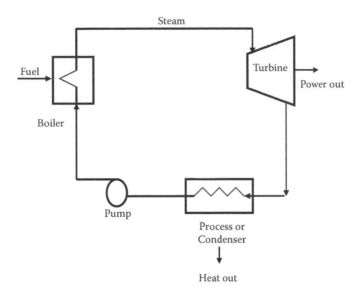

FIGURE 11.5
Simple steam turbine power cycle.

- Extraction turbines have openings in their casings for the extraction of a portion of the steam at some intermediate pressure for process or building heating.
- Back-pressure turbines exhaust the entire flow of steam to the process or facility at the required pressure.

DID YOU KNOW?

Steam turbine life is extremely long. Some steam turbines have been in service for over 50 years. Overhaul intervals are measured in years. When properly operated and maintained (including proper control of boiler water chemistry), steam turbines are extremely reliable. They require controlled thermal transients as the massive casing heats up slowly and differential expansion of the parts must be minimized.

Steam turbines are one of the most versatile and oldest prime mover technologies still in commercial production. In use for about 100 years, steam turbines replaced reciprocating steam engines due to their higher efficiencies and lower costs. Conventional steam turbines can range from 50 kW to several hundred megawatts for large utility power plants. Steam turbines are widely used for CHP applications.

Gas Turbine Technologies

Combustion turbines, or gas turbines, have been used for power generation for decades and are often the technology of choice for new electric generation in the United States and much of the world due to their low capital cost, low maintenance, and low emissions.

Turbine technology was developed in the 1930s as a means of propulsion for jet aircraft. Use of turbines for power generation began in the 1940s and 1950s, but it was not until the early 1980s that improvements in turbine efficiency and reliability resulted in increased utilization for power production. The gas turbine is an internal combustion engine that operates with rotational rather than reciprocating motion. Turbines can be fueled by natural gas or biogas and are used in a broad scope of applications, including electric power generation, gas pipeline compressors, and various industrial applications requiring shaft power. Although many newer turbines are large utility units, manufacturers are producing smaller and more efficient units that are well suited to distributed generation applications. Turbines range in size from 30 kW (microturbines) to 250 MW (large industrial units). Gas turbines can be used in a variety of configurations:

- *Simple-cycle operations*—A single gas turbine produces power only.
- *CHP operations*—A simple-cycle gas turbine with a heat recovery/heat exchanger recovers the heat from the turbine exhaust and converts it to useful thermal energy, usually in the form of steam or hot water.
- *Combined-cycle operation*—High-pressure steam generated from recovered exhaust heat is used to create additional power using a steam turbine. Some combined-cycle turbines extract steam at an intermediate pressure for use in industrial processes, making them combined-cycle CHP systems.

An illustration of the configuration of a gas turbine is shown in Figure 11.6. As shown in the figure, gas turbine power generation systems use the Brayton cycle and consist of a compressor to compress the air to high pressure, a combustor chamber operating at high pressure, the gas turbine itself, and the generator. The turbine section compresses one or more sets of turbine blades that extract mechanical energy from the hot combustion products. Some of that energy is used to power the compressor stage; the remaining energy is available to drive an electric generator or other mechanical load. The compressor and all of the turbine blades can be on one shaft, or there can be two shafts—one for the compressor and the turbine stages that drive it and a second for the turbine stages that produce useful output. To inject the fuel into the pressurized combustion chamber, the fuel must also be pressurized. A low-Btu gas—like most biogas—will require only a small pump, whereas

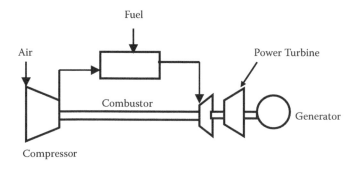

FIGURE 11.6
Components of a simple-cycle gas turbine. (Adapted from Energy and Environmental Analysis, Inc., *Gas-Fired Distributed Generation Technology Characterization: Microturbines*, NREL/TP-620-34783, prepared for National Renewable Energy Laboratory, Golden, CO, 2003.)

high-Btu gas (greater than about 1000 Btu/scf) requires a small compressor. Theoretical turbine efficiency is a function of turbine inlet temperature and pressure ratio across the power turbine, with higher levels of both factors leading to higher efficiency. Inlet temperature is limited by the ability of the turbine blades to operate at that temperature over the lifetime of the turbine.

There are some notable differences in gas turbine performance with biogas fuels. For example, a biomass gas turbine system requires landfill gas, anaerobic digester gas, or a biomass gasifier to produce the biogas for the turbine. This biogas must be carefully filtered of particulate matter to avoid damaging the blades of the gas turbine. Additionally, because a typical biomass gasifier produces a low-Btu biogas (e.g., 100 Btu/scf), the fuel compressor must be sized to handle about 10 times the gas flow compared to natural gas to provide the same Btu to the combustor. These flow requirements make the shaft power to the fuel compressor comparable to the power of the air compressor, thereby reducing the turbine's efficiency. In a conventional gas turbine, the turbine is designed to handle about 10% more flow (standard cubic feet per minute) than the air compressor. In a low-Btu gas turbine, the turbine must be designed to handle about twice the flow of the air compressor. In addition, the air-to-fuel ratio is lower for biogas than for natural gas, so not all of the compressed air is needed. Some of the compressed air can be redirected to provide energy to the air compressors for pressurized direct gasifiers or to help compress the biogas in atmospheric gasifiers. Without removal of the excess air, the capacity of the turbine would be significantly reduced.

Because of all the modifications required, existing natural gas turbines cannot easily be retrofitted to operate on low-Btu biogas (<300 Btu/scf). Gas turbines designed for low-Btu biogas generally cost at least 50% more than natural gas turbines on a per-kilowatt basis. Many gas turbine manufacturers offer gas turbine products that operate on medium-Btu landfill and wastewater treatment gas at equipment costs that are slightly higher than natural gas turbines, assuming that the gas is properly treated and cleaned. Non-fuel O&M costs will increase for gas turbines using low- and medium-Btu biogas due to increased cleaning and more frequent maintenance intervals.

Although a number of gas turbines have been studied and tested for low-Btu biogas use modification and integration, commercial experience is very limited. Some biomass gasifiers produce syngas for co-firing in integrated coal gasification combined-cycle power plants, and other biomass gasifiers have been built with the intention of ultimately integrating them with gas turbines; in the interim, biogas is being used as a supplement fuel.

Microturbines

Microturbines are small electricity generators that burn gaseous and liquid fuels to create high-speed rotation that turns an electrical generator. Today's microturbine technology is the result of development work in small stationary and automatic gas turbines, auxiliary power equipment, and turbochargers, much of which was pursued by the automotive industry beginning in the 1950s. Microturbines entered field testing around 1997 and began initial commercial service in 2000. Microturbines available and in development range in size from 30 to 250 kW, while conventional gas turbine sizes ranges from 500 kW to 250 MW. Microturbines run at high speeds and, like larger gas turbines, can be used in power-only generation or in combined heat and power (CHP) systems. They are able to operate on a variety of fuels, including natural gas, sour gases (high sulfur, low Btu content), and liquid fuels such as gasoline, kerosene, and diesel fuel/distillate heating oil. In resource recovery applications, they burn waste gases that would otherwise be flared or released directly into the atmosphere. Microturbines generally have lower electrical efficiencies than similarly sized reciprocating engine generators and large gas turbines; however, because of their design simplicity and relatively few moving parts, microturbines offer the potential for reduced maintenance compared to reciprocating engines.

Microturbines usually have an internal heat-recovery heat exchanger called a *recuperator*. In typical microturbines, the inlet air is compressed in a radial compressor and then preheated in the recuperator using heat from the turbine exhaust. Heated air from the recuperator is mixed with fuel in the combustor and ignited. The hot combustion gas is then expanded in one or more turbine sections, producing rotating mechanical power to drive the compressor and the electric generator. In single-shaft models, a signal expansion turbine turns both the compressor and the generator. Two-shaft models use one turbine to drive the compressor and a second turbine to drive the generator, with exhaust from the compressor turbine powering the generator turbine. The exhaust of the power turbine is then used in the recuperator to preheat the air from the compressor.

The basic components of a microturbine are shown in Figure 11.7. The heart of the microturbine is the compressor–turbine package, which is most commonly mounted on a single shaft along with the electric generator. Because the turbine shaft rotates at a very high speed, the electric output of the generator must be processed to provide 60-Hz power (the frequency standard in the United States). The single shaft is supported by two (or more) high-speed bearings. Because single-shaft turbines have only one moving part, they have the potential for low maintenance and high reliability. In two-shaft versions of the microturbine, the turbine on the first shaft drives only the compressor while a power turbine on a second shaft drives a gearbox and conventional electrical generator producing 60-Hz power. The two-shaft design has more moving parts but does not require sophisticated power electronics to convert high-frequency alternating current (AC) power output to usable 60-Hz power. Microturbines require gaseous fuel to be supplied in the range of 64 to 100 psig or above. Rotary vane, scroll, and screw compressors have been used to boost fuel gas pressure at the site to the pressure needed by the microturbine; however, this further reduces the efficiency of the system.

In CHP operation, a second heat-recovery heat exchanger—the exhaust gas heat exchanger—can be used to transfer remaining energy from the microturbine exhaust to a hot water system. Recuperated microturbines have lower temperature exhaust than simple-cycle turbines; however, exhaust heat at low temperatures can be used for a variety of different applications, including process or space heating, heating potable water, driving absorption chillers, or regenerating desiccant dehumidification equipment. Some

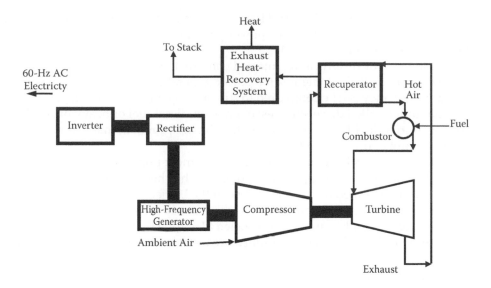

FIGURE 11.7
Microturbine-based CHP system (single-shaft design). (Adapted from Energy and Environmental Analysis, Inc., *Gas-Fired Distributed Generation Technology Characterization: Microturbines*, NREL/TP-620-34783, prepared for National Renewable Energy Laboratory, Golden, CO, 2003.)

microturbine-based CHP applications have the ability to bypass their recuperator to adjust their thermal-to-electric ratio or do not use recuperators at all. The temperature of the exhaust from these microturbines is much higher (up to 1200°F); thus, more and higher temperature heat is available for recovery.

Microturbines have demonstrated that they can handle landfill and wastewater treatment gas and, in some cases, low-Btu biogas reasonably well because of their simple design. No major modifications are needed, but in addition to the power required by the fuel gas compressor, there is a small reduction in power output (10 to 15%) when running on landfill or digester gas. With both factors considered, a 15 to 20% increase in price per kilowatt would be expected for microturbines operating on landfill or digester gas compared to the price for natural-gas-fired units of the same size. Maintenance costs would also increase 30 to 40% due to shorter maintenance intervals and increased inspections (Resource Dynamics Corp., 2004).

DID YOU KNOW?

Microturbines produce thermal output at temperatures in the range of 400 to 600°F, suitable for supplying a variety of building thermal needs.

Reciprocating Internal Combustion Engine Technologies

Reciprocating internal combustion engines are a widespread and well-known technology. North American production exceeds 35 million units per year for automobiles, trucks, construction and mining equipment, marine propulsion, lawn care, and a diverse set of power generation applications. A variety of stationary engine products are available for a range of power generation market applications and duty cycles, including standby and emergency power, peaking service, intermediate and base load power, and CHP. Reciprocating internal combustion engines are available for power generation applications in sizes ranging from a few kilowatts to more than 5 MW.

The two basic types of reciprocating internal combustion engines are *spark ignition* (SI) and *compression ignition* (CI). SI engines for power generation use natural gas as the preferred fuel, although they can be configured to run on propane, gasoline, biogas, or landfill gas. CI engines (often called diesel engines) operate on diesel fuel or heavy oil, or they can be set up to run in a duel-fuel configuration that burns primarily natural gas or biogas with a small amount of diesel pilot fuel.

Although diesel engines have historically been the most popular type of reciprocating IC engine for power generation applications, their use has been increasingly restricted to emergency standby or limited duty-cycle service in the Untied States and other industrialized nations because of air emission concerns, particularly associated with NO_x and particulate matter. Consequently, a natural-gas-fueled SI engine that could also run on biogas is now the engine of choice for the higher-duty-cycle stationary power market (more than 500 hr/yr).

Reciprocating internal combustion engine technology has improved dramatically over the past three decades, driven by economic and environmental pressure for power density improvements (more output power per unit of engine displacement), increased fuel efficiency, and reduced emissions.

Computer systems have greatly advanced reciprocating engine design and control, accelerating advanced engine designs and enabling more precise control and diagnostic monitoring of the engine process. Stationary engine manufacturers and worldwide engine research and development firms continue to drive advanced engine technology, including accelerating the diffusion of technology and concepts from the automatic market to the stationary market.

A biogas-fired reciprocating engine system will encounter many of the same operating issues as a biogas-fired gas turbine:

- Landfill gas, an anaerobic digester, or a biomass gasifier is needed to produce the biogas fuel for the engine.
- The biogas must be carefully filtered of particulate matter to avoid damaging the engine.
- The engine must be derated for burning low-Btu biogas rather than natural gas.

The engines will require modification to accommodate higher flow rates and impurities; however, required modifications to reciprocating engines are achieved more easily. In most cases, more filtration devices and new manifolds are all that is required to accommodate medium-Btu gases such as landfill and digester gas, typically adding about 5% to the cost of a natural gas engine. In addition, the lower heating values of biogas result in about a 15% decrease in power output compared to a natural gas engine, further increasing the overall equipment cost on a per-kilowatt basis. Maintenance issues associated with biogas use in reciprocating engines include increased wear and tear, more cleaning, and up to eight times more frequent oil changes. Total non-fuel O&M costs for a biogas engine are approximately 60 to 70% higher than for a natural gas engine (Resource Dynamics Corp., 2004).

The recovery of heat from a reciprocating IC engine is more complex but more flexible than from a gas turbine. Heat can be recovered not only from the exhaust but also from the jacket water and then engine oil. The high-temperature heat source is the engine exhaust, at 600 to 1200°F. Depending on the design, between 1000 and 2200 Btu can be recovered from the exhaust per kilowatt of engine shaft power. The jacket water leaves the engine at about 200°F. As much as 4000 Btu/kWh of heat can be recovered from the jacket water, depending on the system design, but 2500 Btu/kWh is more typical. The heat from the engine exhaust is used to heat the jacket water before it is sent to the heat exchanger. If the heat demand is less than the heat produced by the CHP system, some of the jacket water is shunted to the excess heat exchanger, where the heat is dumped to the atmosphere. After moving through the heat exchangers, the jacket water is pumped through the oil cooler heat exchanger (slightly heating the jacket water) and back into the engine. In a separate circuit, the engine lube oil is pumped from the oil pan through the oil cooler and back into the engine. Only 300 to 900 Btu/kWh can be recovered from the engine lube oil. Another heat source is turbocharger intercooling and aftercooling, which may be either separate or part of the jacket cooling system; the three potential heat loops offer an opportunity to design the heat recovery to most closely match the heat load of the site.

A large number of gas internal combustion engines are operating on medium-Btu gas from landfills, wastewater treatment plants, and some installations at animal feedlots. Major engine manufacturers offer engine configuration packages and ratings specifically for medium-Btu gas. Additionally, some modular biomass gasification development and demonstration projects fire a low-Btu biogas in reciprocating internal combustion engines. These systems require a supplementary liquid or gaseous fuel for start-up.

Fuel Cell Systems

It is important to note that fuel cells and fuel cell feedstock technology are in the early stages of development and commercial introduction. The point is that long-term experience on both fuel cells and biogas has been limited, making it difficult to estimate what impacts the use of biogas would have on overall equipment and maintenance costs. It is probably safe to say, however, that the operating costs of biogas-operated units would likely cost slightly more than natural gas versions and have a small decline in output. Maintenance would also likely be higher as biogas with more impurities might require increased cleaning and maintenance of the fuel gas reformer. It is likely that both equipment and maintenance costs of a biogas-fueled fuel cell would be at least 10% higher than a comparable natural-gas-fueled system. As the reader reviews the information on fuel cells and their efficacy with regard to biogas feed systems, it is important to remember that biogas-fed fuel cells are a work in progress, and the jury is still out on their effective use.

With regard to the term *cell*, depending on your education level and social, cultural, or economic background, this term may conjure up images that are as diverse in variety as the colors, sizes, and shapes of lightning bolts (which, by the way, are a huge potential source of renewable energy): plant cells, animal cells, cell structure, cell diagram, cell membranes, human memory, cell theory, cell walls, cell parts, cell functions, honeycomb cells, prison cells, electrolytic cells (for producing electrolysis), aeronautical gas cells (contained in a balloon), ecclesiastical cells, or even cell phones. Fuel cells, however, are similar to batteries in that both produce direct current (DC) through an electrochemical process without direct combustion of a fuel source. The battery, of course, is a type of cell—an electric cell, electrochemical cell, galvanic cell, voltaic cell—often referred to as just a plain old battery. This type of cell, no matter what we call it, is a device that generates electrical energy from chemical energy, usually consisting of two different conducting substances placed in an electrolyte. When we talk about electricity generated by a battery, we use the term loosely, as a battery does not produce or actually generate electricity; rather, it stores electricity.

Moreover, whereas a battery delivers power from a finite amount of stored energy, fuel cells can operate indefinitely given a continuous fuel source. Simply, a fuel cell is a device that uses electrochemical principles—two electrodes (a cathode and anode) pass charged ions in an electrolyte, and a catalyst enhances the process—to convert external supplies of hydrogen and oxygen into electricity, heat, and hot water with virtually no emissions. Fuel cells offer the potential for clean, quiet, and efficient power generation. Because the fuel is not combusted, but instead reacts electrochemically, virtually no air pollution is associated with their use. Fuel cells have been under development for over 35 years as the power source of the future. Moreover, before moving on to our basic discussion of fuel cells and their associated terminology and applications, it is important to point out that, although fuel cells are not topics of discussion anywhere near as common as those other types of cells (e.g., cell phones), we predict that the day is coming when we will refer to fuel cells as commonly as we mention cell phones. It will be the fuel cell that powers our lives, just as the cell phone powers our communication today.

Open Cells vs. Closed Cells

Batteries store electrical energy chemically, contrary to the popular belief that they make electrical energy; hence, they are a thermodynamically open system. By contrast, fuel cells are different from conventional electrochemical cell batteries in that they consume reactant from an external source, which must be replenished (Science Reference Services, 2011); hence, they are a thermodynamically open system. A fuel cell is an electrochemical cell that converts a source fuel into an electrical current; this fuel can be hydrogen gas or a hydrogen-rich liquid, as well as hydrocarbons, alcohols, chlorine, or chlorine dioxide, among others (Meibuhr, 1966). The hydrogen gas or hydrogen-rich liquid fuel is reacted or triggered in the presence of an electrolyte (an oxidant), usually oxygen from the air, to produce electricity, heat, and water. The reactants flow into the cell, and the reaction products flow out of it, while the electrolyte remains with it (see Figure 11.8). Fuel cells can operate continuously as long as the reactant and oxidant flows are maintained.

Fuel Cell Operation

Figure 11.8 shows a generalized representation of the fuel cell process. Fuel cells produce direct current electricity through an electrochemical process, much like a standard battery. Again, however, unlike a standard battery a fuel supply continuously replenishes the fuel cell. The reactants, most typically hydrogen and oxygen gas, are fed into the fuel cell reactor, and power is generated as long as these reactants are supplied. The hydrogen (H_2) is typically generated from a hydrocarbon fuel such as natural gas or liquefied petroleum gas (LPG), and the oxygen (O_2) is from ambient air. Fuel cell systems designed for distributed generation (DG) applications are primarily natural gas or LPG fueled systems. Each fuel cell system consists of three primary subsystems: (1) the fuel cell stack, which generates direct current electricity; (2) the fuel processor, which converts the natural gas into a hydrogen-rich feed stream; and (3) the power conditioner, which processes the electric energy into alternating current or regulated direct current.

Figure 11.8 illustrates the electrochemical process of a typical single-cell, acid-type fuel cell. As shown, a fuel cell consists of a cathode (positively charged electrode), an anode (negatively charged electrode), an electrolyte, and an external load. The anode provides an interface between the fuel and the electrolyte, catalyzes the fuel reaction, and provides a path through which free electrons conduct to the load via the external circuit. The cathode

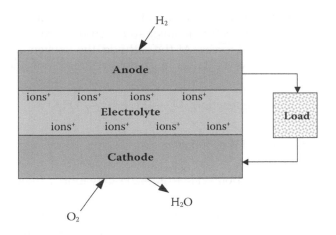

FIGURE 11.8
Generalized block diagram of the fuel-cell process.

provides an interface between the oxygen and the electrolyte, catalyzes the oxygen reaction, and provides a path through which free electrons conduct from the load to the oxygen electrode via the external circuit. The electrolyte, an ionic conductive (nonelectrically conductive) medium, acts as the separator between hydrogen and oxygen to prevent mixing and the resultant direct combustion. It completes the electrical circuit of transport ions between the electrodes.

The hydrogen and oxygen are fed to the anode and cathode, respectively. The hydrogen and oxygen gases do not directly mix, and combustion does not occur; instead, the hydrogen oxidizes one molecule at a time, in the presence of a catalyst. Because the reaction is controlled at the molecular level, there is no opportunity for the formation of NO_x and other pollutants.

At the anode, the hydrogen gas is electrochemically dissociated (in the presence of a catalyst) into hydrogen ions (H^+) and free electrons (e^-):

$$\text{Anode reaction: } 2H_2 \rightarrow 4H^+ + 4e^-$$

DID YOU KNOW?

Distributed generation (DG), sometimes referred to as on-site generation or customer-owned generation, involves locating small electrical power generation units close to the point of use. Several benefits are gained through DG applications, including (Massey, 2009):

- Decreased electric utility bills
- Increased reliability of electrical power
- Improved payback of required generation systems
- Making power available to sell to the power grid
- Generating environmentally friendly power (low emission profiles)

The electrons flow (current flow) out of the anode through an external electrical circuit. The hydrogen ions flow into the electrolyte layer and eventually to the cathode, driven by both concentration and potential forces. At the cathode, the oxygen gas is electrochemically combined (in the presence of a catalyst) with the hydrogen ions and free electrons to generate water:

$$\text{Cathode reaction: } O_2 + 4H^+ + 4e^- \rightarrow 2H_2O$$

The overall reaction in a fuel cell is as follows:

$$\text{Net fuel cell reaction: } 2H_2 + O_2 \rightarrow 2H_2O \text{ (vapor) + Energy}$$

The amount of energy released is equal to the difference between the Gibbs free energy (i.e., the maximum amount of non-expansion work that can be extracted from a closed system) of the product and the Gibbs free energy of the reactants.

When generating power, electrons flow through the external circuit, ions flow through the electrolyte layer, and chemicals flow into and out of the electrodes. Each process has natural resistances, and overcoming these reduces the operational cell voltage below the theoretical potential. There are also irreversibilities that impact actual open circuit potentials. These irreversibilities are changes in the potential energy of the chemical that are not reversible through the electrochemical process. Typically, some of the potential energy is converted into heat even at open circuit conditions when current is not flowing. A simple example is the resistance to ionic flow through the electrolyte while the fuel cell is operating.

This potential energy "loss" is really a conversion to heat energy, which cannot be reconverted into chemical energy directly within the fuel cell. The electrical power generated by the fuel cell is the product of the current measured in amps and the operational voltage (i.e., Ohm's law, where Power = Voltage × Current). Based on the application and economics, a typical operating fuel cell will have an operating voltage of between 0.55 and 0.80 volts. The ratio of the operating voltage and the theoretical maximum of 1.48 volts represents a simplified estimate of the stack (number of fuel cells connected into one unit) electrical efficiency on a higher heating value (HHV) basis.

As mentioned, resistance heat is also generated along with the power. Because the electric power is the product of the operating voltage and the current, the quantity of heat that must be removed from the fuel cell is the product of the current and the difference between the theoretical potential and the operating voltage. In most cases, the water produced by the fuel cell reactions exits the fuel cell as vapor; therefore, the 1.23-volt lower heating value (LHV) theoretical potential is used to estimate sensible heat generated by the fuel cell electrochemical process.

The overall electrical efficiency of the fuel cell is dependent on the amount of power drawn from it. The maximum thermodynamic efficiency of a hydrogen fuel cell is the ratio of the Gibbs free energy and the heating value of the hydrogen. The Gibbs free energy decreases with increasing temperatures, because the product water produced at the elevated temperature of the fuel cell includes the sensible heat of that temperature, and this energy cannot be converted into electricity without the addition of a thermal energy conversion cycle (such as a steam turbine). Therefore, there is an inverse relationship between the maximum efficiency of a pure fuel cell system and temperature; that is, efficiency decreases with increasing temperature.

Fuel Cell Stacks

Practical fuel cell systems require voltages higher than 0.55 to 0.80. To deliver the desired amount of voltage (energy), fuels cells can be combined (stacked) in series and parallel circuits. When cells are connected in series, they yield higher voltage; connecting them in parallel allows a higher current to be supplied. Typically, there are several hundred cells in a single cell stack. Increasing the active area of individual cells manages current flow. Cell area can range from 100 cm^2 to over 1 m^2, depending on the type of fuel cell and application power requirements. Although this stacking procedure and the subsequent reaction give off a lot of heat (exothermic reaction), the heat can be recovered for uses such as space heating, process heating, and domestic hot water.

Fuel Processors

In distributed generation applications, the most viable fuel cell technologies (at present) use natural gas as the fuel source for the system. To operate on natural gas or other fuels, fuel cells require a fuel processor or reformer, a device that converts the fuel into the hydrogen-rich gas stream. While adding fuel flexibility to the system, the reformer also adds significant cost and complexity. The three primary types of reformers are steam reformers, autothermal reformers, and partial oxidation reformers. The fundamental differences are the source of oxygen used to combine with the carbon within the fuel to release the hydrogen gases and the thermal balance of the chemical process. Steam reformers use steam, while partial oxidation units use oxygen gas, and autothermal reformers use both steam and oxygen.

Steam reforming is extremely endothermic and requires a substantial amount of heat input. Autothermal reformers typically operate at or near the thermal neutral point and therefore do not generate or consume thermal energy. Partial oxidation units combust a portion of the fuel (i.e., partially oxidize it), releasing heat in the process. When integrated into a fuel cell system that allows the use of anode off-gas, a typical natural gas reformer can achieve conversion efficiencies in the range of 75 to 90% LHV, with 83 to 85% being an expected level of performance. These efficiencies are defined as the LHV of hydrogen generated divided by the LHV of the natural gas consumed by the reformer. Some fuel cells can function as internally steam reforming fuel cells. Because the reformer is an endothermic catalytic converter and the fuel cell is an exothermic oxidizer, the two combine into one with mutual thermal benefits. More complex than a pure hydrogen fuel cell, these types of fuel cells are more difficult to design and operate. Although combining two catalytic processes is difficult to arrange and control, these internally reforming fuel cells are expected to account for a significant market share as fuel-cell-based DG becomes more common.

Power Conditioning Subsystem

The fuel cell generates direct current (DC) electricity, which requires conditioning before serving a DG application. Depending on the cell area and number of cells, the direct current electricity is approximately 200 to 400 volts per stack. If the system is large enough, stacks can operate in series to double or triple individual stack voltages. Because the voltage of each individual cell decreases with increasing load or power, the output is considered an unregulated voltage source. The power conditioning subsystem boosts the output voltage to provide a regulated higher voltage input source to an electronic inverter. The inverter then uses a pulse width modulation technique at higher frequencies to generate

TABLE 11.7

Characteristics of Fuel Cells

Fuel-Cell Type	Electrolyte	Current Uses
Metal hydride	Aqueous alkaline solution	Commercial/research
Electrogalvanic	Aqueous alkaline solution	Commercial/research
Direct formic acid	Polymer membrane (ionomer)	Commercial/research
Zinc–air battery	Aqueous alkaline solution	Mass production
Microbial	Polymer membrane/humic acid	Research
Upflow microbial	—	Research
Regenerative	Polymer membrane (ionomer)	Commercial/research
Direct borohydride	Aqueous alkaline solution	Commercial
Alkaline	Aqueous alkaline solution	Commercial/research
Direct methanol	Polymer membrane (ionomer)	Commercial/research
Reformed methanol	Polymer membrane (ionomer)	Commercial/research
Direct–ethanol	Polymer membrane (ionomer)	Research
Proton exchange membrane	Polymer membrane (ionomer)	Commercial/research
RFC–redox	Liquid electrolytes w/redox shuttle and polymer membrane (ionomer)	Research
Phosphoric acid	Molten phosphoric acid (H_3PO_4)	Commercial/research
Molten carbonate	Molten alkaline carbonate	Commercial/research
Tubular solid oxide	O^{2-}-conducting ceramic oxide	Commercial/research
Protonic ceramic	H^+-conducting ceramic oxide	Research
Direct carbon	Several different	Commercial/research
Planar solid oxide	O^{2-}-conducting ceramic oxide	Commercial/research
Enzymatic biofuel	Any that will not denature the enzyme	Research
Magnesium air	Saltwater	Commercial/research

stimulated alternating current (AC) output. The inverter controls the frequency of the output, which can be adjusted to enhance power factor characteristics. Because the inverter generates alternating current within itself, the output power is generally clean and reliable. This characteristic is important to sensitive electronic equipment in premium power applications. The efficiency of the power conditioning process is typically 92 to 96% and is dependent on system capacity and input voltage–current characteristic.

Types of Fuel Cells

Fuel cells come in many varieties or types (see Table 11.7 and Table 11.8); however, they all work in the same general manner. Table 11.7 lists 22 types of fuel cells according to their electrolytic composition. In addition to typing a fuel cell according to its electrolyte or ion conduction material, it is common practice (at present) to type or classify fuel cells into the five groups shown in Table 11.8 according to their consideration for distributed generation applications. These five types of fuel cells are (1) phosphoric acid (PAFC), (2) proton exchange membrane (PEMFC), (3) molten carbonate (MCFC), (4) solid oxide (SOFC), and (5) alkaline (AFC). Each type is distinguished by the electrolyte and operating temperatures. Operating temperatures range from near ambient to 1800°F, and electrical generating efficiencies range from 30 to over 50% HHV. As a result, they can have different performance characteristics, advantages, and limitations and therefore will be suited to distributed generation applications in a variety of approaches.

TABLE 11.8

Characteristics of Major Fuel Cell Types

	Proton Exchange Membrane (PEMFC)	Alkaline (AFC)	Phosphoric Acid (PAFC)	Molten Carbonate (MCFC)	Solid Oxide (SOFC)
Type of electrolyte	H^+ ions (with anions bound in polymer membrane)	OH^- ions (typically aqueous KOH solution)	H^+ ions (H_3PO_4 solutions)	CO_3^- ions (typically molten $LiKaCO_3$ eutectics)	O^- ions (stabilized ceramic matrix with free oxide ions)
Typical construction	Plastic, metal, or carbon	Plastic, metal	Carbon, porous ceramics	High-temperature metals, porous ceramic	Ceramic, high-temperature metals
Internal reforming	No	No	No	Yes; good temperature match	Yes; good temperature match
Oxidant	Air to O_2	Purified air to O_2	Air to enriched air	Air	Air
Operational temperature	150–180°F (65–85°C)	190–500°F (90–260°C)	370–410°F (190–210°C)	1200–1300°F (650–700°C)	1350–1850°F (750–1000°C)
DG system-level efficiency (percent HHV)	25 to 35%	32 to 40%	35 to 45%	40 to 50%	45 to 55%
Primary contaminant sensitivities	CO, sulfur, NH_3	CO, CO_2, sulfur	CO < 1%, sulfur	Sulfur	Sulfur

Source: Energy Nexus Group, *Biomass Combined Heat and Power Catalog of Technologies*, prepared for U.S. Environmental Protection Agency, Climate Protection Partnership Division, Washington, D.C., 2002.

Fuel Cell Type Description

Two of the fuel cell types shown in Table 11.8, proton exchange membrane fuel cell (PEMFC) and phosphoric acid fuel cell (PAFC), have acidic electrolytes and rely on the transport of H^+ ions. Two others, alkaline fuel cell (AFC) and molten carbonate fuel cell (MCFC), have basic electrolytes that rely on the transport of OH^- and CO_3^{2-} ions, respectively. The fifth type, solid oxide fuel cell (SOFC), is based on a solid-state ceramic electrolyte in which oxygen ions (O^{2-}) are the conductive transport ion. Each fuel cell type operates at an optimum temperature, which is a balance between the ionic conductivity and component stability. These temperatures differ significantly among the five basic types. The proton-conducting fuel cell type generates water at the cathode and the anion-conducting fuel cell type generates water at the anode. A basic description of the five basic types of fuel cell under consideration for distributed generation applications is provided below:

- *PEMFC (proton exchange membrane fuel cell or polymer electrolyte membrane)* was developed by NASA and was used in the 1960s for the first manned spacecraft. The PEMFC uses a solid polymer electrolyte and operates at low temperatures (~200°F). The PEMFC transforms the chemical energy liberated during the electrochemical reaction of hydrogen and oxygen to electrical energy, as opposed to the direct combustion of hydrogen and oxygen gases to produce thermal energy. Over the past several years, the PEMFC has received significant media coverage due to the large auto industry investment in the technology. Due to their modularity and potential for simple manufacturing, reformer/PEMFC systems or residential DG applications have attracted considerable development capital. PEMFCs have high power density and can vary their output quickly to meet demand. This type of fuel cell is highly sensitive to CO poisoning.

- *AFC (alkaline fuel cell)*, also known as the Bacon fuel cell after it British inventor, was developed later by NASA for use on the Apollo spacecraft and on the space shuttles. AFC technology uses alkaline potassium hydroxide as the electrolyte. The primary advantages of AFC technology are improved performance (electrical efficiencies approaching 70% HHV), use of non-precious metal electrodes, and the fact that no unusual materials are needed. The primary disadvantage is the tendency to absorb carbon dioxide, converting the alkaline electrolyte to an aqueous carbonate electrolyte that is less conductive. The attractiveness of AFCs has declined substantially with improvements in PEMFC technology.

- *PAFC (phosphoric acid fuel cell)* is a type of fuel cell that uses liquid phosphoric acid as an electrolyte. The electrodes are made of carbon paper coated with a finely dispersed platinum catalyst, which make them expensive to manufacture. The current 200-kW product has a stack lifetime of over 40,000 hours and commercially based reliabilities in the range of 90 and 95%.

- *MCFC (molten carbonate fuel cell)* uses an alkali metal carbonate (Li, Na, K) as the electrolyte and has a developmental history that dates back to the early part of the 20th century. Due to its operating temperature range of 1100 to 1400°F, the MCFC holds promise in both combined heat and power and distributed generation applications. This type of fuel cell can be internally reformed, can operate at high efficiencies (50% HHV), and is relatively tolerant of fuel impurities. Government and industry research and development programs during the 1980s and 1990s resulted in several individual pre-prototype system demonstrations.

- *SOFC (solid oxide fuel cell)* is a class of fuel cell characterized by the use of a solid, nonporous oxide material as the electrolyte. Several SOFC units up to 100 kW in size and based on a concentric tubular design were built and tested by Siemens Westinghouse Power Generation (known today as Siemens Power Generation, Inc.). In addition, many companies are developing planar SOFC designs, which offer higher power densities and lower costs than the tubular design, but these have yet to achieve the reliability of the tubular design. Despite relative immaturity, the SOFC offers several advantages (high efficiency, stability, and reliability, as well as high internal temperatures) that have attracted development support. The SOFC has projected service electric efficiencies of 45 to 60% and higher for large hybrid, combined-cycle plants. Efficiencies for smaller SOFC DG units are expected to be about 50%. Stability and reliability of the SOFC are due to an all-solid-state ceramic construction. Test units have operated in excess of 10 years with acceptable performance. The high internal temperatures of the SOFC are both an asset and a liability. As an asset, high temperatures make internal reforming possible. As a liability, these high temperatures add to materials and mechanical design difficulties, thus reducing stack life and increasing costs. Although SOFC research has been ongoing for 30 years, costs of these stacks are still comparatively high.

Hydrogen Fuel Cell*

Containing only one electron and one proton, hydrogen (H) is the simplest element on Earth. Hydrogen is a diatomic molecule—each molecule has two atoms of hydrogen (which is why pure hydrogen is commonly expressed as H_2). Although abundant on Earth as an element, hydrogen combines readily with other elements and is almost always found as part of another substance, such as water hydrocarbons or alcohols. Hydrogen is also found in biomass, which includes all plants and animals.

Hydrogen is an energy carrier, not an energy source. Hydrogen can store and deliver usable energy, but it does not typically exist by itself in nature; it must be produced from compounds that contain it. Hydrogen can be produced using diverse, domestic resources including nuclear, natural gas, and coal, as well as biomass and other renewables, including solar, wind, hydroelectric, or geothermal energy. This diversity of domestic energy sources makes hydrogen a promising energy carrier and important to our nation's energy security. It is expected and desirable for hydrogen to be produced using a variety or resources and process technologies (or pathways). The U.S. Department of Energy (DOE) is focusing on hydrogen production technologies that result in near-zero net greenhouse gas emissions and use renewable energy sources, nuclear energy, and coal (when combined with carbon sequestration). To ensure sufficient clean energy for our overall energy needs, energy efficiency is also important.

For hydrogen to be successful in the marketplace, it must be cost competitive compared with the available alternatives. In the light-duty vehicle transportation market, this competitive requirement means that hydrogen would have to be available untaxed at $2 to $3/ GGE (gasoline gallon equivalent). This price would result in hydrogen fuel cell vehicles having the same cost to the consumer on a cost-per-mile-driven basis as a comparable conventional internal combustion engine or hybrid vehicle.

The various process technologies that can be used to produce hydrogen include thermal (natural gas reforming, renewable liquid and bio-oil processing, and biomass and coal gasification), electrolytic (water splitting using a variety of energy resources), and photolytic (splitting water using sunlight via biological and electrochemical materials). The DOE is engaged in research and development of a variety of hydrogen production technologies. Some are further along in development than others; some can be cost competitive for the transition period (beginning in 2015), but others are considered longer term technologies that will not be cost competitive until after 2030.

Hydrogen can be produced in large, centralized facilities (50 to 300 miles from point of use), in smaller semi-central facilities (located within 25 to 100 miles of use), or where it is needed (near or at point of use) and then distributed. Infrastructure is required to move hydrogen from the location where it is produced to the dispenser at a refueling station or stationary power site. Infrastructure includes the pipelines, trucks, railcars, ships, and barges that deliver fuel, as well as the facilities and equipment needed to load and unload them. Delivery technology for hydrogen infrastructure is currently available

---

* Material in this section is based on EERE, *Fuel Cell Technologies Program: Hydrogen Production*, U.S. Department of Energy, Energy Efficiency & Renewable Energy, Washington, D.C., 2008 (http://www1.eere.energy.gov/ hydrogenandfuelcells/production/basics.html); EERE, *Fuel Cell Technologies Program: Hydrogen Storage*, U.S. Department of Energy, Energy Efficiency & Renewable Energy, Washington, D.C., 2008 (http://www1.eere. energy.gov/hydrogenandfuelcells/storage/basics.html?m=1&); EERE, *Fuel Cell Technologies Program: Fuel Cells*, U.S. Department of Energy, Energy Efficiency & Renewable Energy, Washington, D.C., 2008 (http://www1. eere.energy.gov/hydrogenandfuelcells/fuelcells/basics.html?m=1&).

commercially, and several U.S. companies deliver bulk hydrogen today. Some of the infrastructure is already in place because hydrogen has long been used in industrial applications, but it is not sufficient to support widespread consumer use of hydrogen as an energy carrier. Because hydrogen has a relatively low volumetric energy density, its transportation, storage, and final delivery to the point of use comprise a significant cost and result in some of the energy inefficiencies associated with using it as an energy carrier.

Options and trade-offs for hydrogen delivery from central, semi-central, and distributed production facilities to the point of use are complex. The choice of a hydrogen production strategy greatly affects the cost and method of delivery. For example, larger, centralized facilities can produce hydrogen at relatively low costs due to economies of scale, but the delivery costs for centrally produced hydrogen are higher than the delivery costs for semi-central or distributed production options (because the point of use is farther away). In comparison, distributed production facilities have relatively low delivery costs, but the hydrogen production costs are likely to be higher—lower volume production means higher equipment costs on a per-unit-of-hydrogen basis.

Key challenges to hydrogen delivery include reducing delivery cost, increasing energy efficiency, maintaining hydrogen purity, and minimizing hydrogen leakage. Further research is needed to analyze the trade-offs between the hydrogen production options and the hydrogen delivery options taken together as a system. Building a national hydrogen delivery infrastructure is a big challenge. It will take time to develop and will likely include combinations of various technologies. Delivery infrastructure needs and resources will vary by region and type of market (e.g., urban, interstate, or rural). Infrastructure options will also evolve as the demand for hydrogen grows and as delivery technologies develop and improve.

Hydrogen Storage

Storing enough hydrogen on board a vehicle to achieve a driving range of greater than 300 miles is a significant challenge. On a weight basis, hydrogen has nearly three times the energy content of gasoline (120 MJ/kg for hydrogen vs. 44 MJ/kg for gasoline); however, on a volume basis the situation is reversed (8 MJ/liter for liquid hydrogen vs. 32 MJ/liter for gasoline). On-board hydrogen storage in the range of 5 to 13 kg H_2 is required to encompass the full platform of light-duty vehicles.

Hydrogen can be stored in a variety of ways, but for hydrogen to be a competitive fuel for vehicles the hydrogen vehicle must be able to travel a distance comparable to that of conventional hydrocarbon-fueled vehicles. Hydrogen can be physically stored as either a gas or a liquid. Storage as a gas typically requires high-pressure tanks (5000 to 10,000 psi). Storage of hydrogen as a liquid requires cryogenic temperatures because the boiling point of hydrogen at one atmosphere pressure is –252.8°C. Hydrogen can also be stored on the surfaces of solids (by adsorption) or within solids (by absorption). In adsorption, hydrogen is attached to the surface of a material either as hydrogen molecules or as hydrogen atoms. In absorption, hydrogen is dissociated into hydrogen atoms, and the hydrogen atoms are then incorporated into the solid lattice framework. Hydrogen storage in solids may make it possible to store large quantities of hydrogen in smaller volumes at low pressures and at temperatures close to room temperature. It is also possible to achieve volumetric storage densities greater than liquid hydrogen because the hydrogen molecule is dissociated into atomic hydrogen within the metal hydride lattice structure.

Finally, hydrogen can be stored through the reaction of hydrogen-containing materials with water (or other compounds such as alcohols). In this case, the hydrogen is effectively stored in both the material and in the water. The terms *chemical hydrogen storage* and *chemical hydrides* are used to describe this form of hydrogen storage. It is also possible to store hydrogen in the chemical structures of liquids and solids.

How Hydrogen Fuel Cells Work

The fuel cell uses the chemical energy of hydrogen to cleanly and efficiently produce electricity, with water and heat as byproducts. Fuel cells are unique in terms of variety of their potential applications; they can provide energy for systems as large as a utility power station and as small as a laptop computer. Fuel cells have several benefits over conventional combustion-based technologies currently used in many power plants and passenger vehicles. They produce much smaller quantities of greenhouse gases and none of the air pollutants that create smog and cause health problems. If pure hydrogen is used as a fuel, fuel cells emit only heat and water as byproducts.

DID YOU KNOW?

Hydrogen fuel cell vehicles (FCVs) emit approximately the same amount of water per mile as vehicles using gasoline-powered internal combustion engines (ICEs).

A hydrogen fuel cell is a device that uses hydrogen (or hydrogen-rich fuel) and oxygen to create electricity by an electrochemical process. A single fuel cell consists of an electrolyte and two catalyst-coated electrodes (a porous anode and cathode). Again, while there are different fuel cell types, all fuel cells work similarly:

- A fuel such as hydrogen is fed to the anode where a catalyst separates the negatively charged hydrogen electrons from the positively charged ions (protons).
- At the cathode, oxygen combines with electrons and, in some cases, with species such as protons or water, resulting in water or hydroxide ions, respectively.
- For polymer electrolyte membrane and phosphoric acid fuel cells, protons move through the electrolyte to the cathode to combine with oxygen and electrons, producing water and heat.
- For alkaline, molten carbonate, and solid oxide fuel cells, negative ions travel through the electrolyte to the anode where they combine with hydrogen to generate water and electrons.
- The electrons from the anode cannot pass through the electrolyte to the positively charged cathode; they must travel around it via an electrical circuit to reach the other side of the cell. This movement of electrons is an electrical current.

The hydrogen fuel cell example above was deliberately highlighted here to make a point. Even though direct methanol fuel cells are under development, at the present time fuel cells require hydrogen for operation, but it is generally impractical to use hydrogen directly as a fuel source; instead, it is extracted from hydrocarbon fuels or biogas feed using a reformer. The *reformers* produce or increase the concentration of hydrogen and decrease the concentration of gas species toxic to the fuel cell. In all three types of reformers (particle oxidation, autothermal reformation, and preferential oxidation), fuel processing techniques use some of the energy contained in the fuel to convert the hydrocarbons to hydrogen and CO. The reforming process is often performed at elevated pressure to allow a smaller equipment footprint.

Stirling Heat Engine

Simply, the Stirling heat engine consists of an isothermal compression, an isometric heat addition, an isothermal expansion, and an isometric heat rejection. It operates by cyclic compression and expansion of air or other gas, the working fluid, at different temperature levels such that there is a net conversion of heat energy to mechanical work. Like a steam engine, the heat in a Stirling engine flows in and out through the engine wall. Like the internal combustion engine, the Stirling engine is a reciprocating engine; however, unlike the internal combustion engine, the Stirling engine is an externally heated engine. The Stirling engine is noted for its high efficiency (resulting from heat regeneration), quiet operation, and the ease with which it can use almost any heat source (such as fossil fuels, solar, nuclear, and waste heat), but the Stirling engine is particularly well suited to biomass fuels. This compatibility with renewable energy sources has become increasingly significant as the price of conventional fuels rises and in light of concerns about climate change and peak oil. An example of this compatibility with renewable energy sources is evident when a Stirling engine is placed at the focus of parabolic mirrors where it can convert solar energy to electricity with an efficiency better than nonconcentrated photovoltaic cells and compared to concentrated photovoltaics. As an aside, it is interesting to note that the Stirling engine was designed in 1816 as a safer alternative to the steam engines of the time, but it now is finding use in modern renewable energy and CHP systems.

Chapter Review Questions

11.1   Why does society need fuel cells, and what can they be used for?

11.2   Are fuel cells a viable form of renewable energy? Explain.

11.3   _____ is the simplest element on Earth.

11.4   What three technologies are used to produce hydrogen?

11.5   Does the Stirling engine have a renewed future?

References and Recommended Reading

ACAA. (2009). *ACAA 2009 CCP Report*, American Coal Ash Association, Aurora, CO (http://acaa.affiniscape.com/displaycommon.cfm?an=1&subarticlenbr=3).

Allen, H.L., Fox, T.R., and Campbell, R.G. (2005). What's ahead for intensive pine plantation silviculture? *Southern Journal of Applied Forestry*, 29:62–69.

Anon. (2007). BQ-9000 inductee earns accreditation, certification, *Biodiesel Magazine*, January 24.

Antares Group, Inc. (2003). *Assessment of Power Production at Rural Utilities Using Forest Thinnings and Commercially Available Biomass Power Technologies*, prepared for the U.S. Department of Agriculture, U.S. Department of Energy, and National Renewable Energy Laboratory, Washington, D.C.

Arthur D. Little, Inc. (2001). *Aggressive Growth in the Use of Bioderived Energy and Products in the United States by 2010*, Final Report, Arthur D. Little, Inc., Cambridge, MA.

Ashford, R.D. (2001). *Ashford's Dictionary of Industrial Chemicals*, 2nd ed., Wavelength Publications, London.

ASTM. (2010). *ASTM C-618: Standard Specification for Coal Fly Ash and Raw or Calcined Natural Pozzolan for Use in Concrete*, ASTM International, West Conshohocken, PA.

Aylott, M.J. (2008). Yield and spatial supply of bioenergy poplar and willow short-rotation coppice in the U.K., *New Phytologist*, 178(2):358–370.

Bain, R. (2006). *Biomass Gasification*, presented at the USDA Thermochemical Conversion Workshop, Pacific Northwest National Laboratory, September 6, Richmond, WA.

Baize, J. (2006). *Bioenergy and Biofuels*, Agricultural Outlook Forum, February 17, Washington, D.C.

Baker, J.B. and Broadfoot, W.M. (1979). *A Practical Field Method of Site Evaluation for Commercially Important Southern Hardwoods*, General Technical Report SO-26, U.S. Department of Agriculture, Forest Service, South Forest Experiment Station, New Orleans, LA.

Barbir, F. (2005). *PEM Fuel Cells: Theory and Practice*, Elsevier, New York.

Bender, M. (1999). Economic feasibility review for community-scale farmer cooperatives for biodiesel, *Bioresource Technology*, 70:81–87.

Cheng, S. and Zhu, S. (2009). Lignocellulosic biorefinery, *BioResources*, 4(2):456–457.

Christi, Y. (2007). Biodiesel from microalgae, *Biotechnology Advances*, 25:294–306.

Coyle, W. (2007). The future of biofuels: a global perspective, *Amber Waves*, November (http://www.ers.usda.gov/AmberWaves/november07/features/biofuels.htm).

Curtis, W., Ferland, C., McKissick, J., and Barnes, W. (2003). *The Feasibility of Generating Electricity from Biomass Fuel Sources in Georgia*, University of Georgia, Center of Agribusiness and Economic Development, Athens.

Dickman, D. (2006). Silviculture and biology of short-rotation woody crops in temperate regions: then and now, *Biomass and Bioenergy*, 30:696–705.

EERE. (2008a). *Biomass Program*, U.S. Department of Energy, Energy Efficiency and Renewable Energy, Washington, D.C. (http://www1.eere.energy.gov/biomass/feedstocks_types.html).

EERE. (2008b). *Fuel Cell Technologies Program: Hydrogen Production*, U.S. Department of Energy, Energy Efficiency & Renewable Energy, Washington, D.C. (http://www1.eere.energy.gov/hydrogenandfuelcells/production/basics.html).

EERE. (2008c). *Fuel Cell Technologies Program: Hydrogen Storage*, U.S. Department of Energy, Energy Efficiency & Renewable Energy, Washington, D.C., 2008 (http://www1.eere.energy.gov/hydrogenandfuelcells/storage/basics.html?m=1&).

EERE. (2008d). *Fuel Cell Technologies Program: Fuel Cells*, U.S. Department of Energy, Energy Efficiency & Renewable Energy, Washington, D.C., 2008 (http://www1.eere.energy.gov/hydrogenandfuelcells/fuelcells/basics.html?m=1&).

EG&G Technical Services, Inc. (2004). *Fuel Cell Technology Handbook*, 7th ed., Contract No. DE-AM26-99FT40575, U.S. Department of Energy, Office of Fossil Energy, National Energy Technology Laboratory, Morgantown, WV.

EIA. (2006). *Annual Energy Outlook 2007 with Projections to 2030*, DOE/EIA-0383(2007), U.S. Department of Energy, Energy Information Administration, Office of Integrated Analysis and Forecasting, Washington, D.C.

Eisenberg, A. and Kim, J.S. (1998). *Introduction to Ionomers*, Wiley, New York.

Elliott, D.C., Fitzpatrick, S.W., Bozell, J.J. et al. (1999). Production of levulinic acid and use as a platform chemical for derived products, in Overend, R.P. and Chornet, E., Eds., *Biomass: A Growth Opportunity in Green Energy and Value-Added Products, Proceedings of the Fourth Biomass Conference of the Americas*, Elsevier Science, Oxford, U.K., pp. 595–600.

Energy and Environmental Analysis, Inc. (2003). *Gas-Fired Distributed Generation Technology Characterization: Microturbines*, NREL/TP-620-34783, prepared for National Renewable Energy Laboratory, Golden, CO.

Energy Nexus Group. (2002). *Technology Characterization: Fuel Cells*, prepared for U.S. Environmental Protection Agency, Climate Protection Partnership Division, Washington, D.C.

Energy Products of Idaho. (2011). *Fluidized-Bed Combustors*, Energy Products of Idaho, Coeur d'Alene, ID (http://www.energyproducts.com/fluidized_bed_combustors.htm).

EPRI. (1997). *Renewable Energy Technology Characterizations*, TR-109496, U.S. Department of Energy, Electric Power Research Institute, Washington, D.C.

Fahey, J. (2001). Shucking petroleum, *Forbes Magazine*, 168(13):206–208.

Farrell, A.E. and Gopal, A.R. (2008). Bioenergy research needs for heat, electricity, and liquid fuels, *MRS Bulletin*, 33:373–387.

Fehrs, J. and Donovan, C. (1999). *Co-Firing Wood in Coal-Fired Industrial Stoker Boilers: Strategies for Increasing Co-Firing in New York and the Northeast*, prepared by C.T. Donovan Associates for Northeast Regional Biomass Program, Coalition of Northeastern Governors (CONEG) Policy Research Center, Washington, D.C., and New York State Energy Research and Development Authority, Albany.

Gentille, S.B. (1996). *Reinventing Energy: Making the Right Choices*, American Petroleum Institute, Washington, D.C.

Goldstein, L., Hedman, B., Knowles, D., Freedman, S.I., Woods, R., and Schweizer, T. (2003). *Gas-Fired Distributed Energy Resource Technology Characterizations*, National Renewable Energy Laboratory, Golden, CO.

Graham, R., Nelson, R., Sheehan J., Perlack, R., and Wright L. (2007). Current and potential U.S. corn stover supplies, *Agronomy Journal*, 99:1–11.

Haas, M., McAloon, A., Yee, W., and Foglia, T. (2006). A process model to estimate biodiesel production costs, *Bioresource Technology*, 97:671–678.

Harrar, E.S. and Harrar, J.G. (1962). *Guide to Southern Trees*, 2nd ed., Dover, New York.

Harrell, G. (2002). *Steam System Survey Guide*, ORNL/TM-2001/263, Oak Ridge National Laboratory, Knoxville, TN (www1.eere.energy.gov/industry/bestpractices/pdfs/steam_survey_guide.pdf).

Hartman, E. (2008). A promising oil alternative: algae energy, *The Washington Post*, January 6 (http://www.washingtonpost.com/wp-dyn/content/article/2008/01/03/AR2008010303907.html).

Hileman, B. (2003). Clashes over agbiotech, *Chemical & Engineering News*, 81:25–33.

Hoffman, J. (2001). BDO outlook remains healthy, *Chemical Market Reporter*, 259(14):5.

Hoogers, G. (2003). *Fuel Cell Technology Handbook*, CRC Press, Boca Raton, FL.

Hughes, E. (2000). Biomass cofiring: economics, policy and opportunities, *Biomass and Bioenergy*, 19:45–465.

ILSR. (2002). *Accelerating the Shift to a Carbohydrate Economy: The Federal Role*, Executive Summary of the Minority Report of the Biomass Research and Development Technical Advisory Committee, Institute for Local Self-Reliance, Washington, D.C.

Kagan, J. (2010). *Third and Fourth Generation Biofuels: Technologies, Markets, and Economics Through 2015*, Greentech Media, Boston, MA (http://www.greentechmedia.com/research/report/third-and-fourth-generation-biofuels).

Kantor, S.L., Lipton, K., Manchester, A., and Oliveira, V. (1997). Estimating and addressing America's food losses, *Food Review*, 20(1):3–11.

Klass, D.I. (1998). *Biomass for Renewable Energy, Fuels, and Chemicals*, Academic Press, San Diego, CA.

Larminie, J. and Dicks A. (2003). *Fuel Cell Systems Explained*, 2nd ed., John Wiley & Sons, New York.

Lee, S. (1996). *Alternative Fuels*, Taylor & Francis, Boca Raton, FL.

Markarian, J. (2003). New additives and basestocks smooth way for lubricants, *Chemical Market Reporter*, April 28.

Massey, G.W. (2009). *Essentials of Distributed Generation Systems*, Jones and Bartlett Publishers, Sudbury, MA.

McDill, S. (2009). Can algae save the world—again? *Reuters*, February 20 (http://www.reuters.com/article/2009/02/10/us-biofuels-algae-idUSTRE5196HB20090210?pageNumber=2&virtualBrandChannel=0).

McGraw, L. (1999). Three new crops for the future, *Agricultural Research Magazine*, 47(2):17.

McKeever, D.B. (1998). Wood residual quantities in the United States, *BioCycle: Journal of Composting and Recycling*, 39(1):65–68; as cited in Antares Group, Inc., *Assessment of Power Production at Rural Utilities Using Forest Thinnings and Commercially Available Biomass Power Technologies*, prepared for the U.S. Department of Agriculture, U.S. Department of Energy, and National Renewable Energy Laboratory, Washington, D.C., 2003.

McKeever, D.B. (2004). Inventories of woody residues and solid wood waste in the United States, 2002, in *Proceedings of the Ninth International Conference on Inorganic-Bonded Composite Materials Conference*, October 10–13, Vancouver, British Columbia.

McKendry, P. (2002). Energy production from biomass. Part I. Overview of biomass, *Bioresource Technology*, 83:37–46.

Meibuhr, S.G. (1966). Review of United States fuel-cell patents issued from 1860 to 1947, *Electrochimica Acta*, 11:1301–1308.

Miles, P.D. (2004). *Fuel Treatment Evaluator: Web-Application Version 1.0*, U.S. Department of Agriculture, Forest Service, North Central Research Station, St. Paul, MN (http://ncrs2/4801/fiadb/fueltreatment.fueltreatmentwc.asp).

Miles, T.R., Miles, Jr., T.R., Baxter, L.L., Jenkins, B.M., and Oden, L.L. (1993). Alkali slagging problems with biomass fuels, in *First Biomass Conference of the Americas: Energy, Environment, Agriculture, and Industry*, Vol. 1, National Renewable Energy Laboratory, Burlington, VT, pp. 406–421.

Morey, R.V., Tiffany, D.G., and Hartfield D.L. (2006). Biomass for electricity and process heat at ethanol plants, *Applied Engineering in Agriculture*, 22:723–728.

NRDC. (2001). *Reducing America's Oil Dependence*, Natural Resources Defense Council, New York (http://www.nrdc.org/air/energy/fensec.asp).

NREL. (2009). *Energy Storage: Ultracapacitors*, National Renewable Energy Laboratory, Golden, CO (http://www.nrel.gov/vehiclesandfuels/energystorage/ultracapacitors.html?print).

ODOE. (2010). *Bioenergy in Oregon: Biomass Technologies*, Oregon Department of Energy, Salem (http://www.oregon.gov/ENERGY/RENEW/Biomass/biofuels.shtml).

Perlin, J. (2005). *A Forest Journey: The Story of Wood and Civilization*, Countryman Press, Woodstock, VT.

Polagye, B., Hodgson, K., and Malte, P. (2007). An economic analysis of bioenergy options using thinning from overstocked forests, *Biomass and Bioenergy*, 31:105–125.

Resource Dynamics Corp. (2004). *Combined Heat and Power Market Potential for Opportunity Fuels*, Distributed Energy Program Report, prepared for U.S. Department of Energy, Energy Efficiency and Renewable Energy, Washington, D.C.

Rossell, J.B. and Pritchard, J.L.R., Eds. (1991). *Analysis of Oilseeds, Fats and Fatty Foods*. Elsevier, London.

Sauer, P. (2000). Domestic spearmint oil producers face flat pricing, *Chemical Market Reporter*, 258(18):14.

Schenk, P., Thomas-Hall, S., Stephens, R., Marx U., Mussgnug, J., Posten, C., Kruse, O., and Hankamer, B. (2008). Second generation biofuels: high-efficiency microalgae for industrial production, *BioEnergy Research*, 1(1):20–43.

Science Reference Services. (2011). *Batteries, Supercapacitors, and Fuel Cells*, Library of Congress, Washington, D.C. (http://www.oc.gov/rr.scietech/tracer-bullets/batteriestb.html#scompe).

Sedjo, R. (1997). The economics of forest-based biomass supply, *Energy Policy*, 25(6):559–566.

Segall, L. (2010). Ex-Shell president sees $5 gas in 2012, CNNMoney.com, December 27 (http://money.cnn.com/2010/12/27/markets/oil_commodities/index.htm).

Silva, B. (1998). Meadowfoam as an alternative crop, *AgVentures*, 2(4):28.

Singhal, S.C. and Kendall, K. (2003). *High-Temperature Solid Oxide Fuel Cells: Fundamentals, Design and Applications*, Elsevier, New York.

Sioru, B. (1999). Process converts trash into oil, *Waste Age*, 30(11):20.

Smith, W.B. et al. (2004). *Forest Resources of the United States, 2002*, General Technical Report NC-241, U.S. Department of Agriculture, Forest Service, North Central Research Station, St. Paul, MN.

Spellman, F.R. (2009). *Handbook of Water and Wastewater Treatment Plant Operations*, 2nd ed., CRC Press, Boca Raton, FL.

Thompson, P. (2008). *Wolf Thoughts* [blog], http://wolfatd.blogspot.com/.

Tillman, D. (2000). Biomass cofiring: the technology, the experience, the combustion consequences, *Biomass and Bioenergy*, 19:365–384.

Uhlig, H. (1998). *Industrial Enzymes and Their Applications*, John Wiley & Sons, New York.

USDA. (2003). A Strategic Assessment of Forest Biomass and Fuel Reduction Treatments in Western States, U.S. Department of Agriculture, Forest Service, Research and Development, Washington, D.C. (www.fs.fed.us/research/pdf/Western_final.pdf).

USDA. (2004). *Timber Products Output Mapmaker Version 1.0*, U.S. Department of Agriculture, Forest Service, Washington, D.C.

USDOE. (2003). *Industrial Bioproducts: Today and Tomorrow*, U.S. Department of Energy, Washington, D.C.

USDOE. (2006). *Breaking the Biological Barriers to Cellulosic Ethanol: A Joint Research Agenda*, Report from the December 2005 Workshop, DOE-SC-0095, U.S. Department of Energy, Office of Science, Washington, D.C.

USDOE. (2007a). *Understanding Biomass: Plant Cell Walls*, U.S. Department of Energy, Washington, D.C.

USDOE. (2007b). *Hydrogen, Fuel Cells & Infrastructure Technologies Program: Multi-Year Research, Development, and Demonstration Plan*, U.S. Department of Energy, Washington, D.C. (www.eere.energy.gov/hydrogenandfuelcells/mypp/pdfs/production.pdf).

USEPA. (1979). *Process Design Manual: Sludge Treatment and Disposal*, EPA/625/625/1-79-011, U.S. Environmental Protection Agency, Office of Research and Development, Washington, D.C.

USEPA. (2004a). *Clean Watersheds Needs Survey 2004, Report to Congress*, U.S. Environmental Protection Agency, Washington, D.C. (http://water.epa.gov/scitech/datait/databases/cwns/toc.cfm).

USEPA. (2004b). Overview of biogas technology, in Roos, K.F., Martin, Jr., J.B., and Moser, M.A., Eds., *AgSTAR Handbook*, U.S. Environmental Protection Agency, Washington, D.C. (http://www.epa.gov/agstar/documents/chapter1.pdf).

USEPA. (2006). *Biosolids Technology Fact Sheet: Multi-Stage Anaerobic Digestion*, EPA/832-F-06-031, U.S. Environmental Protection Agency, Office of Water, Washington, D.C.

USEPA. (2007a). *Fuel Economy Impact Analysis of RFG*, U.S. Environmental Protection Agency, Washington, D.C. (http://www.epa.gov/oms/rfgecon.htm).

USEPA. (2007b). *Biomass Combined Heat and Power Catalog of Technologies*, U.S. Environmental Protection Agency, Washington, D.C.

USEPA. (2008). *Climate Leaders Greenhouse Gas Inventory Protocol Core Module Guidance: Direct Emissions from Stationary Combustion Sources*, EPA/430-K-08-003, U.S. Environmental Protection Agency, Washington, D.C.

Valigra, L. (2000). Tough as soybeans, *The Christian Science Monitor*, January 20.

Vielstich, W., Gasteiger, H.A., and Yokokawa, H., Eds. (2009). *Handbook of Fuel Cells: Advances in Electrocatalysis, Materials, Diagnostics and Durability*, Vol. 6, Wiley, New York.

Walsh, M. et al. (1999). *Biomass Feedstock Availability in the United States: 1999 State Level Analysis*, Oak Ridge National Laboratory, Knoxville, TN.

Wang, S. and Baxter, L. (2007). *Comprehensive Investigation of Biomass Fly Ash in Concrete: Strength, Microscopy, Quantitative Kinetics, and Durability*, paper presented at ACERC Annual Conference, February 28, Brigham Young University, Provo, UT.

WAOB. (2008). *World Agricultural Supply and Demand Estimates*, U.S. Department of Agriculture, Washington, D.C.

Wilhelm, W.W., Johnson, J.M.F., Karlen, D.L., and Lightle, D.T. (2002). Corn stover to sustain soil organic carbon further constrains biomass supply, *Agronomy Journal*, 99:1665–1667.

Wiltsee, G. (1998). *Urban Wood Waste Resource Assessment*, NREL/SR-570-25918, National Renewable Energy Laboratory, Golden, CO.

Wiltsee, G. (2000). *Lessons Learned from Existing Biomass Power Plants*, Subcontract No. AXE-8-18008, prepared for National Renewable Energy Laboratory, Golden, CO.

Wood, M. (2002). Desert shrub may help preserve wood, *Agricultural Research Magazine*, 50(4):10–11.

Wright, L., Boundy, B., Perlack, B., Davis, S., and Saulsbury, B. (2006). *Biomass Energy Data Book*, U.S. Department of Energy, Energy Efficiency and Renewable Energy, Washington, D.C.

12

Forest Biomass Removal: Environmental Impact

He took also of the seed of the land, and planted it in a fruitful field; he placed it by great waters, and set it as a willow tree.

And it grew, and became a spreading vine of low stature, whose branches turned toward him, and the roots thereof were under him: so it became a vine, and brought forth branches, and shot forth sprigs.

There was also another great eagle with great wings and many feathers: and, behold, this vine did bend her roots toward him, and shot forth her branches toward him, that he might water it by the furrows of her plantation.

It was planted in a good soil by great waters, that it might bring forth branches, and this might bear fruit that it might be a goodly vine.

Say thou, Thus saith the Lord God; Shall it prosper? Shall he not pull up the roots thereof, and cut off the fruit thereof, that it wither? It shall wither in all the leaves of her spring, even without great power or many people to pluck it up by the roots thereof.

Yea, behold, being planted, shall it prosper? shall it not utterly wither, when the east wind toucheth it? It shall wither in the furrows where it grew.

—**Ezekiel 17:5–10**

… like trees in autumn shedding their leaves, going to dust like beautiful days to night, proclaiming as with the tongues of angels the natural beauty of death.

—**John Muir (*The Century*, 1903)**

Warm Fire

One of my all-time favorite places to visit and explore is Jacob Lake, Arizona. It is a small unincorporated community on the Kaibab (pronounced "kie-bab") Plateau in Coconino County, at the junction of U.S. Route 89A and State Route 67. Known as the "Gateway to the North Rim of the Grand Canyon" and named after the Mormon explorer Jacob Hamblin, it is the starting point of Route 67, the only paved road leading to the North Rim of the Grand Canyon, 44 miles to the south.

The quaint settlement of Jacob Lake, with its local inn, cabins, restaurant, lunch counter, gift shop, bakery, horseback riding center, campground, general store, vast forest area, and a U.S. Forest Service visitor center, is one of those out-of-the-way, looked-forward-to, stopping-off places that I always enjoy—a forest oasis, for sure. Although the North Rim of the Grand Canyon with its Bright Angle Point Trail, Cliff Springs Trail, and much longer North Kaibab Trail (to the Canyon floor and 14 miles across to the South Rim) are the main draws to this particular area, it is the 44 mile scenic drive through the priceless heritage forest from Jacob Lake to the North Rim that many travelers come to appreciate. Along this paved highway, situated at roughly 8000 feet above sea level in a large ponderosa pine forest, giving way in places to aspen, spruce, and fir, the trees, grazing meadows, and wildlife are breathtaking, soothing to the eye, and warming to the heart at the same time. Home to the endangered Kaibab squirrel, this much traveled road is also home to mule deer, elk, coyotes, porcupines, bobcats, numerous bird species, horned lizards, mountain lions, and buffalo.

The drive along Route 67 today is different from what it was in the past (and maybe this is good, because differences make one take notice). Nature has stepped in and made a few changes through the years. On the afternoon of June 8, 2006, a lightning storm swept across the Kaibab Plateau. One of the high-voltage, sky/earth encounters set a tree on fire south of Jacob Lake. Local fire crews could have responded immediately, but that is not what happened. Instead, the National Forest Service decided to let the fire burn. They named it "Warm Fire," and designated it a *Wildland Use Fire* (WUF). "Wildland" can have several different definitions, but for our purposes we define it here as a natural environment that has not been significantly modified by human activity—a last truly wild natural area.

Forest Service management chose to let the naturally ignited fire burn because they understood that fire plays a critical role in wildlands by recycling nutrients, regenerating plants, and reducing high concentrations of fuels that contribute to disastrous wildland fires. U.S. land managers recognize the role that controlled wildland fires play in ecosystems, and through careful planning they can manage naturally occurring fires, such as Warm Fire, for resource benefit. It is important to point out that human-caused fires are never used for resource benefit; they are always declared unwanted wildland fires that must be suppressed using the most cost-effective means to protect lives, property, and the environment.

Wildland fire use, then, is the management of naturally ignited wildland fire to accomplish resource management objectives for specific areas. The three primary objectives for allowing wildland fire use are

- Provide for the health and safety of firefighters and the public.
- Maintain the natural ecosystems of a given area and allow fire to play its natural role in those ecosystems.
- Reduce the risks and consequences of unwanted fire.

The old idiom, "The best laid plans of mice and men oft go astray," is apropos here, though. Why? Consider the following account based on my investigation (and several regulatory agency reports) of the incident at Kaibab Plateau.

According to various private, public, personal, and U.S. Forest Service accounts, after the lightning struck and ignited the tree, the fire smoldered in the decaying leaves and branches (duff) covering the forest floor and crept around for a few days. At first glance all appeared well or under control, but then the wind picked up and the fire jumped Route 67 on June 15. It jumped it again 2 days later and consumed 750 acres, but there were no apparent signs of panic yet.

On June 18, Warm Fire had grown to over 3000 acres, and overnight it doubled, to over 6000 acres. By June 22, the fire had grown to over 10,000 acres, but still no one panicked. The mindset? Let it burn! And burn it did. On June 23, strong winds came blowing hard and howling loudly out of the west and Warm Fire blew up; overnight it grew to over 15,000 acres. On June 24, the Kaibab National Forest Fire Service came to the conclusion that enough was enough and it was time for human intervention (we humans are good at that). By the time action was taken on June 26, Warm Fire had grown to 30,000 acres. By the next day, it was 60,000 acres, primarily due to the back fires set by professional firefighters who were trying to contain the fire. These same professionals, using abundant ground and air resources, contained, controlled, and mostly extinguished Warm Fire by July 4. The cost to Kaibab Forest was immeasurable. Many critics felt that the fire was not a cleanser but instead was a destroyer.

In February 2007, the Kaibab Ranger District posted on its website a rather frank report entitled "Warm Fire Assessment: Post-Fire Conditions and Management Considerations." According to the Ranger District, the following types of damage occurred:

- *Financial*—Total cost of Warm Fire was estimated at $70 million.
- *Soil erosion*—The intense burn denuded vast acreages. Rain and snow runoff incised deeper channels and accelerated soil loss. Known archaeological sites were eroded away and partially lost.
- *Water quality*—Ash and sediment were transported in Marble and Kanab Creeks to the Colorado River.
- *Heritage sites*—Many historic, cultural sites were damaged or destroyed by Warm Fire.
- *Forest vegetation*—More than 60% of the Warm Burn experienced 100% mortality to the trees.
- *Fire hazard*—Surface and fine fuels were partially consumed, but the addition of dead snags increased coarse fuels.
- *Wildlife*—A large population of the Mexican spotted owl was destroyed, as was habitat for the northern goshawk.

DID YOU KNOW?

The Warm Fire event affected the habitat of the northern goshawk, and it will take a long time, perhaps over 100 years, for a viable nesting habitat to develop again in fire-decimated areas of the Kaibab Plateau. Important prey species that lived on the Kaibab Plateau included the American robin, Stellar's jay, hairy woodpecker, northern flicker, red-naped and Williamson's sapsuckers, chipmunks, golden-mantled ground squirrels, cottontail rabbits, Kaibab squirrel, and red squirrel (Reynolds et al., 1992; Wiens et al., 2006).

After Warm Fire, many private organizations, think tanks, and environmental thinkers jumped all over the National Fire Service for letting Warm Fire cause the damage it did. Although natural burns in forest areas may be at odds with images of Smokey Bear and blackened forests, the effects of fires are temporary. During several walk-arounds of hundreds of acres scorched by the Warm Fire area along Route 67, I have concluded that life, represented by various wildlife species and the obvious growth of tree seedlings and other undergrowth, is apparent throughout, and the burned area is recovering.

Based on personal observation, study, experience, and assessment, it is my judgment that instead of pointing a finger of blame toward those who made the decision to let Warm Fire burn, the emphasis should be on gathering lessons learned and making sure they are shared with those who might face a similar situation in the future. The rapid recovery of vegetation and the apparent ability of most species of wildlife, especially mule deer, to thrive on Warm Fire burned areas and the high-quality habitat provided during post-fire recovery suggest that, in this instance, fire actually enhanced the habitat for many plants and animals on the Kaibab Plateau.

DID YOU KNOW?

In the immediate aftermath of Warm Fire, it was rare to hear that resonant chuckling or deep churring call of the Kaibab squirrels; normally, before the fire, they could be expected to be expertly hiding in the branches of trees. In the burned areas, though, there were no tree branches left, just charred sentinels without arms. Kaibab squirrel mortality from the fire was likely high because the crown fire moved rapidly through the area.

Forest Biomass Waste and Purpose-Grown Material Utilization

The Warm Fire event recounted above points to the need to protect our forests. Dense growth has limited the size and resiliency of trees in some forested areas of the United States. In the Cascade region of Oregon and Washington, for example, the health of large areas of forestland has deteriorated. Similar conditions exist in forests throughout the western United States. In many areas, the natural ecosystem has been significantly altered, creating a high risk of intense wildfire. Moreover, in dry interior western U.S. forests, biomass accumulates faster than it decomposes. According to an Oregon Department of Energy study (Sampson et al., 2001), 39 million acres (about 30%) of National Forest land in the West is threatened by unnatural fuel accumulations.

Because of fire suppression and logging practices, the condition of the forest in these overgrown areas is not natural. Selective thinning would improve the general health of the remaining trees and reduce the risk of fire. With less competition for nutrients and water, the remaining trees would have a better chance of maturing into old-growth stands. The surplus biomass obtained from thinning unnaturally overgrown forest areas is a large renewable energy resource. Carefully planned forest thinning activities can preserve wildlife habitat and minimize soil erosion so forest biomass can be used in a sustainable manner.

Thinning forests to protect against fire and disposing of woody biomass—material from trees and woody plants, including limbs, tops, needles, leaves, and other woody parts that are byproducts of forest management, ecosystem restoration, or hazard fuel reduction treatments—for the production of renewable energy are environmentally friendly practices. The first part of this chapter focuses on the environmental benefits of processing woody biomass for energy production. The second part of the chapter focuses on the environmental impact of utilizing purpose-grown biomass fuels (e.g., poplar and willow trees) and whole tree harvesting, with particular emphasis on the impact on soil productivity.

Benefits of Woody Biomass Use for Energy Production*

In addition to using the removal of woody biomass to protect forests as described above, other major environmental benefits can be derived from using forest woody biomass for energy production. These are described in the following.

Reducing Greenhouse Gases: Carbon Dioxide

Increasing the use of forest-derived biomass power will improve the health of our nation's forests and reduce the amount of greenhouse gases released to the atmosphere. Carbon dioxide (CO_2), methane, nitrous oxide, and certain other gases are called *greenhouse gases* because they trap heat in the Earth's atmosphere. The global concentration of CO_2 and other greenhouse gases is increasing. A natural greenhouse effect of trace gases and water vapor warms the atmosphere and makes the Earth habitable; however, human-caused greenhouse gas emissions are having an effect on regional climate and weather patterns. Actual climate change effects (e.g., rate and magnitude) are not yet clear.

Trees and plants remove carbon from the atmosphere through photosynthesis, forming new biomass as they grow. Carbon is stored in biomass. When biomass is burned, carbon returns to the atmosphere in the form of CO_2. There is no net increase in atmospheric CO_2 if the new growth of plants and trees fully replaces the supply of biomass consumed for energy; however, if the collection or processing of biomass consumes any fossil fuel, then additional biomass would have to be grown to offset the carbon released from the fossil fuel; the balance must be maintained. Unfortunately, the combustion of natural gas, coal, and petroleum fuels for energy adds CO_2 to the atmosphere without a counterbalance for removing it. Using biomass fuels instead of fossil fuels may reduce the risks of adverse climate change from greenhouse gas emissions.

Reducing Greenhouse Gases: Methane

Methane (CH_4) is emitted from a variety of both human-related (anthropogenic) and natural sources. Compared to CO_2, methane has 21 times the global warming potential. Methane is released by the natural decomposition of organic materials, especially in wetlands, and by other sources such as gas hydrates, permafrost, termites, oceans, freshwater bodies, non-wetland soils, and wildfires. Human-related activities that release significant quantities of methane to the atmosphere include fossil fuel production, animal husbandry, concentrated animal feeding operations (CAFOs), rice cultivation, biomass burning, and waste management activities. It is estimated that 60 to 80% of the global methane emissions are related to human-related activities (USEPA, 2010). Using biomass-derived methane to produce useful energy consumes methane and reduces the risk to the environment that would otherwise result from natural decomposition. In addition, generating electricity with biomass-derived methane fuel can offset power produced from fossil fuels and reduce the net CO_2 emissions for electrical power generation.

---

* Much of the information in this section is based on ODOE, *Biomass Energy and the Environment*, Oregon Department of Energy, Salem, 2008 (http://www.oregon.gov/ENERGY/RENEW/Biomass/Environment.shtml).

Keeping Waste Out of Landfills

Using urban wood waste for fuel reduces the volume of waste that otherwise would be buried in landfills. The ash residue that remains after combustion of waste wood is less than 1% of the volume of the wood waste consumed. Uncontaminated ash can be used as a soil amendment to add minerals and to adjust soil acidity.

Reducing Air Pollution

Field burning of agricultural residue emits particulate matter and other air pollutants. Due to air quality concerns, many states have enacted regulations designed to reduce the amount of open field burning of grass seed straw. Grass seed straw and other agricultural residues are potential biomass fuels. These materials are suitable as fuel for appropriately designed combustion boilers to produce heat, steam, or electrical power. They are also potential feedstock for conversion to ethanol. Removing biomass from forested areas where an excess of dead wood has accumulated reduces not only the risk of forest fires but also smoke emissions from forest fires and slash burning, thus protecting air quality.

Reducing Acid Rain (Acid Deposition) and Smog

Let's talk about the acid rain problem for a moment. Consider the following: In the evening, when you stand on your porch and look out over your terraced lawn and that flourishing garden of perennials during a light rainfall, you probably feel a sense of calm and relaxation that's difficult to describe—but not hard to accept. Maybe it's the sound of raindrops falling on the roof of the porch, the lawn, the sidewalk, and the street and that light wind blowing through the boughs of the evergreens that are soothing you. Whatever it is that makes you feel this way, rainfall is a major ingredient. But those who are knowledgeable or trained in environmental science might take another view of such a seemingly peaceful event. They might wonder to themselves whether the rainfall is as clean and pure as it should be. Is this actually just rainfall—or is it rain carrying acids as strong as lemon juice or vinegar and capable of harming both living and nonliving things such as trees, lakes, and buildings? This may seem strange to some folks who might wonder why anyone would be concerned about such off-the-wall matters.

Such a concern was unheard of before the Industrial Revolution, but today the purity of rainfall is a major concern for many people, especially with regard to its acidity. Most rainfall is slightly acidic because of decomposing organic matter, the movement of the sea, and volcanic eruptions, but the principal factor is atmospheric carbon dioxide, which causes carbonic acid to form. *Acid rain* (pH <5.6) is produced by the conversion of the primary pollutants sulfur dioxide and nitrogen oxides to sulfuric acid and nitric acid, respectively. These processes are complex, depending on the physical dispersion processes and the rates of the chemical conversions.

Contrary to popular belief, acid rain is not a new phenomenon nor does it result solely from industrial pollution. Natural processes—volcanic eruptions and forest fires, for example—produce and release acid particles into the air, and the burning of forest areas to clear land in Brazil, Africa, and other countries also contributes to acid rain; however, the rise in manufacturing that began with the Industrial Revolution literally dwarfs all other contributions to the problem. The main culprits are emissions of sulfur dioxide from the burning of fossil fuels, such as oil and coal, and nitrogen oxide, formed mostly from internal combustion engine emissions, which is readily transformed into nitrogen dioxide. These mix in the atmosphere to form sulfuric acid and nitric acid.

In dealing with atmospheric acid deposition, the Earth's ecosystems are not completely defenseless; they can deal with a certain amount of acid through natural alkaline substances in soil or rocks that buffer and neutralize acid. Highly alkaline soil (limestone and sandstone) in the American Midwest and southern England provides some natural neutralization; however, areas with thin soil and those laid on granite bedrock have little ability to neutralize acid rain. Scientists continue to study how living beings are injured or even killed by acid rain. This complex subject has many variables. We know from various episodes of acid rain that pollution can travel over very long distances. Lakes in Canada and New York are feeling the effects of coal burning in the Ohio Valley. For this and other reasons, the lakes of the world are where most of the scientific studies have taken place. In lakes, the smaller organisms often die off first, leaving the larger animals to starve to death. Sometimes the larger animals (e.g., fish) are killed directly; as lake water becomes more acidic, it dissolves heavy metals, leading to toxic and often lethal concentrations. Have you ever wandered up to the local lake shore and observed thousands of fish belly-up? Not a pleasant sight or smell, is it? Loss of life in lakes also disrupts the system of life on the land and the air around them. In some parts of the United States, the acidity of rainfall has fallen well below 5.6. In the northeastern United States, for example, the average pH of rainfall is 4.6, and rainfall with a pH of 4.0, a level 1000 times more acidic than distilled water, has occurred.

Despite intensive research into most aspects of acid rain, there are still many areas of uncertainty and disagreement. That is why progressive, forward-thinking countries emphasize the importance of further research into acid rain, and that is why the 1990 Clean Air Act was strengthened to initiate a permanent reduction in SO_2 levels. One of the interesting features of the Clean Air Act is that it allowed utilities to trade allowances within their systems or buy and sell allowances to and from other affected sources. Each source must have sufficient allowances to cover its annual emissions. If not, the source is subject to excess emissions fees and a requirement to offset the excess emissions in the following year. The 1990 law also included specific requirements for reducing emissions of nitrogen oxides for certain boilers.

Efficient combustion of biomass results in lower emissions of SO_2 and the production of fewer organic compounds that cause smog compared to emissions from facilities that burn coal or oil. Co-firing biomass with coal can reduce SO_2 and NO_x emissions at coal-fired power plants. The level of NO_x emissions from biomass combustion facilities depends on the design of the facility and the nitrogen content of the feed stock. Pollution control equipment can further reduce NO_x and particulate emissions.

Woody Biomass Collection or Conversion Effects on Plant Resources

A large body of research has demonstrated that, in general, thinning of densely stocked stands of conifers such as ponderosa pine and Douglas fir improves the vigor of trees that are left in the stand (residual trees) by reducing competition for water and soil nutrients. Trees that are more vigorous are also less susceptible to insect attack. Indeed, improved tree health and growth and resistance to insect attack are often key aspects of the rationale for thinning treatments, but thinning prescriptions involve more complex harvesting procedures than clearcutting and can result in damage to residual trees. Opening up the stand can also make residual trees more susceptible to windthrow (tree blowdown) and increase wind speeds, which in turn can affect fire behavior.

DID YOU KNOW?

Because fire is not a uniform disturbance, forests can be maintained through attributes of both plant resistance and resilience; for example, because of the thickness of its bark, mature *Pinus ponderosa* is resistant to fire and can generally tolerate relatively high temperatures from moderate fires (Bailey and Whitham, 2002).

Forests in many inland western ponderosa-pine-dominated forests have changed from the open, low-density stands encountered by Europeans when they first came to this country to closed, high-density stands, a change that has been detrimental to the vigor of old-growth trees. Stone et al. (1999) examined whether the vigor of old-growth, presettlement trees could be improved by restoring the original stand structure through thinning of smaller trees that established after settlement. This treatment resulted in the following changes in presettlement trees and their environment in the first year following thinning: an increase in volumetric soil water content between May and August, an increase in predawn xylem water potential in July and August, a decrease in midday xylem water potential in June and August, an increase in net photosynthetic rate in August, an increase in foliar nitrogen concentration in July and August, and an increase in bud and needle size. These results show that the thinning restoration treatment improved the condition of presettlement ponderosa pines by increasing canopy growth and the uptake of water, nitrogen, and carbon.

Kolb et al. (1998) compared foliar physiology and several measures of tree resistance to insect attack among ponderosa pine trees growing in thinned stands. The study area was a second-growth forest area in northern Arizona where four different density treatments (6.9, 18.4, 27.6, 78.2 m^2 ha^{-1}) had been experimentally maintained by frequent thinnings for 32 years before measurement began in 1994. Most of the physiological characteristics measured were affected by the basal area treatments. As stand basal area increased from 6.9 to 78.2 m^2 ha^{-1}, predawn water potential, midday water potential, net photosynthetic rate, resin production, phloem thickness, and foliar toughness decreased. Foliar nitrogen concentration was greatest in trees in the intermediate basal area treatments. Results indicated that the physiological condition of second-growth ponderosa pine can be improved by reducing stocking levels and that dense stocking levels increase tree stress and decrease tree resistance to insect attack.

Sala et al. (2005) measured soil water and nitrogen availability, physiological performance, and wood radial increment of second-growth ponderosa pine trees in the Bitterroot National Forest, Montana, 8 and 9 years after four treatments: (1) thinning only, (2) thinning followed by prescribed fire in spring, (3) thinning followed by prescribed fire in fall, and (4) untreated controls. Trees of similar size and canopy condition in the three thinned treatments (with and without fire) displayed higher leaf-area-based photosynthetic rates, stomatal conductance, and mid-morning leaf water potential in June and July, as well as higher wood radial increments relative to trees in control units. Results suggest that, despite minimal differences in soil resource availability, trees in managed units where the basal area was reduced had improved gas exchange and growth compared with trees in unmanaged units. Prescribed fire (spring or fall) in addition to thinning had no measurable effect on the mid-term physiological performance and wood growth of second-growth ponderosa pine.

Thinned and unthinned stands of lodgepole pine in eastern Oregon were evaluated by Mitchell et al. (1983) to determine their vigor and susceptibility to mountain pine beetle attack. Comparisons of stem growth per square meter of crown leaf showed that thinning from below improved the vigor or residual trees and reduced beetle attack. Beetle mortality was significant in unthinned and lightly thinned stands where annual stemwood growth of residual trees averaged less than 80 g/m^2 of foliage. Stands where stemwood growth in residual trees was around 100 g/m^2 were beginning to suffer beetle attacks. No mortality was observed in heavily thinned stands where stemwood growth in residual trees exceeded 120 g/m^2. These findings suggest that lodgepole pine can be managed through stocking control to avoid mountain pine beetle attack.

Understory composition and structure in thinned and unthinned Douglas fir/western hemlock stands on 32 western Oregon sites were studied by Bailey and Tappeiner (1998). These stands had regenerated naturally after timber was harvested between 1880 and 1940 and were thinned between 1969 and 1984. Commercially thinned stands had 8 to 60% of their volume removed 10 to 24 years before the study. Undisturbed old-growth Douglas fir stands were compared on 20 paired sites. Conifer regeneration density and frequency were strongly related to the volume removed and to the stand density index (and other measures of overstory density) just after thinning. In thinned stands, the density of small trees (intermediate crown class overstory trees and advanced regeneration) was 159/ha, significantly greater than in unthinned stands (90/ha), but not significantly different from that of old growth (204/ha). The live crown ratio of these trees in thinned stands (66%) was greater than in unthinned (4%) and old-growth (48%) stands. Bailey and Tappeiner (1998) concluded that thinning young Douglas fir stands will hasten development of multistory stands by recruitment of conifer regeneration in the understory and by enabling the survival of small overstory trees and growth of advance understory regeneration.

In northeastern Oregon, McIver et al. (2003) examined fuel reduction by mechanical thinning and removal in mixed-conifer stands. The experiment compared a single-grip harvester coupled with either a forwarder or a skyline yarding system and unharvested control sites. Both extraction systems achieved nearly equivalent (~46%) mass fuel reduction. Of the seedling trees examined, 32% had noticeable damage after harvest, including bole wounds (38.9% of damaged stems), bark scraping (35.0%), wrenched stems (28.9%), broken branches (26.5%), broken terminal leaders (15.4%), and crushed foliage (4.1%). More damage occurred to residual large trees than to seedlings.

Treatment Effects on Non-Tree Plant Understory Species

Opening up densely stocked stands increases understory plant biomass and biodiversity, which in turn increases habitat heterogeneity for wildlife, but newly available niches may also be colonized by invasion. This section discusses these and other effects of thinning and removal of forest biomass on plant communities in the forest understory. Non-tree species (e.g., shrubs, fungi, native mosses, lichens, ferns, herbs) with commercial, social, or cultural value are affected by forest management practices, including thinning and selection harvesting. In addition to the social value of these species, many of them play important ecological roles in forest communities by contributing to biodiversity and long-term ecosystem productivity and by underpinning populations of mammals and birds. Understory species are also an important aesthetic component of forests (Kerns et al., 2003).

From the end of World War II until fairly recently, clear-cut logging and even-age management dominated forest practices in forests west of the Cascade Range, with the primary objectives of timber production and maintaining vigorous crop trees. With this management legacy, the understory in managed stands is generally an unintentional byproduct of timber management. Since the early 1990s, there has been an increasing focus on alternative silvicultural systems and forest practices that embrace a broader range of values, including biodiversity and forest structural complexity (Kerns et al., 2003).

Generally, overstory stand structure strongly influences understory plant communities by controlling the amount of light that penetrates the canopy. Other factors that influence the composition of forest understories include the disturbance that originated the stand, the degree of biotic legacies (e.g., downed logs, roots, seeds, surviving vegetation) retained following the disturbance, and the rapidity with which trees established on the site and formed a dense canopy.

Active management prescriptions that remove woody biomass can directly or indirectly alter rates and patterns of succession (see Case Studies 12.1 and 12.2) among understory species. If small conifer biomass is removed from the understory, the newly available habitat may be recolonized by conifer seedlings, but it is often colonized by other species—native and exotic. Removing biomass from the overstory influences understory species distribution and abundance by increasing light availability—a major limiting resource for most photosynthetic understory species in coniferous forests.

Case Study 12.1. Ecological Succession*

Ecosystems can and do change. For example, if a fire devastates a forest, it will grow back, eventually, because of ecological succession. Ecological succession is the observed process of change (a normal occurrence in nature) in the species structure of an ecological community over time; that is, a gradual and orderly replacement of plant and animal species takes place in a particular area over time. The result of succession is evident in many places—in an abandoned pasture, in any lake or any pond. Succession can even be seen where weeds and grasses grow in the cracks in a tarmac, roadway, or sidewalk. Additional specific examples of observable succession include the following:

1. Consider a red pine planting area where the growth of hardwood trees (including ash, poplar, and oak) occurs. The consequence of this hardwood tree growth is increased shading by the shade-tolerant hardwood seedlings and subsequent mortality of the sun-loving red pines. The shaded forest floor conditions generated by the pines prohibit the growth of sun-loving pine seedlings but allow the growth of the hardwoods. The consequence of the growth of the hardwoods is the decline and senescence of the pine forest.

2. Consider raspberry thickets growing in sunlit forest sections beneath gaps in the canopy generated by wind-thrown trees. Raspberry plants require sunlight to grow and thrive. Beneath a dense shade canopy, particularly of red pines but also dense stands of oak, there is not sufficient sunlight for the raspberry's survival; however, anywhere a tree has fallen the raspberry canes can proliferate into

---

* This case study is based on Spellman, F.R., *Ecology for the Non-Ecologist*, Government Institutes Press, Lanham, MD, 2008.

dense thickets. Within these raspberry thickets, by the way, are dense growths of hardwood seedlings. The raspberry plants provide a protected nursery for these seedlings and prevent a major browser of tree seedlings, the white tail deer, from eating and destroying the trees. By giving these trees a shaded haven in which to grow, the raspberry plants are setting up conditions for a future tree canopy that will extensively shade the future forest floor and consequently prevent the future growth of more raspberry plants!

Succession usually occurs in an orderly, predictable manner. It involves the entire system. The science of ecology has developed to such a point that ecologists are now able to predict several years in advance what will occur in a given ecosystem. For example, scientists know that if a burned-out forest region receives light, water, nutrients, and an influx or immigration of animals and seeds, it will eventually develop into another forest through a sequence of steps or stages.

Two types of ecological succession are recognized by ecologists: primary and secondary. The particular type that takes place depends on the condition at a particular site at the beginning of the process. Primary succession, sometimes called *bare-rock succession*, occurs on surfaces such as hardened volcanic lava, bare rock, and sand dunes, where no soil exists and where nothing has ever grown before. Obviously, in order to grow, plants need soil; thus, soil must form on the bare rock before succession can begin. Usually this soil formation process results from weathering. Atmospheric exposure—weathering, wind, rain, and frost—forms tiny cracks and holes in rock surfaces. Water collects in the rock fissures and slowly dissolves the minerals out of the rock's surface. A pioneer soil layer is formed from the dissolved minerals and supports such plants as lichens. Lichens gradually cover the rock surface and secrete carbonic acid, which dissolves additional minerals from the rock. Eventually, the lichens are replaced by mosses. Organisms called *decomposers* move in and feed on dead lichen and moss. A few small animals such as mites and spiders arrive next. The result is what is known as a *pioneer community*. The pioneer community is defined as the first successful integration of plants, animals, and decomposers into a bare-rock community (Miller 1988).

After several years, the pioneer community builds up enough organic matter in its soil to be able to support rooted plants such as herbs and shrubs. Eventually, the pioneer community is crowded out and is replaced by a different environment. This, in turn, works to thicken the upper soil layers. The progression continues through several other stages until a mature or climax ecosystem is developed, several decades later. It is interesting to note that in bare-rock succession, each stage in the complex succession pattern dooms the stage that existed before it. According to Tomera (1990), "Mosses provide a habitat most inhospitable to lichens, the herbs will eventually destroy the moss community, and so on until the climax stage is reached."

Secondary succession is the most common type of succession. Secondary succession occurs in an area where the natural vegetation has been removed or destroyed but the soil is not destroyed; for example, succession that occurs in abandoned farm fields, known as *old-field succession*, illustrates secondary succession. An example of secondary succession can be seen in the Piedmont region of North Carolina. Early settlers of the area cleared away the native oak–hickory forests and cultivated the land. In the ensuing years, the soil became depleted of nutrients, reducing the fertility of the soil. As a result, farming ceased in the region a few generations later, and the fields were abandoned. Some 150 to 200 years after abandonment, the climax oak–hickory forest was restored.

Case Study 12.2. From Lava Flow to Forest: Primary Succession*

Probably the best example of primary succession occurred, and is still occurring, on the Hawaiian Islands. One of the most striking aspects of a newly formed lava flow is its barren and sterile nature. The process of colonization of new flows begins almost immediately as certain native organisms specially adapted to the harsh conditions begin to arrive from adjoining areas. A wolf spider and cricket may be the first to take up residence, consuming other invertebrates that venture onto the forbidding new environment. The succession process relies heavily on adjacent ecosystems. A steady rain of organic material, seeds, and spores slowly accumulates in cracks and pockets along with tiny fragments of the new lava surface. Some pockets of this infant soil retain enough moisture to support scattered òhià seedlings and a few hardy ferns and shrubs. Over time, the progeny of these colonizers and additional species from nearby forests form an open cover of vegetation, gradually changing the conditions to those more favorable to other organisms. The accumulation of fallen leaves, bark, and dead roots is converted by soil organisms into a thin but rich organic soil. A forest can develop in wet regions in less than 150 years.

On Hawaiian lava flows, primary succession proceeds rapidly on wet windward slopes, but more slowly in dry areas. The influence of moisture can be seen on the Kona side, where the same flow can support a forest along the Belt Highway but be nearly barren near the dry coast. Except for the newer flows and disturbed areas, the windward surfaces of Kilauea are heavily forested, but the leeward slope is barren or sparsely vegetated.

All the undisturbed flows on Kilauea, Mauna Loa, and Hualaiai volcanoes are young enough to be in some degree of primary succession, and the patterns and relative age of lava flows are reflected in the maturity of vegetation. Only a few of the newest flows on the dry upper slopes of dormant Mauna Kea are young enough to reflect primary succession. Extinct Kohala volcano is too old to produce such flows, and vegetation differences reflect rainfall amounts and disturbance.

On wetter slopes of Hualaiai and Mauna Loa, younger flows stand out against a more uniform, older background, as the surfaces are recovered by lava at rates of only 20 to 40% a century. Small and more active, Kilauea renews about 90% of its surface in the same time period, and the resulting pattern is a patchwork of flows and vegetated remnants (kipuka). The many younger flows rely on the older kipuka to provide sources of plants and animals.

The native forest ecosystems have adapted to the overpowering nature of volcanic eruptions by being able to quickly recolonize from the many kipuka around new flows; however, the added losses due to forest clearing and alien invasion provide additional threats to which the native biota is not adapted. If too many of the native forest areas are cleared or taken over by introduced organisms, natural succession may not be able to provide a replacement native ecosystem on the younger flows. The continuing primary succession process may be already partially interrupted in low Puna, where so much of the native forest has been cleared for development and where colonizers from nearby areas are mostly introduced organisms.

Understory Species Response

Most thinning operations are low- to medium-intensity disturbances compared to high-intensity disturbances such as clear-cut logging and stand-replacing fires. Thinning can increase microhabitats, creating new germination sites and small openings in the canopy.

---

* This case study is based on USGS, *From Lava Flow to Forest: Primary Succession*, U.S. Geological Survey, Menlo Park, CA, 1999 (http://hvo.wr.usgs.gov/volcanowatch/1999/99_01_21.html).

Light, water, nutrient availability, and soil temperatures may increase; therefore, thinning favors species the can rapidly colonize or expand into newly available resources, either by seeds or by vegetative propagation.

Understory species response can be influenced by thinning intensity, frequency, stand age when thinned, uniformity of thinning, and operational disturbance. Uniform thinning leaves evenly spaced trees and usually a compositionally simple understory. Irregular or variable density thinning that creates openings and tree patches of different sizes can increase understory biodiversity. Understory response to thinning, especially by shrubs, is typically correlated with the amount of canopy removed. With very light thinning, impacts of the initial disturbance may outweigh the benefits of making more resources available to understory plants. Moreover, increased wind and sunlight caused by heavy thinning with even spacing can reduce moisture availability.

The introduction or spread of invasive exotic species is a major concern with thinning or woody biomass removal. Greater frequency and abundance of exotics in thinned compared to unthinned stands has been reported in several studies; these effects can last for decades and lead to significant management costs in their own right. The degree to which thinning favors desirable native or invasive exotic species depends to a large extent on the seed species present in the soil and on how the seed bed is treated after thinning. The bottom line on variable density thinning prescriptions is that it shows promise for increasing the biodiversity of understory plants, fungi, native mosses, lichens, ferns, and herbs (Kerns et al., 2003).

Metlen et al. (2004) evaluated the understory response to fuel treatments in northeastern Oregon ponderosa pine–Douglas fir forests. Treatments included no management control, prescribed fall burning (burn), low thinning (thin), and low thinning followed by prescribed fall burning (thin/burn), replaced four times in a completely randomized design. Treatment effects were observed three seasons after thinning. Species richness of understory vegetation was significantly lower in the thin than in the control, but Shannon–Weaver's index of diversity (a commonly used index for measuring diversity at the ecosystem level) was not affected by fuel reduction treatments. Graminoid (grasses and grasslike plants such as sedges and rushes) cover was influenced by treatment, forb (herbaceous flowering plants that at not graminoids) cover was reduced in treatments that included thinning, and shrub and total cover were reduced in treatments that included burning. Individual species responded to treatment in a manner consistent with their life history characteristics. Prairie Junegrass cover increased in those treatments that included burning, but the cover of other graminoid species was not significantly influenced. This treatment significantly lowered elk sedge and total cover but did not strongly influence the cover of other species. Resilience of plant community diversity to fire and the consistent effect of burning on individual species demonstrate their adaptation to frequent low-intensity fire and the subsequent moderate impact of low thinning and fall prescribed burning on understory vegetation.

Metlen and Fiedler (2006) evaluated the effects of no action, thin-only, spring burning, and thinning followed by spring burning on the understory plant community in a second-growth western Montana ponderosa pine–Douglas fir forest that initiated after harvest in

DID YOU KNOW?

Metlen et al. (2004) suggested that modest treatments (probably tied to the disturbance history of the sites) would likely elicit only moderate responses from the understory vegetation.

the early 1990s and has not burned since. Treatments were implemented at an operational scale (~22 acres). Data were collected before and in each of three years after treatment, at two spatial scales: plot (1000 m²) and quadrat (1 m²). Treatments differentially impacted the understory plant community, most dramatically in the thin/burn. The burn-only treatment initially reduced understory richness and cover but by year three all active treatments increased plot-scale understory richness relative to pretreatment and the control. Forbs, both native and exotic, were the most responsive lifeform and increased in richness and cover after thinning, with the greatest response in the thin/burn. Increased native species richness was not detected at the quadrat scale in any treatment but was significant at the plot scale in numerous combinations of treatments and years. Short-term reduction in shrub richness and abundance after burning was detected at the quadrat scale. Sapling density was reduced in all active treatments. Active treatments can create more open overstories and increase understory diversity at the stand level, but a mix of treated and untreated areas will likely maximize heterogeneity and diversity at the landscape scale.

SHANNON–WEAVER INDEX

The Shannon–Weaver index, sometimes referred to as the Shannon–Wiener index (both misnomers, as it is correctly referred to as the Shannon index), is one of several diversity indices used to measure diversity in categorical data (Krebs, 1989). It is the entropy of the distribution—that is, information about the distribution's disorder—its unpredictability or more precisely uncertainty; it treats species as symbols and their relative population sizes as the probability. This index has the advantage of taking into account the number of species and the evenness of the species. The index is increased either by having additional unique species or by having a greater evenness. Typically, the value of the index ranges from 1.5 (low species richness and evenness) to 3.5 (high species evenness and richness), though values beyond these limits may be encountered. The Shannon index is particularly useful when comparing similar ecosystems or habitats, as it can highlight one example being richer or more even than another.

In the dry coniferous forests of central Washington, an assessment was made of the prescribed fire and thinning treatment effects on understory vegetation species richness, cover, and species composition. The thinning and prescribed fire treatments were applied to 12 large (10 ha) management units; the understory vegetation was surveyed before treatment and during the second growing season after treatment. Many understory vegetation traits changed significantly, regardless of treatment. Changes were often proportional to pretreatment conditions. In general, cover declined and species richness increased. Thinning followed by prescribed fire increased species richness, particularly where species richness was initially low. Thinning alone had a similar but lesser effect. Forb richness was increased by thinning, and shrub richness was increased by the combined thin/burn treatment, but graminoid (grass) richness was unaffected. Exotic species cover and richness also increased in the thin/burn treatment, but constituted only a very small portion of the total understory (Dodson et al., 2008).

Understory plant cover was not affected by treatments, but did decline from pre- to post-treatment sampling, with cover losses highest in areas where cover was high prior to treatment. Forb cover increased in the thin/burn treatment where forb cover was low initially. Burning reduced graminoid cover with or without thinning. Species composition varied within and among treatment units but was not strongly or consistently affected by treatments. Thinning and burning treatments had mostly neutral to beneficial effects on understory vegetation, with only minor increases in exotic species; however, the pre-treatment condition had strong effects on understory dynamics and also modified some responses to treatments. It was determined that the maximum benefit of restoration treatments appears to be where understory richness is lower prior to treatment; Dodson et al. (2008) suggested that restoration efforts should focus on these areas.

In a different study, Korb et al. (2001) investigated the inoculum potential for arbuscular mycorrhizal (AM) fungi, which help plants capture nutrients such as phosphorus and micronutrients in the soil, and ectomycorrhizal (EM) fungi, which are important for soil chemistry of vascular plants, in thinned and uncut control stands in a northern Arizona ponderosa pine forest. Three stands of each treatment were sampled by collecting soil cores along 10 randomly chosen transects within each stand. The relative amount of infective propagules of AM fungi was significantly higher in samples collected from thinned stands than controls. Conversely, the relative amount of infective propagules of EM fungi in samples collected from thinned stands decreased slightly compared to controls but this difference was not significant. These preliminary results indicate that population densities of AM fungi can rapidly increase following restoration thinning in northern Arizona ponderosa pine forests, which may have important implications for restoring the herbaceous understory of these forests because most understory plants depend upon AM associations for normal growth.

On 32 western Oregon sites, the understory composition and structure in thinned and unthinned Douglas fir–western hemlock stands were studied by Bailey and Tappeiner (1998). These stands had regenerated naturally after timber was harvested between 1880 and 1940 and were thinned between 1969 and 1984. Commercially thinned stands had 8 to 60% of their volume removed 10 to 24 years before the study. Undisturbed old-growth Douglas fir stands were compared to 20 paired sites. There was significantly less tall shrub cover in unthinned stands than in either thinned or old-growth stands, which did not differ. Thinned stands had the most low shrub cover. Salal and bracken fern cover was greater in thinned stands than in the other stand types, but there was no difference in sword fern and Oregon grape cover. The leaf area index in thinned stands (6.6) was not significantly different from that in unthinned (6.8) and old-growth stands (7.1); however, shrubs had more leaf area in the thinned stands. Bailey and Tappeiner (1998) concluded that thinning Douglas fir stands will help develop the shrub layer by increasing tall shrub stem density and cover of some low shrubs.

Treatment Effects on Conifer Root Diseases

Rippy et al. (2005) reviewed current knowledge regarding the effects of thinning on conifer root diseases and some of the processes by which these effects can occur, as well as the influence of fuel treatments, including thinning, on conifer root diseases in the Inland West. The authors noted that the knowledge base regarding interactions of fuel treatments and diseases is limited. Moreover, interactions among trees, root pathogens, and the environment are complex and vary by forest structure, stand, and land use history; habitat type; species composition; soil characteristics; bark beetle populations; and activity of other forest insects and pathogens. Based on their research, however, they concluded that conifer root diseases are a natural and necessary part of forest ecosystems, and they play important, beneficial roles in the ecological processes of succession, decomposition, and fires in forests with moderate- to high-intensity fire regimes. Native root diseases usually do not cause permanent loss of large stands or threaten the existence of host tree species; however, shifts in stand composition and other natural and human-caused disturbances can increase the prevalence of root diseases. Mid-seral tree species such as white pine, western larch, and ponderosa pine tend to be root disease tolerant but have declined over much of the inland West. In northern Idaho, the combined effects of blister rust, selective harvesting, planting practices, and fire exclusion have reduced the stand representation of these species but have increased the presence of root-disease-susceptible Douglas fir and true firs. In some situations, changes in root rot dynamics may influence forest growth and succession for centuries.

Fuel treatments can influence root disease in complex ways that should be considered at the ecosystem level. Successful use of thinning and biomass removal to lower the risk of severe wildfire depends, in part, on the impacts of such treatments on levels of root disease that contribute to subsequent fuels accumulation. Tree mortality may result directly from root disease or from indirect consequences of infection, such as susceptibility to bark beetle attack. Killed trees can persist as standing snags or windthrown logs, so treatments that cause increase in root disease mortality will likely increase the accumulation of wood debris. A single type of fuel treatment is not appropriate for all forest conditions and root diseases, so the silvicultural prescription should be determined on a site-by-site basis, with knowledge of diseases that are present and where they are located in the stand. Some fuel treatments may increase root disease incidence, so managers should consider what levels of mortality and growth loss will be acceptable. The type of treatment can often be selected to match the overall objectives for a stand while addressing root rot dynamics and possible detrimental effects on a site.

Tree thinning can generate stumps and residual root systems that are susceptible to colonization by root-rot pathogens. Stumps, roots, and slash can serve as nutrient substrates that increase growth rates and distribution of root-rot fungi. Thinning-associated wounds on retained trees can provide vectors for new root disease infections. When mid-seral-stage tree species (e.g., pine and larch) are removed in thinning operations, the resulting stand may have a higher proportion of late-seral-stage tree species that are more susceptible to root rot, such as grand fir or Douglas fir. Thinning may increase the damage from some root diseases, such as Armillaria root rot, laminated root rot, and *Annosus* root disease. Incidence of black stain root disease and *Schweinitzii* root and butt rot may be less affected by thinning if specific guidelines are followed. Soil compaction may predispose tree roots to damage by some root diseases. For all root diseases, steps should be taken to avoid wounding trees during harvesting and removal to reduce future impacts on stand health. Thinning treatments should also favor the species most tolerant to root disease and be timed to avoid problems with bark beetles.

Root disease effects may be considered beneficial or detrimental depending on social values, landowner goals, and management objectives for the stand. Impacts generally viewed as negative include increase in fuels, reduced timber volume, damage to buildings from wind-thrown trees, and potential personal injury. Conversely, root-disease-induced changes in forest structure such as canopy openings, reduced stand density, and increases in snags and downed large wood may produce benefits such as forage for large game animals, habitat for small animals and cavity-nesting birds, greater plant biodiversity, sources for specialty forest products, and reduced likelihood of severe wildfire.

Treatment Effects on Wildfire Behavior

Research has been conducted to investigate relationships between fuel treatment and the effects of subsequent wildfires. The primary rationale, of course, for treatments that reduce wildfire fuel loadings is that doing so will reduce the intensity and severity of the effects of wildfires when they do occur. Research done by Raymond and Peterson (2005) quantified the relationship between fuels and fire severity using prefire surface and canopy fuel data and fire severity data after a wildfire in a mixed-evergreen forest of southwestern Oregon with a mixed-severity fire regime (Biscuit Fire). Modeled fire behavior showed that thinning reduced canopy fuels, thereby decreasing the potential for crown fire spread. Potential for crown fire initiation remained fairly constant despite reductions in ladder fuels, because thinning increased surface fuels, which contributed to greater surface fire intensity. Thinning followed by underburning reduced canopy, ladder, and surface fuels, thereby decreasing surface fire intensity and crown fire potential; however, crown fire is not a prerequisite for high fire severity. Damage to and mortality of overstory trees in the wildfire were extensive despite the absence of crown fire. Mortality was most severe in thinned treatments (80 to 100%), moderate in untreated stands (53 to 54%), and the least severe in the thinned and underburned treatment stands (5%). Thinned treatments had higher fine-fuel loading and more extensive crown scorch, suggesting that greater consumption of fine fuels contributed to higher tree mortality. Fuel treatments intended to minimize tree mortality will be most effective if both ladder and surface fuels are treated. Raymond and Peterson (2005) also noted that

> ... comparing fire severity at the scale of a few hectares is informative for evaluating the relative effects of three management options (no action, thinning, and thinning followed by burning) on fuels and fire severity at small scales. However, further inference is limited by the small spatial scale of this study relative to the spatial scale of the disturbance being studied. The Biscuit Fire burned a large area, created a mosaic of low, moderate, and high fire severity patches, and when the study area is evaluated in the context of the large fire, then each treatment is simply one patch within the mosaic. At this scale, factors in addition to forest structure, such as topography, weather, and climate may control the size and relative abundance of patches within the burn mosaic.

DID YOU KNOW?

The southern Oregon forest type studied by Raymond and Peterson (2005) is not typical of eastside forests in Oregon that are usually considered for fuel treatments. Much of the understory involved in Biscuit Fire was evergreen hardwoods that typically do not function as fuel ladders (Brown, 2008a).

Agee and Lolley (2006) modeled wildfire behavior in a series of prescribed fire and low thinning treatments (four control, four thin, two burn, and two thin/burn units) applied to dry forests dominated by ponderosa pine and Douglas fir in the eastern Washington Cascades. Low thinning focused on smaller, commercial-sized trees and was designed to reduce basal area to 10 to 14 $m^2 ha^{-1}$ in a non-uniform pattern to mimic natural stand patterns and increase resistance to bark beetle attack. Yarding was done by helicopter; branches and tops were left onsite. Nonmerchantable small trees were felled by hand on thinned units. Burning was a low-intensity spring burn, but coverage was spotty, ranging from 23 to 51%, and was considered ineffective in reducing fuels at the time of application by management and research personnel. Both thinning and burning affected vegetation and fuel variables. Thinning reduced canopy closure, canopy bulk density, and basal area and increased canopy base height. Burning had no influence on these canopy variables. Thinning increased 10-hr timelag (0.62 to 2.54 cm) fuels. Burning decreased 1-hr timelag (0 to 0.62 cm) and 10-hr

BISCUIT FIRE

Biscuit Fire, located in southern Oregon and Northern California, began on July 13, 2002, and eventually covered 499,965 acres. Estimated to be one of Oregon's largest in recorded history, Biscuit Fire encompassed most of the Kalmiopsis wilderness. The boundary of Biscuit Fire stretched from 10 miles east of the coastal community of Brookings, Oregon; south into northern California; east to the Illinois Valley; and north to within a few miles of the Rogue River. The fire burned in a mosaic pattern; approximately 20% of the area burned lightly, and less than 25% of the vegetation was killed. Another 50% of the area burned very hot, and more than 75% of the vegetation was killed. Many acres of critical habitat for wildlife burned, and the late-seral and old-growth stands that remain are very precious. Recovery from the fire area is progressing. Flowers, brush, and hardwoods are resprouting amidst the burned trees. In the Cave Junction area, the salmon have returned (USFS 2006).

timelag fuels, forest floor depth and mass, and increased fuel-bed depth. Differences in fuel properties did not translate into differences in simulated wildfire behavior and tree mortality. Thinning did increase potential surface fire flame length under 97 percentile weather, and active crown fire potential decreased on thinned units, but basal area survival did not significantly differ between treatments under 80 and 97 percentile weather. The scale at which data are presented has a large influence on interpretation of results; for example, torching fire behavior, expressed as an average at the unit level, was low, but 17% of individual plots (about 30 plots total per unit) across all treatments did exhibit potential torching behavior. Agee and Lolley (2006, pp. 155–156) stated that "thinning increased crowning index and therefore reduced crown fire hazard. But whatever the crowning index it must be compared to expected windspeeds in the area in order to identify real risk. Arbitrary thresholds that define low, moderate, and high crowning potential across a state or region need to have a more site-specific interpretation biased on local wind data."

Fire effects in treated and untreated stands in four ponderosa pine sites in Montana, Washington, California, and Arizona were quantitatively examined by Pollet and Omi (2002). Fuel treatments included prescribed fire only, whole-tree thinning, and thinning followed by prescribed fire. On-the-ground fire effects were measured in adjacent treated and untreated forests. The researchers developed *post facto* fire severity and stand structure measurement techniques to complete field data collection. They found that crown fire severity was mitigated in stands that had some type of fuel treatment compared to stands without any treatment. At all four of the sites, the fire severity and crown scorch were significantly lower at the treated sites. Results from this research indicated that fuel treatments removing small-diameter trees may be beneficial for reducing crown fire hazards in ponderosa pine sites. Differences in fire severity and crown scorch were less extreme at sites treated with prescribed fire than sites treated with thinning. Apparently, mechanical fuel treatments allowed for more precise and controlled results compared to prescribed fire. Mechanical fuel treatments may specify the exact number of post-treatment residual trees per hectare. By contrast, prescribed fire fuel treatment often varies across a stand and results in less precise stand structure changes.

Pollet and Omi (2002) noted that, while topography and weather may play a greater role than fuels in governing fire behavior, fuels are the only leg of the fire environment triangle (fuels, topography, weather) that land managers are able to manipulate. However, the authors cautioned that, in extreme weather conditions (e.g., drought and high winds) fuel treatments may do little to mitigate fire spread or severity. They also suggested focusing programs, funding, and management attention where the wildfire risk is greatest: wildland–urban interface areas, tree plantations, critical watersheds, and habitats for threatened and endangered species. They argued that, in order to lower the probability of a severe wildfire over a landscape, the entire landscape should be analyzed to determine the most appropriate scales and locations for fuel treatments. Intensively treating most of the landscape may not be necessary (or feasible) but treating strategically located stands for fuel treatment or treating strips of fuels may be beneficial for reducing severe wildfire potential across a large area.

Treatment Effects on Snags and Coarse Woody Debris

Stephens and Moghaddas (2005) studied the impacts on a Sierra Nevada mixed-conifer forest of several replicated fuel treatments on snag and coarse woody debris (CWD) quantity and structure. These treatments included prescribed fire, commercial thinning (crown thinning and thinning from below) followed by rotary mastication of understory trees, mechanical followed by prescribed fire, and control. Density of snags greater than 15 cm DBH (diameter at breast height) in decay class 1 significantly increased in fire-only and mechanical-plus-fire treatments compared with mechanical-only and control treatments. Snag volumes (m^3 ha^{-1}) were not significantly different among treatments for all decay classes. The density, percent cover, and volume of CWD in decay classes 1 and 2 were not significantly altered by any treatment when aggregated across all diameter classes; however, the volume of CWD in decay class 3 was significantly reduced in the fire-only treatment when compared to controls. The density and volume of CWD in class 4 were significantly reduced in mechanical-plus-fire and fire-only treatments when compared with controls and mechanical-only treatments. The authors concluded that retaining large CWD levels may benefit some wildlife species in the short term but can increase fire hazards and make fire control more difficult. High overall fuel loads also increase the probability of snag and CWS consumption when an area inevitably burns. Influences of altering snag and CWD characteristics should be analyzed in the context of long-term forest management goals, including reintroduction of fire as an ecosystem process and creation of forest structures that can incorporate wildfire without tree mortality outside a desired range.

Woody Biomass Collection or Conversion Effects on Wildlife

Depending on local conditions and the specific prescriptions used, thinning and removal of biomass from dense, overstocked stands can often improve wildlife habitat, but not all species benefit and some may be negatively impacted. This section discusses general issues that land managers might wish to consider and provides summaries of what is known about effects on individual species and wildlife groups. To have a discussion about wildlife, even at the periphery of the topic, it is important to provide a working definition of *wildlife*. Generally, wildlife refers to all species of amphibians, birds, fishes, mammals, and reptiles occurring in the wild, where *in the wild* is interpreted as being undomesticated and free roaming in a natural environment. For use in this book, wildlife is defined as all undomesticated animal life (Helms, 1998); forest wildlife species, specifically, are those whose primary habitat includes trees as the dominant form of plant life (Patton, 1992). The forest, of course, is composed of trees in various combinations of size and density. The basic descriptor of forest structure is the stand (Patton, 2011), which is "a community of trees possessing sufficient uniformity in age, spatial arrangement, or conditions, to be distinguishable from adjacent communities and forming a silvicultural or management entity" (Ford-Robertson, 1983, p. 254).

With regard to managers predicting the effects of fuel treatments on wildlife, this is a difficult undertaking due to the paucity of information for most species and wide variability in habitat needs and responses. Predictions may be possible, however, after considering an animal's ecology and then using available information in a conceptual framework that includes: (1) species distribution and abundance, (2) migratory and dispersal characteristics, (3) habitat requirements and preferences, and (4) potential responses to changes in habitat (Pilliod, 2004). There are opportunities to jointly improve wildlife habitat and reduce wildfire risk; for example, "low elevation, dry forests appear to offer the clearest opportunities for thinning—in conjunction with prescribed fire—to contribute to restoration of wildlife habitat while making forests more resistant to uncharacteristically severe fire" (Brown, 2000, p. 25).

To reduce fire risk, the priorities are to reduce surface and ladder fuels and raise the bottom of the live canopy (Agee et al., 2000; van Wagtendonk, 1996). Thinning is most apt to be appropriate where understory trees are sufficiently large or dense that attempting to kill them with fire would run a high risk of also killing overstory trees (Arno et al., 1997; Christensen, 1988; Fule et al., 1997; Moore et al., 1999; Stephenson, 1999). Using prescribed fire alone can be desirable in that it provides the full range of ecological effects of fire; however, fire is an imprecise tool, and a chainsaw or harvester can provide much more control over which trees are actually killed (Swezy and Agee, 1991; Thomas and Agee, 1986). Brown (2000) cautioned that, "Although anecdotal evidence, computer modeling, and common sense provide considerable support for the premise that thinning can reduce fire risk and restore habitat, there is precious little empirical scientific research on the subject and utility and caution should be the order of the day" (p. 25).

A primary concern of wildlife managers is species extirpations caused by habitat fragmentation. Active management of small-diameter stands can directly address both habitat fragmentation and habitat maintenance/restoration issues (Lehmkuhl et al., 2002). Forested landscapes in the interior Pacific Northwest are dominated by small-diameter forests as a result of previous management and wildfire in the early 20th century. Old forest habitats have declined and become fragmented, and some wildlife species associated with them have become threatened or endangered. In general, fewer wildlife species use middle-aged, pole-sapling, or young forests, where dense canopy structures result in relatively low understory vegetation diversity, compared to older forests (Lehmkuhl et al., 2002). Active management (thinning) in these middle-aged, small-diameter stands can diversify understory habitat and accelerate development of old forest characteristics, such as large-diameter trees and patchy understories, that are currently lacking.

In western dry coniferous forest, Pilliod et al. (2006) found tremendous gaps in the information necessary to evaluate the effects of fuel treatments on the majority of terrestrial wildlife and invertebrates in ponderosa pine, dry-type Douglas fir, lodgepole pine, and mixed-conifer forests. Differences in study location, fuel treatment type and size, and pre- and post-treatment habitat conditions resulted in considerable variability in species response. A species may respond positively to fuel reduction in one situation and negatively in another. There is a great need for long-term observational studies on the effects of a range of fuel reduction treatments at multiple spatial scales (stand or larger). Until more complete information on many species becomes available, management activities that allow retention of critical habitat elements are warranted, particularly for elements that are slow to recover such as large-diameter downed wood and snags.

Despite these information gaps, Pilliod et al. (2006) identified a few consistent patterns. Fire-dependent species, species preferring open habitats, and species associated with early successional vegetation or that consume seeds and fruit usually benefit from fuel treatments. Conversely, species that prefer closed-canopy forests or dense understories and species that are closely associated with large snags or down logs that may be removed by fuel treatments will likely be negatively affected. Some habitat loss may persist for only a few months or years, such as understory vegetation and litter that recover quickly. But lost large snags and downed wood—important habitat elements for many wildlife and invertebrate species—may take decades to recover and thus represent some of the most important habitat elements to conserve. Managers should retain refugia (areas that have escaped ecological change) of untreated stands and critical habitat elements, particularly slow-to-recover features such as large-diameter downed wood and snags, to increase habitat heterogeneity and benefit the greatest number of species over time.

Because, in most cases, conserving habitat heterogeneity also conserves biodiversity it might be prudent for fuel treatment planners to coordinate with biologists to plan treatments that, over time, create a mosaic of forest structures and conditions that approximate natural disturbance patterns. Such actions could be expected to support greater species diversity than large, homogeneous stands given the same treatment. Treating fuels in habitat patches adjacent to untreated patches that are occupied by a given wildlife species may increase the rate of colonization and recovery compared to restoring areas at random (Pilliod et al., 2006).

A species' response to habitat changes from fuel treatments depends on habitat elements that the species needs to survive and reproduce and on how the treatment affects these habitat elements. Potential effects of a fuel treatment on a particular species should be considered within the context of that species' distribution and abundance, migratory and dispersal characteristics, habitat association, and potential responses to habitat changes. If a species is widely distributed, localized fuels projects may have relatively minor effects on population viability, depending on the species' ability to recolonize from surrounding areas. But, for species with a limited distribution, especially in cases where an entire sub-population of a rare species lies within the treatment area, it may be necessary to protect specific habitat components or leave untreated refugia within project boundaries. Species responses may also vary over time; for example, the population of a small mammal species that requires shrub cover to avoid predators may decline following treatment but then later exceed pretreatment levels when shrubs recover and food resources increase from greater light, herbaceous growth and seed production resulting from opening up the forest canopy (Pilloid et al., 2006).

Treatment Effects on Forest Carnivores

Forest carnivores native to western dry coniferous forests include mountain lion, bobcat, wolf, coyote, black bear, grizzly bear, fisher, and American marten. Wolverine and lynx generally prefer cooler, higher elevation forests. American marten primarily inhabit older forests of Engelmann spruce and subalpine fir in northeastern Oregon but occasionally occur in drier stands of fir and lodgepole pine with high densities of downed wood and snags (Bull et al., 2005a). Most forest carnivores are relatively rare, and there are substantial information gaps on their responses to tree harvest. Because most have relatively large ranges, few forest carnivore species will be affected by stand-level fuel projects, but some could be affected by a loss of denning habitat and changes in prey populations due to cumulative effects of past management, past disturbances, and larger scale projects (Pilliod et al., 2006). Both marten and fisher are sensitive to loss in canopy cover and are strongly associated with downed wood cover. These are both indicator species that can complement ecosystem-level conservation planning by revealing thresholds in habitat area and landscape connectivity (Carroll et al., 2001).

Black bears are more common than most other forest carnivores, and their diet and habitat may be influenced by fuels projects. Black bears use areas with abundant downed wood and dense thickets of shrubs and smaller trees adjacent to or within mature forests. About 25% of the black bear's diet can consist of insects (mainly ants and yellowjackets) obtained primarily from downed logs (Bull et al., 2001), so decreases in downed wood can make less of this food available to the bears. Black bears also strip bark from smaller trees and feed on sapwood. In a western larch, lodgepole pine, and Engelmann spruce forest in northwestern Montana, the rate of black bears feeding on sapwood was 5 times higher in thinned stands where bears mostly selected 5- to 10-inch larch trees compared with adjacent unthinned stands (Mason and Adams, 1989). The health and condition of residual trees may have attracted the bears. Treated stands may also provide dependable food sources for bears when fruit, mast, grass, and herbaceous plant production increase after prescribed fire, but fuel treatments may reduce the amount of escape cover, perhaps the most critical component of black bear habitat (Hamilton, 1981). Sites used by black bears for traveling and resting typically have high stem density and dense canopies, presumably for security. Bears use large-diameter hollow logs for denning (Bull et al., 2000) but these may be lost in thinning operations.

> **DID YOU KNOW?**
>
> Omnivores, black bears will feed opportunistically on whatever food is available. Plants make up 75% of a black bear's diet. Naturally, they will consume berries, fruit, nuts, insects, bird eggs, small mammals, and carrion. When black bears emerge from their winter dens they primarily eat newly emergent skunk cabbage, forbs, grasses, tubers, bulbs, and insects. They may also feed on carrion, such as deer carcasses.

Treatment Effects on Ungulates (Hooved Mammals)

North American ungulate species, including deer, elk, moose, mountain goats, and bighorn sheep generally utilize open areas for foraging and forested areas for cover. Ungulates such as deer and elk use dense thickets of shrubs and trees as thermal cover, to hide from predators, for daybeds, and for fawning. All ungulates require an area of abundant forage, such as grasses, forbs, and shrubs. Both of these habitat types, and their proximity to each other, are important. In Douglas fir and mixed-conifer forests in western Montana, elk generally remain within 200 meters of foraging areas during summer. Thinning alone or combined with prescribed fire generally increases forage quantity and quality for elk and other ungulates (Demarais and Krausman, 2000; Huffman and Moore, 2004); however, retaining patches of dense cover (greater than 40% closure of midstory canopies) is beneficial for providing security and escape cover for mule deer (Chambers and Germaine, 2003).

Thinning large areas of forest cover may force wintering ungulates to lower elevations, resulting in increased conflict and damage losses or neighbors. Subsequent green-up of thinned areas may attract ungulates for longer periods during the spring and summer, increasing the potential for damage to adjacent property of crops. Integrating wildlife needs with fuel reduction can be achieved through intensive planning and close cooperation among resource disciplines (Henjum, 2006).

In central Oregon, thinning dense second-growth trees and mowing or burning of understories dominated by fire-sensitive bitterbrush have successfully reduced fire risk, but the treatments often stimulate fire-dependent species such as rabbitbrush, snowbrush, and green leaf manzanita (Riegel, 2006). This shift in species composition reduces bitterbrush cover and abundance—the primary browse for mule deer.

Long et al. (2008a) investigated the effects of fuel reduction on the quantity and quality of forage available to elk in the Starkey Experimental Forest and Range in northeastern Oregon. From 2001 to 2003, 26 stands of true fir and Douglas fir were thinned and burned; 27 similar stands were left untreated as experimental controls. (The effects of thinning only or thinning and removal of thinned biomass were not tested separately.) The percentage of cover, percent of dry matter digestibility, and percent of nitrogen (%N) of 16 important forage species and genera were estimated in treatment and control stands in the spring (May and June) and summer (July and August) of 2005 and 2006. The quantity and quality of forage were lower in summer than spring in both stand types. Total cover of forage was higher in treatment stands during spring and higher in control stands during summer. For grasses, %N was higher in control than in treatment stands, whereas digestibility did not differ between stand types. For forbs, neither index of forage quality differed between stand types. When treated stands were separated out by years since burning, %N and digestibility of forbs and %N of graminoids increased from 2 to 5 years following treatment, and by the fifth year after burning had exceeded maximum values observed in control stands in both seasons. Due to the interacting effects of fuel reduction and season on forage characteristics, treated stands provided better spring foraging for elk, whereas control stands provided better summer foraging. Long et al. (2008a) suggested that maintaining a mosaic of burned and unburned (late successional) habitat may be of greater benefit to elk than burning a large proportion of a landscape. Where elk and mule deer coexist, a combination of thin/burn fuel treatments may benefit both (Long et al., 2008b).

Treatment Effects on Small Mammals

Small mammals require cover and food resources (Chambers, 2002). Shrubs, downed wood, and snags provide important cover from predators, so loss of these habitat elements may negatively impact some small mammal species, but species that prefer open habitats can benefit from food resources provided by fruit-producing shrubs, grasses, and forbs that may establish after fuel treatments. Small mammals seem to recolonize disturbed

---

DID YOU KNOW?

Temporal trends in mammal abundance and diversity are generally similar for both harvested and burned forest stands, with some differences occurring in the initiation stage, 0 to 10 years after the disturbance (Fisher and Wilkinson, 2005).

areas soon after the disturbance, although diversity and species dominance differ as succession progresses. Generalist species typically dominate in early successional stages, and specialist species dominate later (Fisher and Wilkinson, 2005).

Small mammal species vary in habitat preference and thus in response to different types of fuel treatments. In a lodgepole pine–mixed-conifer forest in northeastern Oregon, a commercial thinning designed as a fuel treatment resulted in an increase in chipmunks and a decrease in red-backed voles, red squirrels, and snowshoe hares one year after thinning (Bull and Blumton, 1999). Hares continued to use small patch-cut stands (10-m-wide circular cuts interspersed in unthinned stands comprising 30 to 40% of the stand), but avoided stands thinned to a uniform 4.2-m spacing, which suggests that retaining unthinned patches may help maintain hare populations in fuel treatment areas until understory vegetation recovers (Ausband and Baty, 2005; Shick, 2003; Sullivan and Sullivan, 1988).

DID YOU KNOW?

Forest floor litter is important to some small mammals, especially in westside Oregon forests. Fuel reduction treatments, primarily fire, are likely to reduce the presence or depth of litter.

Small mammals that prefer high canopy closure may be adversely affected by thinning. Thinning will likely have a drying effect on high-canopy, high-density stands of grand fir, potentially having a negative effect on northern flying squirrel populations. Northern flying squirrel abundance in dry Douglas fir–ponderosa pine–western larch forests in northeastern Oregon decreased 1 to 2 years after a thinning treatment (Bull et al., 2004), possibly due to habitat changes and decreases in truffles, the primary food of northern flying squirrels and other small mammals (Lehmkuhl et al., 2004). Thinned and burned stands are likely to be poor bushy-tailed woodrat habitat in eastern Washington dry forests due to woodrats' association with abundant large snags, mistletoe brooms, and soft log cover (Lehmkuhl et al., 2006).

In a mixed-conifer forest of the southern Sierra Nevada of California, Meyer et al. (2005) examined the short-term impacts of prescribed burning (no burn and burn), mechanical thinning (no thin, light thin, and heavy thin), and combinations of these treatments on production of truffles and their consumption by lodgepole chipmunks. Truffle frequency, biomass, and species richness were lower in thinned or burned plots than controls, as was frequency and richness of truffles in the diet of lodgepole chipmunks. These impacts also were lower in heavily thinned and thinned and burned plots than in those that were only burned. These results suggest that either thinning or burning can reduce short-term truffle production and consumption, and potentially the dispersal of ectomycorrhizal spores by small mammals. Moreover, truffles decreased with treatment intensity, suggesting that heavy thinning and higher burn intensity, particularly when applied together, can significantly affect short-term truffle abundance and small mammal consumption. Impacts to truffles will also impact northern flying squirrels.

DID YOU KNOW?

Small mammals are most abundant immediately after a disturbance and decrease as the stand ages. Lynxes and hares utilize mid-successional stands but are rarely found in young or old stands (Fisher and Wilkinson, 2005).

Converse et al. (2006) examined short-term patterns in small-mammal responses to mechanical thinning, prescribed fire, and mechanical thinning/prescribed fire combination treatments at eight different study areas across the United States by modeling taxa-specific densities and total small-mammal biomass as functions of treatment types and study area effects. Small-mammal taxa examined included deer mice, golden-mantled ground squirrels, and all *Peromyscus* (deer mouse) and *Tamias* (chipmunk) species. The top-ranked model of total small-mammal biomass was one with biomass varying only with treatment (i.e., treated vs. untreated), not by treatment type or study area. Individual species and taxa appear to have variable responses to the various types of fuel treatment in different areas: however, total small-mammal biomass appears generally to increase after any type of fuel reduction. Converse et al. (2006) found that predicting responses of a particular small-mammal species to treatment is difficult given current information, but they indicated that it is reasonable to expect total small-mammal biomass to increase with thinning, prescribed fire, or a combination of thinning and prescribed fire treatments. Adaptive management policies may be necessary to help reduce uncertainty as to which treatments are locally optimal for meeting management objectives for small-mammal populations.

Edge and Manning (2002) reported preliminary results of a study that examined the effects on small mammals of three types of treatments for managing fuel loadings: (1) piling, (2) piling and burning, and (3) lopping and scattering. Small mammals were sampled before (1999) and after (2000) treatments. Fifteen small-mammal species were captured during the 2000 sampling. The deer mouse and western red-backed vole were most abundant and were used for analysis. The authors found no evidence that the mean survival of either species was affected by fuel treatments, but they suspected that intense solar radiation reaching the forest floor following thinning resulted in severe environmental conditions that masked treatment effects on small mammal survival. Effects of thinning on the small-mammal community are indicated by the approximately 10% decline in mean total number of small mammals captured across all treatments. Results are preliminary, but suggest interactions among fuel treatments, regional climate, and opening of the forest canopy following thinning. Slash management may affect small mammals differently elsewhere in the Pacific Northwest, where solar radiation is less extreme (Edge and Manning, 2002).

Conservation of small mammals requires knowledge of ecologically meaningful spatial scales (e.g., individual or populations) at which species respond to habitat heterogeneity. Between July and October of 1998, Manning and Edge (2004) sampled small mammals, understory vegetation, and downed wood at multiple scales (trap sites, 1-hectare forest patches, and stands) in two managed Douglas fir forests in western Oregon. The objectives were to determine if downed wood or understory vegetation varied among or within forest patches or among forest stands and whether variation in survival of small mammals coincided with the scale at which these varied. Understory vegetation explained most variation within patches but did not vary among patches or stands. Survival of the two most abundant species, the deer mouse and creeping vole, also varied within patches by differing among individual home ranges and was most related to downed wood volume (cubic meters per 0.01 hectare) and herb and grass cover. Survival of deer mice was explained by a function of downed wood within individual home ranges. Survival of creeping voles was dependent on a function of downed wood with home ranges and was highest in home ranges lacking downed wood. Results demonstrate that these species may not be generalists, as previously suspected, but rather specialists tied to specific amounts of particular habitat components within home ranges.

Because of fires and intensive logging practices, young forests dominate much of the Pacific Northwest landscape. Most young stands were reforested with Douglas fir at high densities. Researchers have proposed thinning these stands as a means to improve habitats for vertebrates, but the effects of thinning intensity on forest-floor small mammals are not well understood. From 1994 to 1996, Suzuki and Hayes (2003) conducted experimental and retrospective studies using pitfall trapping to assess effects of thinning intensity on abundance and reproduction of small mammals in Oregon Coast Range Douglas fir forests. They investigated short-term effects of thinning stands to moderate and low tree densities on small mammals during the first 2 years following thinning and potential long-term effects of thinning by comparing relative abundance and reproductive performance of small mammals in previously thinned (7 to 24 years prior to the study) and unthinned stands. Among 12 small mammal species examined in the experiment, the number of captures increased for four and decreased for one within 2 years of thinning. Responses were similar between moderately and heavily thinned stands. Among nine species examined in the retrospective study, the number of captures was greater for five and lower for none in previously thinned compared to unthinned stands. Furthermore, the total number of small mammals captured was higher in previously thinned than in unthinned stands. The effects of thinning on two species—creeping voles and Pacific jumping mice—were consistent in the short and long term. Captures for both species increased in the first 2 years following thinning and were greater in stands thinned 7 to 24 years previously than unthinned stands. Western red-backed vole captures decreased within 2 years of thinning but were similar in stands thinned 7 to 24 years previously and in unthinned stands. The reproductive performance of deer mice and creeping voles improved following thinning in the short term. In the retrospective study, the reproductive performance of western red-backed voles was higher in thinned than in unthinned stands. Overall, thinning did not have substantial detrimental effects on any species investigated and had positive effects on several. Suzuki and Hayes (2003) suggested that thinning is a viable option to enhance habitat quality for several species of forest-floor small mammals in densely stocked, young Douglas fir stands.

DID YOU KNOW?

The Oregon Coast Range is a mountain range in the Pacific Coast Ranges physiographic region along the Pacific Ocean. The range runs north and south and extends over 200 miles (321.9 km) from the Columbia River in the north, at the border of Oregon and Washington, to the middle fork of the Coquille River in the south. It is 30 to 60 miles (50 to 100 km) wide and averages around 1500 feet (457.2 meters) in elevation above sea level.

Treatment Effects on Bats

Bats inhabit stands with adequate roosting habitat (large trees and snags) and plentiful food (flying insects). Few data exist on the direct effects of fuel treatments on bats, but some inferences can be made based on their known habitats. Long-legged myotis, silver-haired bats, and other bat species roost under the bark of tall, large-diameter trees or in cavities of large snags (Betts, 1998; Ormsbee and McComb, 1998; Rabe et al., 1998; Vonhof and Barclay, 1996). In South Dakota, silver-haired bats roosted primarily under loose bark, in tree crevices, and in woodpecker cavities in ponderosa pine snags averaging about

---

DID YOU KNOW?

The long-legged myotis is most common in woodland and forest habitats above 1200 m (4000 ft). The long-legged myotis feeds on flying insects, primarily moths. Flight of this species is strong and direct. It feeds at fairly low heights (3 to 5 meters; 10 to 15 feet) over water, close to trees and cliffs, and in openings in woodlands and forests (Black 1974; Whitaker et al., 1977, 1981).

---

39 cm DBH (Mattson et al., 1996). These stands averaged 21 snags per hectare. As long as large snags and trees are protected, thinning may have minimal or even positive effects on bat populations depending on initial site conditions and history of land use (Boyles and Aubrey, 2006; Patriquin and Barclay, 2003; Schmidt, 2003) whereas loss of these habitat features may be detrimental (Chambers et al., 2002).

In the Oregon Coast Range, bat activity in Douglas fir stands was highest in old-growth stands, lowest in unthinned second-growth stands (50 to 100 years old), and intermediate in thinned second-growth stands (Humes et al., 1999). Elsewhere in Oregon, old-growth Douglas fir–western hemlock, ponderosa pine–sugar pine, and true fir stands were also preferred by hoary and silver-haired bats compared to younger stands (Perkins and Cross, 1988). Bat activity was higher in old growth stands of Douglas fir than in mature and young stands in the southern Washington Cascades and Oregon Coast Range (Thomas, 1988). This preference for old growth might be explained by the high density of large-diameter snags used as roosts; open spaces among the trees, which allow bats to gain flight when leaving their roosts; or other factors. Whether or not bats also prefer old-growth stands in the interior dry coniferous forests of the west remains unknown.

Loeb and Waldrop (2008) tested the effects of thinning and burning on bat foraging and commuting activity in pine stands in South Carolina's Clemson Experimental Forest. They also tested whether vertical use of stands varied with treatment and whether activity of three common species varied among treatments. Twelve stands dominated by loblolly and shortleaf pine at least 14 ha in size were selected. Three replicates of four treatments were installed: (1) prescribed burn, (2) thin only, (3) thin and burn, and (4) no treatment (control). Bat activity was sampled in 2001 and 2002. Big brown bats, eastern red bats, and eastern pipistrelles were the most frequently recorded species. In 2001, overall activity was significantly greater in thin-only stands compared to control stands; activity in the prescribed burn and thin and burn stands was intermediate. Activity was also greater in treated stands than in control stands in 2002, but the difference was not statistically significant. None of the treatments affected vertical use of the stands. Activity of big brown bats and red bats was significantly higher in thinned stands than in control or burned stands, but activity of the genus of pipistrelles bats did not vary among treatments. Results suggest that treatments that reduce clutter, particularly thinning, increase the suitability for pine stands for bats' foraging and commuting activity in managed pine forests in the South.

DID YOU KNOW?

Bats, arboreal sciurids (rodents such as squirrels), and mustelids (carnivores such as weasels) increase in abundance with stand age and are most abundant in old growth (Fisher and Wilkinson, 2005).

Treatment Effects on Birds in General

Bruns (1960) summarized the role of birds in the forests:

> Within the community of all animals and plants of the forest, birds form an important factor. The birds generally are not able to break down an insect plague, but their function lies in preventing insect plagues. It is our duty to preserve birds for esthetic as well as economic reasons ... where nesting chances are diminished by forestry work. ... It is our duty to further these biological forces and to conserve or create a rich and diverse community. By such a prophylactic ... the forests will be better protected than by any other means.

Factors that influence bird species richness and diversity in a stand include the structure and composition of living and dead vegetation. In western dry coniferous forests, bird community composition depends on the diversity of habitats available, proximity to water, fire history, and silvicultural legacy (Brawn and Balda, 1988; Finch et al., 1997; Hejl et al., 2002). As with other wildlife, the bird species most likely to be affected by fuel treatments are those whose nesting and foraging habitats are associated with the fuels being removed or created and species that either prefer or avoid disturbed areas. Because of their great mobility as adults, population-level responses of birds to fuel treatments are strongly influenced by their distribution and abundance in the surrounding landscape, but pretreatment conditions will also influence response. Reported effects of fuel treatments on birds are somewhat inconsistent. Stand-scale effects may differ from those at the landscape or regional scale (Kotliar et al., 2002), but it is possible to make some general predictions. Thinning conducted during the nesting season is more likely to result in a high mortality of nestlings, especially for species that nest on the ground or in shrubs and small trees (Smith, 2000). Bird species that nest on the ground or in shrubs are likely to increase in abundance after fuel reduction (DeGraaf et al., 1991; Hagar et al., 1996; Provencher et al., 2002; Simon et al., 2000; Wilson et al., 1995). If conducted prior to nesting season, fuel treatments are likely to reduce nestling habitat for ground- and shrub-nesting species, but shrubs and ground cover lost during treatment will likely recover within a few years (Pilliod et al., 2006).

Treatment Effects on Raptors

Raptors are birds of prey that hunt for food primarily on the wing, using their keen senses, especially vision; they include hawks, eagles, buzzards, harriers, kites, and old-world vultures. More open forest understories may benefit hawks, owls, and eagles that prey on small mammals and birds in open forests and small clearings because prey species that have less cover are more easily captured (Pilliod et al., 2006); however, some raptor species and some small mammal and avian prey prefer closed canopy forests and thus avoid stands that have been treated to reduce fuels. Any fuel reduction treatments that reduce the density of pole-sized to mature trees is likely to negatively impact accipiter hawks (i.e., a group of birds of prey in the family Accipitridae, many of which are named as goshawks and sparrowhawks). All three species, especially sharp-shinned and Cooper's hawks, are closely associated with very dense stands (Moore and Henny, 1983; Reynolds and Wight, 1978; Reynolds et al., 1992;).

Removal of trees with dwarf mistletoe brooms during thinning will likely be detrimental to raptor species that nest in the brooms, including the great gray owl, long-eared owl, great horned owl, northern spotted owl, northern goshawk, Cooper's hawk, and red-tailed hawk (Bull et al., 1997). Northern goshawks prefer closed canopy forests of large-diameter trees with relatively open understories (Reich et al., 2004). Management recommendations for sustaining northern goshawk habitat and prey include thinning from below to achieve non-uniform spacing of large trees, a maximum of 30 to 50% canopy opening, and various slash treatments (Reynolds et al., 1992; Squires and Reynolds, 1997).

The endangered Mexican spotted owl and northern spotted owl both occur in dry forest environments. Management guidelines for both species specify little active management in defined areas. By reducing the risk of stand-replacing fires in these defined areas, restoration treatments outside of these areas should benefit spotted owls over time (Beier and Maschinski, 2003). Variable-density thinning may accelerate development of northern spotted owl habitate and dense prey populations (Carey, 2001; Carey and Wilson, 2001; Carey et al., 1999a,b; Muir et al., 2002), especially when snags, capacity trees, and large downed wood are conserved (Bunnell et al., 1999; Carey, 2002; Carey et al., 1999a,b), although this treatment may be more appropriate in mixed-conifer stands with mixed-severity fire regimes than in drier pine, low-intensity fire regimes (Lehmkuhl et al., 2006). Fuel treatments may reduce northern spotted owl habitat quality, but this should be weighed against the risk of stand-replacing fires and complete loss of habitat over large areas (Everett et al., 1997).

DID YOU KNOW?

Most of us are familiar with Christmas mistletoe and perhaps have experienced a sweet holiday smooch or two under the mistletoe, but mistletoe is important in other vital ways, as it provides essential food, cover, and nesting sites for an amazing number of critters in the United States and elsewhere. The thing that all mistletoes have in common is this: They all grow as parasites on the branches of trees and shrubs. In fact, the scientific name for American mistletoe, *Phoradendron*, is Greek for "thief of the tree." The plant is aptly named; it begins its life as a handily sticky seed that often hitchhikes to a new host tree on a bird beak or feather or on mammal fur. The plant's common name—mistletoe—is derived from early observations that the plant would often appear in places where birds had left their droppings. *Mistel* is the Anglo-Saxon word for "dung," and *tan* is the word for "twig" (*misteltan* was the Old English word for the plant). Thus, mistletoe means "dung-on-a-twig." Rule of thumb: A high abundance of dwarf mistletoe in a forest means that many kinds and numbers of birds inhabit that forest (USGS, 2009).

Treatment Effects on Cavity-Nesting Birds

Primary cavity nesters excavate nest cavities in large trees and large snags. These cavities also provide nesting, roosting, and shelter habitat for secondary cavity-nesting birds and some mammals. Preference for large snags for nesting has been well documented for the pileated woodpecker, hairy woodpecker, northern flicker, and William's sapsucker in dry forests in northeastern Oregon (Bull, 1986), woodpeckers in ponderosa pine forests in central Oregon (Bate, 1995; Dixon, 1995), and cavity nesters in ponderosa pine forests after wildfires in southwestern Idaho (Saab and Dudley, 1998; Saab et al., 2002) and northern Arizona (Chambers and Mast, 2005).

If fuel treatments remove snags, a loss of nesting habitat for primary and secondary cavity-nesting birds might be expected for many years. Most studies report that thinning or thinning and burning fuel treatments result in reduced populations of cavity nesters due to loss of dead trees used for nesting and roosting. Preliminary results of research in southeastern British Columbia show that thinning in dry Douglas fir and ponderosa pine forests resulted in a decline in snag densities, cavity-nesting bird densities, and species richness of cavity nesters in the first two breeding season after thinning (Machmer, 2002). Treatments that do not remove large snags may improve ponderosa pine habitat for breeding Lewis's woodpeckers by decreasing canopy cover, as open canopies are preferred habitat for this species (Saab and Vierling, 2001).

When we think about a forest or a particular forest stand we generally picture live trees and understory growth in our minds, but it is important to note that in addition to live trees providing food, homes, and special-use sites for many animal species, use of the trees continues after they die. Dead trees provide different types of wildlife homes and food than when they are alive and healthy, and they serve an important ecological function in forests (Franklin and Forman, 1987). Recognition that snags or dying trees are important components of the forest is not new (Patton, 2011). In light of recognizing the importance of snags, it is also important to recognize that maintaining a minimum number of snags per unit area in a ponderosa pine or Douglas fir stand may be insufficient given the ephemeral nature of snags, which continually change in height, decay state, and presence of sapwood decay organism (Parks et al., 1999; Zack et al., 2002).

Research in dry forests in Idaho suggests that most cavities are only used once by seven of the eight woodpecker species that were studied (Saab et al., 2004), so a continual supply of snags is necessary to maintain new nesting sites over time. Responses of secondary cavity nesters to fuel treatments in ponderosa pine and Douglas fir forests will likely follow a pattern similar to primary cavity nesters—some positive, some negative, and some neutral (Medin and Booth, 1989). Clutch size and number of nestlings of the western bluebird, a secondary cavity nester, did not differ between control stands and stands thinned and burned in ponderosa pine forests in Arizona (Germaine and Germaine, 2002). Blue birds had a higher probability of successfully fledging young in treated stands but with greater risk of parasitic infestations, the ultimate effects of which on post-fledging survival are unknown.

In northeastern Oregon, 65% of pileated woodpecker foraging occurs in downed wood and snags. Bull et al. (2005b) compared the abundance of logs, snags, stumps, and woodpecker foraging in mixed-conifer (ponderosa pine–Douglas fir–grand fir) stands that had undergone (1) prescribed burning after mechanical fuel treatment, (2) mechanical fuel treatment without prescribed burning, or (3) no treatment. Pileated woodpecker foraging was significantly higher in untreated or mechanically treated stands. Ants, the primary prey of pileated woodpeckers, were also significantly more abundant in these stands. Pileated woodpecker foraging in mechanical removal treatments was less common than in control treatments but significantly higher than in prescribed burn treatments. Prescribed burning did not allow the same degree of control in retaining coarse woody debris as mechanical fuel treatment.

In 2004 and 2005, Lyons et al. (2008) examined the short-term response of cavity-nesting birds in dry conifer forests of Washington to fuel treatment to determine if mechanical thinning or prescribed burning or a combination of the two would alter foraging tree selection. Results suggested that fuel treatments had a positive impact. Chickadees selected for large-diameter, live Douglas fir trees in treated stands; nuthatches selected for large-diameter, ponderosa pine in treated stands; and woodpeckers selected for large-diameter, ponderosa pine snags in thinned and thinned/burned stands. Birds were more likely to forage in treated stands, an outcome that was strongest in stands that received both thinning and burning. Bird groups selected trees at least 1.6 times as large in diameter in treated stands compared to control stands. Thinning and burning may enhance foraging habitat for bark-gleaning species as a whole. Results support treatment design considerations including small tree removal. Canopy opening increases herbaceous and bare ground cover, resulting in improved invertebrate assemblages and thus improved forage abundance, in addition to promoting the retention of large trees and snags (>40 cm DBH) that provide important foraging substrate and nesting habitat.

McIver et al. (2003) examined fuel reduction by mechanical thinning and removal in mixed-conifer stands in northeastern Oregon in an experiment that compared a single-grip harvester coupled with either a forwarder or a skyline yarding system and unharvested control sites. Both extraction systems achieved nearly equivalent (~46%) mass fuel reductions. Mean log length was lower in harvested units compared to unharvested controls, but this did not decrease occupation of logs by ants or the activities of woodpeckers feeding on them.

Treatment Effects on Other Birds

Hayes et al. (2003) experimentally manipulated stands to evaluate influences of two thinning intensities on populations of diurnal breeding birds in western Oregon. They conducted point counts of birds seven times each year in 1994 (before treatment) and from 1995 to 2000 (after treatment). Of the 22 species for which they obtained adequate data, nine species decreased and eight species increased in thinned stands relative to controls. There was no strong evidence that thinning influenced numbers of five species. Of the 17 species that responded to thinning, the magnitude of response of eight species varied with thinning intensity. For each of these species, response was greatest in more heavily thinned stands. No species was extirpated from stands following thinning but detections of Hutton's vireos, golden-crowned kinglets, brown creepers, black-throated gray warblers, and varied thrushes decreased to less than half of detections in controls in one or more treatment types. This evidence suggests that thinning may significantly reduce the numbers of detections for the species that were studied.

In contrast, American robins, Townsend's solitaires, and Hammonds's flycatchers were rare or absent in controls but regularly present in thinned stands, and detections of western tanagers, evening grosbeaks, and hairy woodpeckers increased by threefold or more in thinned stands relative to controls. Only Pacific-slope flycatchers, warbling vireos, and western tanagers showed strong evidence of temporal trends in response. For these species, differences between numbers in controls and treated stands became more extreme through time. Findings suggest that thinning densely stocked conifers in landscapes dominated by younger stands enhances habitat suitability for several species of birds, but that some unthinned patches and stands should be retained to provide refugia for species that are impacted by thinning.

Hagar et al. (2004) studied forest songbird response to three different intensities and patterns of thinning in 40-year-old stands dominated by Douglas fir in the Oregon Cascades. They compared changes in songbird density 2 years before and 4 years after experimental thinning, between each thinning treatment and the control. Species richness and density of ten species increased following thinning. Four additional species were detected with higher frequency in thinned stands. The density of five species decreased following thinning but no species disappeared. Commercial thinning rapidly promotes diversity of breeding songbirds in young, conifer-dominated stands, but Hagar et al. (2004) suggest using a variety of thinning intensities and patterns, from no thinning to very widely spaced residual trees, to maximize avian diversity at the landscape scale and structural diversity both within and among stands.

In another study, Hagar et al. (1996) compared the abundance and diversity of breeding and winter birds in commercially thinned and unthinned 40- to 55-year-old Douglas fir stands in the Oregon Coast Range. Breeding bird abundance was greater in thinned stands. Bird species richness was correlated with habitat patchiness and densities of hardwoods, snags, and conifers. During the breeding season, Hammond's flycatchers, hairy woodpeckers, red-breasted nuthatches, dark-eyed juncos, warbling vireos, and evening grosbeaks were more abundant in thinned than unthinned stands. Pacific-slope flycatchers were more abundant in unthinned stands. Golden-crowned kinglets, gray jays, and black-throated gray warblers were more abundant in unthinned than thinned stands, but these patterns were inconsistent between seasons, regions, or years. Stand-scale habitat features were associated with the abundance of 18 bird species.

Siegel and DeSante (2003) compared avian community composition and nesting success in thinned and unthinned stands of commercially managed Sierran mixed-conifer forest. They conducted point counts and monitored 537 active nests of 37 species on 10 study plots during three consecutive breeding seasons in the northern Sierra Nevada. All 10 study plots had a similar long-term management history that included fire suppression and single-tree selection logging, but five plots also underwent combined commercial and biomass thinning 5 to 8 years prior to the study. Pooling species by nest substrate, Siegel and DeSante (2003) found that detections of ground-nesting bird species were similar on thinned and unthinned plots, but they detected canopy-, cavity-, and especially shrub-nesting species much more frequently on the thinned plots. Nest success rates were not statistically different between thinned and unthinned plots for ground-, shrub-, canopy-, or cavity-nesting species. Thinned stands were characterized by significantly less canopy cover, significantly lower density of small and medium conifers, and significantly greater understory cover and deer brush cover than the unthinned stands. Siegel and DeSante (2003) surmised that thinning stimulated vigorous shrub growth and concluded that forest conditions associated with a relatively open-canopy and a well-developed shrub understory are highly beneficial to numerous breeding bird species in the Sierran mixed-conifer community, including many species that may not nest or forage in the understory. Forest thinning that promotes vigorous shrub growth may correlate with increased abundance of nesting birds, at least in stands affected by historical fire suppression and single-tree selection logging.

Treatment Effects on Reptiles

Reptile diversity is generally lower in forests than in deserts, grasslands, and chaparral. Few liquid and snake species occupy western closed canopy coniferous forests except for perhaps the rubber boa, but reptile species do inhabit specific forest patches, such as wetlands, meadows, and rock outcroppings that provide shelter, microclimates, and prey (Heatwole, 1977; Lillywhite et al., 1977). Examples include the western fence lizard (also known as the Blue Belly), eastern fence lizard, sagebrush lizard, ornate tree lizard, western skink, northern alligator lizard, garter snake, mountain king snake, racer snake, gopher snake, and western rattlesnake. Very little is known about the effect of thinning on reptiles, and Pilliod et al. (2006) found no studies specific to dry coniferous forests. Reptile species that prefer forest floor cover might decrease when such habitat is lost, but most reptile species would probably benefit from a reduction in shrubs, ground vegetation, and litter cover and depth (Germaine and Germaine, 2002; Knox et al., 2001; Mushinsky, 1985; Singh et al., 2002). Many lizard species prefer snags and downed wood over live trees (James and M'Closkey, 2003), so fuel treatments that leave snags and downed wood onsite may improve habitat for these species.

DID YOU KNOW?

The rubber boa snake is the northernmost relative of the giant boa constrictor and anaconda of South America. Like those snakes, it kills prey by squeezing it to death with its muscular body and swallows the meal whole. It gets its name from its rubbery appearance, the result of small, smooth scales and somewhat loose skin. Uniformly colored on its back, the adult ranges from tan to brown to olive green. Rubber boas are one of the smallest of the boa family, reaching 14 to 28 inches long (Curtis, 2006).

Treatment Effects on Amphibians

Most forest amphibians use upland habitats at various times during the year, depending primarily on the availability of moist duff and litter and rotting downed wood. Unlike reptiles, amphibians' response to reduced canopy cover will likely be negative due to warmer, drier conditions created in the understory vegetation, downed wood, litter, and soil (McGraw, 1997; Meyer et al., 2001; Pough et al., 1987). Most terrestrial salamanders require moist soils or decomposing wood to maintain water balance; therefore, dry conditions usually result in suppressed populations (Bury and Corn, 1988; DeMaynadier and Hunter, 1995). Frogs and toads may be less affected by environmental changes associated with fuel treatments because they tend to travel at night and during rain events, their greater mobility compared to salamanders, and their close association with wetlands (Constible et al., 2001; Pilliod et al., 2003). Still, species that frequently occupy terrestrial habitats, including many salamanders, boreal toads, and tree forms, may be killed during fuel treatments or find post-treatment conditions unsuitable (Pilliod et al., 2003).

Pilliod et al. (2006) found only one study that addressed the effects of thinning on amphibians in dry coniferous forests. In a western Montana lodgepole pine forest, 70% fewer long-toed salamanders were captured in selectively harvested stands compared to unharvested stands, a difference that was attributed to a reduction in overstory canopy and subsequent temperature increases (Naughton et al., 2000). The importance of overstory canopy has been demonstrated for other amphibian species in the central Oregon Coast Range (Martin and McComb, 2003).

DID YOU KNOW?

Associations of the salamander with fire appear in the writings of da Vinci, Pliny, Aristotle, Ray Bradbury, and the Talmud. Legends about salamanders no doubt arose from the tendency of many salamanders to dwell inside rotting logs. When such logs were thrown onto a fire, the salamanders attempted to escape from the log, contributing to the belief that salamanders were created from flames—a belief that gave the creature its name, as *salamander* is Greek for "fire lizard" (Ashcroft, 2002).

INVERTEBRATE TROPHIC WEBS

All organisms, alive or dead, are potential sources of food for other organisms. All organisms that share the same general type of food in a food chain are said to be at the same *trophic level* (nourishment or feeding level). Since green plants use sunlight to produce food for animals, they are *producers*, or the first trophic level. Herbivores, which eat plants directly, are *primary consumers*, or the second trophic level. Carnivores are flesh-eating consumers; they include several trophic levels, from the third level on up. At each transfer, a large amount of energy (about 80 to 90%) is lost as heat and wastes. Thus, nature normally limits food chains to four or five links; however, in aquatic ecosystems, "food chains are commonly longer than those on land" (Dasmann 1984). The aquatic food chain is longer because several predatory fish may be feeding on the plant consumers. Even so, the built-in inefficiency of the energy transfer process prevents development of extremely long food chains.

Tomera (1989) described a simple food chain that can be seen in a prairie dog community: "The grass in the community manufactures food. The grass is called a *food producer*. The grass is eaten by a prairie dog. Because the prairie dog lives directly off the grass, it is termed a *first-order consumer*. A weasel may kill and eat the prairie dog. The weasel is, therefore a predator and would be termed a *second-order consumer*. The second-order consumer is twice removed from the green grass. The weasel, in turn, may be eaten by a large hawk or eagle. The bird that kills and eats the weasel would therefore be a *third-order consumer*, three times removed from the grass. Of course, the hawk would give off waste materials and eventually die itself. Wastes and dead organisms are then acted on by decomposers" (p. 50). Only a few simple food chains or trophic webs are found in nature. Thus, when attempting to identify the complex food relationships among many animals and plants within a community, it is useful to create a *food web*. The fact is that most simple food chains are interlocked; this interlocking of food chains forms a food web. A food web can be characterized as a map that shows what eats what (Miller, 1988). Most ecosystems support a complex food web. A food web involves animals that do not feed on one trophic level; for example, humans feed on both plants and animals. The point is that an organism in a food web may occupy one or more trophic levels. Trophic level is determined by an organism's role in its particular community, not by its species. Food chains and webs help to explain how energy moves through an ecosystem.

An important trophic level of the food web that has not been discussed thus far is comprised of the *decomposers* (bacteria, mushrooms, etc.). The decomposers feed on dead plants or animals and play an important role in recycling nutrients in the ecosystem. Note that there is no waste in ecosystems. All organisms, dead or alive, are potential sources of food for other organisms.

Invertebrate trophic webs can be staggeringly complex. Consider an example from the Sierra Nevada (Sugden, 2000). The western sand wasp, which functions as both a predator and pollinator, preys on the big black horse fly, which in turn preys on humans and other large mammals. Via such predation, the Bodega gnat benefits from reduced competition with the big black horse fly for mammalian blood meals. Eggs of the big black horse fly are parasitized by a scelionid wasp.

Larvae of Edwards' cuckoo wasp attack the larvae of the western sand wasp. The cuckoo wasp does not get a free lunch, as it is in turn parasitized by Sacker's velvet ant. This food web could be expanded almost indefinitely. There is a perception that tropical food webs are more complex than those of temperate regions, but this is not true (Schoenly et al., 1991); this relatively small extract of a Sierran web indicates the depth of temperature trophic complexity.

Concerns have been expressed that fuel treatments may contribute fine sediment to streams because of increased surface runoff (Elliot et al., 1999; Robichaud, 2000; Robichaud and Waldrop, 1994). Sedimentation reduces egg and tadpole survivorship of some stream-breeding amphibians that lay eggs and rear tadpoles under rocks or in spaces in stream cobbles (Corn and Bury, 1989; Gillespie, 2002; Spellman, 1996).

Treatment Effects on Invertebrates in General

Why consider conserving insects? ... In short, they make ecosystems tick. ... They cannot be ignored.

—Samways (1994)

Invertebrates are "webmasters" that govern ecosystem function to a degree far out of proportion to their biomass.

—Coleman and Hendrix (2002)

Invertebrates are critical in decomposition, nutrient cycling, physical processes, and disturbance regimes.

—Meffe and Carroll (1997)

Insects are the "glue and building blocks" of ecosystems.

—Janzen (1987)

Insects are important ecosystem regulators via the action of many species as powerful herbivores.

—Schowalter (2000)

Although generally minute, most ecological interactions, functions and flows must pass through various invertebrate taxa.

—Bailey and Whitham (2002)

Fuel treatment effects on terrestrial invertebrates in western dry coniferous forests—insects, spiders, mites, scorpions, centipedes, millipedes, isopods, worms, snails, and slugs—are probably as diverse as the group itself. Invertebrates comprise over half the animal diversity in forests (Niwa et al., 2001), occupy all forested habitats, and have varied functional roles, including detritovores, predators, herbivores, and pollinators. Many invertebrates are specifically associated with habitat elements targeted in fuel treatments (e.g., shrubs, snags, litter, and duff). Most invertebrates occupy distinctly different habitats during their life cycles; for example, some species live below ground when young and above ground as adults or feed on vegetation as immatures and then on flower nectar as adults. Forest invertebrates are short lived and either have small dispersal ranges or are sedentary in one or more life stages, so fuel treatments can potentially affect local populations through direct mortality depending on the season, type, and size of treatment. Some

DID YOU KNOW?

The "stealth" trophic function of invertebrates is decomposition. As an example, millipedes in Big Bend National Park have been found to be key detritivores (Crawford, 1976) that make nutrients available to other organisms (Holmquist, 2004).

invertebrate species can potentially explode in population, but many are scarce and recognized as threatened, endangered, or of conservation concern. Across invertebrate groups and for diverse ecosystems, the provision of refugia (leaving untreated areas from which populations can recolonize) is widely recommended to minimize the effects of direct mortality and accelerated recovery (Pillion et al., 2006).

Relatively few studies have examined fuel treatment effects on invertebrates in dry coniferous forests, a significant limitation in understanding and predicting the response of species in this group—native and nonnative—to such treatments. Threats to forest health posed by outbreaks, particularly of some bark beetles and defoliators, are somewhat unique to invertebrates. Silvicultural practices such as precommercial thinning that improve stand health have long been assumed to reduce the risk and impact of these disturbance agents (Gast et al., 1991), but the use of fuel treatments to achieve the combined goals of fuel and insect management has not been widely applied or studied, with the exception of thinning residue treatments aimed at reducing surface fuels and population build-ups of some bark and woodboring insects (Pilliod et al., 2006).

McIver et al. (2003) examined fuel reduction by mechanical thinning and removal in mixed-conifer stands in northeastern Oregon in an experiment that compared a single-grip harvester coupled with either a forwarder or a skyline yarding system and unharvested control sites. Both extraction systems achieved nearly equivalent (~46%) mass fuel reductions. Light displacement of the soil resulted in a short-term increase in the abundance of soil microarthropods. The effects of compaction on litter microarthropods were more persistent, with lower numbers in compacted litter a year after harvest.

Treatment Effects on Invertebrate Detritivores

Invertebrates of the forest soil and floor are crucial to decomposition and nutrient cycling and include detritivores—earthworms, land snails and slugs, and arthropods, including millipedes, isopods, mites, and springtails—and species active in decompositions of dead downed wood and snags, such as termites, beetles, and ants. Few studies have especially examined the effects of fuel treatments on detritivores and decomposition in western dry coniferous forests. Researchers (e.g., Niwa et al., 2001) have hypothesized that thinning is likely to have significant negative effects, at least in the short term, on invertebrates of the soils and organic layers through soil compaction and disruption or loss of organic layers. Compaction will depend on soil type and thinning treatment.

Soil organisms may be more protected from treatment effects than those in litter layers; for example, long-term (16 to 41 years after treatment) effects of thinning on densities of microinvertebrates (including oribatid, mesotigmid, and protigmatid mites) were found in the litter of late-successional white fir and Douglas fir forests in the southwestern Oregon Cascades. No differences were found in mite *Collembola* densities in the upper 5 cm of soil between thinned and unthinned stands (Peck and Niwa, 2004). Tenebrionid beetles— forest floor scavengers—were found at higher richness and diversity in ponderosa pine stands of northern Arizona 13 to 14 years after fuel treatments (thinning alone and thinning with prescribed burning) compared to untreated stands (Chen et al., 2006). Studies in other ecosystems have shown that fuel treatments can negatively affect soil and litter invertebrates directly through mortality and loss of food and cover, and many have prolonged recovery periods (Hanula and Wade, 2003). Better information is needed on the effects of changes in soil and litter invertebrates on ecosystem functions (e.g., nutrient cycling) (Pilliod et al., 2006).

Treatment Effects on Predatory Invertebrates

Predatory invertebrates are found in all forest habitats. Forest floor predators are frequently studied, especially spiders and carabid beetles. These invertebrates usually recover from fuel treatments more quickly than those in the soil, particularly those with greater mobility and ability to disperse. They may also benefit from greater habitat diversity created by fuel treatments. One study on response to thinning (>16 years after treatment), focused

THE MILLIPEDE AND THE BEETLE*

The remarkable relationship between some invertebrate predators and their prey influences the rate at which litter is decomposed on the forest floor. The polydesmid millipede *Harpaphe hydeniana*, for example, is an important forest detritivore. It and similar millipedes are the major consumers of coniferous and deciduous leaf litter in western coniferous forests, where earthworms are virtually absent. They chew up and digest leaves, producing feces that are a food resource accessible to many other small arthropods, fungi, and bacteria. Five adults can consume more than 95% of a full gallon can of coniferous or deciduous litter in 2-1/2 weeks. *Harpaphe* and other polydesmid millipedes secrete cyanide and are usually conspicuously colored to warn away would-be predators. The defensive reaction of *Harpaphe* is to roll up into a ball to protect its neck. In the immature stage it is unpigmented and probably confined underground to the lower humus layers. When about half grown, it becomes pigmented and moves to the surface, where it feeds on litter. *Promecognathus laevissimus* is an unusual carabid beetle in that it feeds only on polydesmid millipedes, which are often much larger than itself. The feeding behavior of carabid beetles can be summarized as follows: Approaching to within a distance of less that 1 cm, the beetle pounces on its prey, straddling and lining up parallel to the millipede's body. Moving to one end of the body, the predator constantly opens and closes its mandibles around the millipede. If *Promecognathus* chances to reach the posterior end, it quickly turns about-face and moves up to the head. Upon reaching the head, it plunges its long, scimitar-shaped mandibles down around the neck, severing or crushing the ventral nerve cord and paralyzing its prey. The beetle returns to feed on the paralyzed millipede for several days, emptying the body contents and breaking off segment as it works down the body, until only a pile of segment rings remains. It may cache its prey under a shelter between meals.

---

* From Parsons, G.L. et al., *Invertebrates of the H.J. Andrews Experimental Forest, Western Cascade Range, Oregon. V: An Annotated List of Insects and Other Arthropods*, General Technical Report PNW-GTR-290, U.S. Department of Agriculture, Forest Service, Pacific Northwest Research Station, Portland, OR, 1991.

on spiders and carabid beetles in late-successional white fir and Douglas fir forests in the Cascade Mountains of Oregon (Peck and Niwa, 2004). No differences were found in overall abundance or species richness of carabids between thinned and unthinned stands, but some individual species were more or less abundant in thinned stands. Spider abundance and species richness were significantly higher in thinned stands. Generally, hunting spiders were more numerous in thinned stands, whereas sheet-web-building spiders were more numerous in unthinned stands. Site management history and size of treatment area were important factors in moderating changes in diversity and abundance of leaf litter arthropods in mixed-conifer stands in the Sierra Nevada Mountains of California one year after fire and fire surrogate treatments (prescribed burning, overstory thinning with understory mastication, and combined thinning and burning) (Apigian et al., 2006).

In a longer term study in northern Arizona ponderosa pine stands, the ground beetle species assemblage did not vary between thinned (4 to 10 years after treatment) and untreated stands, was richer in thinned stands that were subsequently (3 to 4 years) treated with prescribed fire, and was highest in stands 2 years following wildfire (Villa-Castillo and Wagner, 2002). High diversity in burned stands was attributed to temporary colonization by open-area species, pyrophilous species, and recolonization from refugia. Another study in northern Arizona ponderosa pine stands found consistently higher richness and diversity of carabids more that 13 years following fuel treatments (Chen et al., 2006). Carabids may provide a useful indicator of ponderosa pine stand conditions following fuel treatments or wildfire.

Treatment Effects on Invertebrate Herbivores and Pollinators

Depending on the timing of fuel treatments, invertebrate herbivores and pollinators, such as moths and butterflies, that inhabit and feed on live vegetation during some life stage can be immediately affected through direct mortality or loss of food or cover. Longer term, these invertebrates may benefit from changes in structural diversity resulting from fuel treatments that increase the amount of light reaching foliage and the forest floor. Two years after thinning that removed 50% of basal area in young ponderosa pine in central Oregon, the abundance of pandora moths (which feed as larvae on pine foliage) was not affected, but adult emergence and egg hatch occurred 7 to 10 days earlier in thinned stands, presumably due to increased solar radiation and temperature (Ross, 1995).

In a northern Arizona ponderosa pine–oak forest, adult butterfly abundance and species richness were greater with a combination of thinning and prescribed fire than in paired control sites. Elevated response up to 2 years after treatment corresponded with significantly greater light intensity in the treated site, although larval host and nectar plant richness did not change (Waltz and Covington, 2004). Huntzinger (2003) reported similar results after prescribed fire (1 to >15 years) in late-successional reserve Douglas fir and hardwoods in the Siskiyou Mountains of Oregon and mid- to late-seral pines and firs in western Sierra Nevada in California. Nonnative plant species invasions following prescribed fire and other disturbances could have secondary negative impacts on invertebrate herbivores and pollinators through the loss of host plants.

Treatment Effects on Bark- and Wood-Boring Insects

Species such as root and bark beetles and wood borers that benefit from stress or weakened defenses of living host trees usually respond positively in the short term to disturbance created by thinning and prescribed fire, but treatment timing affects response.

Some bark beetles and wood borers increased significantly in abundance and diversity after prescribed fire only (170% increase), partial harvest and prescribed fire (54% increase), and partial harvest only (45% increase) compared to untreated controls (up to a year after treatment) in interior Douglas fir and ponderosa pine stands in British Columbia (Machmer, 2002). Several other studies (e.g., Apigian et al., 2006; Hanula et al., 2002) have produced similar results, but thinning and prescribed fire treatments are not always examined separately. In prescribed fire studies, attack success has often been higher in injured trees when beetle populations were low (e.g., Elkin and Reid, 2004). At high population levels, injured and uninjured trees are usually equally likely to be successfully attacked. Over a 3-year period in northern Arizona ponderosa pine, there was no evidence of an outbreak of endemic bark beetles regardless of treatment history including untreated, thinned (4 to 11 years before), thinned followed by prescribed fire (3 to 4 years later), or stand replacement wildfire (2 years before) (Sanchez-Martinez and Wagner, 2002). These results are counter to the hypothesis that dense stands should be more susceptible to beetle infestation.

All about Soil*

In this text, not only is the environmental impact of the removal of woody biomass fuels important to our discussion but we must also consider and discuss the impact of whole-tree harvesting on soil productivity. Thus, to gain a better understanding of the effects of woody biomass fuel and whole-tree harvesting on forest soil productivity, we must have a solid understanding of soils basics. Weekend gardeners tend to think of soil as the first few inches of the surface of the Earth—the thin layer that has to be weeded and that provides a firm foundation for plants. But, the soil actually extends from the surface down to the hard rocky crust of the Earth. Soil represents a zone of transition, and, as in many of nature's transition zones, the soil is the site of important chemical and physical processes. In addition, because plants need soil to grow, it is arguably the most valuable of all the mineral resources on Earth (Beazley, 1992). Before we begin a journey that takes us through the territory that is soil and examine soil from micro to macro levels, we need to stop for a moment and discuss why, beyond the obvious reason, soil is so important to us, to our environment, to our very survival. Is soil really that big a deal? Is it that important? Do we need to even think about soil? Yes, yes, and yes. Soil is all of these things and more.

Functions of Soil

We normally relate soil to our backyards, to farms, to forests, or to a regional watershed. We think of soil as the substance upon which plants grow. Soils play other roles, though. They have six main functions important to us: (1) soil is a medium for plant growth, (2) soils regulate our water supplies, (3) soils are recyclers of raw materials, (4) soils provide a habitat for organisms, (5) soils are used as an engineering medium, and (6) soils provide materials. Let's take a closer look at each of these functions of soils.

---

* This section is based on material presented in Spellman, F.R., *The Science of Environmental Pollution*, 2nd ed., CRC Press, Boca Raton, FL, 2008.

Plant Growth Medium

We are all aware of the primary function of soil: Soil serves as a plant growth medium, a function that becomes more important with each passing day as Earth's population continues to grow. Although it is true that soil is a medium for plant growth, soil itself is actually alive as well. Soil exists in paradox: We depend on soil for life, and at the same time soil depends on life. Its very origin, its maintenance, and its true nature are intimately tied to living plants and animals. What does this mean? Let's take a look at how one renowned environmental writer, whose elegant prose brought this point to the forefront, explained this paradox:

> The soil community ... consists of a web of interwoven lives, each in some way related to the others—the living creatures depending on the soil, but the soil in turn is a vital element of the earth only so long as this community within it flourishes (Carson, 1962, p. 56).

What Rachel Carson said is important—the meaning is clear. The soil might say to us if it could, "Don't kill off the life within me and I will do the best I can to provide life that will help to sustain your life." What we have here is a trade-off—one vitally important to soil and to ourselves. Remember that most of Earth's people are tillers of the soil; the soil is their source of livelihood, and those soil tillers provide food for us all.

As a plant growth medium, soil provides vital resources and performs important functions for the plant. To grow in soil, plants must have water and nutrients. Soil provides these. To grow and to sustain growth, a plant must have a root system. Soil provides pore spaces for roots. To grow and maintain growth, a plant's roots must have oxygen for respiration and carbon dioxide exchange and ultimate diffusion out of the soil. Soil provides the air and pore spaces (the ventilation system) for this. To continue to grow, a plant must have support. Soil provides this support.

If a plant seed is planted in a soil and is exposed to the proper amount of sunlight for growth to occur, the soil still must provide nutrients through a root system that has space to grow, a continuous stream of water (about 500 g of water are required to produce 1 g of dry plant material) for root nutrient transport and plant cooling, and a pathway for both oxygen and carbon dioxide transfer. Just as important, soil water provides the plant with the normal fullness or tension (turgor) it needs to stand—the structural support it needs to face the sun for photosynthesis to occur.

As well as the functions stated above, soil is also an important moderator of temperature fluctuations. If you have ever dug in a garden on a hot summer day, you probably noticed that the soil was warmer (even hot) on the surface but much cooler just a few inches below the surface.

Regulator of Water Supplies

When we walk on land, few of us probably realize that we are actually walking across a bridge. This bridge (in many areas) transports us across a veritable ocean of water below us, deep—or not so deep—under the surface of the Earth. Consider what happens to rain. Where does the rain water go? Some, falling directly over water bodies, become part of the water body again, but an enormous amount falls on land. Some of the water, obviously, runs off—always following the path of least resistance. In modern communities, stormwater runoff is a hot topic. Cities have taken giant steps to try to control runoff, to send the runoff where it can be properly handled to prevent flooding. Let's take a closer look at precipitation and the sinks it pours into, then relate this usually natural operation to soil water. We begin with surface water, then move on to that ocean of water below the soil's surface: groundwater.

Surface water (water on the Earth's surface as opposed to subsurface water, or groundwater) is mostly a product of precipitation: rain, snow, sleet, or hail. Surface water is exposed or open to the atmosphere and results from overland flow, the movement of water on and just under the Earth's surface. This overland flow is the same thing as surface runoff, which is the amount of rainfall that passes over the Earth's surface. Specific sources of surface water include rivers, streams, lakes, impoundments, shallow wells, rain catchments, tundra ponds, or meskegs (peat bogs).

Most surface water is the result of surface runoff. The amount and flow rate of surface runoff are highly variable. This variability stems from two main factors: (1) human interference (influences) and (2) natural conditions. In some cases, surface water runs quickly off land. Generally, this is undesirable from a water resources standpoint because the water does not have enough time to infiltrate into the ground and recharge groundwater aquifers. Other problems associated with quick surface water runoff are erosion and flooding. Probably the only good thing that can be said about surface water that quickly runs off land is that it does not have enough time (normally) to become contaminated with high mineral content. Surface water running slowly off land may be expected to have all the opposite effects.

Surface water travels over the land to what amounts to a predetermined destination. What factors influence how surface water moves? Surface water's journey over the face of the Earth typically begins at its drainage basin, sometimes referred to as its *drainage area*, *catchment*, or *watershed*. For a groundwater source, this is known as the *recharge area*, the area from which precipitation flows into an underground water source. The area of a surface water drainage basin is usually measured in square miles, acres, or sections. If a city takes water from a surface water source, essential information for the assessment of water quality includes how large the drainage basin is and what lies within it.

We know that water does not run uphill; instead, surface water runoff (like the flow of electricity) follows along the path of least resistance. Generally speaking, water within a drainage basin will naturally be shunted by the geological formation of the area toward one primary watercourse (a river, stream, creek, brook) unless some manmade distribution system diverts the flow. Various factors directly influence the surface water's flow over land:

- *Rainfall duration*—Length of the rainstorm affects the amount of runoff. Even a light, gentle rain will eventually saturate the soil if it lasts long enough. When the saturated soil can absorb no more water, rainfall builds up on the surface and begins to flow as runoff.
- *Rainfall intensity*—The harder and faster it rains, the more quickly soil becomes saturated. With hard rains, the surface inches of soil quickly become inundated; with short, hard storms, most of the rainfall may end up as surface runoff because the moisture is carried away before significant amounts of water are absorbed into the ground.
- *Soil moisture*—Obviously, if the soil is already laden with water from previous rains, the saturation point will be reached sooner than if the soil were dry. Frozen soil also inhibits water absorption; up to 100% of snow melt or rainfall on frozen soil will end up as runoff because frozen ground is impervious.

- *Soil composition*—Runoff amount is directly affected by soil composition. Hard rock surfaces will shed all rainfall, obviously, but so will soils with heavy clay composition. Clay soils possess small void spaces that swell when wet. When the void spaces close, they form a barrier that does not allow additional absorption or infiltration. On the opposite end of the spectrum, coarse sand allows easy water flowthrough, even in a torrential downpour.

- *Vegetation cover*—Runoff is limited by ground cover. Roots of vegetation and pine needles, pine cones, leaves, and branches create a porous layer (sheet of decaying natural organic substances) above the soil. This porous organic sheet (ground cover) readily allows water into the soil. Vegetation and organic waste also act as a cover to protect the soil from hard, driving rains. Hard rains can compact bare soils, close off void spaces, and increase runoff. Vegetation and ground cover work to maintain the infiltration and water-holding capacity of the soil. Note that vegetation and ground cover also reduce evaporation of soil moisture as well.

- *Ground slope*—Flat-land water flow is usually so slow that large amounts of rainfall can infiltrate the ground. Gravity works against infiltration on steeply sloping ground, where up to 80% of rainfall may become surface runoff.

- *Human influences*—Various human activities have a definite impact on surface water runoff. Most human activities tend to increase the rate of water flow; for example, canals and ditches are usually constructed to provide steady flow, and agricultural activities generally remove ground cover that would work to retard the runoff rate. On the opposite extreme, manmade dams are generally built to retard the flow of runoff.

Human habitations, with their paved streets, tarmac, paved parking lots, and buildings create surface runoff potential, because so many surfaces are impervious to infiltration. All these surfaces hasten the flow of water, and they also increase the possibility of flooding, often with devastating results. Because of urban increases in runoff, a whole new industry has developed: stormwater management. Paving over natural surface acreage has another serious side effect. Without enough area available for water to infiltrate the ground and percolate through the soil to eventually reach and replenish (recharge) groundwater sources, those sources may eventually fail, with devastating impact on the local water supply.

Now let's shift gears and take a look at groundwater. Water falling to the ground as precipitation normally follows three courses. Some runs off directly to rivers and streams, some infiltrates to ground reservoirs, and the rest evaporates or transpires through vegetation. The water in the ground is essentially invisible and may be thought of as a temporary natural reservoir (ASTM, 1969; Spellman, 2008). Almost all groundwater is in constant motion toward rivers or other surface water bodies. Groundwater is defined as water below the Earth's crust but above a depth of 2500 feet; thus, if water is located between the Earth's crust and the 2500 foot level, it is considered usable (potable) freshwater. In the United States, it has been estimated that at least 50% of total available freshwater storage is in underground aquifers (Kemmer, 1979).

In this text, we are concerned with that amount of water retained in the soil to ensure plant life and growth. Recall that earlier we stated that 500 g of water are required to produce 1 g of dry plant material. Note that about 5 g of this water become an integral part of the plant. Unless rainfall is frequent, you don't have to be a rocket scientist to figure out that the ability of soil to hold water against the force of gravity is very important. Thus, one of the vital functions of soil is to regulate the water supply to plants.

Recycler of Raw Materials

Imagine what it would be like to step out into the open air and be hit by a stench that not only would offend your olfactory sense but could almost reach out and grab you. Imagine looking out upon cluttered fields in front of your domicile and seeing nothing but stack upon stack upon stack of the sources of the horrible, putrefied, foul, decaying, gagging, choking, retching stench. We are talking about plant and animal remains and waste, mountains of it, reaching toward the sky and surrounded by colonies of flies of all varieties. "Ugh," you say. Well, if it were not for the power of the soil to recycle waste products, then this scene or something very much like it would be possible. Of course, ultimately this scenario would be impossible because under such conditions there would be no more life to die and stack up.

Soil is a recycler—probably the premier recycler on Earth. The simple fact is that if it were not for the incredible recycling ability of soil, plants and animals would have run out of nourishment long ago. Soil recycles in other ways; for example, consider the geochemical cycles (i.e., the chemical interactions between soil, water, air, and life on Earth) in which soil plays a major role.

Soil possesses the incomparable ability and capacity to assimilate great quantities of organic wastes, turn them into beneficial organic matter (humus), and then convert the nutrients in the wastes to forms that can be utilized by plants and animals. In turn, the soil returns carbon to the atmosphere as carbon dioxide, where it again will eventually become part of living organisms through photosynthesis. Soil performs several different recycling functions—most of them good, some of them not so good. Not so good? Yes. Consider one recycling function of soil that may not be so good. Soils have the capacity to accumulate large amounts of carbon as soil organic matter which can have a major impact on global changes such as the greenhouse effect.

Habitat for Soil Organisms

> Life not only formed the soil, but other living things of incredible abundance and diversity now exist within it; if this were not so the soil would be a dead and sterile thing

> —Carson (1962, p. 53)

One thing is certain, most soils are not dead and sterile things. The fact is, a handful of soil is an ecosystem. It may contain up to billions of organisms, belonging to thousands of species. Table 12.1 lists a few (very few) of these organisms. Obviously, communities of living organisms inhabit the soil. What is not so obvious is that they are as complex and intrinsically valuable as are those organisms that roam the land surface and waters of Earth.

Engineering Medium

We usually think of soil as being firm and solid (solid ground, *terra firma*). As solid ground, soil is usually a good substrate upon which to build highways and structures; however, not all soils are firm and solid. Some are not as stable as others. Whereas construction of buildings and highways may be suitable in one location on one type of soil, it may be unsuitable in another location with different soil. To construct structurally sound and stable highways and buildings, construction on soils and with soil materials requires knowledge of the diversity of soil properties. Note that working with manufactured building materials that have been engineered to withstand certain stresses and forces is much different than

TABLE 12.1

Soil Organisms (Representative Sample)

Actinomycetes	Mice, moles, voles
Algae	Microorganisms (protists)
Ants	Millipedes and centipedes
Arthropod animals	Mites
Bacteria	Nematodes
Crustacea	Nonarthropod animals
Diplopoda	Protozoa
Diptera	Rabbits, gophers, squirrels
Earthworms and potworms	Springtails
Fungi	Vertebrates
Harvestman	

working with natural soil materials, even though engineers have the same concerns about soils as they do with manmade building materials (concrete and steel). It is much more difficult to evaluate the ability of a soil to resist compression or to remain in place and to determine its bearing strength, shear strength, and stability than it is to make these same determinations for manufactured building materials.

Soil Basics

Any fundamental discussion about soil should begin with a definition of what soil is. The word *soil* is derived through Old French from the Latin *solum*, which means "floor" or "ground." A more concise definition is made difficult by the great diversity of soils throughout the globe. In general, however, we can describe soil as a mixture of air, water, mineral matter, and organic matter; the relative proportions of these four components greatly influence the productivity of soils. Keep in mind that the four major ingredients that make up soil are not mixed or blended like cake batter. Instead, a major and critically important constituent of soil is the pore spaces, which are vital to air and water circulation, as they provide space for roots to grow and microscopic organisms to live. Without sufficient pore space, soil would be too compacted to be productive. Ideally, the pore space will be divided roughly equally between water and air, with about one-quarter of the soil volume consisting of air and one-quarter consisting of water. The relative proportions of air and water in a soil typically fluctuate significantly as water is added and lost. Compared to surface soils, subsoils tend to contain less total pore space, less organic matter, and a larger proportion of micropores, which tend to be filled with water.

Soil air circulates through soil pores in the same way air circulates through a ventilation system. Only when the pores (the ventilation ducts) become blocked by water or other substances does the air fail to circulate. Although soil pores normally connect to interface with the atmosphere, soil air is not the same as atmospheric air. It differs in composition from place to place. Soil air also normally has a higher moisture content than the atmosphere. The content of carbon dioxide (CO_2) is usually higher and that of oxygen (O_2) lower than the accumulations of these gases in the atmosphere.

It is important to state again that only when soil pores are occupied by water or other substances does air fail to circulate in the soil. For proper plant growth, this is of particular importance, because in soil pore spaces that are water dominated, air oxygen content is low and carbon dioxide levels are high, which restricts plant growth.

The presence of water in soil (often reflective of climatic factors) is essential for the survival and growth of plant and other soil organisms. Soil moisture is a major determinant of the productivity of terrestrial ecosystems and agricultural systems. Water moving through soil materials is a major force behind soil formation. Along with air, water, and dissolved nutrients, soil moisture is critical to the quality and quantity of local and regional water resources.

Mineral matter varies in size and is a major constituent of nonorganic soils. Mineral matter consists of large particles (rock fragments), including stones, gravel, and coarse sand. Many of the smaller mineral matter components are made of a single mineral. Minerals in the soil (for plant life) are the primary source of most of the chemical elements essential for plant growth. Soil organic matter consists primarily of living organisms and the remains of plants, animals, and microorganisms that are continuously broken down (biodegraded) in the soil into new substances that are synthesized by other microorganisms. These other microorganisms continually use this organic matter and reduce it to carbon dioxide via respiration until it is depleted, making repeated additions of new plant and animal residues necessary to maintain soil organic matter (Brady and Weil, 1996).

Now that we have defined soil, let's take a closer look at a few of the basics pertaining to soil and some of the common terms used in any discussion related to soil basics. Soil is the layer of bonded particles of sand, silt, and clay that covers the land surface of the Earth. Most soils develop in multiple layers. The topmost layer, topsoil, is the layer of soil moved in cultivation and in which plants grow. This topmost layer is actually an ecosystem composed of both biotic and abiotic components—inorganic chemicals, air, water, and decaying organic material that provides vital nutrients for plant photosynthesis, as well as living organisms. Below the topmost layer is the subsoil, the part of the soil below the plow level, usually no more than a meter in thickness. Subsoil is much less productive, partly because it contains much less organic matter. Below that is the parent material, the unconsolidated (and more or less chemically weathered) bedrock or other geologic material from which the soil is ultimately formed. The general rule of thumb is that it takes about 30 years to form 1 inch of topsoil from subsoil; it takes much longer than that for subsoil to be formed from parent material, with the length of time depending on the nature of the underlying matter (Franck and Brownstone, 1992).

Physical Properties of Soil

Five major physical properties of soil are of interest. They are soil texture, slope, structure, organic matter, and soil color. Soil texture, or the relative proportions of the various soil separates in a soil, is a given and cannot be easily or practically changed significantly. It is determined by the size of the rock particles (sand, silt, and clay particles) or the soil separates within the soil. The largest soil particles are gravel, which consists of fragments larger than 2.0 mm in diameter. Particles between 0.05 and 2.0 mm are classified as sand. Silt particles range from 0.002 to 0.05 mm in diameter, and the smallest particles (clay particles) are less than 0.002 mm in diameter. Clays are composed of the smallest particles, but these particles have stronger bonds than silt or sand; once broken apart, though, they erode more readily. Particle size has a direct impact on erosion. Rarely does a soil consist of only one single size of particle; most are a mixture of various sizes.

The slope (or steepness of the soil layer) is another given, important because the erosive power of runoff increases with the steepness of the slope. Slope also allows runoff to exert increased force on soil particles, which breaks them apart more readily and carries them farther away.

Soil structure (tilth) should not be confused with soil texture—they are different. In fact, in the field, the properties determined by soil texture may be considerably modified by soil structure. Soil structure refers to the combination or arrangement of primary soil particles into secondary particles (units or peds). Simply stated, soil structure refers to the way various soil particles clump together. Clusters of soil particles, called *aggregates*, can vary in size, shape, and arrangement; they combine naturally to form larger clumps called *peds*. Sand particles do not clump because sandy soils lack structure. Clay soils tend to stick together in large clumps. Good soil develops small friable (easily crumbled) clumps. Soil develops a unique, fairly stable structure in undisturbed landscapes, but agricultural practices break down the aggregates and peds, lessening erosion resistance.

The presence of decomposed or decomposing remains of plants and animals (organic matter) in soil helps not only fertility but also soil structure—especially the ability of water to store water. Live organisms such as protozoa, nematodes, earthworms, insects, fungi, and bacteria are typical inhabitants of soil. These organisms work to either control the population of organisms in the soil or to aid in the recycling of dead organic matter. All soil organisms, in one way or another, work to release nutrients from the organic matter, changing complex organic materials into products that can be used by plants.

Just about anyone who has looked at soil has probably noticed that soil color is often different from one location to another. Soil colors range from very bright to dull grays, to a wide range of reds, browns, blacks, whites, yellows, and even greens. Soil color is dependent primarily on the quantity of humus and the chemical form of iron oxides present.

Soil scientists use a set of standardized color charts (the *Munsell Color System*) to describe soil colors. They consider three properties of color—hue, value, and chroma—in combination to come up with a large number of color chips to which soil scientists can compare the color of the soil being investigated.

Soil Separates

Soil particles are divided into groups (*soil separates*) based on their size—sand, silt, and clay— by the International Soil Science Society System, the U.S. Public Roads Administration, and the U.S. Department of Agriculture. In this text, we use the classification established by the U.S. Department of Agriculture (USDA). The size ranges in these separates reflect major changes in how the particles behave and in the physical properties they impart to soils. Table 12.2 lists the separates, their diameters, and the number of particles in 1 g of soil (according to the USDA).

Sand, one of the individual-sized groups of mineral soil particles, ranges in diameter from 2 mm to 0.05 mm and is divided into five classes (see Table 12.2). Sand grains are more or less spherical (rounded) in shape, with variable angularity, depending on the extent to which they have been worn down by abrasive processes such as rolling around in flowing water during soil formation.

Sand forms the framework of soil and gives it stability when in a mixture of finer particles. Sand particles are relatively large, which allows voids that form between each grain to also be relatively large. This promotes free drainage of water and the entry of air into the soil. Sand is usually composed of a high percentage of quartz; because quartz is most resistant to weathering, its breakdown is extremely slow. Many other minerals are found in sand, depending on the rocks from which the sand was derived. In the short term (on an annual basis), sand contributes little to plant nutrition in the soil; however, in the long term (thousands of years of soil formation), soils with a lot of weatherable minerals in their sand fraction develop a higher state of fertility.

TABLE 12.2

Characteristics of Soil Separates

Separate	Diameter (mm)	Number of Particles per Gram
Very coarse sand	2.00–1.00	90
Coarse sand	1.00–0.50	720
Medium sand	0.50–0.25	5700
Fine sand	0.25–0.10	46,000
Very fine sand	0.10–0.05	722,000
Silt	0.05–0.002	5,776,000
Clay	Below 0.002	90,260,853,000

Silt (essentially microsand), though spherically and mineralogically similar to sand, is smaller, too small to be seen with the naked eye. It weathers faster and releases soluble nutrients for plant growth faster than sand. Too fine to be gritty, silt imparts a smooth feel (like flour) without stickiness. The pores between silt particles are much smaller than those in sand (sand and silt are just progressively finer and finer pieces of the original crystals in the parent rocks). In flowing water, silt is suspended until it drops out when flow is reduced. On the land surface, silt, if disturbed by strong winds, can be carried great distances and is deposited as loess.

The clay soil separate is (for the most part) much different from sand and silt. Clay is composed of secondary minerals that were formed by the drastic alteration of the original forms or by the recrystallization of the products of their weathering. Because clay crystals are platelike (sheeted) in shape they have a tremendous surface area-to-volume ratio, giving clay a tremendous capacity to adsorb water and other substances on its surfaces. Clay actually acts as a storage reservoir for both water and nutrients. There are many kinds of clay, each with different internal arrangements of chemical elements which give them individual characteristics.

Soil Formation

Everywhere on Earth's land surface is either rock formation or exposed soil. When rocks formed deep in the Earth are thrust upward and exposed to the Earth's atmosphere, the rocks adjust to the new environment, and soil formation begins. Soil is formed as a result of physical, chemical, and biological interactions in specific locations. Just as vegetation varies among biomes, so do the soil types that support that vegetation. The vegetation of the tundra and rain forest differ vastly from each other and from vegetation of the prairie and coniferous forest; soils differ in similar ways.

In the soil-forming process, two related, but fundamentally different, processes are occurring simultaneously. The first is the formation of soil parent materials by weathering of rocks, rock fragments, and sediments. This set of processes is carried out in the zone of weathering. The end point is producing parent material for the soil to develop in and is referred to as C horizon material. It applies in the same way for glacial deposits as for rocks. The second set of processes is the formation of the soil profile by soil-forming processes, which gradually change the C horizon material into A, E, and B horizons.

Soil development takes time and is the result of two major processes: weathering and morphogenesis. Weathering (the breaking down of bedrock and other sediments that have been deposited on the bedrock by wind, water, volcanic eruptions, or melting glaciers) happens physically, chemically, or a combination of both. Physical weathering involves the breaking down of rock primarily by temperature changes and the physical action of water, ice, and wind. When a geographical location is characterized as having an arid desert biome, the repeated exposure to very high temperatures during the day followed by low temperatures at night causes rocks to expand and contract and eventually to crack and shatter. At the other extreme, in cold climates rock can crack and break as a result of repeated cycles of expansion of water in cracks and pores during freezing and contraction during thawing. Another type of physical weathering occurs when various vegetation types spread their roots and grow, and the roots can exert enough pressure to enlarge cracks in solid rock, eventually splitting and breaking the rock. Plants such as mosses and lichens also penetrate rock and loosen particles.

Bare rocks are also subjected to chemical weathering, which involves chemical attack and dissolution of rock. Accomplished primarily through oxidation via exposure to oxygen gas in the atmosphere, acidic precipitation (after having dissolved small amounts of carbon dioxide gas from the atmosphere), and acidic secretions of microorganisms (bacteria, fungi, and lichens), chemical weathering speeds up in warm climates and slows down in cold ones. Physical weathering and chemical weathering do not always (if ever) occur independently of each other; instead, they normally work in combination.

The final stages of soil formation consist of the processes of morphogenesis, or the production of a distinctive soil profile with its constituent layers or horizons. The soil profile (the vertical section of the soil from the surface through all its horizons, including C horizons) gives the environmental scientist critical information. When properly interpreted, soil horizons can provide warnings about potential problems in using the land and can tell much about the environment and history of a region. The soil profile allows us to describe, sample, and map soils.

Soil horizons are distinct layers, roughly parallel to the surface, which differ in color, texture, structure, and content of organic matter. The clarity with which horizons can be recognized depends upon the relative balance of the migration, stratification, aggregation, and mixing processes that take place in the soil during morphogenesis. In podzol-type soils (formed mainly in cool, humid climates), striking horizonation is quite apparent; in vertisol-type soils (soils high in swelling clays), the horizons are less distinct. When horizons are studied, they are given a letter symbol to reflect the genesis of the horizon.

Certain processes work to create or destroy clear soil horizons. Processes that tend to create clear horizons by vertical redistribution of soil materials include the leaching of ions in soil solutions, movement of clay-sized particles, upward movement of water by capillary action, and surface deposition of dust and aerosols. Clear soil horizons are destroyed by mixing processes that occur because of organisms, cultivation practices, creep processes on slopes, frost heave, and swelling and shrinkage of clays—all part of the natural soil formation process.

Soil Classification

> Classification schemes of natural objects seek to organize knowledge so that the properties and relationships of the objects may be most easily remembered and understood for some specific purpose. The ultimate purpose of soil classification is maximum satisfaction of human wants that depend on use of the soil. This requires grouping soils with similar properties so that lands can be efficiently managed for crop production. Furthermore, soils that are suitable or unsuitable for pipelines, roads, recreation, forestry, agriculture, wildlife, building sites, and so forth can be identified.
>
> —Foth (1978, p. 255)

When people become ill, they may go to a doctor to seek a diagnosis of what is causing the illness and perhaps a prognosis regarding how their illness will progress. What do diagnosis and prognosis have to do with soil? Actually, quite a lot. The diagnostic techniques used by a physician to identify the causative factors leading to a particular illness are analogous to the soil practitioner using diagnostic techniques to identify a particular soil. Sound farfetched? It shouldn't, because it isn't. Soil scientists must be able to determine the type of soil they study or work with.

Determining the type of soil makes sense, but what does prognosis have to do with all this? Soil practitioners must be able to identify or classify a soil type, because this information allows them to correctly predict how a particular pollutant will react or respond when spilled in that type of soil. The fate of the pollutant is important in determining the possible damage inflicted on the environment—soil, groundwater, and air—because ultimately a spill could easily affect all three. Thus, the soil practitioner not only must use diagnostic tools in determining soil type but also must be familiar with the soil type to determine how a particular pollutant or contaminant will respond when spilled in that soil type.

Let's take a closer look at the genesis of soil classification. From the time humans first advanced from hunter–gatherer status to cultivators of crops, they noticed differences in productive soils and unproductive soils. The ancient Chinese, Egyptians, Romans, and Greeks all recognized and acknowledged the differences in soils as media for plant growth. These early soil classification practices were based primarily on texture, color, and wetness.

Soil classification as a scientific practice did not gain a foothold until the later 18th and early 19th centuries, when the science of geology was born. This is when such terms (with an obvious geological connotation) as *limestone soils* and *lake-laid soils*, as well as *clayey* and *sandy soils*, came into being. The Russian scientist Dokuchaev was the first to suggest, in the late 1800s, that soils were natural bodies, and he developed a generic classification of soils that was later expanded. The system was based on the theory that each soil has a definite form and structure (morphology) related to a particular combination of soil-forming factors. This system was used until 1960, when the USDA published its original *Soil Classification: A Comprehensive System*. This classification system placed major emphasis on soil morphology and gave less emphasis to its genesis or soil forming factors as compared to previous systems. In 1975, this text was replaced by *Soil Taxonomy: A Basic System of Soil Classification for Making and Interpreting Soil Surveys*, now in its second edition (USDA, 1999). *Soil Taxonomy* classifies objects according to their natural relationships, and soils are classified based on measurable properties of soil profiles.

Note that no clear delineation or line of demarcation can be drawn between the properties of one soil and those of another. Instead, a gradation (sometimes quite subtle, like comparing one shade of white to another) occurs in soil properties as one moves from one soil to another. Brady and Weil (2007, p. 58) noted that, "The gradation in soil properties can be compared to the gradation in the wavelengths of light as you move from one color to another. The changing is gradual, and yet we identify a boundary that differentiates what we call 'green' from what we call 'blue.'" To properly characterize the primary characteristics of a soil, a soil must be identified down to the smallest three-dimensional characteristic sample possible; however, to accurately perform a particular soil sample characterization, a sampling unit must be large enough so the nature of its horizons can be studied and the range of its properties identified. The *pedon* (rhymes with head-on) is this unit. The pedon is roughly polygonal in shape and designates the smallest characteristic unit that can still be called a soil.

Because pedons occupy a very small space (from approximately 1 to 10 m²), they cannot, obviously, be used as the basic unit for a workable field soil classification system. To solve this problem, a group of pedons, termed a *polypedon*, is of sufficient size to serve as a basic classification unit (or, as it is commonly referred to, a *soil individual*). In the United States, these groupings have been called a *soil series*.

There is a difference between *a* soil and *the* soil. This difference is important in the soil classification scheme. A soil is characterized by a sampling unit (pedon), which as a group (polypedons) form a soil individual. The soil, on the other hand, is a collection of all of these natural ingredients and is distinguishable from other bodies such as water, air, solid rock, and other parts of the Earth's crust. By incorporating the difference between *a* soil and *the* soil, a classification system has been developed that is effective and has been widely used.

Diagnostic Horizons and Temperature and Moisture Regimes

Soil taxonomy uses a strict definition of soil horizons called *diagnostic horizons*, which are used to define most of the orders. Two kinds of diagnostic horizons are recognized: surface and subsurface. The surface diagnostic horizons are *epipedons* (Greek *epi*, "over"; *pedon*, "soil"). The epipedons include the dark (organic rich) upper part of the soil, the upper eluvial horizons, and sometimes both. Soils beneath the epipedon horizons are *subsurface diagnostic horizons*. Each of these layers is used to characterize different soils in soil taxonomy.

In addition to using diagnostic horizons to strictly define soil horizons, soil moisture regime classes can also be used. A soil moisture regime refers to the presence of plant-available water or groundwater at a sufficiently high level. The control section of the soil (ranging from 10 to 30 cm for clay and from 30 to 90 cm for sandy soils) designates that section of the soil where water is present or absent during given periods in a year. The control section is divided into upper and lower sections. The upper portion is defined as the depth to which 2.5 m³ of water will penetrate within 24 hours. The lower portion is the depth that 7.5 m³ of water will penetrate.

Six soil moisture regimes are identified:

- *Aridic*—Characteristic of soils in arid regions
- *Xeric*—Characteristic of having long periods of drought in the summer
- *Ustic*—Soil moisture generally high enough to meet plant needs during growing season
- *Udic*—Common soil in humid climatic regions
- *Perudic*—An extremely wet moisture regime annually
- *Aquic*—Soil saturated with water and free of gaseous oxygen

Table 12.3 lists the moisture regime classes and the percentage distribution of areas with different soil moisture regimes.

In soil taxonomy, several soil temperature regimes are also used to define classes of soils. These soil temperature regimes, shown in Table 12.4, are based on mean annual soil temperature, mean summer temperature, and the difference between mean summer and winter temperatures.

TABLE 12.3

Soil Moisture Regimes

Moisture Regime	Percent of Global Area Occupied (%)
Aridic	35.9
Xeric	3.5
Ustic	18.0
Udic	33.1
Perudic	1.0
Aquic	8.3

Source: Adapted from Eswaran, H., *Pedologie*, 43, 19–39, 1993.

TABLE 12.4

Soil Temperature Regimes

Soil Temperature Regimes (Mean Annual Temperature)	Percent of Global Area Occupied (%)
Pergelic (0°C)	10.9
Cryic (0–8°C)	13.5
Frigid (0–8°C)	1.2
Mesic (8–15°C)	12.5
Thermic (15–22°C)	11.4
Hyperthermic (>22°C)	18.5
Isofrigid (0–8°C)	0.1
Isomesic (8–15°C)	0.3
Isothermic (15–22°C)	2.4
Isohyperthermic (>22°C)	26.0
Water (NA)	1.2
Ice (NA)	1.4

Source: Adapted from Eswaran, H., *Pedologie*, 43, 19–39, 1993.

Forest Ecosystem Mineral Cycles and Their Dynamics*

To gain an understanding of nutrient distribution, mineral cycles, and their dynamics in the forest ecosystem, one needs to have an understanding of the key terms discussed herein and also of a basic conceptual diagram commonly used to illustrate the forest ecosystem nutrient cycle.

---

* Much of this section is based on Farve, R. and Napper, C., *Biomass Fuels & Whole Tree Harvesting Impacts on Soil Productivity—Review of Literature*, U.S. Department of Agriculture, Forest Service, National Technology & Development Program, Washington, D.C., 2009; Waring, R. and Running, S.W., *Forest Ecosystems: Analysis at Multiple Scales*, 3rd ed., Elsevier, Burlington, MA, 2007.

Overview of Forest Ecosystem Nutrient Distribution and Cycling

To understand the role of dead trees and the dead wood cycle, you must learn and understand the key terms used. Here are some of the key terms you will encounter in this section:

- *Coarse woody debris* (CWD)—Dead standing and downed pieces larger than 3 inches in diameter, which corresponds to the size class that defines large wood fuel (Harmon et al., 1986). Some ecologists include woody material larger than 1 inch in diameter as CWD. Coarse woody debris is an important component in the structure and functioning of ecosystems. A dead tree, from the time it dies until it is fully decomposed, contributes to many ecological processes as a standing snag and fallen woody material lying on and in the soil. Fire, insects, pathogens, and weather are responsible for the decomposition of dead organic matter and the recycling of nutrients (Olson, 1963; Stoszek, 1988). Fire directly recycles the carbon of living and dead vegetation. The relative importance of fire and biological decomposition depends on site and climate (Harvey, 1994). In cold or dry environments, biological decay is limited, which allows accumulation of plant debris. Fire plays a major role in recycling organic matter in these environments. Without fire in these ecosystems, nutrients are tied up in dead woody material for a long period. Fire, insects, and diseases perform similar roles in that they both create and consume CWD and smaller dead woody material.
- *Dead wood cycle*—Though this term is rarely used in the literature, it refers to the process of tree death, tree fall, and decay in the forested ecosystem. The cycle begins with live healthy trees and ends with their incorporation into the soil organic horizon or aquatic environment (Lofroth, 1998).
- *Dead wood facultative*—Organism that may use but does not require some component of the dead wood cycle for some or all of its life history (Lofroth, 1998).
- *Dead wood obligate*—Organism that requires some component of the dead wood cycle for some or all of its life history. The conservation of populations of such species can be considered dependent on the presence of dead wood (Lofroth, 1998).
- *Fine fuel/slash*—The debris left behind following forest harvesting, including woody material of any size (McRae et al., 1979; Trowbridge et al., 1987).
- *Large organic debris* (LOD)—Essentially, coarse woody debris in aquatic ecosystems. LOD is downed woody material, including tree boles, limbs, and rooting structures. Van Sickle and Gregory (1990) defined it as material greater than 10 cm in diameter and greater than 1.5 m long. Bisson et al. (1987) used a limit of 10 cm in diameter but did not specify a minimum length (Lofroth, 1998).
- *Snags*—Dead standing trees, typically with a specified lower size limit. A range of desirable snag depending on bird species has been suggested for forests in the Northern Rocky Mountains. About 25% of the bird species in the Northern Rocky Mountains are cavity nesters (Bull et al., 1997). Cavity-nesting birds and bats mostly utilize standing snags, especially those of large diameters. Smaller snags can be important for foraging. Salvage logging can enhance habitat for some species but diminish it for others, resulting in a shift in diversity but not in richness (McIver and Starr, 2001).

- *Wildlife tree*—A tree that provides present or future valuable habitat of the conservation or enhancement of wildlife (Guy and Manning, 1995). Wildlife trees may be distinguished by attributes such as structure, age, abundance, location, and surrounding habitat features. They range from live and healthy trees to decayed stubs and, as such, include snags (Lofroth, 1998).

Nutrients enter the forest ecosystem as atmospheric inputs, and they are exported from the ecosystem in a number of ways (e.g., gas emissions, fire). Within this system, nutrients are subjected to several interrelated processes: plant processes, litter processes, soil processes, intrasystem recycling, and internal geologic processes. The following discussion briefly summarizes these processes, many of which are covered in greater detail later in the text.

Within the forest ecosystem, nutrients are cycled in two separate but interconnected nutrient reserves or pools: the aboveground plant pool and the belowground soil pool. The aboveground plant nutrient pool is created when inorganic nutrients are taken up by trees from the soil and converted into an organic form by plants for metabolism. Once taken up by trees, these nutrients are retained and recycled within the tree (more or less efficiently depending on the species of tree), thereby creating an above-the-ground reservoir of nutrients. The obvious advantage of this strategy is that by recycling and reusing nutrients, the tree reduces its dependency on belowground nutrients that may be in short supply. As the tree discards leaves, branches, or bark or it dies, the tree's organic nutrients are returned to the soil, where they are eventually converted back to an inorganic form by soil organisms. These nutrients remain in the soil pool for eventual reuse by trees or are leached out of the forest ecosystem.

Mineral Cycle Dynamics

Plants acquire most of their nutrients from the soil in solution. Depending on the species, they require some nutrients in large quantities (macronutrients) and other nutrients in trace amounts. Most forest practitioners agree that the sustainability and soil productivity of forest ecosystems are largely determined by the availability of inorganic (water-soluble) nitrogen (N), phosphorus (P), potassium (K), and calcium (Ca) (Swank and Waide, 1980; Waring and Running, 2007).

Inorganic Nutrients Entering the Soil Pool

Nearly all of the nitrogen and sulfur in the forest ecosystem initially enter from the atmosphere. Magnesium, sodium, calcium, and potassium also enter from the atmosphere, but their primary entrance is by weathering of the underlying mineral rock. All atmospheric elements enter the ecosystem via rain or snow (wetfall) or deposit via gravity (dryfall). Nitrogen is a key nutrient that is required in large amounts by plants. Even

---

DID YOU KNOW?

Nitrogen fixation accounts for only a fraction of the total N in forest ecosystems. Plants acquire most of their nitrogen N from organic material that has been converted into an inorganic form by soil microbes. The organic N comes from decomposed plant material that accumulated in the soil from prior plant growth.

though nitrogen makes up over 78% of the atmosphere in the gaseous form, it is unavailable to plants. As such, even though abundant, it is not unusual for nitrogen to be in short supply in the soils of most terrestrial ecosystems (LeBauer and Treseder, 2008; Vitousek and Howarth, 1991).

Gaseous N only can be made available to plants by the actions of N-fixing soil organisms (i.e., organisms that can fix or incorporate N in a chemical compound that can be used by plants). N-fixing organisms exist in the soil either free living or in symbiotic associations of certain forest shrubs and trees with bacteria (*Rhizobium*) in the root nodules. These soil organisms possess an enzyme that can convert gaseous N_2 into an inorganic form, thereby making it available for plant uptake. In the forest ecosystem, N fixation can occur in tree-loving fungi (e.g., brown rot fungus) that decompose coarse woody debris and in the rhizomes of certain forest plants and shrubs. In the forest, N fixation is highly variable seasonally and also changes with stand development.

Except for N, most nutrients trace their origin in the forest ecosystem from the mineral composition of the underlying rocks. The forest soil's fertility, texture, and buffering capacity to changes in pH are determined by the type and age of the parent material from which the soil is derived. The soil's parent material, therefore, is the main natural source of most nutrients entering the forest ecosystem (Buol et al., 2003).

As rock is uplifted and exposed near the surface, it undergoes chemical or mechanical (physical) weathering. Simply, weathering (which projects itself on all surface material above the water table) is the general term used for all the ways in which a rock may be broken down. Factors that influence weathering include the following:

- *Rock type and structure*—Each mineral contained in rocks has a different susceptibility to weathering. A rock with bedding planes, joints, and fractures provides pathways for the entry of water, leading to more rapid weathering. Differential weathering (where rocks erode at different rates) can occur when rock combinations consist of some rocks that weather faster than more resistant rocks.

- *Slope*—On steep slopes, weathering products may be quickly washed away by rains. Wherever the force of gravity is greater than the force of friction holding particles upon a slope, these tend to slide downhill.

- *Climate*—Higher temperatures and high amounts of water generally cause chemical reactions to run faster. Rates of weathering are higher in warmer than in colder dry climates.

- *Animals*—Rodents, earthworms, and ants that burrow into soil bring material to the surface where it can be exposed to the agents of weathering.

- *Time*—Rates of weathering depend on slope, climate, and animals.

Although we consider the physical and chemical weathering processes to be separate, it is important to recognize that they work in tandem to break down rocks and minerals to smaller fragments.

Physical (Mechanical) Weathering

Physical weathering involves the disintegration of a rock by physical processes. These include freezing and thawing of water in rock crevices, disruption by plant roots or burrowing animals, and the changes in volume that result from chemical weathering with the rock. These and other physical weathering processes are discussed below.

- *Development of joints*—Joints are one way in which rocks yield to stress. Joints are fractures or cracks that occur when the rocks on either side of a fracture have not undergone relative movement. Joints form as a result of expansion due to cooling or relief of pressure as overlying rocks are removed by erosion. They form free spaces in the rock that allow other agents of chemical or physical weathering to enter (unlike faults, which show offset across the fracture). They play an important part in rock weathering as zones of weakness and water movement.
- *Crystal growth*—As water percolates through fractures and pore spaces it may contain ions that precipitate to form crystals. When crystals grow they can cause the necessary stresses needed for mechanical rupturing of rocks and minerals.
- *Heat*—It was once thought that daily heating and cooling of rocks was a major contributor to the weathering process. This view is no longer shared by most practicing geologists; however, it should be pointed out that sudden heating of rocks from forest fires may cause expansion and eventual breakage of rock.
- *Biological activities*—Plant and animal activities are important contributors to rock weathering. Plants contribute to the weathering process by extending their root systems into fractures and growing, causing expansion of the fracture. Growth of plants and their effects are evident in many places, such as when they are planted near cement work (streets, brickwork, and sidewalks). Also, animals burrowing in rock cracks can break the rock.
- *Frost wedging*—Frost wedging is often produced by the freezing and thawing of water in rock pores and fissures. Expansion of water during freezing causes the rock to fracture. Frost wedging is more prevalent at high altitudes, where there may be many freeze–thaw cycles. One classic and striking example of weathering of Earth's surface rocks by frost wedging can be seen in the formation of hoodoos in Bryce Canyon National Park, Utah. "Although Bryce Canyon receives a meager 18 inches of precipitation annually, it's amazing what this little bit of water can do under the right circumstances" (NPS, 2008). Approximately 200 freeze–thaw cycles occur annually in Bryce. During these periods, snow and ice melt in the afternoon and water seeps into the joints of the Bryce or Claron Formation. When the sun sets, temperatures plummet and the water refreezes, expanding up to 9% as it becomes ice. This frost wedging process exerts tremendous pressure or force on the adjacent rock and shatters the weak rock. The assault from frost wedging is a powerful force, but, at the same time, rainwater (the universal solvent), which is naturally acidic, slowly dissolves away the limestone, rounding off the edges of fractured rocks and washing away the debris. Small rivulets of water run down Bryce's rim, forming gullies. As gullies are cut deeper, narrow walls of rock known as fins begin to emerge. Fins eventually develop holes known as windows. Windows grow larger until their roofs collapse, creating hoodoos. As old hoodoos age and collapse, new ones are born.

Chemical Weathering

Chemical weathering involves the decomposition of rock by chemical changes or solution. Rocks that are formed under conditions found deep within the Earth are exposed to quite different conditions when uplifted onto the surface; for example, temperatures and pressures are lower on the surface, and copious amounts of free water and oxygen are available. The chief chemical weathering processes are oxidation, carbonation, hydration, and solution in water above and below the surface. The main agent responsible for chemical weathering reactions is not water movement but water and the weak acids formed in water. The acids formed in water are solutions that have abundant free H^+ ions. The most common weak acid that occurs in surface waters is carbonic acid (H_2CO_3), which is produced when atmospheric carbon dioxide dissolves in water; it exists only in solution. Hydrogen ions are quite small and can easily enter crystal structures, releasing other ions into the water.

$$H_2O + CO_2 \quad \rightarrow \quad H_2CO_3 \quad \rightarrow \quad H^+ + HCO_3^-$$

Water + Carbon Dioxide → Carbonic Acid → Hydrogen Ion + Bicarbonate Ion

Chemical weathering breaks rocks down by adding or removing chemical elements and changing them into other materials. Again, as stated, chemical weathering consists of chemical reactions, most of which involve water. Types of chemical weathering include

- *Hydrolysis*, which is a water–rock reaction that occurs when an ion in the mineral is replaced by H^+ or OH^-
- *Leaching*, which occurs when ions are removed by dissolution into water
- *Oxidation*, which is a result of oxygen being plentiful near the Earth's surface and reacting with minerals to change the oxidation state of an ion
- *Dehydration*, which occurs when water or a hydroxide ion is removed from a mineral
- *Complete dissolution*

DID YOU KNOW?

Nitrogen fixation accounts for only a fraction of the total nitrogen in forest ecosystems. Plants acquire most of their nitrogen from organic material that has been converted into an inorganic form by soil microbes. Organic nitrogen comes from decomposed plant material that accumulated in the soil from prior plant growth.

Most soils of the Northeast, Midwest, and Pacific Northwest are in the tens of thousands of years old and still contain much weatherable material to supply nutrients. Some soils of the upper coastal plain of the Southwestern United States are much older (by 104 times) and lack weatherable nutrients (Stone, 1979).

Nutrients (Organic and Inorganic) Leaving the Forest Ecosystem

Soluble inorganic nutrients are easily leached and often lost to the ecosystem. This is especially true of excessively drained (coarse) soils, which, as a result, tend to be less fertile (productive) than finer soils. Organic (insoluble) nutrients are less likely to be lost from the soil pool when compared to soluble inorganic nutrients. The exception is organic potassium, which is highly soluble.

Plant Processes

The plant processes of nutrient uptake, nutrient storage, internal nutrient recycling, and return of nutrients to the soil as litter ensure that nutrients essential for forest plant growth and health requirements are met.

Plant Uptake

A compound synthesized by a living organism and containing carbon is simplistically defined as an *organic compound*. An *inorganic compound*, in general, is considered a compound that is not generated by a life force and typically does not contain carbon, especially complex carbon. Organic compounds are generally insoluble, but many inorganic compounds are soluble. This is also a key characteristic of organic vs. inorganic nutrients. That is, organic nutrients are generally large, insoluble, complex polymers that typically are not taken up by roots. Inorganic nutrients, however, are less complex, soluble compounds that are readily available for plant uptake. The important inorganic nutrients of nitrogen, potassium, phosphorus, and calcium are made available to plants largely in solution (i.e., in water).

Inorganic nutrients are taken up by plants via their roots and mycorrhizae (symbiotic units that assist roots in nutrient and water uptake). The nutrient demand of the plant depends on its life-cycle stage and the species. Older mature trees typically require more nutrients than seedlings or saplings. The primary method of movement of most nutrients into the plant is via mass flow. Nutrients typically enter the plant passively, especially if they are found in sufficient concentrations in the soil; however, some plants can actively transport nutrients across root membranes if in lower concentrations.

Soil water (and therefore critical nutrients) is critical to plant survival. The movement of soil water to plants is greatly affected by soil pore size and volume. The principal source of water for plants is soil macropores (i.e., voids >14 μm in radius), where the water is loosely held. (Note that sandy soils have more pore space than clay.) As water is depleted from the macropores, they can become partially recharged from water bound in the more closely packed micropores; however, water can eventually become so depleted and the water–soil affinity so great that plants cannot extract water from the soil, thereby making it unavailable for plant uptake. Any activity that compacts the soil and alters pore size tends to reduce the availability of water and nutrients for plant uptake. This is especially true for clay soils.

Storage and Recycling

When plants have taken in enough nutrients to meet their current metabolic requirements, they typically take on an additional "luxury" consumption of nutrients and store the excess in the foliage. For most trees, after leaves, the next highest nutrient concentrations occur in twigs and small branches, then larger branches, then the bole bark, and finally the bole wood. Because different tree species and tree tissues can have significantly different nutrient demands, only gross generalizations can be made about where nutrients are stored in tissues while being recycled in the plant nutrient pool. Each species of tree and each part of a tree (leaves, bark, branch, trunk) retain different amounts of nutrients. The distribution of nutrients in western forests shows a similar pattern; Table 12.5 shows the distribution of nutrients in Douglas fir as reported by Pang et al. (1987). In general, the highest concentration of aboveground nutrients of most species, however, occurs in the foliage and branches and the lowest concentrations occur in bole wood.

TABLE 12.5

Aboveground Macronutrient Distribution (Percent of Mean Concentration) in Tree Components of 34-Year-Old Douglas Fir

Tree Component	Macronutrient Concentration (%)				
	N	P	Ca	K	Mg
Current foliage	29	26	14	29	28
Old foliage	26	39	29	25	27
Current twigs	21	17	14	19	20
Branches	10	7	17	11	10
Bark	7	7	10	12	8
Dead branches	6	2	15	2	6
Wood	2	1	1	2	1

Source: Pang, P.C. et al., *Canadian Journal of Forest Research*, 17, 1379–1384, 1987.

DID YOU KNOW?

Belowground, large-diameter roots of some species can store considerable amounts of nutrients. Nutrients stored in roots typically support root expansion but can also be mobilized to support leaf expansion.

Plants have evolved a more or less (depending on tree species) efficient means to retain and recycle nutrients that they have acquired from the soil (i.e., a plant pool). Nutrients stored in the leaves and foliage can be distributed to other aboveground organs (especially rapidly elongating shoots) when root uptake is inadequate to meet current demands. Prior to seasonal shedding, nutrients from leaves are typically relocated to perennial foliage tissues.

In general, evergreen species (such as conifers) have lower nutrient loss rates to the soil pool than deciduous species. This reduced loss appears to be the result of a general low nutrient content and longer lifespan of evergreen tissue compared to deciduous tissue. There is no evidence that evergreen species are more efficient at resorption of nutrients from leaves prior to leaf abscission. Compared to deciduous species, evergreen leaves decompose more slowly and thereby release few nutrients to the soil pool (Aerts, 1995). Some suggest that this results in a positive feedback that keeps evergreen soils relatively infertile, which in turn favors evergreens over deciduous species (Aerts, 1995, 1996; Hopper et al., 2000). The nutrient requirement, uptake, and return to the soil are summarized in Table 12.6.

Nitrogen appears to be the most efficiently recycled or retained nutrient prior to seasonal leaf abscission. Several studies have indicated that at least half or better of all available nitrogen in the ecosystem appears to be located and retained in the plant pool (Aerts, 1996; Helmisaari, 1995; Whittaker et al., 1979). The only nutrient that could be considered to be leached easily from living tissues (mostly stomata guard cells of the leaf) to the soil is potassium, which is highly soluble. The other nutrients are much less prone to leaching loss. The general pattern of loss via leaching appears to be potassium > phosphorus > nitrogen > calcium (Waring and Running, 2007).

TABLE 12.6

Nutrient Requirement, Uptake, and Return
of Deciduous Forests vs. Coniferous Forests

Element	Process	Deciduous (kg/ha/yr)	Coniferous (kg/ha/yr)
Nitrogen (N)	Requirement	94	39
	Uptake	70	39
	Return	57	30
Potassium (K)	Requirement	46	22
	Uptake	48	25
	Return	40	20
Calcium (Ca)	Requirement	54	16
	Uptake	84	35
	Return	67	29
Magnesium (Mg)	Requirement	10	4
	Uptake	13	6
	Return	11	4
Phosphorus (P)	Requirement	7	4
	Uptake	6	5
	Return	4	4

Source: Adapted from Cole, D.W. and Rapp, M., in Reichle, D.E., Ed., *Dynamic Properties of Forest Ecosystems*, Cambridge University Press, Cambridge, U.K., 1980, pp. 341–409.

Return to Litter

The major route of plant pool nutrient return to the soil is through litterfall (especially leaves and branches). Annually, significant amounts of leaves, twigs, branches, and fruit fall to the forest floor. If foliage falls to the forest floor early in the season (i.e., before the normal leaf abscission process), a significant nutrient concentration is returned to the soil. Because coarse woody debris (large branches, bole bark) typically has few stored nutrients, this component (from dying or diseased trees) accounts for a smaller fraction of nutrient loss from the plant pool. Litterfall collects on top of the mineral soil of the forest floor. Chemical, physical, and mechanical processes are necessary to break down and decompose the litterfall to a state where nutrients can enter the upper soil layers and become incorporated into the mineral soil.

DID YOU KNOW?

In general, nitrogen, phosphorus, and calcium move from the plant pool to the soil pool through litterfall. Potassium enters by throughfall and leaching. Magnesium is intermediate and varies with forest sites (Cole and Rapp, 1980).

Soil and Litter Processes

Organic Material Entering the Soil Pool

When plant material has fallen to the forest floor it can be considered part of the soil nutrient pool. Leaf litter, of course, is a large part of the fallen plant material. Leaf litter on the forest floor can often be distinguished as three layers: a top layer of nondecomposed litter, a middle layer of fragmented organic material, and a bottom humus layer. (In the humus, the organic litterfall has decomposed into a more inorganic form.) Below the litterfall (commonly called the O horizon) lies the mineral soil, which sometimes has clearly distinguished A, B, and C horizons. As chemical, physical, and biological processes decompose and mineralize the nutrients of litterfall into a soluble, inorganic form, the A, B, and C horizons have decreasing organic content. Most plants get their annual nutrient requirement from the nutrients that are made available through the processes of decomposition and mineralization. The process by which organic material is decomposed, mineralized and incorporated into the mineral soil is discussed below.

Fragmentation and Mixing

Litterfall is physically cut, fragmented, comminuted, and mixed into lower layers by a wide variety of forest soil invertebrates (e.g., roundworms, insects, microarthropods, segmented worms). They range in size and diversity from microscopic nematodes to large millipedes and earthworms. Temperature and soil pH tend to dictate what species are typically present in the leaf litter. In general, the total biomass of these soil animals increases by a factor of 6 from boreal (taiga) to tropical forests.

Microbial Decomposition

Decomposition describes the process whereby fragmented organic litter is broken down (i.e., consumed) by soil bacteria and fungi. The time it takes to decompose organic litter (decomposition rates) or the time it takes for nutrients to move through the nutrient pool (mean residence time) can be used to quantify the fate of nutrients in the soil pool. Soil animals and fungi attack cell walls, cellulose, resins, and lignin of litterfall and break them apart. (Lignin is most resistant to decomposition; only a few fungi can break down this compound.) Because the activity of soil animals and fungi is influenced by temperature and moisture content, decomposition rates typically increase exponentially with increased temperatures, but no simple temperature (or moisture) function provides a reliable predictor of decomposition rates, as the chemical composition of the litter also generally includes decomposition. A summary of decomposition rates for general forest types by climate is presented in Table 12.7.

Another metric used to measure the amount of time nutrients remain in the soil pool is *mean residence time* (MRT), which is a generic term used to determine how long a material, once added to a pool, remains in the pool. In nutrient cycling, MRT reflects how long a nutrient remains in the soil pool; it is the ratio of the total nutrient pool size to the rate at which nutrients are added and removed from the pool over time. An estimate of the MRT for minerals held as a pool on the forest floor can be obtained by determining the ratio of all forest floor mass to annual litterfall (assuming the forest flow is in a steady state). The MRT is the number of years that nutrients from the litterfall remain part of the forest floor biomass pool. A general pattern for MRT across biomes is 20 for boreal forest, 10 for temperate evergreen forests, 4 for temperate deciduous forest, and 2 for tropical deciduous forests (Waring and Running, 2007).

TABLE 12.7

Decomposition Rates of Foliage and Coarse Wood by Climate

Climate (Biome)	Forest Type	Tissue Decomposed (Mass Lost per Year)	
		Foliage	Coarse Wood
Boreal (Taiga)	Deciduous	0.39–0.7	0.02–0.3
	Evergreen	0.22–0.45	—
Cold temperate	Deciduous	0.28–0.85	—
	Evergreen	0.14–0.89	—
Warm temperate	Deciduous	0.44–2.46	0.03–0.27
	Evergreen	0.16–0.75	0.04
Tropical	Deciduous	0.62–4.16	—
	Evergreen	0.16–2.81	0.12–0.48

Source: Adapted from Waring, R.H. and Running, S.W., *Forest Ecosystems: Analysis at Multiple Scales*, 3rd ed., Academic Press, San Diego, CA, 2007.

Mineralization Process

Mineralization describes the specific hydrolysis and oxidation process whereby carbon is released (as CO_2) from organic material; this results in the conversion of organic nutrients to an inorganic, soluble form that is again available for uptake by plants. Mineralization typically is the result of soil microbes (especially bacterial) consuming organic material in the soil to extract the carbon for their use. The waste product of the microbe is the inorganic nutrient. Because the mineralization of nitrogen has been studied extensively, it provides a good example of the mineralization process. Simplistically, the pathway begins with soil microbial consumption of organic N. The microbes use the carbon for their metabolism and excrete N in an inorganic form—ammonium (NH_4^+). Other microbes can consume ammonium, break it down further, and convert it into inorganic nitrate (NO_3^-). The inorganic N

DID YOU KNOW?

The ratio of carbon to nitrogen is important. Plant and animal residues that have a C:N ratio of 30:1 or over have too little N to allow for rapid decomposition; therefore, microorganisms will take ammonium and nitrate out of the soil to fuel decomposition. This depletes the soil of nitrate and ammonium. Plant and animal residues with low C:N ratios (20:1 or less) have sufficient N for microorganisms to decompose the residues without taking nitrate and ammonium from the soil.

High C:N Ratio

Decomposition slower
Microorganisms deplete soil of nitrate and amonnium

Low C:N Ratio

Rapid decomposition (N higher)
Microorganisms are satisfied with plant N

excreted by soil microbes is thereby made available for uptake by the roots of aboveground plants. Plants can take up either inorganic form (i.e., NH_4^+ or NO_3^-); however, some plants prefer one over the other, and temperature and pH can affect the uptake rates (Barber, 1995). Soil microbes can store organic nitrogen and digest it at a later time. In that manner, organic nitrogen is immobilized until released later by the microbes. As a result, the levels of ammonium and nitrate do not remain constant in the soil.

The rate of mineralization (i.e., how long organic nutrients remain in the soil pool in an organic form) depends on the number of soil organisms and the rate at which they consume organic material. As such, mineralization rates and the amount of available inorganic nutrients depend on how favorable soil conditions (i.e., soil chemistry, temperature, moisture, aeration) are for soil biota. For most temperate forests, this causes the soil to accumulate a very large reservoir of organic nutrients (especially N) which can eventually be mineralized and made available to plants (Alban, 1982; Cole and Rapp, 1980).

Quantity of Nutrients in Soil Pool vs. Plant Pool

Most of the nutrients within the forest ecosystem are located in the soil pool. The results of several studies indicate that within the forest ecosystem only about 10 to 20% of the total nitrogen, calcium, or magnesium and about 20 to 50% of phosphorus occur in aboveground vegetation (Alban et al., 1978; Cole and Rapp, 1980; Helmisaari, 1995; Pastor and Bockheim, 1984). Potassium (K) is responsible for a large number of leaf cellular and plant physiological activities, such as photosynthesis, stomatal regulation, phloem transport, and enzyme activation (Fisher and Binkley, 2000). As such, compared to other macronutrients, it is typically more abundant in the plant pool (as opposed to the soil pool) than other macronutrients.

Alban et al. (1978) showed the relationship of nutrients in the aboveground tree (plant pool) and soil layer (soil pool) of three trees in Minnesota (Table 12.8). For these tree species, phosphorus and potassium were located primarily in the plant pool. Pastor and Bockheim (1984) studied nutrient cycling and distribution in an aspen–mixed hardwood forest in Wisconsin and reported nutrient plant pool/soil pool percentages for aspen that were somewhat higher for phosphorus and calcium and lower for potassium than Alban et al. (1978). The role of aboveground quaking aspen trees as a nutrient sink of calcium has emerged over the past few years from several other studies. It has been well documented that aspens take up large amounts of calcium from the soil pool and retain this nutrient in the perennial tissues (plant pool). This distribution of the nutrient capital of calcium in aboveground vegetation has also been reported for white spruce and other deciduous species such as a sugar maple and big-tooth aspen.

TABLE 12.8

Percentage of Nutrients in Aboveground Tree vs. Soil Layer (0–36 cm)

Element	Aspen	White Spruce	Red Pine
N	18	15	13
P	54	84	54
K	98	77	58
Ca	32	26	7
Mg	22	12	18

Source: Adapted from Alban, D.H. et al., *Canadian Journal of Forest Research*, 8, 290–299, 1978.

> **DID YOU KNOW?**
>
> The nutrient capital of the soil pool is the nutrients that are available for use by aboveground vegetation depending on internal ecosystem dynamics (i.e., the interchange between sequestered organic nutrients and available inorganic nutrients). The nutrient capital increases when inputs to the system exceed outputs and decreases when outputs exceed inputs (Grigal and Bates, 1992).

Aboveground/Belowground Interactions

Soil organisms have long been known to influence the aboveground plant. Recent research, however, is shedding light on the effect of this aboveground plant on the belowground organisms. Recent studies have indicated that plants can create positive feedback relative to nutrient cycling by regulating their uptake of nutrients, by use and loss of nutrients, and by influencing herbivores and belowground soil microbes. These interactions may be functionally important for an ecosystem-level understanding of ecosystem maintenance and stability.

A general principle emerging from recent research is that plant species adapted to fertile soils tend to be much different than plants adapted to infertile conditions. By the quantity and quality of biomass the plant produces under these conditions, the plant appears to have a significant effect on soil organisms belowground. Plants from infertile soils tend to have slower growth rates, smaller leaves, and fewer nutrients stored in foliage. These characteristics create positive feedbacks that affect belowground soil animals that are involved with nutrient cycling and thereby can further reduce nutrient available (Hobbie, 1992).

In contrast, on fertile soils, a positive feedback loop is begun when these soils make more nutrients available to the aboveground plant. The result is that the plant produces more biomass or stores more nutrients in the aboveground biomass (via luxury consumption). This relatively higher nutrient biomass is returned to the soil as high-nutrient litter, which promotes higher mineralization rates by soil organisms. Also, high-quality aboveground biomass supports more animal herbivores that can return nutrients in their waste products. The high-nutrient litter and animal waste tend to increase mineralization rates in the soil, in turn increasing the availability of nutrients for plant uptake (i.e., increase soil fertility)—thus completing the loop (Hobbie, 1992; Wardle et al., 2004).

Forest Soil Fertility and Productivity

This section focuses on the importance of key macronutrients to the forest ecosystem. To minimize confusion, the key terms of *soil fertility*, *soil productivity*, *forest productivity*, and *forest sustainability* are defined here:

- *Soil fertility* is the soil characteristic that allows soil to support abundant plant life. Fertile soil is rich in available, necessary macronutrients (N, K, P, Ca) and has sufficient trace elements (micronutrients). Additionally, the soil texture allows for water retention and drainage to support abundant vegetation. To be fertile, soil also must contain a store of organic material and a relatively neutral pH.

- *Soil productivity* is the capacity of soil, in its normal environment, to support plant growth. In the forest, typically it is reflected in the growth of forest vegetation.

- *Forest productivity* is the amount of biomass (or merchantable wood volume) produced annually (Edmonds et al., 1989).

- *Forest sustainability* is a term that is defined differently depending on the eye beholding the forest (Burger and Kelting, 1998); however, it is generally agreed that sustainable forest management includes management that is economically feasible, environmentally sound (i.e., maintains forest biodiversity), and socially desirable (i.e., politically acceptable).

Soil Nitrogen

Because N is essential to many biological functions of plants and its entry into the ecosystem relies on N-fixing organism and recycling of organic material within the soil, it is not surprising that it is typically a limiting factor in most terrestrial ecosystems. Globally, N limitation constrains the productivity of most terrestrial ecosystems (LeBauer and Treseder, 2008; Vitousek and Howarth, 1991). The amount of N required for tree growth is greater than any other mineral nutrient and therefore is usually the most limiting nutrient in forest soils—especially in the West (Binkley, 1991; Edmonds et al., 1989). With regard to the supply and demand for N over time, in the early stages of stand development (especially after a harvest), crop trees have a modest N demand, but N demand soon overtakes the available soil nutrient supply. Intermediate-aged and older stands typically exhibit a deficiency in nutrients, especially nitrogen. The discussion below focuses on the response of forest stands to the addition of nutrients (fertilization) to improve soil fertility.

DID YOU KNOW?

One method to determine whether an element is limiting plant productivity is to add it to the soil and document the increased productivity.

Nitrogen in Western Forest Soils

The absence of N is considered to be the major growth-limiting factor that limits forest productivity in the Pacific Northwest (Edmonds et al., 1989). Douglas fir is known to consistently respond to N fertilization, and Sitka spruce also appears to benefit from increased N. Other Pacific Northwest and Rocky Mountain forest trees have inconsistently responded to N, but they have demonstrated improved productivity when fertilizers included N + P (Binkley, 1991) and N + K or N + S (Garrison et al., 2000). N deficiency in boreal and cool temperate western forests is caused by frequent fires, low N mineralization rates due to low soil temperatures, excessive accumulation of decaying wood (high C:N ratios), and reduced aboveground litter fall from evergreens (Kimmins, 1996).

Nitrogen in Eastern Forest Soils

In the northcentral states, jack pine, lodgepole pine, and black spruce are known to respond positively to N and N + P + K combinations (Brockley, 2005; Newton and Amponsah, 2006). Many northeastern forests are considered nitrogen rich or at a point of nitrogen saturation. The cause of this excess N in the forest ecosystem is much speculated (e.g., air pollution), as well as its effect (i.e., contributing to forest decline.

Coniferous vs. Deciduous Forest Soils

Conifers (as well as most evergreen trees) are better adapted to N-deficient soils. Conifers have a lower N requirement (and greater N-use efficiency) than deciduous trees primarily because of their slower growth rates, smaller leaves, and fewer nutrients stored in foliage. As such, conifers can thrive in soils that would not support deciduous forests. Although phosphorus, calcium, and potassium are not required to the extent of nitrogen, they are nevertheless important to plant health and vigor.

- *Phosphorous* (P)—Phosphorus typically is added to forest soils with nitrogen as a fertilizer in the Pacific Northwest and Southeast. In the Southeast, P deficiencies in the poorly drained soils of the lower Coastal Plain and certain sites with old, ancient soils (sites that have weathered in place for several hundred thousand years) in the upper Gulf Coastal Plain are well known (Allen, 1987; Bengtson, 1979; Huntington et al., 2000).

- *Calcium* (Ca)—Calcium is seldom a limiting factor for most forest soils; however, because it is disproportionately distributed in perennial tissue (branches and stems) of certain deciduous tree species (most importantly aspen and to a lesser extent deciduous species such as sugar maple, beech, yellow birch, and spruce), Ca is considered susceptible to depletion from the soil under intensive harvesting conditions (Federer et al., 1989; Huntington et al., 2000; Silkwood and Grigal, 1982; Wilson and Grigal, 1995).

- *Potassium* (K)—Potassium is one of the most common alkali metals and the seventh most abundant element in the Earth's curst. It is typically weathered as a soluble cation and readily moves into the soil solution. K is commonly made available in the soil by the weathering of parent material containing potassium feldspar, biotite, and muscovite. Organic K is highly soluble and readily lost from living plants and dead plant material. Relative to macronutrients, K can rapidly cycle through the forest ecosystem. For this reason, movement of K in space and time through the forest ecosystem is of general concern in maintaining long-term forest productivity (growth and health). Fertilization studies suggest that many forests respond favorably to increases in K. Potassium concentrations in the foliage of ponderosa pine and Douglas fir in the Inland Northwest (northern Idaho and Montana) do not change after applications of K fertilizers. Compared to N-only fertilization, applications of N + K treatments are believed to decrease tree mortality by reducing loss to mountain pine beetles and Armillaria root rot. N-only fertilization also tends to reduce foliar concentrations of K (Garrison-Johnston et al., 2000, 2003, 2005; Mandzak and Moore, 1994).

Armillaria Root Rot*

Armillaria root rot disease has been reported in nearly every state in the continental United States. Hosts include hundreds of species of trees, shrubs, vines, and forbs growing in forests, along roadsides, and in cultivated areas. The disease is caused by fungi that

---

* This section is based on Williams, R.E. et al., *Armillaria Root Disease*, Forest Insect and Disease Leaflet 78, U.S. Department of Agriculture, 1986.

live as parasites on living host tissue or as saprophytes (feed on dead organic materials) on dead woody material. The fungus most often identified as causing the disease is *Armillaria mellea*. Recent research, however, indicates that several different but closely related species are involved; still, the generic term "Armillaria" is used to refer to this group.

These fungi are natural components of forests, where they live on the coarse roots and lower stems of conifers and broad-leaved trees. As parasites, the fungi cause mortality, wood decay, and growth reduction. They infect and kill trees that have been already weakened by competition, other pests, or climatic factors. This type of activity occurs throughout the United States—especially in deciduous forests of the east. The fungi also infect healthy trees, either killing them outright or predisposing them to attacks by other fungi or insects. Such behavior typically occurs in the relatively dry, inland coniferous forests of the western United States.

The disease has several common names. *Shoestring root rot* refers to the rootlike fungal structures (rhizomorphs) that spread the fungi. The names *honey mushroom, honey agaric, mushroom root rot*, or *toadstool disease* refer to the mushrooms produced. Conifers often respond to infection by producing a copious flow of resin, hence the names *resin glut* and *resin flow*. When oaks are the common host, *Armillaria* is often called the *oak fungus*. Because these fungi commonly inhabit roots, their detection is difficult unless characteristic mushrooms are produced around the base of the tree or symptoms become obvious in the crown or on the lower stem.

Crown symptoms on conifers and broad-leaved trees vary somewhat. Generally, however, the foliage thins and discolors, turning yellow, then brown; branches die back; and shoot and foliar growth are reduced. On large, lightly infected, or vigorous trees, crown symptoms develop over a number of years until the trees die. Conifers, particularly Douglas fir and western larch, frequently produce a larger than normal crop of cones, known as *stress cones*, shortly before they die. On small, extensively infected, or low-vigor trees, crown symptoms develop rapidly; the foliage quickly discolors, and the tree often dies within a year. On such trees, premature foliage loss and reduced shoot and foliar growth may not be apparent.

Trees affected by prolonged drought or attacked by rodents, bark beetles, or other fungi, particularly other root pathogens, can produce crown symptoms similar to those caused by *Armillaria*. Thus, additional evidence, often found on the roots and on the lower stem, is needed to diagnose the disease. On most conifers, the infected portions of the lower stems are somewhat enlarged and exude large amounts of resin. Infected portions of the roots frequently become heavily encrusted with resin, soil, and sometimes fungal tissue. In contrast, infected portions of broad-leaved trees sometimes develop sunken cankers covered with loose bark or bark infiltrated with gum and other exudates, but most often these cankers are inconspicuous or absent.

If *Armillaria* is present, removing bark that covers infections will expose the characteristic, white mycelial mats or the rhizomorphs that grow between the wood and the bark. The white mycelial mats are marked by irregular, fanlike striations; hence, they are often referred to as mycelial *fans*. The thick mats decompose, leaving impressions on the resin-impregnated inner bark. Rhizomorphs growing beneath the bark are flat, black to reddish brown, and up to 0.20 inch (5 mm) wide. They have a compact outer layer of dark mycelium and an inner core of white mycelium. Rhizomorphs also grow through the soil. Except for being cylindrical and about half as wide, subterranean rhizomorphs are similar to those produced beneath the bark.

Mushrooms, the reproductive stage of these fungi, confirm the present of *Armillaria*. The short-lived mushrooms may be found growing in clusters around the bases of infected trees or stumps. They are produced sporadically in later summer or autumn and are most abundant during moist periods. Mushrooms of the different species vary somewhat but generally have yellow or brown stalks about 2 inches (5 cm) long, and a ring is sometimes found around the stalk just below the gills. The stalks have honey-yellow caps, 2 to 5 inches (5 to 12.5 cm) across. The upper side of the cap may be slightly sticky and dotted with dark-brown scales; underneath, the cap has light-colored gills, which produce millions of light yellow to white spores.

Armillaria causes a white rot of infected wood. When wood first begins to decay, it looks fairly water soaked, then it turns light brown. In the advanced stages of decay, wood becomes light yellow or white and may be marked by numerous black lines. Advanced decay is spongy in hardwoods but often stringy in conifers. In live trees, stem decay, referred to as *butt rot*, is confined largely to the inner woody tissues. Butt rot seldom extends more than a few feet above the ground.

Trees of different species and sizes may be killed individually throughout stands. This pattern often occurs in managed stands reforested with species unsuited to the site but may also occur in unmanaged stands. *Armillaria* also kills trees—primarily conifers—in a pattern of progressively expanding disease centers. These centers develop in managed or unmanaged stands and vary from small areas affecting several trees to areas of up to 1000 acres (400 ha). Within disease centers and on their expanding margins, trees in varying stages of decline are normally present. One or all species and sizes of conifers may be affected.

Armillaria may live for decades in coarse woody material. From its food source, the fungi spread to living hosts. Spread occurs when rhizomorphs, growing through the soil, contact uninfected roots or when uninfected roots contact infected ones. Rhizomorphs can grow for distances of up to 10 feet (3 m) through the upper soil layers, and they penetrate the roots by a combination of mechanical pressure and enzyme action. Rhizomorph growth and ability to penetrate roots depend upon the specific fungus, the type and amount of the food source, the soil environment, and the host species. When uninfected roots contact infected ones, the fungal mycelium invades uninfected roots without forming rhizomorphs. Such spread is common in dense stands where root contact is frequent. Vigorously growing trees often confine the fungi to localized lesions and limit their spread up the roots by secreting resin and rapidly forming callus tissues, but when infected trees are in a weakened condition *Armillaria* spreads rapidly through the roots. If the growth of the tree improves, fungal growth is checked. Such interaction occurs throughout the life of an infected host until it outgrows the fungi or the fungi reach the root collar, girdle the stem, and kill the tree.

When infected live trees are cut, *Armillaria* rapidly spreads into the uncolonized parts of roots and stump. As a result, the food source increases and may be responsible for initiating new disease centers. Outright mortality is the most frequently observed result of infection; it can be a problem in timber stands, recreation areas, or orchards. On the other hand, mortality can improve resource values—particularly in dense, young coniferous forests of the western United States. Infection also results in growth reduction and wood decay. Growth reduction often goes undetected or is ascribed to other agents and thus is probably underestimated. Likewise, decay extends only a few feet into the lower stem and will often go unnoticed until the tree fails or is cut. Tree failures are significant hazards in recreation and urban areas.

Because these fungi are indigenous to many areas and live on a wide variety of plants and woody material, their eradication or complete exclusion is not feasible; management should be directed toward limiting disease buildup or reducing its impact. Where individual trees are of high value, chemical fumigants, including chloropicrin, methyl bromide, and carbon disulfide, can reduce the infection level. These fumigants are applied in and around the base of infected stems or in holes left after trees have been uprooted.

Cultural management shows promise for dealing with *Armillaria* in commercial forests. Management considerations include: (1) reforesting stands with a mixture of species ecologically suited to the site and not obviously infected by *Armillaria*; (2) maintaining vigorous tree growth without causing undue damage to soils; (3) minimizing stress to and wounding of crop trees; and (4) reducing the food source by uprooting infected or susceptible root system and stumps. Where infection is limited, integrating the first three considerations into management prescriptions may be adequate. Where infection levels are high, such as in root disease centers, all four approaches may be used. Stumps and roots should be removed in a zone extending at least 33 feet (10 m) beyond the visible margin of the disease center because root systems in this area are likely infected. Sometimes other pests or stand conditions may be more significant than *Armillaria*. A thorough evaluation of existing or potential pest activity, site and stand characteristics, and the feasibility of the various options available should always be made before selecting a management alternative.

Summary of Nutrient Cycling under Natural Conditions

The following is a list of a few key points discussed to this point relative to nutrient cycling within the unharvested forest ecosystem, especially focusing on those issues that are likely to be of concern for the harvested forest—many of these concerns are addressed in sections that follow.

- Nutrients naturally enter the forest ecosystem from the atmosphere or mineral weathering; nutrients exit primarily by leaching into groundwater.
- The key macronutrients for the forest ecosystem are nitrogen, phosphorus, potassium, and calcium.
- Trees (and all plants) uptake nutrients in mostly soluble, inorganic form from the soil. These inorganic nutrients are converted to an organic form by the plant for metabolism.
- Most of the tree's nutrients are distributed in the foliage (leaves, twigs, and branches). The bark and wood typically have fewer nutrients than the foliage.
- Depending on the tree species, nutrients are more or less efficiently recycled within the plant pool.
- Evergreen trees (especially conifers) require fewer amounts of nutrients than deciduous species and, as a result, can exist on relatively infertile soils.
- As the tree discards leaves, branches, or bark or dies, the tree's organic nutrients are returned to the soil.
- Organic material returned to the soil is decomposed and the nutrients are mineralized (i.e., converted to an inorganic form) by soil organisms depending on the physical conditions of the soil (e.g., moisture, temperature, aeration).
- Within the forest ecosystem, two pools of nutrients exist: a relatively large belowground soil pool and a smaller aboveground plant pool.
- The soil pool has several orders of magnitude of nutrients more than the plant pool. The soil pool consists mostly of unavailable organic nutrients and, to a much lesser extent, available inorganic nutrients. The total soil nutrient capital is largely unavailable for immediate plant use.

- Inorganic (available) nutrients comprise only a small fraction of the total nutrient capital in the soil pool.

- Nitrogen is required in large amounts by plants and is typically a limiting nutrient in most terrestrial ecosystems. Phosphorus is sometimes a limiting nutrient, especially in areas of ancient, well weathered soils of the Southeast. Potassium deficiencies are linked to increased tree mortality, especially from mountain pine beetles.

- Aspen and several eastern broadleaf deciduous species are atypical of most tree species in that they accumulate a high percentage of calcium in the aboveground tissue. Intensive harvest of these species has the potential to result in a significant loss of soil calcium over time.

Woody Biomass Collection or Conversion Effects on Soil Resources

One of the primary concerns cited by conservation groups is the effect of thinning and biomass removal activities on forest soils. Thinning and removal of small-diameter wood and slash generally requires lighter equipment than traditional logging products, but active harvesting of any kind entails some degree of soil impact. A key consideration is weighing the physical soil impacts of management activities to remove small-diameter wood against impacts that can result from uncharacteristically severe wildfires that may occur if stands are left untreated.

Removing thinnings and slash from a stand rather than leaving it onsite to decompose can also affect soil chemistry. Decomposing wood helps replenish soil nutrients, but managers must consider how leaving excess forest biomass on the forest floor affects fuel loadings and wildfire risk, as uncharacteristically severe wildfire can also negatively impact soil qualities. Whether or not removing biomass negatively affects soil fertility and the growth of residual plants depends on the overall soil nutrient budget in the stand and which nutrients are limiting.

A primary challenge with understory thinning, and a major concern to those who are concerned about forest soil degradation, is that the low value of the wood usually targeted for removal encourages the use of low-cost harvesting methods, such as the use of ground-based equipment (Brown, 2000). Soil not only is the fundamental source of productivity in forest ecosystems but also strongly influences hydrologic function and water quality. Soil compaction, which can take decades to reverse, reduces plant growth and also inhibits infiltration of water, which increases erosion, sedimentation, and spring runoff. Fire can also adversely affect forest soils, but Brown (2000) suggested that these effects are relatively short-lived. To maintain ecological integrity during thinning operations, it is essential to employ low-impact equipment and use it properly. Establishing standards to keep soil compaction, disturbance, and puddling to less than 10% of a project area and monitoring to ensure that these standards are met would be a significant and feasible improvement over past practices (Brown, 2000).

Using a replicated field experiment in Montana, Gundale et al. (2005) studied the responses, 1 and 3 years after forest restoration treatment, of the physical, chemical, and biological properties of soil to the treatments of thinning, controlled burning, or thinning combined with controlled burning. Individual restoration treatments were implemented in 9-ha units. Treatments had a pronounced effect on the depth of the soil organic (O) horizons. Both the burn and thin/burn treatments had thinner O horizons, whereas the thin-only treatment had a thicker O horizon compared to the control. By year 3, the O horizons remaining diminished in both burn treatments, but the thin-only treatment did not differ from the control, suggesting that significant settling and decomposition had occurred. During year 1, both burn treatments had significantly more exposed mineral soil compared to the thin-only and control, but these differences were no longer detectable by year 3. No differences in soil bulk density were detected among the treatments. Most likely this was the result of harvest operations conducted on frozen and snow-covered soil using harvesting techniques designed to minimize soil compaction. A high C:N substrate decomposed more rapidly in both burn treatments relative to the unburned treatments. Treatments had no immediate effect on the soil microbial community.

Moghaddas and Stephens (2008) evaluated mechanical fuel treatment effects on soil compaction in a managed Sierra Nevada mixed-conifer forest using three treatments: thin only, thin and burn, and an untreated control. To examine impacts of mastication equipment traveling through a stand to reduce fuels, soil sampling was stratified to examine the effects at the treatment unit, skid trail network, and non-skid trail area scales. At all scales, thin-only and thin/burn treatments did not increase soil bulk density compared to the control. At the treatment unit level, soil strength was increased in thin and burn relative to the control, but this was attributed to increased strength in skid trails rather than in the non-skid portion of the stand. Compacting forces of the masticator were buffered by the debris bed it created. No significant compaction due to mastication was observed away from the skid trails. Soil strength appeared to be a more sensitive measure of compaction, although a very weak relationship was observed between soil bulk density and soil strength. Despite frequent stand entries prior to these fuel treatments, the cumulative extent of detrimental compaction was not increased as a result of thin-only or thin/burn treatments. Mean soil strength in skid trails was consistently greater than in on-skid trail areas to a depth of nearly 60 cm. Measures to avoid creation of new skid trails will help curtail increased soil compaction in managed forest stands, particularly in areas that may require repeated fuel treatments to remain effective.

In mixed-conifer stands in northeastern Oregon, McIver et al. (2003) examined fuel reduction by mechanical thinning and removal in an experiment that compared a single-grip harvester coupled with either a forwarder or a skyline yarding system and unharvested control sites. Both extraction systems achieved nearly equivalent (~46%) mass fuel reduction. Of 37 logged hectares, 1.4% (0.5 ha) of the soil area was compacted, mostly within forwarder units, log landings, and tails close to landings. Displaced soil varied from 5 to 43% area among units and was located within trails or in intertrail areas between the trails.

A major public concern with any natural or human-made forest disturbance activity is the potential for forest stream pollution by excessive sedimentation. For sound management of forest resources, adequate erosion simulation tools are needed. The Water Erosion Prediction Project (WEPP) watershed model (applicable to agriculture, rangelands, and forests), a physically based erosion prediction software program developed by the U.S. Department of Agriculture, has proved useful. This has especially been the case in areas where Hortonian flow dominates, such as modeling erosion from roads, harvested units, or areas burned by wildfire or prescribed fire (Dun et al., 2009). (Hortonian flow occurs when water flows horizontally across land surfaces when rainfall has exceeded the infiltration capacity.)

Using the WEPP model, Elliot and Miller (2002) compared erosion rates from fuel management operations, including roads, to erosion following wildfire for climates across the western United States. Forest sediment yields were estimated with the Disturbed WEPP online forest erosion prediction interface, an adjunct to the WEPP model that allows users to easily describe numerous disturbed forest and rangeland erosion conditions. The interface presents the probability of a given level of erosion conditions occurring the year following a disturbance. All scenarios assumed that if no harvesting occurred then the area would progress to the point of high-severity fire. No scenarios were examined for undisturbed conditions (neither harvesting nor fire occurrences). All thinning and prescribed fire simulations retained 85% surface cover as recommended by most Forest Service regional soil quality guidelines. The wildfire scenario assumed a 45% cover. Results indicated that erosion from fuel treatments, including thinning and prescribed fire, is less than from wildfire, even when road erosion rates are included. Forest erosion rates from the wildfire scenario were predicted to be about 40 times the erosion rates from prescribed fire with buffers. Erosion due to thinning was predicted to be about 70% that of prescribed fire, or 1% that of wildfire.

Treatment Effects on Limiting Nutrients

Among the factors affecting growing conditions in forestlands, the one that, if increased, will result in the greatest corresponding increase in productivity of the stand is considered to be the critical or "most limiting factor." Nitrogen (N) is a critical limiting nutrient that regulates plant productivity and the cycling of other essential elements in forests. Johnson and Curtis (2001) performed a meta-analysis of forest management effects on soil carbon (C) and nitrogen (N). Results indicated that forest harvesting, on average, had little or no effect on soil C and N, but significant effects of harvest type and species were noted. Leaving residues on site (sawlog harvesting) caused an 18% increase in soil C and N, whereas residue removal (whole-tree harvesting) caused a 6% decrease compared to controls. The positive effect of sawlog harvesting appeared to be restricted to coniferous species, although reasons for this were not clear. Conversely, several studies from widely varied conditions clearly showed that residues had little or no effect on soil C or N in hardwood or mixed forests. Johnson and Curtis (2001, p. 235) discussed harvesting residues:

> Several studies found that soil C and N temporarily increase after sawlog harvesting, apparently a result of residues becoming incorporated into the soil. The general trends found in a number of these studies, however, are consistent with the concept of high C:N ratio residues becoming incorporated into soils over the short term, with soil C re-equilibrating to lower levels and to C:N ratios more similar to background as time passes. This raises other questions regarding C balances. Specifically, what is the long-term fate of the residues? They remain part of the O [organic] horizon for long periods in some cool coniferous forests but in warmer hardwood forests they rapidly decompose. If leaving residues on site has no long-term positive effect on mineral soil C, removing residues for biomass burning may be more C efficient (by offsetting fossil fuel combustion) than leaving them on site. Conversely, nutrients left behind in residues may result in long-term carbon gains in aboveground vegetation and cause residue removal to be less C efficient. If the latter is true, how do the C and economic costs of fertilization compare with the costs of leaving residues on site? Our analysis in this study can only provide partial answers to these questions. Indeed, the answers to these questions, while very important to both intensive forestry and the global C issue, surely vary substantially by site and probably defy generalization.

Nutritional Value of Coarse Woody Debris

Prescott and Laiho (2002) assessed the nutritional significance of coarse woody debris (CWD) in lodgepole pine, white spruce–lodgepole pine, and subalpine fir–Engelmann spruce forests in the Rocky Mountains of southern Alberta, Canada. Mass loss and changes in carbon (C), nitrogen (N), and phosphorus (P) concentrations in decomposing log segments were measured over a 14-year period. Organic matter input was measured for 10 years for CWD, 1 year for ground vegetation, and 5 years for other aboveground litter types. Carbon, N, and P releases from decomposing litter were simulated for a period of 40 years to determine the relative contribution of each aboveground litter type, including CWD. After 14 years, pine log segments had lost 71% of their dry mass; spruce and fir lost 38% and 40%, respectively. The nitrogen content of the logs increased in pine, changed little in spruce, and decreased by almost 30% in fir logs. Phosphorus accumulated in decaying log segments of all three species, especially fir logs, in which P content was nearly five times the initial content after 14 years. Tree species with the lowest initial concentration had the greatest relative accumulation; thus, wood decay organisms may compete with vegetation for limiting nutrients in these forests. The proportion of CWD in aboveground litter input was 19% in the pine site, 3% at the spruce site, and 24% at the fir site. The contribution of CWD to N and P release was 2% or less, except at the fir site, where CWD released 5% of the N. Prescott and Laiho (2002) stated that their estimates were only valid for the forests they studied and at the time of measurements, but the sites represent a variety of common forest types and stand development stages: a dense self-thinning lodgepole pine stand, a mature post-thinning spruce stand, and an old-growth fir–spruce stand. Their findings indicate that CWD does not appear to make a significant contribution to N and P cycling in these forests and may actually compete with vegetation for limiting nutrients. The authors argued that guidelines for management of CWD should be based on management objectives related to other potential values (e.g., wildlife habitat) rather than its role in N and P cycling.

Whole-Tree Harvest Effects on Forest Soil Productivity

The studies cited in the following discussion typically involve investigations whereby the nutrients of aboveground biomass were determined empirically, and the expected impact of intensive (nutrient draining) harvesting on the ecosystem was calculated or predicted as changes in soil nutrient availability and productivity over time. These analyses typically used nutrient budgets and simulation models to forecast changes in nutrient pool sizes (i.e., plant and soil pools) and transfer rates between the pools. These typically assume that the nutrient pool sizes and transfer rates between nutrient pools are the prime regulators of nutrient availability and forest productivity. Forest harvesting (nutrient removal) is then used to predict future nutrient availability and soil productivity within the ecosystem. The major shortcomings of these forecasts and simulations are that few forest ecosystems have been studied in sufficient detail to accurately predict nutrient transfers between pools, and the rates of key ecosystem processes involved in nutrient cycling (e.g., decomposition, weathering, nutrient mobilization/immobilization) are poorly understood. As a result, the predictions of the consequences of various harvesting regimes may be grossly over- or understated. Nevertheless, up until long-term studies began in the 1990s, these forecasts were the best available information that resource specialists had to assess soil productivity impacts from intensive harvesting.

Nutrient Capital and Nutrient Mining

The terms *nutrient capital* and *nutrient mining* will be used extensively in this discussion of the effects of whole-tree harvesting. *Nutrient capital* refers to the nutrients in the ecosystem that are available to plants depending on the internal dynamics of the ecosystem. If inputs of nutrients exceed the outputs from the ecosystem, then the nutrient capital is expected to increase. The initial nutrient capital of a site is considered to be the amount of nutrient stored in the soil prior to a disturbance to the ecosystem. *Nutrient mining* describes the rate of removal of nutrients that exceeds the rate of replenishment into the forest ecosystem. There has been concern that whole-tree harvesting results in significant nutrient mining of the nutrient capital of the forest ecosystem (Grigal and Bates, 1992).

General Impacts of Whole-Tree Harvesting

Because most nutrients of the plant (aboveground) pool are in the branches and foliage, a whole-tree harvest can remove as much as three times the nutrients as compared to a conventional bole-only harvest (Alban et al., 1978; Johnson et al., 1982; Phillips and Van Lear, 1984). However, because the belowground soil nutrient pool contains most of the nutrient capital of a forest ecosystem (by several orders of magnitude), in general, the removal of the whole tree during timber harvesting should result in only a small percentage of nutrient loss from the forest ecosystem. In general, simulation and nutrient budget studies predict that the longer the interval between rotations, the more likely nutrients lost from whole-tree harvesting are naturally replenished to preharvest conditions (Aber et al., 1978; Kimmins, 1977; Swank and Waide, 1980). Mann et al. (1988) compared preharvest and postharvest nutrient budgets for bole-only vs. whole-tree harvest in sites in seven states (Tennessee, Washington, Maine, Connecticut, New Hampshire, South Carolina, and Florida). That analysis concluded that whole-tree harvests remove significantly more nutrients than bole-only harvests from the ecosystem. In general, phosphorus and calcium (at eastern hardwood sites) are the nutrient most likely lost (mined) if not replaced by natural mineral weathering between rotations. Yanai (1998) provided an analysis of whole-tree harvesting effects of phosphorus cycling using nutrient budgets and measurements of uptake of phosphorus in vegetation regrowth 2 years after the harvest. Impacts to nutrient supply (i.e., loss of phosphorus) within the forest ecosystem were considered small or negligible. Predictions of impacts from future harvest rotations were not made because the rates of key ecosystem processes (e.g., mineralization/immobilizations, decomposition, mineral weathering) that input phosphorus to the ecosystem are unknown and difficult to estimate.

Mining of Calcium from the Soil Pool

Of the key macronutrients, calcium is consistently predicted to be susceptible to nutrient mining, as soil calcium levels are typically low (but not typically limiting) at many forest locations, and calcium accumulation in the aboveground pool of several tree species tends to be significant. Nutrient budget and simulation studies from the late 1970s through the later 1980s began to identify calcium as the nutrient most susceptible to being mined from the soil pool of eastern forest soils. Researchers began to document that certain deciduous trees (e.g., aspen, white spruce, sugar maple, beech, yellow birch) stored a significant amount of calcium in their aboveground perennial tissue (especially branches and stems) which, in effect, operated as a nutrient sink for calcium. Even though calcium is seldom a limiting factor for most forest soils, under whole-tree harvesting scenarios, it is consistently predicted to be a nutrient that might be mined from the soil pool and lost from the ecosystem.

Long-Term Validation Studies of Whole-Tree Harvest Effects

By the mid-1980s, researchers began to critically examine and question the predictions and forecasts made in the previous decade of intensive harvesting vs. soil productivity. Researchers began to notice that the nutrient deficiencies that these studies predicted (which relied on nutrient budgets and model simulations) were not being observed on forest sites. Johnson (1983), in particular, critically questioned the oft-cited generalizations and predictions of nutrient deficiencies made by the previous forecasts. Dyck and Mees (1990) discussed the shortcomings of model simulations and forests based on chronosequence and retrospective studies. A chronosequence study involves research on a series of stands of various ages but with all stands having essentially the same treatment (Turner, 1981), and retrospective research involves examining previous treatments of stands and inferring impacts. Even though these studies provided a rapid means of assessing long-term effects of a treatment, a researcher has no control over the previous management practice and has to accept a host of simplifying assumptions (e.g., sites are assumed to have similar slopes, soils, natural pathogens). Dyck and Mees (1990) stressed that these studies are merely a convenient substitute for a more controlled (and time-consuming) long-term empirical study. As such, several researchers (Dyck and Bow 1992; Dyck and Mees, 1990; Powers et al., 1990; Smith, 1986) began to call for long-term empirical validation studies to validate the actual impacts of intensive harvesting on soil and forest productivity. Even though these studies require much more time to obtain definitive results, the researcher has control over key factors such as design and application of the treatments.

Long-Term Soil Productivity Program

The Long-Term Soil Productivity (LTSP) program began in 1989 and has developed into a national program for the Forest Service with the goal of examining the long-term consequences of forest harvesting on forest soil productivity. The partnerships and affiliations have expanded into Canada. The more than 100 LTSP core and affiliated sites comprise the world's largest coordinated research network addressing applied issues of forest management and sustained productivity. The LTSP program is based on the principle that the net primary productivity of a forest site is directed by physical, chemical, and biological soil processes that are readily affected by forest management practices. The key properties of the soil that are most affected by management are the soil porosity and organic matter of a site; therefore, the LTSP program targets porosity and organic matter for large-scale, long-term experiments.

The LTSP program has forest types, age classes, and conditions that are likely to come under active forest management. Plots are 0.4 ha in size with comparable soil and stand variability. Sites were sampled to quantify pretreatment standing biomass and nutrient levels in the overstory, understory, and forest floor. The program is planned to extend to include a physical rotation of about 20 years for tropical forests and as long as 80 years for boreal forests (Powers, 2006). Powers et al. (2005) drew the following conclusions based on analysis of 10-year old LTSP sites:

- Complete removal of surface organic matter from the forest floor leads to significant universal declines in soil carbon (C) and reduced availability of nitrogen (N).

- Loss of soil C and N had no general effect on standing forest biomass, except for aspen stands of the Great Lakes.

- Soil bulk density was increased by compaction treatments; increases were greater for soils with initial low-to-moderate densities.
- Forest productivity impacts from compaction depended on soil texture and presence of an understory.

Powers et al. (2005) acknowledged that because only a third of the LTSP installations had reached 10 years of age, it was possible that the above-mentioned trends may change as more sites reached the decade age.

Other Long-Term Studies

The effects of whole-tree vs. bole-only harvesting on a mixed-oak forest on the Oak Ridge Reservation, near Oak Ridge, Tennessee, was first reported as a 1979 clear-cut in a nutrient budget analysis by Johnson et al. (1982). The study area was reexamined in 1995 by Johnson and Todd (1998). In that analysis, Johnson and Todd (1998) compared the nutrient budget in 1995 with the budget projections of 1979. The reevaluation found that 15 years after treatment:

- Soil carbon was the same for bole-only and whole-tree harvesting.
- Soil calcium was greater in the bole-only area.
- Soil carbon and nitrogen increased in both areas over time.
- Foliar calcium, magnesium, and potassium levels were higher in the bole-only area than the whole-tree harvested area.
- Foliar nitrogen and phosphorus levels were not significantly different in both areas.
- Soil-exchangeable calcium did not decline in whole-tree harvesting as might be expected from predictions. Calcium from nonexchangeable reserves (i.e., deep soil reserves or bedrock) probably replenished exchangeable reserves.

Johnson and Todd (2007) concluded that nutrient budget analyses may have, in general, greatly overestimated the rates at which soil calcium would be depleted and greatly underestimated rates of nitrogen accumulation. They suggest that incomplete knowledge of weathering, deep rooting, and nitrogen fixation should cause researchers to be cautious about making long-term predictions without long-term studies.

Johnson et al. (1988) reported on the changes over an 11-year period of monitoring nutrient distribution in the forest and soils of the Walker Branch watershed, another site near Oak Ridge, Tennessee. The area was not the subject of an empirical study to validate the effects of intensive harvesting vs. soil productivity effects, but the area had been allowed to revert into a forest since 1942 from its previous use as a woodland pasture. As such, the soils and forests are considered far from having achieved a steady-state condition. The long-term monitoring provided some insight into the nutrient dynamics of certain forest ecosystems on eastern forest soils that have been previously disturbed. The 11-year period (1972 through 1982) of the monitoring study of vegetation, litter, and soil (surface and subsoil) nutrient contents revealed a general decline in fertility, especially for calcium and magnesium, in areas of low soil calcium and magnesium reserves.

Richter et al. (1994) reported on long-term monitoring of the Calhoun Experimental Forest in Union County, South Carolina, that had been allowed to revert to forest since about 1957. The previous 150+ years of afforestation for agricultural use of the area are known to have severely affected soil productivity. The long-term (1962 to 1990) monitoring results of soils, nutrient removals, and nutrient dynamics are similar to Johnson et al. (1998), especially in terms of the declines in soil calcium and magnesium in previously afforested eastern forest. Monitoring results suggest that multiple rotations are not sustainable on these impoverished soils (Richter et al., 1994).

The Hubbard Brook Experimental Forest in New Hampshire has been the site of three decades of biogeochemistry study of the flux of water and nutrients through an aggrading northern hardwood forest ecosystem (Likens and Bormann, 1995). The experimental forest (established in 1955) covers 7810 acres (3160 ha) of six contiguous watershed ecosystems where studies have been conducted on the interaction of the forest nutrient and hydrologic cycle. Of the hundreds of publications generated from the biogeochemistry studies at the experimental forest, several have monitored the long-term effects of harvesting on specific and general nutrient cycles.

Woody Biomass Collection or Conversion Effects on Water Quality

Undisturbed watersheds have little erosion, but natural forests have natural disturbances, including wildfire and large floods, with return periods that range from decades to centuries. When wildfires or floods occur, there will be significant upland erosion, as well as sediment deposition and movement in forest streams. Thus, long-term natural background sediment yields from watersheds are a combination of low levels of erosion from undisturbed forests plus added erosion from occasional disturbances. Activities such as thinning generally cause some level of disturbance. Erosion rates associated with these activities are generally much lower than rates from a wildfire, landslide, or flooding but may occur more frequently (Elliot and Robichaud, 2005). Estimates of long-term average annual sediment delivery for different types of disturbance in northern Rocky Mountain forest ecosystems were provided by Elliot and Robichaud (2005):

Wildfire:	0.15 Mg/ha = 0.0669 tons/acre
Prescribed fire:	0.001 Mg/ha = 0.0005 tons/acre
Thinning or logging:	0.005 Mg/ha = 0.0022 tons/acre
Roads (assume 2.5% of watershed):	0.125 Mg/ha = 0.0558 tons/acre

Woody Biomass Collection or Conversion Effects on Air Quality

The effects of noncommercial forest biomass removal and utilization on air quality is a complex topic that ranges from local-level smoke management concerns to national- and international-scale issues of carbon budgets, climate change, and energy policy. Finding economically viable ways to remove huge numbers of small, nonmerchantable trees from the landscape is the central dilemma facing managers as they try to implement hazardous fuel treatments. One option is to simply cut these trees and leave them on the forest floor; however, doing so often increases fire hazards and the severity of pest insect outbreaks.

Historically, this material was burned in either prescribed fires or in uncharacteristically severe wildfires, but high fuel loadings, air-quality restrictions, short windows of appropriate weather, and risk of escaped fire are some factors that limit application of prescribed fire.

Within its natural range, wildfire has a number of beneficial effects on forest ecosystems, and both prescribed fires and wildland fires are viable and useful tools for managers. On the other hand, open burning of forest biomass—whether in wildfires, prescribed fires, or slash burning—can have detrimental impacts on forest ecosystems and produce large amounts of visible smoke and particulates and significant quantities of nitrogen oxides, carbon monoxide, and hydrocarbons that contribute to the formation of atmospheric ozone. Wildfires also emit substantial quantities of carbon dioxide (CO_2) as well as methane and other trace gases (McNeil Technologies, 2003; Morris, 1999) and can impact human health. Quantification of emissions is difficult because of extreme variability in fuels, burning practices, and environmental conditions (Morris, 1999).

Converting biomass waste into energy is one promising way to sustainably address some of these problems. Use of biomass waste as power plant fuel vastly reduces the smoke and particulate emissions associated with its disposal and significantly reduces the amounts of carbon monoxide, nitrogen oxides, and hydrocarbons released to the atmosphere. By one estimate, if nonmerchantable forest thinnings were consumed in biomass power boilers instead of open burning, nitrogen oxide emissions could be reduced by 64% and particulate matter could be reduced by 97% (Antares Group, 2003).

Particulate matter emissions from open burning depend on the amount of fuel and type of fire, generally ranging from 25 to 40 pounds per ton of fuel burned. Estimates of fuel burned per acre range from 11.0 tons for a prescribed fire in a low-density stand to 79.5 tons during a high-intensity wildfire in a high-density stand (Sampson et al., 2001). Resulting emissions therefore could range from 275 to over 3000 lb per acre burned. Estimates of emissions from biomass power include 0.22 to 0.3 lb per MMBTU of fuel input (USEPA,

---

DID YOU KNOW?

Ozone and forest health variables are measured by the Forest Inventory and Analysis (FIA) program of the U.S. Forest Service (USFS, 2011). The Ozone Indicator is one of the primary parameters measured. Ground-level ozone, an air pollutant, is known to interact with forest ecosystems. Ozone pollution has been shown to reduce tree growth, alter species composition, and predispose trees to insect and disease attack. Ozone also causes direct foliar injury to many plant species. Affected leaves are often marked with discoloration and lesions, and they age more rapidly than normal leaves. This approach is known as *biomonitoring* and the plant species used are known as *bioindicators*. The U.S. Forest Service, forest biomass researchers, and others use information about ozone bioindicator plants to answer the following questions about the effects of air quality on forest ecosystems:

- Is regional air quality (specifically, ozone pollution) changing over time? If so, is it improving or deteriorating?
- In what percentage of a region or forest type are ozone bioindicator plants indicating the possibility of air pollution impacts on forest health (e.g., biodiversity, growth increment, crown condition, or damage)?

1995). This equates to 3.7 to 5.1 lb per ton of fuel, or 9 to 20% of the emissions from open burning. Emission levels for ethanol plants are expected to be similar or less than biomass power facilities (McNeil Technologies, 2003).

In three eastern Oregon counties, an assessment of the potential for utilization of forest biomass was made. The use of forest biomass was compared to the use of coal because both are solid fuels that employ similar technologies (McNeil Technologies, 2003). The use of biomass fuels produces lower emissions than coal-fired plants, so to the extent that biomass replaces coal use air quality will benefit. Biomass is lower in sulfur than is most U.S. coal. Typical biomass contains 0.05 wt% to 0.20 wt% sulfur on a dry basis. Biomass sulfur content translates to about 0.12 to 0.50 lb of sulfur dioxide per million BTU. Using biomass to generate power typically produces lower sulfur dioxide emissions compared to using coal. Nitrogen oxide emissions should also generally be lower for biomass, due to lower fuel nitrogen content and the higher volatile fraction of biomass vs. coal. The assessment report also addressed the complex issue of the carbon footprint of forest biomass energy in comparison to fossil fuels. Biomass power plants can produce large emissions of CO_2, sometimes even in excess of fossil fuel plants because of lower combustion efficiencies for biomass. The CO_2 released by combustion of forest biomass, however, was removed from the atmosphere in the recent past through photosynthesis and new plant growth will continue to remove CO_2 from the atmosphere after biomass is harvested—it can be said that the carbon is cycling. For this reason, it is often argued that biomass is CO_2 neutral. In practice, the picture is somewhat more complex. Other carbon flows are involved with biomass power production, including CO_2 emissions from fossil fuels burned during harvesting, processing, and transportation operations. Net CO_2 emissions from a biomass power plant are clearly lower than those from a fossil fuel plant, but under current production practices, biomass power is not a net-zero CO_2 process (McNeil Technologies, 2003).

DID YOU KNOW?

Forests take in CO_2 and water, store carbon in wood, and release oxygen. The carbon stored in forests is released back into the atmosphere when trees are burned, such as in forest fires, or when dead trees and leaves decay. Forest management can greatly affect the amount of carbon stored; vigorously growing forests store more carbon than slow growing ones. When trees are made into lumber or paper, some CO_2 is released, but much continues to be stored in the products or eventually in landfills. Substituting wood for nonrenewable materials can also reduce CO_2 in the atmosphere by reducing fossil fuel energy use (USFS, 2009).

Chapter Review Questions

12.1 In your opinion, does the removal of any type of forest biomass have a negative environmental impact? Explain.

12.2 In your opinion, does whole-tree removal from forest stands have a negative or positive impact on soil? Explain.

References and Recommended Reading

Aber, J.D., Botkin, D.B., and Melillo, J.M. (1978). Predicting the effects of different harvesting regimes on forest floor dynamics in northern hardwoods, *Canadian Journal of Forest Research*, 8(3):306–315.

Aerts, R. (1995). The advantages of being evergreen, *Tree*, 10(0):402–407.

Aerts, R. (1996). Nutrient resorption from senescing leaves of perennials: are there general patterns? *Journal of Ecology*, 84:597–608.

Agee, J.K. and Lolley, M.R. (2006). Thinning and prescribed fire effects on fuels and potential fire behavior in an Eastern Cascades Forest, Washington, USA, *Fire Ecology*, 2(2):142–158.

Agee, J.K., Bahro, B., Finney, M.A. et al. (2000). The use of shaded fuel breaks in landscape fire management, *Forest Ecology and Management*, 127(2000):55–66.

Alban, D.H. (1982). Effects of nutrient accumulation by aspen, spruce, and pine on soil properties, *Soil Science Society of America Journal*, 43:593–596.

Alban, D.H., Perala, D.A., and Schlaegel, B.E. (1978). Biomass and nutrient in aspen, pin, and spruce stands on the same soil type in Minnesota, *Canadian Journal of Forest Research*, 8:290–299.

Allee, W.C. (1932). *Animal Aggregations: A Study in General Sociology*, University of Chicago Press.

Allen, H. (1987). Forest fertilizers: nutrient amendment, stand productivity, and environmental impacts, *Journal of Forestry*, 85(2):37–46.

Allen, H.L., Dougherty, P.M., and Campbell, R.G. (1990). Manipulation of water and nutrients: practice and opportunity in southern U.S. pine forests, *Forest Ecology Management*, 30:437–453.

Anderson, J.G., Ulirich, R.C., Roth. L.I. et al. (1979). Genetic identification of clones of *Armillaria mellea* in coniferous forests in Washington, *Phytopathology*, 69(10):1109–1111.

Antares Group, Inc. (2003). *Assessment of Power Production at Rural Utilities Using Forest Thinnings and Commercially Available Biomass Power Technologies*, Task Order No. TOA KDC-9-29462-19, prepared for U.S. Department of Agriculture, U.S. Department of Energy, National Renewable Energy Laboratory, Golden, CO.

Apigian, K.O., Dahlsten, D.L., and Stephens, S.L. (2006). Fire and fire surrogate treatment effects on leaf litter arthropods in a western Sierra Nevada mixed-conifer forest, *Forest Ecology and Management*, 221:110–122.

Arno, S.F., Smith, J.Y., and Krebs, M.A. (1997). *Old Growth Ponderosa Pine and Western Larch Stand Structure: Influences of Pre-1900 Fires and Fire Exclusion*, U.S. Department of Agriculture, Forest Service, Intermountain Research Station, Fort Collins, CO.

Ashcroft, F. (2002). *Life at the Extremes: The Science of Survival*, University of California Press, Berkeley.

ASTM. (1969). *Manual on Water*, American Society for Testing and Materials, Philadelphia, PA.

Ausband, D.E. and Baty, G.R. (2005). Effects of precommercial thinning on snowshoe hare habitat use during winter in low-elevation montane forests, *Canadian Journal of Forest Research*, 35:1–5.

Bailey, J.D. and Tappeiner, L.C. (1998). Effects of thinning on structural development in 40- to 100-year-old Douglas fir stands in western Oregon, *Forest Ecology and Management*, 108:99–113.

Bailey, J.K. and Whitham, T.G. (2002). Interactions among fire, aspen, and elk affect insect diversity: reversal of a community response, *Ecology*, 83(6):1701–1712.

Ballard, R. (1978). Effects of slash and soil removal on the productivity of second rotation radiata pine on a pumice soil, *New Zealand Journal of Forest Science*, 8:248–258.

Ballard, R. and Gessel, S.P., Eds. (1983). *IUFRO Symposium on Forest Site and Continuous Productivity*, August 22–28, 1982, Seattle, WA, General Technical Report PNW-GTR-163, U.S. Department of Agriculture, Forest Service, Pacific Northwest Forest and Range Experiment Station, Portland, OR.

Barber, S.A. (1995). *Soil Nutrient Bioavailability: A Mechanistic Approach*, John Wiley & Sons, New York.

Bate, L.J. (1995). *Monitoring Woodpecker Abundance and Habitat in the Central Oregon Cascades*, University of Idaho, Moscow.

Beazley, J.D. (1992). *The Way Nature Works*, Macmillan, New York.

Beier, P. and Maschinski, J. (2003). Threatened, endangered, and sensitive species, in Frederici, P., Ed., *Ecological Restoration of Southwestern Ponderosa Pine Forests*, Island Press, Washington, D.C., pp. 306–327.

Bengston, G.W. (1979). Forest fertilization in the United States: progress and outlook, *Journal of Forestry*, 77:222–229.

Berryman, A.A. (1981). *Population Systems: A General Introduction*, Plenum Press, New York.

Berryman, A.A. (1993). Food web connectance and feedback dominance, or does everything really depend on everything else? *Oikos*, 68:13–185.

Berryman, A.A. (1999). *Principles of Population Dynamics and Their Application*, Stanley Thornes Publishers, Cheltenham, U.K.

Berryman, A.A. (2002). *Population Cycles: The Case for Trophic Interactions*, Oxford University Press, London.

Berryman, A.A. (2003). On principles, laws and theory in population ecology, *Oikos*, 103:695–701.

Betts, B.J. (1998). Roost use by mature colonies of silverhaired bats in northeastern Oregon, *Journal of Mammalogy*, 79:643–650.

Binkley, D. (1991). Connecting soils with forest productivity, in Harvey, A.E. and Neuenschwander, L.F., Eds., *Proceedings—Management and Productivity of Western-Montane Forest Soils (April 1990—Missoula, MT)*, General Technical Report INT-280, U.S. Department of Agriculture, Forest Service, Intermountain Research Station, Boise, ID, pp. 66–69.

Birch, T.W. (1996). *Private Forest-Land Owners of the United States (1994)*, Resource Bulletin NE-134, U.S. Department of Agriculture, Forest Service, Northeast Forest Experiment Station, Radnor, PA.

Bisson, P.A., Bilby, R.E., Bryand, M.D. et al. (1987). Large woody debris in forested streams in the Pacific Northwest: past, present and future, in Salo, E.O. and Cundy, T.W., Eds., *Streamside Management: Forestry and Fishery Interactions*, University of Washington Institute of Forestry Resources, Seattle, pp. 143–289.

Black, H.L. (1974). A north temperate bat community: structure and prey populations, *Journal of Mammalogy*, 55:138–157.

Bonner, J.T. (1965). *Size and Cycle*, Princeton University Press, Princeton, NJ.

Boyles, J.G. and Aubrey, D.P. (2006). Managing forests with prescribed fire: implications or cavity-dwelling bat species, *Forest Ecology and Management*, 222:108–115.

Brady, N.C. and Weil, R.R. (1996). *The Nature and Properties of Soils*, 11th ed., Prentice-Hall, Upper Saddle River, NJ.

Brawn, J.D. and Balda, R.P. (1988). The influence of silvicultural activity on ponderosa pine forest bird communities in the southwestern United States, *Bird Conservation*, 3:3–21.

Brockley, R.P. (2005). Effects of post-thinning density and repeated fertilization on the growth and development of young lodgepole pine, *Canadian Journal of Forest Research*, 35:1952–1964.

Brown, L.R. (1994). Facing food insecurity, in Brown L.R. et al., Eds., *State of the World*, W.W. Norton, New York.

Brown, R. (2000). *Thinning, Fire and Forest Restoration: A Science-Based Approach for National Forests in the Interior Northwest*, Defenders of Wildlife, West Linn, OR.

Brown, R. (2008a). *Report: Environmental Effects of Forest Biomass Removal*, Oregon Department of Forestry, Salem.

Brown, R. (2008b). *The Implications of Climate Change for Conservation, Restoration, and Management of National Forest Lands*, Defenders of Wildlife, Washington, D.C.

Bruns, H. (1960). The economic importance of birds in forests, *Bird Study*, 7(4):193–208.

Bull, E.L. (1986). Resource Portioning among Woodpeckers in Northeast Oregon, PhD dissertation, University of Idaho, Moscow.

Bull, E.L. and Blumton, A.K. (1999). *Effect of Fuels Reduction on American Marten and Their Prey*, Research Note PNW-RN-539, U.S. Department of Agriculture, Forest Service, Pacific Northwest Research Station, Portland, OR.

Bull, E.L., Parks, C.G., and Torgersen, T.R. (1997). *Trees and Logs Important to Wildlife in the Interior Columbia River Basin*, General Technical Report PNW-GTR-391, U.S. Department of Agriculture, Forest Service, Pacific Northwest Research Station, Portland, OR.

Bull, E.L., Akenson, J.J., and Henjum, M.G. (2000). Characteristics of black bear dens in trees and logs in northeastern Oregon, *Northwest Science*, 81:148–153.

Bull, E.L., Torgersen, R.T., and Wertz, T.L. (2001). The importance of vegetation, insects, and neonate ungulates in black bear diet in Northeastern Oregon, *Northwest Science*, 75:244–253.

Bull, E.L., Heater, T.W., and Youngblood, A. (2004). Arboreal squirrel response to silvicultural treatments for dwarf mistletoe control in northeaster Oregon, *Western Journal of Applied Forestry*, 19:133–141.

Bull, E.L., Heater, T.W., and Shepherd, J.F. (2005a). Habitat selection by the American marten in northeastern Oregon, *Northwest Science*, 79:37–43.

Bull, E.L., Clark, A.A., and Shepherd, J.F. (2005b). *Short-Term Effect of Fuel Reduction on Pileated Woodpeckers in Northeastern Oregon—A Pilot Study*, Research Paper PNW-RP-564, U.S. Department of Agriculture, Forest Service, Pacific Northwest Research Station, Portland, OR.

Bunnell, F.L., Kremsater, L.L., and Wind, E. (1999). Managing to sustain vertebrate richness in forests of the Pacific Northwest: relationships within stands, *Environmental Review*, 97:97–146.

Buol S.W., Southard, R.J, Graham, R.C., and McDaniel, P.A. (2003). *Soil Genesis and Classification*, 5th ed., Iowa State University Press, Ames.

Burger, J.A. and Kelting, D.L. (1998). Soil quality monitoring for assessing sustainable forest management, in Adams, M.B., Ramakrishna, K., and Davidson, E.A., Eds., *The Contribution of Soil Science to the Development of and Implementation of Criteria and Indictors of Sustainable Forest Management*, SSSA Special Publication 53, Soil Science Society of America, Madison WI.

Bury, R.B. and Corn, P.S. (1988). Response of aquatic and streamside amphibians to timber harvest: a review, in Raedeke, K.J., Ed., *Streamside Management: Riparian Wildlife and Forestry Interactions*, Contribution 59, University of Washington Institute of Forest Resources, Seattle, pp. 165–181.

Calder, W.A. (1983). An allometric approach to population cycles of mammals, *Journal of Theoretical Biology*, 100:275–282.

Calder, W.A. (1996). *Size, Function and Life History*, Dover, Mineola, NY.

Campbell, N.A. and Reece, J.B. (2004). *Biology*, 7th ed., Benjamin Cummings, Upper Saddle River, NJ.

Carey, A.B. (2001). Experimental manipulation of spatial heterogeneity in Douglas-fir forest: effects on squirrels, *Forest Ecology and Management*, 152:13–30.

Carey, A.B. and Wilson, S.M. (2001). Induced spatial heterogeneity in forest canopies: response of small mammals, *Journal of Wildlife Management*, 65:1014–1020.

Carey, A.B., Lippke, B.R., and Sessions, J. (1999a). Intention systems management: managing forest for biodiversity, *Journal Sustainable Forestry*, 9:83–125.

Carey, A.B., Maguire, C.C., Biswell, B.L., and Wilson. T.M. (1999b). Distribution and abundance of Neotoma in Western Oregon and Washington, *Northwest Science*, 73:65–79.

Carroll, C., Noss, R.F., and Paquet, P.C. (2001). Carnivores as focal species for conservation planning in the Rocky Mountain region, *Ecological Applications*, 11(4):961–980.

Carson, R. (1962). *Silent Spring*, Houghton Mifflin, Boston.

Chambers, C.L. (2002). Forest management and the dead wood resource in ponderosa pine forests: effects on small mammals, in Laudenslayer, Jr., W.F. et al., Eds., *Proceedings of the Symposium on the Ecology and Management of Dead Wood in Western Forests*, General Technical Report PSW-GTR-181, U.S. Department of Agriculture, Forest Service, Pacific Southwest Research Station, Albany, CA, pp. 679–693.

Chambers, C.L. and Germaine, S.S. (2003). Vertebrates, in Friederici, P., Ed., *Ecological Restoration of Southwestern Ponderosa Pine Forests*, Island Press, Washington, D.C.

Chambers, C.L. and Mast, J.N. (2005). Ponderosa pine snag dynamics and cavity excavation following wildfire in northern Arizona, *Forest Ecology and Management*, 216:227–240.

Chambers, C.L., Alm, V., Sider, M.S., and Rabe, M.J. (2002). Use of artificial roost by forest-dwelling bats in northern Arizona, *Wildlife Society Bulletin*, 30:1085–1091.

Chambers, J.Q., Higuchi, N., Schimel, J.P., Ferreira, L.V., and Melack, J.M. (2000). Decomposition and carbon cycling of dead trees in tropical forests of the central Amazon, *Oecologia*, 122:380–388.

Chen, Z., Grady K., Stephens, S., Villa-Castillo, J., and Wagner, M.R. (2006). Fuel reduction treatment and wildfire influence on carabid and tenebrionid community assemblages in the ponderosa pine forest of northern Arizona, USA, *Forest Ecology and Management*, 225:168–177.

Christensen, N.L. (1988). Succession and natural disturbance: paradigms, problems, and preservation of natural ecosystems, in Agee, J.K. and D.R. Johnson, Eds., *Ecosystem Management for Parks and Wilderness*, University of Washington Press, Seattle.

Cifelli, E.M., Ed. (1997). *The Collected Poems of John Ciardi*, University of Arkansas Press, Fayetteville.

Clark, L.R., Gerier, P.W., Hughes, R.D., and Harris, R.F. (1967). *The Ecology of Insect Populations*, Methuen, London.

Cole, D.W. and Rapp, M. (1980). Elemental cycling in forested ecosystems, in Reichle, D.E., Ed., *Dynamic Properties of Forest Ecosystems*, Cambridge University Press, Cambridge, U.K., pp. 341–409.

Coleman, D.C. and Hendrix P.F., Eds. (2000). *Invertebrates as Webmasters in Ecosystem*, CABI Publishing, Wallingford, U.K.

Colyvan, M. and Ginzburg, L.R. (2003). Laws of nature and laws of ecology, *Oikos*, 101:649–653.

Constible, J.M., Gregory, P.T., and Anholt, B.R. (2001). Patterns of distribution, relative abundance, and microhabitat use of anurans in a boreal landscape influence by fire and timber harvest, *Ecoscience*, 8:462–470.

Converse, S.J., White, G.C., Farris, K.L., and Zack, S. (2006). Small mammals and forest fuel reduction: national-scale response to fire and fire surrogates, *Ecological Applications*, 16(5):1717–1729.

Corn, P.S. and Bury, R.B. (1989). Logging in western Oregon: responses of headwater habitats and stream amphibians, *Forest Ecology and Management*, 29:19.

Crawford, C.S. (1976). Feeding-season production in the desert millipede *Orthoprous ornatus*, *Oecologia*, 24:265–276.

Curtis, S. (2006). Rubber boa, *Montana Outdoors* (fwp.mt.gov/mtoutdoors/HTML/articles/portraits/rubberboa.htm).

Damuth, J. (1981). Population density and body size in mammals, *Nature*, 290:699–700.

Damuth, J. (1987). Interspecific allometry of population density in mammals and other animals: the independence of body mass and population energy-use, *Biological Journal of the Linnean Society*, 31:193–246.

Damuth, J. (1991). Of size and abundance, *Nature*, 351:268–269.

Dasmann, R.F. (1984). *Environmental Conservation*, John Wiley & Sons, New York.

Davis, G.H. and Pollock, G.L. (2003). Geology of Bryce Canyon National Park, Utah, in Sprinkel, D.A. et al., Eds., *Geology of Utah's Parks and Monuments*, 2nd ed., Utah Geological Association, Salt Lake City.

DeGraaf, R.M., Healy, W.M., and Books, R.T. (1991). Effects of thinning and deer browsing on breeding birds in New England oak woodlands, *Forest Ecology and Management*, 41:179–191.

Demarais, S. and Krausman, P.R. (2000). *Ecology and Management of Large Mammals in North America*, Prentice Hall, Upper Saddle River, NJ.

DeMaynadier, P.G. and Hunter, Jr., M.L. (1995). The relationship between forest management and amphibian ecology: a review of the North American literature, *Environmental Review*, 3:230–261.

Dixon, R.D. (1995). Ecology of White-Headed Woodpeckers in the Central Oregon Cascades, PhD dissertation, University of Idaho, Moscow.

Dodson, E.K., Peterson, D.W., and Harrod, R.J. (2008). Understory vegetation response to thinning and burning restoration treatments in dry conifer forests of the eastern Cascades, USA, *Forest Ecology and Management*, 255(8–9):3130–3140.

Dun, S., Wu, J.Q., Elliot, W.J. et al. (2009). Adapting the Water Erosion Prediction Project (WEPP) model for forest applications, *Journal of Hydrology*, 366:46–54.

Dyck, W.J. and Bow, C.A. (1992). Environmental impacts of harvesting, *Biomass and Bioenergy*, 2:173–191.

Dyck, W.J. and Mees, C.A. (1990). Nutritional consequences of intensive forest harvesting on site productivity, *Biomass*, 22:171–186.

Edge, W.D. and Manning, J. (2002). Response of small mammals to fuel treatments in southwest Oregon, in S.A. Fitzgerald, Ed., *Fire in Oregon's Forests: Risks, Effects, and Treatment Options. A Synthesis of Current Issues and Scientific Literature*, Special Report for Oregon Forest Resources Institute, Portland, OR.

Edmonds, R.L, Binkley, D., Feller, M.C., Sollins, P., Abee, A., and Myrold, D.D. (1989). Nutrient cycling: effects on productivity of northwest forests, in Perry, D.A. et al., Eds., *Maintaining the Long-Term Productivity of Pacific Northwest Forest Ecosystems*, Timber Press, Portland, OR, pp. 17–35.

Elkin, C.M. and Reid, M. (2004). Attack and reproductive success of mountain pine beetles (Coleoptera: Scolytidae) and fire-damaged lodgepole pine, *Environmental Entomology*, 33:1070–1080.

Elliot, W.J. and Miller, I.S. (2002). *Estimating Erosion Impacts from Implementing the National Fire Plan*, paper presented at the 2002 ASAE Annual International Medting/CICR XVth World Congress, July 28–31, 2002, Chicago, IL, American Society of Agricultural Engineers, St. Joseph, MI.

Elliot, W.J. and Robichaud, P. (2005). *Evaluating Sedimentation Risks Associated with Fuel Management*, Fuels Planning: Science Synthesis and Integration, Environmental Consequences Fact Sheet 8, Research Note RMRS-RN-23-08-WWW, U.S. Department of Agriculture, Rocky Mountain Research Station, Fort Collins, CO.

Elliot, W.J., Page-Dumroese, D., and Robichaud, R. (1999). The effects of forest management on erosion and soil productivity, in Lal, R., Ed., *Soil Quality and Soil Erosion*, CRC Press, Boca Raton, FL, 195–208.

Enger, E., Kormclink, J.R., Smith, B.F., and Smith, R.J. (1989). *Environmental Science: The Study of Interrelationships*, William C. Brown, Dubuque, IA.

Eswaran, H. (1993). Assessment of global resources: current status and future needs, *Pedologie*, 43(1):19–39.

Evans, A.M. (2008). *Synthesis of Knowledge from Woody Biomass Removal Case Studies*, Forest Guild, Santa Fe, NM.

Everett, W.J., Schellhaus, D., Spurbeck, D. et al. (1997). Structure of northern spotted owl next stands and their historical conditions on the eastern slope or the Pacific Northwest cascades, USA, *Forest Ecology and Management*, 94:1–14.

Federer, C.A., Hornbeck, J.W., Triton, L.M., Martin, C.W., Pierce, R.S., and Smith, C.T. (1989). Long-term depletion of calcium and other nutrients in eastern U.S. Forests, *Environmental Management*, 13(5):593–601.

Fenchel, T. (1974). Intrinsic rate of natural increase: the relationship with body size, *Oecologia*, 14:317–326.

Finch, D.M., Ganey, J.L. Yong, W., Kimball, R.T., and Sallabanks, R. (1997). Effects and interactions of fire, logging, and grazing, in Block, W.M. and Finch, D.M., Eds., *Songbird Ecology in Southwestern Ponderosa Pine Forests: A Literature Review*, General Technical Report RM-GTR-292, U.S. Department of Agriculture, Forest Service, Rocky Mountain Forest and Range Experiment Station, Fort Collins, CO, pp. 103–136.

Fisher, J.T. and Wilkinson, L. (2005). The response of mammals to forest fire and timber harvest in the North American Boreal forest, *Mammal Review*, 3:51–81.

Fisher, R.F. and Binkley, D. (2000). *Ecology and Management of Forest Soils*, 3rd ed., John Wiley & Sons, New York.

Ford-Robertson, F.C. (1983). *Terminology of Forest Science Technology Practice and Products*, FAO/UFRO Committee on Forestry Bibliography and Terminology, Society of American Foresters, Washington, D.C.

Foth, H.D. (1978). *Fundamentals of Soil Science*, 6th ed., John Wiley & Sons, New York.

Fox, T.R., Allen, H.L., Albaugh, T.J., Rubilar, R., and Carlson, C.A. (2007). Tree nutrition and forest fertilization of pine plantations in the southern United States, *Southern Journal of Applied Forestry*, 31(1):5–11.

Franck, I. and Brownstone, D. (1992). *The Green Encyclopedia*, Prentice Hall, New York.

Franklin, J.F. and Forman, R.T.T. (1987). Creating landscape patterns by forest cutting: ecological consequences and principles, *Landscape Ecology*, 1:5–18.

Fule, P.Z., Covington, W.W., and Moore, M.M. (1997). Determining reference conditions for ecosystem management of southwestern ponderosa pine forests, *Ecological Applications*, 7(3):895–908.

Garrison-Johnston, M.T., Moore, J.A., Shaw, T.M., and Mika, P.G. (2000). Foliar nutrient and tree growth response of mixed-conifer stands to tree fertilization treatments in northwest Oregon and north central Washington, *Forest Ecology and Management*, 132:183–198.

Garrison-Johnston, M.T., Moore, J.A., Cook, S.P., and Niehoff, G.J. (2003). Douglas-fir beetle infestations are associated with certain rock and stand types in the inland northwestern United States, *Environmental Entomology*, 32(6):1364–1369.

Garrison-Johnston, M.T., Shaw, T.M., Mika, P.G., and Johnson, L.T. (2005). *Management of Ponderosa Pine Nutrition Fertilization*, General Technical Report PSW-GTR-198, U.S. Department of Agriculture, Forest Service, Pacific Southwest Research Station, Albany, CA.

Garrison-Johnston, M.T., Lewis, R., and Johnson, L.R. (2007). *Northern Idaho and Western Montana Nutrition Guidelines by Rock Type*, Intermountain Forest Tree Nutrition Cooperative (IFTNC), University of Idaho, Moscow.

Gast, Jr., W.R., Scott, D.W., Schmitt, C. et al. (1991). *Blue Mountains Forest Health Report: New Perspectives in Forest Health*, U.S. Department of Agriculture, Forest Service, Malheur, Umatilla, and Wallowa-Whitman National Forests.

Germaine, H.L. and Germaine, S.S. (2002). Forest restoration treatment effects on the nesting success of western bluebirds (*Sailia mexicano*), *Restoration Ecology*, 110:362–367.

Gillespie, G.R. (2002). Impacts of sediment loads, tadpole density, and food type on the growth and development of tadpoles of the spotted tree frog *Litoria spenceri*: an in-stream experiment, *Biological Conservation*, 106:141–150.

Ginzburg, L.R. (1986). The theory of population dynamics. 1. Back to first principles, *Journal of Theoretical Biology*, 122:385–399.

Ginzburg, L.R. and Colyvan, M. (2004). *Ecological Orbits: How Planets Move and Populations Grow*, Oxford University Press, New York.

Ginzburg, L.R. and Jensen C.X.J. (2004). Rules of thumb for judging ecological theories, *Trends in Ecology and Evolution*, 19:121–126.

Grigal, D.F. (2004). *An Update of Forest Soils: A Technical Paper for a Generic Environmental Impact Statement on Timber Harvesting and Forest Management in Minnesota*, Laurentian Energy Agency, Virginia, MN.

Grigal, D.F. and Bates, P.C. (1992). *Forest Soils: A Technical Paper for a Generic Environmental Impact Statement on Timber Harvesting and Forest Management in Minnesota*, Jaako Poyry Consulting, Tarrytown, NY.

Gundale, M.J., Deluca, T.H., Fiedler, C.E., Ramsey, P.W., Harrington, M.G., and Gannon, J.E. (2005). Restoration treatments in a Montana ponderosa pine forest: effects on soil physical, chemical and biological properties, *Forest Ecology and Management*, 213:25–38.

Guy, S.E. and Manning, E.T. (1995). *Wildlife/Danger Tree Assessor's Course Workbook*, 4th ed., British Columbia Ministry of Forests, Victoria.

Haemig, P.D. (2006). *Laws of Population Ecology*, Ecology Info 23, Ecology Online Sweden (http://www.ecology.info/laws-population-ecology.htm).

Hagar, J.C., McComb, W.C., and Emmingham, W.H. (1996). Bird communities in commercially thinned and unthinned Douglas fir stands of western Oregon, *Wildlife Society Bulletin*, 24(2):353–366.

Hagar, J., Howlin, S., and Ganio, L. (2004). Short-term response of songbirds to experimental thinning of young Douglas-fire forests in the Oregon Cascades, *Forest Ecology and Management*, 199:333–347.

Hamilton, R.J. (1981). Effects of prescribed fire on black bear populations in southern forests, in Wood, G.W., Ed., *Prescribed Fire and Wildlife in Southern Forests*, Clemson University, Belle W. Baruch Forest Science Institute, Georgetown, SC, pp. 129–134.

Hanula, J.L. and Wade, D.D. (2003). Influence of long-term dormant-season burning and fire exclusion on ground-dwelling arthropod populations, in longleaf pine flatwood ecosystems, *Forest Ecology and Management*, 175:163–184.

Hanula, J.L., Meeker, J.R., Miller, D.R., and Barnard, E.L. (2002). Association of wildfire with tree health and numbers of pine bark beetles, reproduction weevils and their associates in Florida, *Forest Ecology and Management*, 170:233–247.

Harmon, M.E. and Sexton, J. (1996). *Guidelines for Measurements of Woody Detritus in Forest Ecosystems*, Publication No. 20, U.S. Long-Term Ecological Research Network Office, University of Washington, Seattle.

Harmon, M.E. et al. (1986). Ecology of coarse woody debris in temperate ecosystems, *Advances in Ecological Research*, 15:133–302.

Harvey, A.E. (1994). Integrated roles for insects, disease and decomposers in fire dominated forests of the inland Western United States: past, present and future forest health, *Journal of Sustainable Forestry*, 2:211–220.

Hayes, J.S., Weikel, J.M., and Huso, M.P. (2003). Response of birds to thinning young Douglas-fir forests. *Ecological Applications*, 13(5):1222–1232.

Heatwole, H. (1977). Habitat selection in reptiles, in Gans, C. and Tinkle, D.W., Eds., *Biology of the Reptilia*. Vol. 7. *Ecology and Behavior*, Academic Presses, New York, pp. 137–155.

Hejl. S.J. Mack, D.E., Young, J.S. Bednarz, J.C., and Hutte, R.L. (2002). Birds and changing landscape patterns in conifer forests of the north-central Rocky Mountains, *Studies in Avian Biology*, 25:113–129.

Helmisaari, H.S. (1995). Nutrient cycling in *Pinus sylvestris* stands in eastern Finland, *Plant and Soil*, 168–169:327–336.

Helms, J.A., Ed. (1998). *The Dictionary of Forestry*, Society of American Foresters, Bethesda, MD.

Henjum, M. (2006). *Wildlife in the WUI: Framing the Issue*, Wildlife and Fire in the Wildland-Urban Interface Conference, Oregon Chapter of the Wildlife Society, April 18–19, Bend, OR.

Hickman, C.P., Roberts, L.S., and Hickman, F.M. (1990). *Biology of Animals*, Times Mirror/Mosby, St. Louis, MO.

Hobbie, S.E. (1992). Effects of plant species on nutrient cycling, *Trends in Ecology and Evolution* (TREE), 7(10):336–339.

Holmquist, J.G. (2004). *Terrestrial Invertebrates: Functional Roles in Ecosystems and Utility as Vital Signs in the Sierra Nevada*, reported submitted to the National Park Service Inventory and Monitoring Program in partial fulfillment of Cooperative Agreement H8R07010001 and Task Agreement J8R07030011.

Hopper, D.U., Bignell, D.E., Brown, V.K. et al. (2000). Interactions between aboveground and belowground biodiversity in terrestrial ecosystems: patterns, mechanisms, and feedbacks, *BioScience*, 50(12):1049–1061.

Hornbeck, J.W. (1988). *Nutrient Cycling: A Key Consideration When Whole-Tree Harvesting*, University of Maine, Maine Agricultural and Forest Experiment Station, Orono, pp. 7–11.

Hornbeck, J.W. and Kropelin, W. (1982). Nutrient removal and leaching from whole-tree harvest of northern hardwoods, *Journal of Environmental Quality*, 11:309–316.

Hubbell, S.P. and Johnson, L.K. (1977). Competition and next spacing in a tropical stingless bee community, *Ecology*, 58:949–963.

Huffman, D.W and Moore, M.M. (2004). Responses of *Fendler ceanothus* to overstory thinning, prescribed fire, and drought in an Arizona ponderosa pine forest, *Forest Ecology and Management*, 198:105–115.

Humes, M.L., Hayes, J.P., and Collopy, M.W. (1999). Bat activity in thinned, unthinned, and old-growth forests in western Oregon, *Journal of Wildlife Management*, 63:553–561.

Huntington, T.G., Hopper, R.P., Olson, C.E., Aulenbach, B.T., Cappellato, R., and Blum, A.E. (2000). Calcium depletion in southeastern United States forest ecosystem, *Soil Science Society of American Journal*, 64:1845–1858.

Huntzinger, M. (2003). Effects of fire management practices on butterfly diversity in forested Western United States, *Biological Conservation*, 113:1–12.

James, S.E. and M'Closkey, R.T. (2003). Lizard microhabitat and fire fuel management, *Biological Conservation*, 114:293–297.

Janzen, D.H. (1987). Insect diversity of a Costa Rican dry forest: why keep it, and how? *Biological Journal of the Linnaean Society*, 30:343–356.

Johnson, D.W. (1983). The effects of harvesting intensity on nutrient depletion in forests, in Ballard, R. and Gessel, S.P., Eds., *IUFRO Symposium on Forest Site and Continuous Productivity*, August 22–28, 1982, Seattle, WA, General Technical Report PNW-GTR-163, U.S. Department of Agriculture, Forest Service, Pacific Northwest Forest and Range Experiment Station, Portland, OR, pp. 157–166.

Johnson, D.W. and Curtis, P.S. (2001). Effects of forest management on soil C and N storage: meta analysis, *Forest Ecology and Management*, 140(15):227–238.

Johnson, D.W. and Todd, D.E. (1998). Harvesting effects on long-term changes in nutrient pools of mixed oak forest, *Soil Science Society of America Journal*, 62:1725–1735.

Johnson, D.W., West, D.C., Todd, D.E., and Mann, L.K. (1982). Effects of sawlog versus whole-tree harvesting on the nitrogen, phosphorous, potassium, and calcium budget of upland mixed oak forest, *Soil Science Society of America Journal*, 46:1353–1363.

Johnson, D.W., Van Miegroet, H., Lindberg, S.E., Todd, D.E., and Harrison, R.B. (1991). Nutrient cycling in red spruce forests of the great Smokey Mountains, *Canadian Journal of Forest Research*, 21:769–787.

Johnson, D.W., Todd, D.E., Trettin, F.F., and Sedinger, J.S. (2007). Soil carbon and nitrogen changes in forests of walker Branch Watershed (1972 to 2004), *Soil Science Society of America Journal*, 71:1639–1646.

Kelly, M.J., D'Amato, A.W., and Barten, P.K. (2008). *Silvicultural and Ecological Considerations of Forest Biomass Harvesting in Massachusetts*, Department of Natural Resources Conservation, University of Massachusetts, Amherst.

Kemmer, F.N. (1979). *Water: The Universal Solvent*, NALCO Chemical Company, Naperville, IL.

Kerns, B.K., Pilz, D., Ballard, H., and Alexander, S.J. (2003). Compatible management of understory forest resources and timber, in Johnson, A.C., Haynes, R.W., and Monserud, R.A., Eds., *Compatible Forest Management: Case Studies from Alaska and the Pacific Northwest*, Kluwer Academic, Norwell, MA, pp. 337–381.

Kimmins, J.P. (1977). Evaluation of the consequences for future tree productivity of the loss of nutrients in whole tree harvesting, *Forest Ecology Management*, 1:169–183.

Kimmins, J.P. (1996). Importance of soil and role of ecosystem disturbance for sustained productivity of cool temperate and boreal forests, *Soil Science Society of American Journal*, 56:1643–1654.

Knox, S.C., Chambers, C., and Germaine, S.S. (2001). Habitat associations of the sagebrush lizard (*Scelopurus graciosus*): potential responses of an ectotherm to ponderosa pine forest restoration treatments, in Vance R.K. et al., Eds., *Ponderosa Pine Ecosystems Restoration and Conservation: Steps toward Stewardship*, U.S. Department of Agriculture, Forest Service, Rocky Mountain Research Station, Ogden, UT, pp. 95–98.

Kolb, T.E., Holmberg, K.M., Wagner, M.R., and Stone, J.E. (1998). Regulation of ponderosa pine foliar physiology and insect resistance mechanisms by basal area treatment, *Tree Physiology*, 18(6):375–381.

Konigsburg, E.M. (1996). *The View from Saturday*, Scholastic, New York.

Korb, J.E., Johnson, N.C., and Covington, W.W. (2001). The effects of restoration thinning on mycorrhizal fungal propagules in a northern Arizona ponderosa pine forest, in Vance, G.K., Covington, W.W., and Edminister, C.B., Eds., *Ponderosa Pine Ecosystems Restoration and Conservation: Steps Toward Stewardship*, Proc. RMRS-P-22, U.S. Department of Agriculture, Forest Service, Rocky Mountain Research Station, Ogden, UT, pp. 74–79.

Kotliar, N.B., Hejl, S.J., Hutto, R.L., Saab, B.A., Melcher, C.P., and McFadzen, M.E. (2002). Effects of fire and post-fire salvage logging on avian communities in conifer-dominated forests of the western United States, *Studies in Avian Biology*, 25:49–64.

Krebs, C. (1989). *Ecological Methodology*, Harper Collins, New York.

Krebs, R.E. (2001). *Scientific Laws, Principles and Theories*, Greenwood Press, Westport, CT.

Laws, E.A. (1993). *Environmental Science: An Introductory Text*, John Wiley & Sons, New York.

LeBauer, D.S. and Treseder, K.T. (2008). Nitrogen limitation of net primary productivity in terrestrial ecosystems is globally distributed, *Ecology*, 89(2):371–379.

Lehmkuhl, J.F., Loggers, C.O., and Creighton, H.H. (2002). Wildlife considerations for small diameter timber harvesting, in Baumgartner, D.M., Johnson, L.R., and DePuit, E.J., Eds., *Proceedings Small Diameter Timber: Resource Management, Manufacturing, and Markets*, February 25–27, 2002, Washington State University Cooperative Extension, Spokane, WA.

Lehmkuhl, J.F., Gould, L.E., Cazares, E., and Hosford, D.R. (2004). Truffle abundance and mycophagy by northern flying squirrels in eastern Washington forests, *Forest Ecology and Management*, 200:49–65.

Lehmkuhl, J.F., Kistler, K.D., and Begley, J.S. (2006). Bushy-tailed woodrat abundance in dry forests of eastern Washington, *Journal of Mammalogy*, 87(2):371–379.

Liebig, J. (1840). *Chemistry and Its Application to Agriculture and Physiology*, Taylor & Walton, London.

Likens, G.E. and Bormann, F.H. (1995). *Biogeochemistry of a Forest Ecosystem*, 2nd ed., Springer-Verlag, New York.

Likens, G.E., Bormann, F.H., Pierce, R.S., and Reiners, W.A. (1978). Recovery of a deforested ecosystem, *Science*, 199(3):492–496.

Lillywhite, H.B., Friedman, G., and Ford, N. (1977). Color matching and perch selection by lizards in recently burned chaparral, *Copeia*, 1977(1):115–121.

Loeb, S.C. and Waldrop, T.A. (2008). Bat activity in relation to fire and fire surrogate treatments in southern pine stands, *Forest Ecology and Management*, 255:3185–3192.

Lofroth, E. (1998). The dead wood cycle, in Voller, J. and Harrison, S., Eds., *Conservation Biology Principles for Forested Landscapes*, British Columbia Ministry of Forests, Vancouver.

Long, R.A., Rachlow, J.L., Kie, J.G., and Vavra, M. (2008a). Fuels reduction in a western coniferous forest: effects on quantity and quality of forage for elk, *Rangeland Ecology and Management*, 61(3):302–313.

Long, R.A., Rachlow, J.L., and Kie, J.G. (2008b). Effects of season and scale on response of elk and mule deer to habitat manipulation, *Journal of Wildlife Management*, 72(5):1133–1142.

Lotka, A.J. (1925). *Elements of Physical Biology*, Williams & Wilkens, Baltimore, MD.

Lyons, A.L., Gaines, W.L., Lehmkuhl, J.F., and Harrod, R.J. (2008). Short-term effects of fire and fire surrogate treatments on foraging tree selection by cavity-nesting birds in dry forests of central Washington, *Forest Ecology and Management*, 255:3203–3211.

Machmer, M. (2002). Effects of ecosystem restoration treatments on cavity-nesting birds, their habitat, and their insectivorous prey in fire-maintained forests of southeastern British Columbia, in *Proceedings of the Symposium of the Ecology and Management of Dead Wood in Western Forests*, General Technical Report PSW-GTR-181, U.S. Department of Agriculture, Forest Service, Pacific Southwest Research Station, Albany, CA, pp. 121–133.

Malthus, T.R. (1798). *An Essay on the Principle of Population*, J. Johnson, London.

Mandzak, J.M. and Moore, J.A. (1994). The role of nutrition in the health of inland western forests, *Journal of Sustainable Forestry*, 2:191–210.

Mann, L.K., Johnson, D.W., West, D.C., Cold, D.W., Hornbeck, J.W., Marin, C.W., Riekerk, H., Smith, C.T., Swank, W.T., Tritton, L.M., and Van Lear, D.H. (1988). Effects of whole-tree and stem-only clearcutting on postharvest hydrologic losses, nutrient capital and regrowth, *Forest Science*, 34(2):412–428.

Manning, L.A. and Edge, D.W. (2004). Small mammal survival and downed wood at multiple scales in managed forests, *Journal of Mammalogy*, 85:87–96.

Martin, K.A. and McComb, B.C. (2003). Amphibian habitat associations at patch and landscape scales in western Oregon, *Journal of Wildlife Management*, 67:672–683.

Mason, A.C. and Adams, D.L. (1989). Black bear damage to thinned timber stands in northwest Montana, *Western Journal of Applied Forestry*, 4:10–13.

Masters, G.M. (1991). *Introduction to Environmental Engineering and Science*, Prentice Hall, Englewood Cliffs, NJ.

Mattson, T.A., Buskirk, S.W., and Stanton, N.L. (1996). Roost sites of the silver-haired bat (*Lasionycteris noctivagans*) in Black Hills of South Dakota, *Great Basin Naturalist*, 56:247–253.

McGraw, R.L. (1997). Timber Harvest Effects on Breeding and Larval Success of Long-Toed Salamanders (*Ambystoma macrodactylum*), PhD dissertation, University of Montana, Missoula.

McIver, J.D. and Starr, I. (2001). A literature review on the environmental effects of postfire logging, *Western Journal of Applied Forestry*, 16(4):159–168.

McIver, J.D., Adams, P.W., Doyal, J.A. et al. (2003). Environmental effects and economics of mechanized logging for fuel reduction in northeastern Oregon mixed-conifer stands, *Western Journal of Applied Forestry*, 18(4):238–249.

McNeil Technologies, Inc. (2003). *Biomass Resource Assessment and Utilization Options for Three Counties in Eastern Oregon*, report prepared for the Oregon Department of Energy, Lakewood.

McRae, D.J., Alexander, M.E., and Stocks, B.J. (1979). *Measurement and Description of Fuels and Fire Behaviour on Prescribed Burns: A Handbook*, Canadian Forestry Service, Great Lakes Forest Research Centre, Ontario.

Medin, D.E. and Booth, G.D. (1989). *Responses of Birds and Small Mammals to Single-Tree Selection Logging in Idaho*, Research Paper INT-RP-408, U.S. Department of Agriculture, Forest Service, Intermountain Research Station, Ogden, UT.

Meffe, G.K. and Carroll, C.R., Eds. (1997). *Principles of Conservation Biology*, 2nd ed., Sinauer Associates, Sunderland, MA.

Metlen, K.L. and Fielder C. (2006). Restoration treatment effects on the understory of ponderosa pine/Douglas-fir forests in western Montana, *USA Forest Ecology and Management*, 222:355–369.

Metlen, K.L., Fielder, C.E., and Youngblood, A. (2004). Understory response to fuel reduction treatments in the Blue Mountains of northeast Oregon, *Northwest Science*, 78(3):175–185.

Meyer, C.L., Sisk, T.D., and Covington, W.W. (2001). Microclimatic changes induced by ecological restoration of ponderosa pine forests in northern Arizona, *Restoration Ecology*, 9:443–452.

Meyer, M.D., North, M.P., and Kelt, D.A. (2005). Short-term effects of fire and forest thinning on truffle abundance and consumption by *Neotamias specious* in the Sierra Nevada of California, *Canadian Journal Forest Research*, 35:1061–1070.

MFRC. (2005). *Sustaining Minnesota Forests Resources: Voluntary Site-Level Forest Management Guidelines for Landowners, Loggers, and Resource Managers*, Minnesota Forest Resources Council, St. Paul, MN.

Miller, G.T. (1988). *Environmental Science: An Introduction*, Wadsworth, Belmont, CA.

Mitchell, R.G., Waring, R.H., and Pitman, G.B. (1983). Thinning lodgepole pine increases tree vigor and resistance to mountain pine beetle, *Forest Science*, 29:204–211.

Moghaddas, E.E., Stephens, Y., and Stephens, S.L. (2008). Mechanized fuel treatment effects on soil compaction in Sierra Nevada mixed-conifer stands, *Forest Ecology and Management*, 255:3098–3106.

Moore, K.R. and Henny, C.J. (1983). Nest site characteristics of three-coexisting *Accipiter* hawks in northeastern Oregon, *Raptor Research*, 17:65–76.

Moore, M.M., Covington, W.W., and Fule, P.Z. (1999). Reference conditions and ecological restoration: a southwestern ponderosa pine perspective, *Ecological Applications*, 9:1266–1277.

Moran, J.M., Morgan, M.D., and Wiersma, H.H. (1986). *Introduction to Environmental Science*, W.H. Freeman, New York.

Morris, G. (1999). *The Value of the Benefits of U.S. Biomass Power*, Green Power Institute, Berkeley, CA, and National Renewable Energy Laboratory, U.S. Department of Energy, Golden, CO.

Morrison, D.J. (1981). *Armillaria Root Disease: A Guide to Disease Diagnosis, Development and Management in British Columbia*, BC-X-23, Canadian Forestry Service, Pacific Forest Research Center, Victoria.

Mowet, F. (1957). *The Dog Who Wouldn't Be*, Willow Books, New York.

Muir, R.S., Mattingly, R.L., and Tappeiner, J.C. (2002). *Managing for Biodiversity in Young Douglas-Fir Forests of Western Oregon*, Biological Science Report USGS/BRD/BSR-2002-0006, U.S Geological Survey, Forest and Rangeland Ecosystem Science Center, Corvallis, OR, 76 p.

Mushinsky, H.R. (1985). Fire and the Florida sandhill herpetofaunal community: with special attention to response of *Cnemidophorus sexlineatus*, *Herpetologica*, 41:333–342.

Naughton, G.P., Henderson, C.B., Foresman, K.R., and McGraw II, R.L. (2000). Long-toed salamanders in harvested and intact Douglas-fir forests of western Montana, *Ecological Applications*, 10:1681–1689.

Newton, P.R. and Amponsah, I.G. (2006). Systematic review of short-term growth response of semi-mature black spruce and jack pine stands to nitrogen-based fertilization treatments, *Forest Ecology and Management*, 237:1–14.

Niwa, C.G., Sandquist, R.E., Crawford, R. et al. (2001). *Invertebrates of the Columbia River Basin Assessment Area*, General Technical Report PNW-GTR-512, U.S. Department of Agriculture, Forest Service, Pacific Northwest Research Station, Portland, OR.

NPS. (2008). *The Hoodoo: Park Planner, Hiking & Shuttle Guide*, National Park Service, U.S. Department of the Interior, Washington, D.C.

ODF. (2008). *Report: Environmental Effects of Forest Biomass Removal*, Office of the State Forester, Oregon Department of Forestry, Salem.

Odum, E.P. (1971). *Fundamentals of Ecology*, Saunders, Philadelphia, PA.

Odum, E.P. (1975). *Ecology: The Link between the Natural and the Social Sciences*, Holt, Rinehart Winston, New York.

Odum, E.P. (1983). *Basic Ecology*, Saunders, Philadelphia, PA.

Olson, J.S. (1963). Energy storage and balance of producers and decomposers in ecological systems, *Ecology*, 44:322–331.

Ormsbee, P.C. and McComb, W.C. (1998). Selection of day roosts by female long-legged myotis in the central Oregon Cascade range, *Journal of Wildlife Management*, 62:596–603.

Pang, P.C., Barclay, H.J., and McCullough, K. (1987). Aboveground nutrient distribution within trees and stands in thinned and fertilized Douglas-fir, *Canadian Journal of Forest Research*, 17:1379–1384.

Parks, C.G., Conklin, D.A., Bednar, L., and Maffei, H. (1999). *Woodpecker Use and Fall Rates of Snags Created by Killing Ponderosa Pine Infected with Dwarf Mistletoe*, Research Paper PNW-RP-515, U.S Department of Agriculture, Forest Service, Pacific Northwest Research Station, Portland, OR.

Pastor, J. and Bockheim, J.G. (1984). Distribution and cycling of nutrients in an Aspen-mixed hardwood-spodosol ecosystem in northern Wisconsin, *Ecology*, 65(2):339–353.

Patriquin, K.J. and Barclay, R.M.R. (2003). Foraging by bats in cleared, thinned and unharvested boreal forest, *Journal of Applied Ecology*, 40:646–657.

Patton, D.R. (1992). *Wildlife Habitat Relationships in Forested Ecosystems*, Timber Press, Portland, OR.

Patton, D.R. (2011). *Forest Wildlife Ecology and Habitat Management*, CRC Press, Boca Raton, FL.

Peck, R.W. and Niwa, C.G. (2004). Longer-term effects of selective thinning on carabid beetle and spiders in the Cascade Mountains of southern Oregon, *Northwest Science*, 78:267–277.

Perkins, J.M. and Cross. S.P. (1988). Differential use of some coniferous forest habitats by hoary and sliver-haired bats in Oregon, *Murrelet*, 69:21–24.

Phillips, D.R. and Van Lear, D.H. (1984). Biomass removal and nutrient drain as affected by total-tree harvest in southern pine hardwood stands, *Journal of Forestry*, 82:547–550.

Pianka, E.R. (1988). *Evolutionary Ecology*, Harper Collins, New York.

Pilliod, D. (2004). *Wildlife Responses to Fuels Treatments: Key Considerations*, Fuels Planning: Science Synthesis and Integration, Environmental Consequences Fact Sheet 4, Research Note RMRSW-RN-23-WWW, U.S. Department of Agriculture, Forest Service, Rocky Mountain Research Station, Fort Collins, CO.

Pilliod, D.S., Bury, R.B., Hyde, E.J., Perl, C.A., and Corn, P.S. (2003). Fire and amphibians in North America, *Forest Ecology and Management*, 178:163–181.

Pilliod, D.S., Bull, E.L., Hayes, J.L., and Wales, B.C. (2006). *Wildlife and Invertebrate Response to Fuel Reduction Treatments in Dry Coniferous Forests of the Western United States: A Synthesis*, General Technical Report RMS-GTR-173, U.S. Department of Agriculture, Forest Service, Rocky Mountain Research Station, Fort Collins, CO.

Pollet, J.P. and Omi, P.N. (2002). Effect of thinning and prescribed burning on crown fire severity in ponderosa pine forests, *International Journal of Wildland Fire*, 11(1):1–10.

Pough, F.H., Smith, E.M., Rhodes, D.H., and Collazo, A. (1987). The abundance of salamanders in forest stands with different histories of disturbance, *Forest Ecology and Management*, 20:1–9.

Powers, R.F. (2002). *Effects of Soil Disturbance on Fundamental, Sustainable Productivity of Managed Forest*, General Technical Report PSW-GRT-183, U.S. Department of Agriculture, Forest Service, Pacific Southwest Research Station, Albany, CA.

Powers, R.F. (2006). Long-term soil productivity: genesis of the concept and principles behind the program, *Canadian Journal of Forest Research*, 36:519–528.

Powers, R.F. and Avers, P.E. (1995). Sustaining forest productivity through soil standards: a coordinated U.S. effort, in Powter, C.B., Abboud, S.A., and McGill, W.B., Eds., *Environmental Soil Science: Anthropogenic Chemicals and Quality Criteria*, Canadian Society of Soil Science, Manitoba, pp. 147–190.

Powers, R.F., Alban, D.H., Miller, R.E. et al. (1990). Sustaining site productivity in North America forests: problems and prospects, in Gessel, S.P., Lacate, D.S., Weetman, G., and Powers, R.R., Eds., *Sustained Productivity of Forest Soils: 7th North American Forest Soils Conference*, July 24–28, 1988, University of British Columbia, Vancouver.

Powers, R.F., Tiarks, A.E., and Boyle, J.R. (1998). Assessing soil quality: practical standards for sustainable forest productivity in the United States, in Adams, M.B., Ramakrishna, K., and Davidson, E.A., Eds., *The Contribution of Soil Science to the Development of and Implementation of Criteria and Indictors of Sustainable Forest Management*, SSSA Special Publication 53, Soil Science Society of America, Madison WI.

Powers, R.F, Scott, D.A., Sanchez, F., Voldseth, R.A., Page-Dumrosese, D., Elioff, J.D., and Stone, D.M. (2005). The North American long-term soil productivity experiment: findings from the first decade of research, *Forest Ecology Management*, 220:31–50.

Prescott, C.E. and Laiho, R. (2002). *The Nutritional Significance of Coarse Woody Debris in Three Rocky Mountain Coniferous Forests*, General Technical Report PSW-GTR-181, U.S. Department of Agriculture, Forest Service, Washington, D.C. (www.fs.fed.us/psw/publications/documents/psw_gtr181/031_Prescott.pdf).

Price, P.W. (1984). *Insect Ecology*, John Wiley & Sons, New York.

Provencher, J., Cobris, N.M., Breman, L.A., Gordon, D.R., and Hardesty, J.F. (2002). Breeding bird response to midstory hardwood reduction in Florida sandhill longleaf pine forests, *Journal of Wildlife Management*, 66:641–661.

Raabe, R.D. (1962). Host list of the root rot fungus, *Armillaria mellea*, *Hilgardia*, 33(2):23–88.

Rabe, M.J., Morrell, T.E., Green, H., deVos, J.C., and Miller, C.R. (1998). Characteristics of ponderosa pine snag roosts used by reproductive bats in northern Arizona, *Journal of Wildlife Management*, 62:612–621.

Raymond, C.L. and Peterson, D.L. (2005). Fuel treatments alter the effects of wildfire in a mixed-evergreen forest, Oregon, USA, *Canadian Journal of Forest Research*, 35:2981–2995.

Reich, R.M., Joy, S.M., and Reynolds, R.T. (2004). Predicting the location of northern goshawk nests: modeling the spatial dependency between nest locations and forest structure, *Ecological Modeling*, 176:109–133.

Reynolds, R.T. and Wight, H.M. (1978). Distribution, density, and productivity of *Accipiter* hawks breeding in Oregon, *Wilson Bulletin*, 90:182–186.

Reynolds, R.T., Graham, R.T., Reiser, M. et al. (1992). *Management Recommendations for the Northern Goshawk in the Southwestern United States*, General Technical Report RM-217, U.S. Forest Service, Rocky Mountain Research Station, Fort Collins, CO.

Richter, D.D., Markewitz, D., Wells, C.G., Allen, H.L., April, R., Hein, P.R., and Urrego, B. (1994). Soil chemical change during three decades in an old-field loblolly pine (*Pinus taeda* L.) ecosystem, *Ecology*, 75:1463–1473.

Riegel, G. (2006). *Fire, Bitterbrush, and Mule Deer: Does Intelligent Design Exist?* Wildlife and Fire in the Wildland-Urban Interface Conference, Oregon Chapter of the Wildlife Society, April 18–19, Bend, OR.

Rippy, R.C., Stewart, J.E., Zambino, P.J., Klopfenstein, N.B., Tirocke, J.M., Kim, M, Thies, M., and Thies, W.G. (2005). *Root Diseases in Coniferous Forest of the Inland West: Potential Implications of Fuels Treatments*, General Technical Report, RMRS-GTR-141, U.S. Department of Agriculture, Forest Service, Rocky Mountain Research Station, Fort Collins, CO.

Robichaud, P.R. (2000). Fire effects on infiltration rates after prescribed fire in northern Rocky Mountain forests, USA, *Journal of Hydrology*, 231–232:220–229.

Robichaud, P.R. and Waldrop, T.A. (1994). A comparison of surface runoff and sediment yields from low- and high-severity site preparation burns, *Water Resources Bulletin*, 30:27–34.

Ross, D. (1995). Short-term impacts of thinning ponderosa pine on pandora moth densities, pupal weights, and phrenology, *Western Journal of Applied Forestry*, 10:91–94.

Saab, V.A. and Dudley, J. (1998). *Responses of Cavity-Nesting Birds to Stand-Replacement Fire and Salvage Logging in Ponderosa Pine/Douglas-Fir Forests of Southwestern Idaho*, Research Paper RMRS-RO-11, U.S. Department of Agriculture, Forest Service, Rocky Mountain Research Station, Fort Collins, CO.

Saab, V.A. and Vierling, K.T. (2001). Reproductive success of Lewis's woodpecker in burned pine and cottonwood riparian forests, *The Condor*, 103:491–501.

Saab, V.A., Brannon, R., Dudley, J., Donohoo, L., Vanderszanden, D., Johnson, V., and Lachowski, H. (2002). Selection of fire-created snags at two spatial scales by cavity-nesting birds, in Shea, P.J., Laudenslayer, Jr., W.F., Valentine, B., Weatherspoon, C.P., and Lisle, T.E., Eds., *Proceedings of the Symposium on the Ecology and Management of Dead Wood in Western Forests*, November 2–4, 1999, Reno, NV, General Technical Report PSW-GTR-181, U.S. Department of Agriculture, Forest Service, Pacific Southwest Research Station, Albany, CA, pp. 835–848.

Saab, V.A., Dudley, J., and Thompson, W.L. (2004). Factors influencing occupancy of nest cavities in recently burned forests, *The Condor*, 10:20–36.

Sala, A., Peters, G.D., McIntyre, L.R., and Harrington, M.G. (2005). Physiological responses of ponderosa pine in western Montana to thinning, prescribed fire and burning season, *Tree Physiology*, 25:339–348.

Sampson, N., Smith, M., and Gann, N. (2001). *Western Forest Health and Biomass Energy Potential*, Oregon Department of Energy, Salem (http://oregon.gov/ENERGY/RENEW/Biomass/forest.shtml).

Samways, M.J. (1994). *Insect Conservation Biology*, Chapman & Hall, London.

Sanchez-Martinez, G. and Wagener, M.R. (2002). Bark beetle community structure under four ponderosa pine forest stand conditions in northern Arizona, *Forest Ecology and Management*, 170:145–160.

Schmidt, C.A. (2003). *Conservation Assessment for the Silverhaired Bat in the Black Hills National Forest South Dakota and Wyoming*, U.S Department of Agriculture, Forest Service, Rocky Mountain Region, Black Hills National Forest, Custer, SD.

Schoenly, K., Beaver, R.A., and Heumier, T.A. (1991). On the trophic relations of insects: a food-web approach, *The American Naturalist*, 137:597–638.

Schowalter, T.D. (2000). Insects as regulators of ecosystem development, in Coleman, D.C. and Hendrix, P.F., Eds., *Invertebrates as Webmasters in Ecosystem*, CABI Publishing, Wallingford, U.K.

Sharov, A. (1992). Life-system approach: a system paradigm in population ecology, *Oikos*, 63:485–494.

Sharov, A. (1996). *What Is Population Ecology?* Department of Entomology, Virginia Tech University, Blacksburg.

Shaw III, C.G. and Roth, L.F. (1978). Control of *Armillaria* root rot in managed coniferous forests: a literature review, *European Journal of Forest Pathology*, 8(3):163–174.

Shick, K. (2003). The Influence of Stand-Level Vegetation and Landscape Composition on the Abundance of Snowshoe Hares (*Lepus americanus*) in Managed Forest Stands in Western Montana, PhD dissertation, University of Montana, Missoula.

Siegel, R.B. and DeSante, D.F. (2003). Bird communities in thinned versus unthinned Sierran mixed conifer stands, *Wilson Bulletin*, 115:155–165.

Silkwood, D.R. and Grigal, D.T. (1982). Determining and evaluation nutrient losses following whole-tree harvesting of aspen, *Soil Science of America Journal*, 46:626–633.

Simon, N.P.P., Schwab, F.E., and Diamond, A.W. (2000). Patterns of breeding bird abundance in relation to logging in western Labrador, *Journal of Forest Research*, 30:257–263.

Singh, S., Symth, A.K., and Blomberg, S.P. (2002). Effect of a control burn on lizards and their structural environment in a eucalypt open-forest, *Wildlife Research*, 23:447–454.

Smith, B.W., Vissage, J.S., Darr, D.R., and Sheffield, R.M. (2001). *Forest Resources of the United States, 1997*, General Technical Report NC-219, U.S. Department of Agriculture, Forest Service, North Central Research Station, St. Paul, MN (http://www.ncrs.fs.fed.us/pubs/gtr/gtr_nc219.pdf).

Smith, B.W. Miles, P.D., Vissage, J.S., and Pugh, S.A. (2003). *Forest Resources of the United States, 2002: A Technical Document Supporting the USDA Forest Service 2005 Update of the RPA Assessment*, General Technical Report GTR-NC-241, U.S. Department of Agriculture, Forest Service, North Central Research Station, St. Paul, MN.

Smith, C.T. (1986). An evaluation of the nutrient removals associated with whole-tree harvesting, in Smith, C.T., Martin, C.W., and Tritton, L.M., Eds., *Proceedings of the 1986 Symposium on the Productivity of Northern Forests Following Biomass Harvesting*, U.S. Department of Agriculture, Forest Service, Northeastern Forest Experiment Station, Upper Darby, PA.

Smith, J.K., Ed. (2000). *Wildland Fire in Ecosystems: Effects of Fire on Fauna*, General Technical Report RMRS-GTR-42, U.S. Department of Agriculture, Forest Service, Rocky Mountain Research Station, Fort Collins, CO.

Smith, R.L. (1974). *Ecology and Field Biology*, Harper & Row, New York.

Spellman, F.R. (1996). *Stream Ecology and Self-Purification*, Technomic Publishing, Lancaster, PA.

Spellman, F.R. (2008). *The Science of Environmental Pollution*, 2nd ed., CRC Press, Boca Raton, FL.

Squires, J. and Reynolds, R. (1997). Northern goshawk, *The Birds of North America*, 298:2–27.

Stephens, S.L. and Moghaddas, J.J. (2005). Fuel treatment effects on snags and coarse woody debris in a Sierra Nevada mixed conifer forest, *Forest Ecology and Management*, 214:53–64.

Stephenson, N.L. (1999). Reference conditions for giant sequoia forest restoration: structure, process, and precision, *Ecological Applications*, 9:1253–1265.

Stone, E.L. (1979). Nutrient removals by intensive harvest: some research gaps and opportunities, in Leaf, A., Ed., *Proceedings on the Symposium on Impacts of Intensive Harvesting on Forest Nutrient Cycling*, August 13–16, 1979, State University of New York, Syracuse, NY, pp. 366–386.

Stone, J.E., Kolb, T.E., and Covington, W.W. (1999). Effects of restoration thinning on presettlement *Pinus ponderosa* in northern Arizona, *Restoration Ecology*, 7(2):172–182.

Stoszek, K.J. (1988). Forest under stress and insect outbreaks, *Northwest Environmental Journal*, 4:247–261.

Sugden, E.A. (2000). Arthropods, in Smith, G., Ed., *Sierra East: Edge of the Great Basin*, University of California Press, Berkeley.

Sullivan, T.P. and Sullivan, D.S. (1988). Influence of stand thinning on snowshoe hare population dynamics and feeding damage in lodgepole pine forest, *Journal of Applied Ecology*. 25:791–805.

Suzuki, N. and Hayes, J.R. (2003). Effects of thinning on small mammals in Oregon coastal forests, *Journal of Wildlife Management*, 67:352–371.

Swank, W.T. and Reynolds, B.C. (1986). Within-tree distribution of woody biomass and nutrients for selected hardwood species, in Brooks, Jr., R., Ed., *Proceedings of the 1986 Southern Forest Biomass Workshop*, June 16–19, Knoxville, TN, pp. 87–91.

Swank, W.T. and Waide, J.B. (1980). Interpretation of nutrient cycling research in a management contest: evaluating potential effects of alternative management strategies on site productivity, in Waring, R.W., Ed., *Forests: Fresh Perspectives from Ecosystem Analysis*, Oregon State University Press, Corvallis, pp. 137–158.

Swezy, D.M. and Agee, J.K. (1991). Prescribed fire on fire-root and tree mortality in old-growth ponderosa pine, *Canadian Journal of Forest Research*, 21:626–634.

Thomas, D.W. (1988). The distribution of bats in different ages of Douglas-fir forests, *Journal of Wildlife Management*, 52:619–626.

Thomas, T.L. and Agee, J.K. (1986). Prescribed fire effects on mixed conifer forest structure at Crater Lake, Oregon, *Canadian Journal of Forest Research*, 16:1082–1087.

Tomera, A.N. (1990). *Understanding Basic Ecological Concepts*, J. Weston Walch, Portland, ME.

Townsend, C.R., Harper, J.L., and Begon, M. (2000). *Essentials of Ecology*, Blackwell Science, Oxford, U.K.

Trettin, C.C., Johnson, D.W., and Todd, Jr., D.E. (1999). Forest nutrient changes and carbon pools at Walker Branch watershed: changes during a 21-year period, *Soil Science Society of America Journal*, 63:1436–1148.

Trowbridge, R., Hakes, B., Macadam, A., and Parminter, J. (1987). *Field Handbook or Prescribed Fire Assessments in British Columbia: Logging Slash Fuels*, British Columbia Ministry of Forests, Victoria.

Turchin, P. (2001). Does population ecology have general laws? *Oikos*, 94:17–26.

Turchin, P. (2003). *Complex Population Dynamics: A Theoretical/Empirical Synthesis*, Princeton University Press, Princeton, NJ.

Turner, J. (1981). Nutrient cycling in an age sequence of western Washington Douglas-fir stands, *Annals of Botany*, 48:159–169.

USDA. (1975a). *Soil Taxonomy: A Basic System of Soil Classification for Making and Interpreting Soil Surveys*, U.S. Department of Agriculture, Natural Resources Conservation Service, Washington, D.C.

USDA. (1975b). *Soil Classification: A Comprehensive System*, U.S. Department of Agriculture, Natural Resources Conservation Service, Washington, D.C.

USDA. (2005). *A Strategic Assessment of Forest Biomass and Fuel Reduction Treatments in Western United States*, General Technical Report RMRS-GTR-149, U.S. Department of Agriculture, Forest Service, Rocky Mountain Research Station, Fort Collins, CO.

USEPA. (1995). *Emissions Factors & AP 42: Compilation of Air Pollution Emission Factors*, Technology Transfer Network Clearinghouse for Inventories & Emissions Factors, U.S. Environmental Protection Agency, Washington, D.C. (http://www.epa.gov/ttn/chief/ap42/index.html).

USEPA. (2010). *U.S. Emissions Inventory 2010: Inventory of U.S. Greenhouse Gas Emissions and Sinks: 1990–2008)*, U.S. Environmental Protection Agency, Washington, D.C.

USFS. (2006). *Rogue River–Siskiyou National Forest*, U.S. Forest Service, Medford, OR (http://www.fs.fed.Us/r6/rogue-siskiyou/biscuit-fire/index.shtml).

USFS. (2009). *Forest Inventory and Analysis National Program*, U.S. Forest Service, Arlington, VA (http://fia.fs.fed.us).

USGS. (2011). *Hawaiian Volcano Observatory*, U.S. Geological Survey (http://hvo.wr.usgs.gov/).

USGS. (2009). *Not Just for Kissing: Mistletoe and Birds, Bees, and Other Beasts*, U.S. Geological Survey (http://www.usgs.gov/newsroom/special/mistleton/).

Van Sickle, J. and Gregory, S.V. (1990). Modeling inputs of large woody debris to steams from falling trees, *Canadian Journal of Forestry Research*, 20:1593–1601.

van Wagtendonk, J.W. (1996). Use of a deterministic fire growth model to test fuel treatments, in *Sierra Nevada Ecosystem Project: Final Report to Congress*. Vol. II. *Assessments and Scientific Basis for Management Options*, Centers for Water and Wildland Resources, University of California, Davis, pp. 1155–1165.

Villa-Castillo, J. and Wagner, M.R. (2002). Ground beetle (Coleoptera: Carabidae) species assemblages as an indicator of forest condition in northern Arizona ponderosa pine forests, *Environmental Entomology*, 31:242–252.

Vitousek, P.M. and Howarth, R.W. (1991). Nitrogen limitation on land and sea—how can it occur? *Biogeochemistry*, 13:87–115.

Vonhof, M.J. and Barclay, R.M.R. (1996). Roost-site selection and roosting ecology of forest-dwelling bats in southern British Columbia, *Canadian Journal of Zoology*, 74:1797–1805.

Wackernagel, M. (1997). *Framing the Sustainability Crisis: Getting from Concerns to Action*, http://www.sdri.ubc.ca/documents/Framing_the_Sustainability_Crisis.doc.

Waltz, A.E.M. and Covington, W.W. (2004). Ecological restoration treatments increase butterfly richness and abundance: mechanisms of response, *Restoration Ecology*, 12:85–96.

Wardle, D.A., Bardgett, R.D., Klironomos, J.N., Setala, H., van der Putten, W.H., and Wall, D.H. (2004). Ecological linkage between aboveground and belowground biota, *Science*, 304:1629–1633.

Wargo, P.M. and Shaw III, C.G. (1985). *Armillaria* root rot: the puzzle is being solved, *Plant Disease*, 69(10):826–832.

Waring, R.H. and Running, S.W. (2007). *Forest Ecosystems: Analysis at Multiple Scales*, 3rd ed., Academic Press, San Diego, CA.

Wessells, N.K. and Hopson, J.L. (1988). *Biology*, Random House, New York.

Whitaker, Jr., J.O., Maser, C., and Keller, L.E. (1977). Food habits of bats of western Oregon, *Northwest Science*, 51:46–55.

Whitaker, Jr., J.O., Maser, C., and Cross, S.P. (1981). Food habits of eastern Oregon bats, based on stomach and scat analyses, *Northwest Science*, 55:281–292.

Whittaker, R.H., Likens, G.E., Bormann, F.H., Eaton, J.G., and Siccama, T. (1979). The Hubbard Brook
    ecosystem study: forest recycling and element behavior, *Ecology*, 60:203–220.
Wiens, D., Reynolds, R.T., and Noon, B.R. (2006). Juvenile movement and natal dispersal of northern
    goshawks in Arizona, *The Condor*, 108:253–269.
Wilson, C.W., Masters, R.E., and Bukenhofer, G.A. (1995). Breeding bird response to pine–grassland
    community restoration for red-cockaded woodpeckers, *Journal of Wildlife Management*, 59:56–67.
Wilson, D.M. and Grigal, D.F. (1995). Effects of pine plantations and adjacent deciduous forest on soil
    calcium, *Soil Sciences Society of America Journal*, 59:1755–1761.
World Commission on Environment and Development. (1987). *Our Common Future*, Oxford
    University Press, New York.
Yanai, R.D. (1998). The effects of whole-tree harvest on phosphorus in northern hardwood forest,
    Forest Ecology and Management, 104:281–295.
Zack, S., George, T.L., and Ladenslayer, Jr., W.F. (2002). *Are There Snags in the System? Comparing
    Cavity Use Among Nesting Birds in "Snag-Rich" and "Snag-Poor" Eastside Pine Forests*, General
    Technical Report PSW-GRT-181, U.S. Department of Agriculture, Forest Service, Pacific
    Southwest Research Station, Albany, CA, pp. 179–191.

Glossary*

Abiotic—Refers to nonliving parts of a physical environment (e.g., light, temperature, soil structure).

Ablation till—A superglacial coarse-grained sediment or till that accumulates as the sub-adjacent ice melts and drains away to finally be deposited on the exhumed sub-glacial surface.

Absorption—(1) Movement of a chemical into a plant, animal, or soil. (2) Any process by which one substance penetrates the interior of another substance; in chemical spill cleanup, this process applies to the uptake of chemical by capillaries within certain sorbent materials.

Absorption units—Devices or units designed to transfer the contaminant from a gas phase to a liquid phase.

Accuracy—Freedom from error or the closeness of a measurement or estimate to the true value. More broadly, it is the degree to which a statement or quantitative result approaches the truth.

Acicular foliage—Needle-shaped leaves.

Acid—A hydrogen-containing corrosive compound that reacts with water to produce hydrogen ions; a proton donor; a liquid compound with a pH less than or equal to 2.

Acid mine drainage—The dissolving and transporting of sulfuric acid and toxic metal compounds from abandoned underground coal mines to nearby streams and rivers when surface water flows through the mines.

Acid rain—Atmospheric precipitation with pH values less than about 5.6, the acidity being due to inorganic acids such as nitric and sulfuric that are formed when oxides of nitrogen and sulfur are emitted into the atmosphere. Any form of precipitation made more acidic from falling through air pollutants (primarily sulfur dioxide) and dissolving them.

Acid soil—A soil with a pH value of <7.0 or neutral. Soils may be naturally acid from their rocky origin or by leaching, or they may become acid from decaying leaves or from soil additives such as aluminum sulfate (alum). Acid soils can be neutralized by the addition of lime products.

Acidic deposition—See *acid rain*.

Acropetal—Developing upward from the base toward the apex so youngest are nearer the apex.

Actinides in the environment—The sources, environmental behavior, and effects of radioactive actinides in the environment.

Adhesion—Molecular attraction that holds the surfaces of two substances (e.g., water and sand particles) in contact.

Adiabatic—Refers to a condition with no loss or gain of heat. When air rises, air pressure decreases and expands adiabatically in the atmosphere; because the air can neither gain nor lose heat, its temperature falls as it expands to fill a larger volume.

---

* Many of the terms listed here are from Burns, R.M. and Honkala, B.H., *Silvics of North America*, Agriculture Handbook 654, U.S. Department of Agriculture, Forest Service, Washington, D.C., 2004 (http://www.na.fs.fed.us/spfo/pubs/silvics_manual/volume_2/glossary/glossary.htm).

Adiabatic lapse rate—The temperature profile or lapse rate; it is used as a basis for comparison for actual temperature profiles (from ground level) and hence for predictions of stack gas dispersion characteristics.

Adsorption—(1) The process by which one substance is attracted to and adheres to the surface of another substance without actually penetrating its internal structure. (2) Process by which a substance is held (bound) to the surface of a soil particle or mineral in such a way that the substance is only available slowly.

Adsorption site density—The concentration of sorptive surface available from the mineral and organic contents of soils. An increase in adsorption sites indicates an increase in the ability of the soils to immobilize hydrocarbon compounds in the soil matrix.

Advective wind—The horizontal air movements resulting from temperature gradients that give rise to density gradients and subsequently pressure gradients.

Adventitious—Plant organs produced in an unusual or irregular position, or at an unusual time of development.

Aeration, soil—The process by which air in the soil is replaced by air from the atmosphere. In a well-aerated soil, the soil air is similar in composition to the atmosphere above the soil. Poorly aerated soils usually contain more carbon dioxide and correspondingly less oxygen than the atmosphere above the soil.

Aerobic—Refers to living in the air; opposite of *anaerobic*.

Aerobic processes—Many biotechnology production and effluent treatment processes are dependent on microorganisms that require oxygen for their metabolism; for example, water in an aerobic stream contains dissolved oxygen, and organisms using the dissolved oxygen can oxidize organic wastes to simple compounds.

Afterripening—Enzymatic process occurring in seeds, bulbs, tubers, and fruit after harvesting; often necessary for germination or resumption of growth.

Aggregates (soil)—Soil structural units of various shapes, composed of mineral and organic material, formed by natural processes, and having a range of stabilities.

Agricultural sources—Organic and inorganic contaminants usually produced by pesticides, fertilizers, and animals wastes, all of which enter water bodies via runoff and groundwater absorption in areas of agricultural activity.

Agronomy—A specialization of agriculture concerned with the theory and practice of field crop production and soil management; the scientific management of land.

Air capacity—Percentage of soil volume occupied by air spaces or pores.

Air layering—Inducing root development on an undetached aerial portion of a plant, commonly by wounding it, treating it with a rooting stimulant, and wrapping it in moisture material under a waterproof covering; the portion so treated is capable of independent growth after separation from the mother plant.

Air pollutants—Generally includes sulfur dioxide, hydrogen sulfide, hydrocarbons, carbon monoxide, ozone, and atmospheric nitrogen, but can include any gaseous substance that contaminates air.

Air pollution—Contamination of atmosphere with any material that can cause damage to life or property.

Air porosity—Proportion of the bulk volume of soil filled with air at any given time or under a given condition, such as a specified moisture potential; usually the large pores.

Air Quality Index—A standardized indicator of the air quality in a given location.

Airborne contaminants—Any contaminant capable of dispersion in air or capable of being carried by air to other locations.

Airborne particulate matter—Fine solids or liquid droplets suspended in the air.

Albedo—The fraction of received radiation reflected by a surface.

Algae—A large and diverse assemblage of eucaryotic organisms that lack roots, stems, and leaves but have chlorophyll and other pigments for carrying out oxygen-producing photosynthesis.

Aliphatic hydrocarbon—Compound comprised of straight-chain molecules as opposed to a ring structure.

Alkalinity—(1) The concentration of hydroxide ions. (2) The capacity of water to neutralize acids because of the bicarbonate, carbonate, or hydroxide content. Alkalinity is usually expressed in milligrams per liter of calcium carbonate equivalent.

Alkanes—A class of hydrocarbons (gas, solid, or liquids depending on carbon content). Its solids (paraffins) are a major constituent of natural gas and petroleum. Alkanes are usually gases at room temperature (methane) when containing fewer than five carbon atoms per molecule.

All-aged—A condition of a forest or stand that contains trees of all or almost all age classes. It is generally a primary stand where individuals have entered at various times when and where space permitted. (See *Uneven-aged*.)

Allele—One of an array of genes possible at a certain position (locus) on a given chromosome.

Allelopathy—The influence of plants, other than microorganisms, upon each other, arising from the products of their metabolism.

Allopatric—Occurring in different areas or in isolation. (See *Sympatric*.)

Alluvial—A type of azonal soil that is highly variable and is classified by texture from fine clay and silt soils through gravel and boulder deposits.

Alluvium—Soil, usually rich in minerals, deposited by water, as in a floodplain.

Alpha pinene—A hydrocarbon of the terpene class occurring in many essential oils; it has a density of about 0.855 and an index of refraction of about 1.465, both at 25°C (77°F).

Amino acids—Building blocks of proteins.

Amoeba (sing.), **Amoebae** (pl.)—One of the simplest living animals, consisting of a single cell and belonging to the protozoa group. The body consists of colorless protoplasm. Its activities are controlled by the nucleus, and it feeds by flowing around and engulfing organic debris. It reproduces by binary fission. Some species of amoebae are harmless parasites.

Anabolism—The process of building up cell tissue, promoted by the influence of certain hormones; the utilization of energy and materials to build and maintain complex structures from simple components; the constructive side of metabolism as opposed to catabolism.

Anaerobic—Refers to not requiring molecular oxygen.

Anaerobic process—Any process (usually chemical or biological) carried out without the presence of air or oxygen, such as in a heavily polluted watercourse that has no dissolved oxygen.

Analysis—The separation of an intellectual or substantial whole into its constituent parts for individual study.

Andesite—An extrusive usually dark grayish rock consisting essentially of oligoclase or feldspar.

Anemophilous—Refers to normally being wind pollinated.

Animal feedlots—A confined area where hundreds or thousands of livestock animals are fattened for sale to slaughterhouses and meat producers.

Animal wastes—Consists of dung (fecal matter) and urine of animals.

Anther—The part of the stamen that develops and contains pollen.

Anthesis—The time at which a flower comes into full bloom.

Anthropogenic sources—Generated by human activity.

Apoenzyme—The protein part of an enzyme.

Apophysis—The rounded, exposed thickening on the scales of certain pine cones (Greek for "offshoot").

Appalachian Highlands—The lands of the Appalachian Mountains extending from central New York south to northeastern Alabama.

Aqueous solution—Solution in which the solvent is water.

Aquifer—Any rock formation containing water. The rock of an aquifer must be porous and permeable to absorb water.

Arbuscula—A small or low shrub having the form of a tree.

Argillite—A compact argillaceous (clayey) rock differing from shale in being cemented by silica and from slate in having no slaty cleavage.

Arillate—Refers to having an aril (an appendage, outgrowth, or outer covering of a seed, growing out from the hilum or funiculus).

Aromatic hydrocarbons—Class of hydrocarbons considered to be the most immediately toxic; they are found in oil and petroleum products and are soluble in water. (Antonym is *aliphatic*.)

Asexual reproduction—Requires one parent cell.

Aspect—The predominant direction of slope of the land. Individual azimuthal map projections are divided into three aspects: the polar aspect, which is tangent at the pole; the equatorial aspect, which is tangent at the equator; and the oblique aspect, which is tangent anywhere else.

Asymptotically unbiased—Estimation bias approaches 0 as sample size approaches population size.

Atmosphere—The layer of air surrounding the surface of the Earth.

Atmospheric dispersion modeling—The mathematical simulation of how air pollutants disperse in the ambient atmosphere.

Atom—A basic unit of physical matter indivisible by chemical means; the fundamental building block of chemical elements that is composed of a nucleus of protons and neutrons, surrounded by electrons.

Atomic number—Number of protons in the nucleus of an atom. Each chemical element has been assigned a number in a complete series from 1 to 100+.

Atomic weight—The mass of an element relative to its atoms.

Attribute—Refers to units classified as having or not having some specific quality.

Auger—A tool used to bore holes in soil to capture a sample.

Automatic samplers—Devices that automatically take samples from a waste stream.

Autotrophic—An organism that can synthesize the organic molecules necessary for growth from inorganic compounds using light energy (photosynthesis) or chemical energy (chemosynthesis).

Auxin—A natural hormone that regulates plant growth, generally identified with β-indolylacetic acid (IAA), a heteroauxin.

Azimuth—The angle measured in degrees between a base line radiating from a center point and another line radiating from the same point. Normally, the base line points north and degrees are measured clockwise from the base line.

Backcross—A cross between a hybrid and either one of its parents.

Baghouse filter—A closely woven bag for removing dust from dust-laden gas streams. The fabric allows passage of the gas with retention of the dust.

Bare rock succession—An ecological succession process whereby rock or parent material is slowly degraded to soil by a series of bioecological processes.

Barrel of oil equivalent (BOE)—The amount of energy contained in a barrel of crude oil, approximately 6.1 GJ (5.8 million Btu), equivalent to 1700 kWh. A *petroleum barrel* is a liquid measure equal to 42 U.S. gallons (35 imperial gallons or 159 liters); about 7.2 barrels are equivalent to 1 metric ton of oil.

Basal area (per site)—The cross-sectional area at breast height of all trees on the site.

Basal area (per tree)—The cross-sectional area at breast height of a tree.

Basalt—A dark gray to black colored, dense to fine-grained igneous rock.

Base—A substance that, when dissolved in water, generates hydroxide (OH^-) ions or is capable of reacting with an acid to form a salt.

Basipetal—Refers to developing toward the base from the apex.

Batture—Land between the river and the manmade levees that border it.

Bearing—A horizontal angle referenced to a quadrant (namely, NE, SE, SW, or NW), measured from north or south to east to west.

Benthos—The term originates from the Greek word for "bottom" and broadly refers to aquatic organisms living on the bottom sediments of the ocean or on submerged vegetation.

Berry—A simple, pulpy fruit of a few or many seeds (but no stones) developed from a single ovary.

Best available technology (BAT)—Essentially, a refinement of best practicable means whereby a greater degree of control over emissions to land, air, and water may be exercised using currently available technology.

Beta phellandrene—A terpene with a density of about 0.84 and an index of refraction of about 1.48, both at 25°C (77°F).

Beta pinene—A terpene with a density of about 0.867 and an index of refraction of about 1.477, both at 25°C (77°F).

Bias—A systematic error introduced into sampling, measurement, or estimation by selecting or flavoring, possibly unintentionally, one outcome or answer over others.

Binomial system of nomenclature—Used to classify organisms; organisms are generally described by a two-word scientific name that includes the *genus* and *species*.

Bioaccumulation—The biological concentration mechanism whereby filter feeders such as limpets, oysters, and other shellfish concentrate heavy metals or other stable compounds present in dilute concentrations in seawater or freshwater.

Biobased product—As defined by the Farm Security and Rural Investment Act (FSRIA), a product determined by the U.S. Secretary of Agriculture to be a commercial or industrial product (other than food or feed) that is composed, in whole or in significant part, of biological products or renewable domestic agriculture materials (including plant, animal, and marine materials) or forestry materials.

Biochemical oxygen demand (BOD)—The amount of oxygen required by bacteria to stabilize decomposable organic matter under aerobic conditions.

Biodegradable—A material capable of being broken down, usually by microorganisms, into basic elements.

Biodegradation—The ability of natural decay processes to break down manmade and natural compounds to their constituent elements and compounds for assimilation in, and by, the biological renewal cycles (e.g., wood being decomposed to carbon dioxide and water).

Bioenergy—Useful, renewable energy produced from organic matter—the conversion of the complex carbohydrates in organic matter to energy. Organic matter may be used directly as a fuel, may be processed into liquids and gases, or may be a residual of processing and conversion.

Biogeochemical cycles—*Bio* refers to living organisms and *geo* to water, air, rocks, or solids. *Chemical* is concerned with the chemical composition of the earth. Biogeochemical cycles are driven by energy, directly or indirectly, from the sun.

Biological oxygen demand (BOD)—The amount of dissolved oxygen taken up by microorganisms in a sample of water.

Biology—The science of life.

Biomass—The total weight of living organisms, including plants and animals, for a given area; usually expressed as kg/ha, lb/acre, g/m^2, etc. For most ecological investigations and for the purposes of this text, biomass is a vegetation attribute that refers to the weight of plant material within a given area. Another commonly used term for biomass is *production*, which refers to how much vegetation is produced on an area. The biomass on a site can be estimated by species (i.e., weight of each individual plant species), or biomass can be estimated in groups such as growth form (trees, grass, forb, or shrubs), plant longevity (annual or perennial), or degree of woodiness (herbaceous or woody).

Biorefinery—A facility that processes and converts biomass into multiple value-added products. These products can range from biomaterials to fuels such as ethanol or important feedstocks for the production of chemicals and other materials. Biorefineries can be based on a number of processing platforms (mechanical, thermal, chemical, or biochemical).

Bioremediation—Any process that uses microorganisms, fungi, green plants, or their enzymes to return a natural environment altered by contaminants to its original condition.

Biosphere—The region of the Earth and its atmosphere in which life exists; an envelope extending from up to 6000 meters above to 10,000 meters below sea level that embraces all life from Alpine plant life to the ocean depths.

Biostimulant—An organic material that, when applied in small quantities, stimulates plant growth and development; an example would be phosphates or nitrates in a water system.

Biota—The animal and plant life of a particular region considered as a total ecological entity.

Biotic—Pertaining to life or specific life conditions.

Biotic index—The diversity of species in an ecosystem is often a good indicator of the presence of pollution; the greater the diversity, the lower the degree of pollution. The biotic index is a systematic survey of invertebrate aquatic organisms that is used to correlate with river quality; it is based on two principles: (1) Pollution tends to restrict the variety of organisms present at a particular point, although large numbers of pollution-tolerant species may persist. (2) In a polluted stream, as the degree of pollution increases, key organisms tend to disappear in the order of stoneflies, mayflies, caddisflies, freshwater shrimp, bloodworms, and tubificid worms.

Bisexual—Having both male and female sexual reproductive structures.

Black liquor—Solution of lignin residues and the pulping chemicals used to extract lignin during the manufacture of paper.

Blastospore—Fungi spores formed by budding.

Board feet—The amount of wood contained in a board 1 inch thick, 12 inches long, and 12 inches wide.

Boiling point—The temperature at which a substance changes from a liquid to a gas.

Bonsai—The culture of miniature potted trees, which are dwarfed by stem and root pruning and controlled nutrition.

Boreal forest—A coniferous forest of the northern hemisphere characterized by evergreen conifers such as spruce, fir, and pine.

Brackish water—Water (nonpotable) containing between 100 and 10,000 ppm of total dissolved solids.

Breast height—The point on a tree stem 1.4 m (4-1/2 feet) above the ground in the United States, New Zealand, Burma, India, Malaysia, South Africa, and some other countries; 1.3 m (4-1/4 feet) above the ground in continental Europe, Great Britain, Australia, Canada, and Mexico.

Browse—The potion or amount of woody plants available for animal consumption; usually the current season's growth of twigs and leaves.

Brush—Stubby vegetation that is not suitable for commercial timber.

Btu—British thermal unit, a measuring unit of heat.

Budding—Type of asexual reproduction in which an outgrowth develops from a cell to form a new individual; most yeasts reproduce in this way.

C unit—Unit of volume, generally supplied to pulpwood, consisting of 100 ft^3 of solid wood.

Calorie—The amount of heat required to raise the temperature of 1 gram of water 1°C.

Campanulate—Bell-shaped.

Canadian Shield—Precambrian nuclear mass centered in Hudson Bay around which, and to some extent upon which, the younger sedimentary rocks have been deposited.

Canopy—The more or less continuous cover of braches and foliage formed collectively by the crowns of adjacent trees.

Cant—A piece of lumber made from a log by removing two or more sides by sawing.

Capsule, bacterial—Organized accumulations of gelatinous material on cell walls.

Carbohydrates—Main source of energy for living things such as sugar and starch.

Carbon adsorption—Process whereby activated carbon, known as the sorbent, is used to remove certain wastes from water by preferentially holding them to the carbon surface.

Carbon cycle—The atmosphere is a reservoir of gaseous carbon dioxide, but to be of use to life this carbon dioxide must be converted into suitable organic compounds ("fixed"), as in the production of plant stems by the process of photosynthesis. The productivity of an area of vegetation is measured by the rate of carbon fixation. The carbon fixed by photosynthesis is eventually returned to the atmosphere as plants and animals die and the dead organic matter is consumed by the decomposer organisms.

Carbon dioxide—A colorless, odorless inert gas that is a byproduct of combustion.

Carbon monoxide—A highly toxic and flammable gas that is a byproduct of incomplete combustion; very dangerous even in very low concentrations.

Carbon sequestration—The process by which atmospheric carbon dioxide is absorbed by trees and other plants through photosynthesis and stored as carbon in biomass (trunks, branches, foliage, and roots), soils, and wood products. Adopting certain agricultural and forestry practices can reduce greenhouse gas emissions to the atmosphere and sequester additional carbon.

Carbonate hardness—Temporary hardwater caused by the presence of bicarbonates; when water is boiled, the bicarbonates are converted to insoluble carbonates that precipitate as scale.

Carr—A deciduous woodland on a permanently wet, organic soil.

Catabolism—In biology, the destructive part of metabolism where living tissue is changed into energy and waste products; the breaking down of complex materials into simpler ones using enzymes and releasing energy.

Catalysis—The acceleration (or retardation) of chemical or biochemical reactions by a relatively small amount of a substance (the catalyst), which itself undergoes no permanent chemical change and which may be recovered when the reaction has finished.

Catalyst—A substance or compound that speeds up the rate of chemical or biochemical reactions.

Catchment—The natural drainage area for precipitation; the collection area for water supplies or a river system. The notional line, or watershed, on surrounding high land defines the area.

Catena—A sequence of different soils, generally derived from similar parent soil material, each of which owes its character to its peculiar physiographic position.

Catkin—A drooping elongated cluster of bracted unisexual flowers found only in woody plants.

Cell—The basic biological unit of plant and animal matter.

Cell (cytoplasmic) membrane—The lipid- and protein-containing, selectively permeable membrane that surrounds the cytoplasm in procaryotic and eucaryotic cells; in most types of microbial cell, the cell membrane is bordered externally by the cell wall. In microbial cells, the precise composition of the cell membrane depends on the species, on growth conditions, and on the age of the cell.

Cell nucleus—Contained within a eucaryotic cell; a membrane-lined body that contains chromosomes.

Cell wall—The permeable, rigid outermost layer of a plant cell composed mainly of cellulose.

Cellular basis of life—Cells are the basis of life. The two types of cells are prokaryotes and eukaryotes.

Chemical bond—A chemical linkage that holds atoms together to form molecules.

Chemical change—A transfer that results from making or breaking chemical bonds.

Chemical equation—A shorthand method for expressing a reaction in terms of written chemical formulas.

Chemical extraction—Process in which excavated contaminated soils are washed to remove contaminants of concern.

Chemical formula—In the case of substances that consist of molecules, the chemical formula indicates the kinds of atoms present in each molecule and the actual number of them.

Chemical precipitation—Process by which inorganic contaminants (heavy metals from groundwater) are removed by the addition of carbonate, hydroxide, or sulfide chemicals.

Chemical process audit/survey—A procedure used to gather information on the type, composition, and quantity of waste produced.

Chemical reactions—A process in which the composition or structure of a chemical is changed to form another substance.

Chemical weathering—A form of weathering brought about by a chemical change in the rocks affected; involves the breakdown of the minerals within a rock and usually produces a claylike residue.

Chemosynthesis—A method of making protoplasm using energy from chemical reactions, in contrast to the use of light energy employed for the same purpose in photosynthesis.

Chlorofluorocarbons (CFCs)—Synthetic chemicals that are odorless, nontoxic, nonflammable, and chemically inert.

Chlorophyll—A combination of green and yellow pigments present in all "green" plants; it captures light energy and enables the plants to form carbohydrate material from carbon dioxide and water in the process known as photosynthesis. Found in all algae, phytoplankton, and almost all higher plants.

Chloroplasts—A structure (or organelle) found within a plant cell containing the green pigment chlorophyll.

Chromosome—A microscopic, usually rodlike body carrying the genes. The number, size, and form of chromosomes are usually constant for each species.

Cilia—Small threadlike organs found on the surface of some cells and composed of contractile fibers that produce rhythmic waving movements. Some single-celled organisms move by means of cilia. In multicellular animals, they keep lubricated surfaces clear of debris. They also move food in the digestive tracts of some invertebrates.

Cirque—A deep, steep-walled basin shaped like a half bowl; found on mountainsides.

Clarification—The process of removing solids from water.

Clay content—The amount of clay (fine-grained sedimentary rock) within a soil.

Clean Air Act—The name given to two acts passed by the U.S. government. The Act of 1963 dealt with the control of smoke from industrial and domestic sources; it was extended by the Act of 1968, particularly to control gas cleaning and heights of stacks of installations in which fuels are burned to deal with smoke from industrial open bonfires. The 1990 Clean Air Act brought wide-ranging reforms for all kinds of pollution from large or small mobile or stationary sources, including routine and toxic emissions ranging from power plants to consumer products.

Clean Water Act—A keystone environmental law passed in 1972 that is credited with significantly cutting the amount of municipal and industrial pollution fed into the nation's waterways. More formally known as the Federal Water Pollution Control Act Amendments, it was derived from a much-amended 1948 law aiding communities in building sewage treatment plants and has itself been much amended, most notably in 1977 and 1987.

Clean zone—That point in a river or stream upstream before a single point of pollution discharge.

Clearcut—The silvicultural system in which the old crop is cleared over a considerable area at one time. Regeneration then occurs from (1) natural seeding from adjacent stands, (2) seed contained in the slash or logging debris, (3) advance growth, or (4) planting or direct seeding. An even-aged forest usually results.

Cleft graft—The stock is cut off and split, and then one or more scions are placed in the split such that the cambium layers of the stock and scion match.

Climate—The composite pattern of weather conditions that can be expected in a given region. Climate refers to yearly cycles of temperature, wind, rainfall, and so on, not to daily variations.

Climax community—The terminal stage of an ecological succession sequence that remains relatively unchanged as long as climatic and physiographic factors remain stable.

Clinal—Sloping.

Clone—Any plant propagated vegetatively and therefore considered a genetic duplicate of its parent.

Coal gasification process—The conversion of coal (via destructive distillation or carbonization) to gaseous fuel.

Coarse woody debris (CWD)—Dead pieces of wood including downed and dead tree and shrub boles, large limbs, and other woody pieces that are severed from their original source of growth or are leaning more than 45 degrees from vertical. For decay classes 1 to 4, CWD transect diameter must be >3.0 inches (7.6 cm); for decay class 5, the transect diameter must be >5.0 inches (12.7 cm).

Coccus (sing.), **cocci** (pl.)—Member of a group of globular bacteria, some of which are harmful to humans.

Codominant—Crown class of species in a mixed crop that are about equally numerous and vigorous; they form part of the upper canopy of a forest and are less free to grow than dominants but freer than intermediate and suppressed trees.

Cofactor—Nonprotein activator that forms a functional part of an enzyme.

Cold-blooded animals—Animals whose body temperature changes with the environment.

Cold front—The leading portion of a cold atmospheric air mass moving against and eventually replacing a warm air mass.

Collenchyma—Flexible, supportive plant tissue, usually of elongated living cells with unevenly thickened walls which are usually interpreted as primary walls.

Colloidal material—A constituent of total solids in wastewater; consists of particulate matter with an approximate diameter range of from 1 millimicron to 1 micron.

Colluvium—Rock detritus and soil accumulated at the foot of a slope.

Color—Physical characteristic of water often used to judge water quality; pure water is colorless.

Composite sample—Sample formed by mixing discrete samples taken at periodic points in time or a continuous proportion of the flow. The number of discrete samples making up the composite depends on the variability of pollutant concentration and flow.

Compound—A substance composed of two or more elements, chemically combined in a definite proportion.

Concentrated solution—Solute in concentration present in large quantities.

Condensation—Air pollution control technology used to remove gaseous pollutants from a waste stream; a process in which the volatile gases are removed from the contaminant stream and changed into a liquid.

Condenser—Air pollution control device used in a condensation method to condense vapors to a liquid phase by either increasing the system pressure without a change in temperature or decreasing the system temperature to its saturation temperature without a pressure change.

Conduction—Heat flow of heat energy through a material without the movement of any part of the material itself.

Confined aquifer—A water-bearing layer sandwiched between two less permeable layers; water flow is restricted to vertical movement only.

Confluence—The point at which two streams merge.

Conformality—Applies to map projections and refers to when at any point the scale is the same in every direction; meridians and parallels intersect at right angles and the shapes of very small areas and angles with very short sides are preserved. The size of most areas, however, is distorted.

Conglomerate—Made up of parts from various sources or of various kinds.

Conidia—Asexual spores borne on aerial mycelia (actinomycetes bacteria).

Consumers—Organisms that cannot produce their own food and eat by engulfing or predigesting the fluids, cells, tissues, or waste products of other organisms.

Continuous variable—A variable expressed in a numerical scale of measurement, where any interval of it can be subdivided into an infinite number of values.

Convection—Method of heat transfer whereby the heated molecules circulate through the medium (gas or liquid).

Coppicing—A traditional method of woodland management where young tree stems are cut down to a low level or sometimes right down to the ground. In subsequent growth years, many new shoots will grow up, and after a number of years the cycle begins again and the coppiced tree is ready to be harvested again. Typically, coppice woodlands are harvested in rotation so a crop is available each year.

Correlation coefficient—A measure of the degree of linear association between two variables that is unaffected by the sizes of scales of the variables.

Corrosive—A substance that attacks and eats away other materials by strong chemical action.

Corymb—A flat-topped floral cluster with outer flowers opening first.

Cotyledon—An embryonic leaf, which often stores food materials; characteristic of seed plants.

Covalent bond—A chemical bond produced when two atoms share one or more pairs of electrons.

Covariance—A variance or measure of association between paired measurements of two variables.

Covariate—A quantitative, often explanatory variable in a model (e.g., regression model). Covariates are often important in improving estimation.

Crown class—Any class into which trees of a stand may be divided based on both their crown development and crown position relative to crowns of adjacent trees. The four classes commonly recognized are dominant, codominant, intermediate, and suppressed.

Cruise—A forest inventory conducted to estimate the quantity of timber on a given area according to species, size, quality, and other characteristics.

Crustacean—One of a class of arthropods that includes crabs, lobsters, shrimps, woodlice, and barnacles.

Culmination of mean annual increment—For a tree or stand of trees, the age at which the average annual increment is greatest. It coincides precisely with the age at which the current annual increment equals the mean annual increment of the stand and thereby defines the rotation of a fully stocked stand that yields the maximum volume growth.

Cultivar—A contraction of "cultivated variety." It refers to a plant type within a particular cultivated species that is distinguished by one or more characters.

Cultural eutrophication—Overnourishment of aquatic ecosystems with plant nutrients resulting from human activities, including agriculture, urbanization, and industrial discharge.

Current annual increment (CAI)—The amount by which the volume of a tree or stand increases in 1 year.

Cushion krummholz—Alpine trees exposed to severe wind conditions are wind-pruned to a cushion-like mat.

Cymose—Bearing a cyme, a more or less flat-topped floral cluster with the central flowers opening first.

Cytochrome—A class of iron-containing proteins important in cell metabolism.

Cytokinins—A class of hormones that promote and control the growth response of plants.

Cytoplasm—The jelly-like matter within a cell.

DBH—The diameter at breast height of a tree.

Decay class—Qualitative assessment of stage of decay (five classes) of coarse wood debris based on visual assessments of the color of the wood, presence or absence of twigs and branches, texture of rotten portions, and structural integrity.

- Decay class 1—Solid wood, recently fallen, bark and twigs present
- Decay class 2—Solid wood, significant weathering, branches present
- Decay class 3—Wood not solid, may be sloughing but nail must still be pounded into tree
- Decay class 4—Wood sloughing or friable, nails may be forcibly pushed into log
- Decay class 5—Wood friable, barely holding shape, nails may be easily pushed into log.

Decomposers—Organisms such as bacteria, mushrooms, and fungi that obtain nutrients by breaking down complex matter in the wastes and dead bodies of other organisms into simpler chemicals, most of which are returned to the soil and water for reuse by producers.

Decomposition—Process whereby a chemical compound is reduced to its component substances. In biology, it is the destruction of dead organisms either by chemical reduction or by the action of decomposers.

Dehisce—Refers to splitting open when ripe, usually along definite lines or sutures to release seeds.

Deliquescent branching—A mode of branching in trees in which the trunk divides into many branches leaving no central axis, as in elms. (See *Excurrent branching*.)

Dendrometer—A device for measuring the diameter of tree stems.

Density—The ratio of the weight of a mass to the unit of volume.

Destructive distillation—The decomposition of wood by heating out of contact with air, producing primarily charcoal.

Developable surface—A simple geometric form capable of being flattened without stretching. Many map projections can then be grouped by a particular developable surface—cylinder, cone, or plane.

Diallel cross—(Complete) A mating design and subsequent progeny test resulting from the crossing of n parents in all possible n^2 combinations including selfs and reciprocals. (Incomplete) A partial sampling; any individual family or type of family may be omitted. In either type of cross, identities of both seed and pollen parents are maintained for each family.

Diatom—Microscopic single-celled alga found in all parts of the world.

Dichogamy—In a perfect flower, maturation of stamens and pistils occurs at different times, thus preventing self-pollination.

Diffusion—(1) Mixing of substances, usually gases and liquids, from molecular motion. (2) The spreading out of a substance to fill a space.

Digestion—Process by which food is broken down into simpler substances.

Dilute solutions—A solution weakened by the addition of water, oil, or other liquid or solid.

Dioecious—Having staminate (male) flowers and pistillate (female) flowers on different plants of the same species.

Diorite—A granular crystalline igneous rock commonly of acid plagioclase and hornblende, pyroxene, or biotite.

Diploid—An organism that has two sets of chromosomes in its cells, paternal and maternal.

Disclimax—A relatively stable ecological community often including kinds of organism foreign to the region and replacing the climax of disturbance.

Discrete variable—Qualitative variables or those represented by integral values or ratios of integral values.

Dispersion—The dilution and reduction of concentration of pollutants in either air or water. Air pollution dispersion mechanisms are a function of the prevailing meteorological conditions.

Dissolved oxygen (DO)—Oxygen dissolved in a stream, river, or lake; amount present is an indication of the degree of health of the body of water and its ability to support a balanced aquatic ecosystem.

DNA (deoxyribonucleic acid)—The double helix of DNA is the unifying chemical of life; its linear sequence defines the diversity of living things.

Dominant—Crown class of trees with crowns extending above the general level of the main canopy of even-aged stands or, in uneven-aged stands, above the crowns of the tree's immediate neighbors; receives full light from above and partly from the sides.

Double sampling—Two levels of sampling where the first level provides information on covariates and the second on the variable of interest to estimate parameters.

Down woody material (DWM)—Woody pieces of trees and shrubs that have been uprooted (no longer supporting growth) or severed from their root system; they are not self-supporting and are lying on the ground. Previously named down woody debris (DWD).

Drainage basin—The geographical region drained by a river or stream.

Duff—The partially decomposed organic matter (litter of leaves, flowers, and fruits) found beneath plants, as on a forest floor.

Dystrophic—Defective nutrition.

Ecological toxicology—The branch of toxicology that addresses the effect of toxic substances, not only on the human population but also on the environment in general, including air, soil, surface water, and groundwater.

Ecology—The study of the interrelationship of an organism or a group of organisms with their environment.

Ecosystem—A self-regulating natural community of plants and animals interacting with one another and with their nonliving environment.

Ecotone—Any zone of intergradation or interfingering, narrow or broad, between continuous plant communities.

Ecotype—A subgroup within a species that is genetically adapted to a habitat type that is different from the habitat type of other subgroups of that species. It normally has a large geographical distribution.

Ectotrophic mycorrhiza—A mycorrhiza growing in a close web on the surface of an associated root; generally formed by basidiomycete fungi.

Edaphic—Pertaining to the soil in its ecological relationships. In general, edaphic refers to characteristics of the soil, including, for example, drainage, texture, or soil chemical properties such as the pH.

Edaphic endemics—Plants or animals endemic to areas of a specific soil type.

Edge effect—The modified environmental conditions or habitat along the margins or edges of rest stands or patches.

Efficient estimator—An estimator that predicts a parameter more reliably than competing estimators where reliability is usually measured by the ratio of the mean square errors of the estimators.

Electrical resistance heating remediation—An *in situ* environmental remediation method that uses the flow of alternating current electricity to heat soil and groundwater and evaporate contaminants.

Electron—A component of an atom; travels in a distant orbit around a nucleus.

Elements—The simplest substance that cannot be separated into more simple parts by ordinary means; a pure substance that cannot be broken down into simpler substances. There are more than 100 known elements.

Emergent vegetation—A subdivision of the littoral zone of a pond; encompasses the shoreline soil area and the immediate shallow water area where emergent plant life can take root under water, grow, and surface above the waterline.

Emerging pollution contaminants—Any synthetic or naturally occurring chemical or any microorganism that is not commonly monitored in the environment but has the potential to enter the environment and cause known or suspected adverse ecological or human health effects. Pharmaceuticals and personal care products (PPCPs) comprise a very broad, diverse collection of thousands of chemical substances, including prescription and over-the-counter therapeutic drugs, fragrances, cosmetics, sunscreen agents, diagnostic agents, nutrapharmaceuticals, biopharmaceuticals, and many others.

Endergonic—A reaction in which energy is absorbed.

Endocarp—The innermost differentiated layer of the pericarp, or fruit wall, as in the stoney part of a drupe.

Endoplasmic reticulum—A membrane network within the cytoplasm of eukaryotic cells involved in the synthesis, modification, and transport of cellular materials; a protein-containing lipid bilayer.

Endosperm—A nutritive tissue in seed plants formed within the embryo sac.

Endotrophic mycorrhiza—A mycorrhiza penetrating into the associated root and ramifying between the cells; generally formed by phycomycete fungi.

Energy—A system capable of producing a physical change of state.

Entropy—A measure of the disorder of a system.

Environment—All the surroundings of an organism, including other living things, climate, soil, etc.; in other words, the conditions affecting development or growth.

Environmental degradation—All the limiting factors that act together to regulate the maximum allowable size or carrying capacity of a population.

Environmental factors—Factors that influence volatilization of hydrocarbon compounds from soils. Environmental factors include temperature, wind, evaporation, and precipitation.

Environmental radioactivity—The study of radioactive material in the human environment.

Environmental science—The study of the human impact on the physical and biological environment of an organism. In its broadest sense, it also encompasses the social and cultural aspects of the environment.

Environmental toxicology—The branch of toxicology that addresses the effect of toxic substances, not only on the human population but also on the environment in general, including air, soil, surface water, and groundwater.

Enzymes—Proteinaceous substances that catalyze microbiological reactions such as decay or fermentation. They are not used up in the process but speed it up greatly. They can promote a wide range of reactions, but a particular enzyme can usually only promote a reaction on a specific substrate.

Epicotyl—The portion of the axis of an embryo or young seedling above the point where the cotyledon is attached.

Epigeal—The part of the seedling above the cotyledon. (See *Hypogeal*.)

Epilimnion—Upper layer of a lake heated by the sun; it is lighter and less dense than the underlying water.

Epiphyte—An organism that grows on another plant but is not parasitic on it.

Equal areas—A map projection is equal area if every part, as well as the whole, has the same area as the corresponding part on the Earth, at the same reduced scale. No flat map can be both equal area and conformal.

Equidistant—Showing true distances only from the center of the projection or along a special set of lines. An azimuthal equidistant map, for example, centered at Washington shows the correct distance between Washington and any other point on the projection (e.g., the correct distance between Washington and San Diego and between Washington and Seattle), but it does not show the correct distance between San Diego and Seattle. No flat map can be both equidistant and equal area.

Estimate—The numerical value calculated from an estimator for a sample.

Estimator—A function of the values in a sample or a formula used for estimating a parameter based on a sample.

Estimator of population mean—The formula used in estimating the population mean from a sample.

Estimator of population variance—The formula used in estimating the population variance from a sample.

Eucaryotic—An organism characterized by a cellular organization that includes a well-defined nuclear membrane; they are found in human and other multicellular organisms (plants and animals), as well as algae and protozoa.

Euphotic—Refers to the surface layer of an ocean, lake, or other body of water through which sufficient sunlight passes to allow photosynthesis.

Eutrophic lake—Lake with a large or excessive supply of plant nutrients (mostly phosphates and nitrates).

Eutrophication—Natural process in which lakes receive inputs of plant nutrients as a result of natural erosion and runoff from the surrounding land basin.

Evaporative emissions—The evaporative emission of fuel from internal combustion systems caused by diurnal losses, hot soak, and running losses.

Evapotranspiration—The process whereby plants lose water to the atmosphere during the exchange of gases necessary for photosynthesis. Water loss by evapotranspiration constitutes a major flux back to the atmosphere.

Even-aged management—The application of a combination of actions that results in the creation of stands in which trees of essentially the same age grow together. The difference in age between trees forming the main canopy level of a stand usually does not exceed 20% of the age of the level of a stand at maturity. Regeneration in a particular stand is obtained during a short period at or near the time when a stand has reached the desired age or size for regeneration and is harvested. Cutting methods producing even-aged stands include clear-cut, shelter wood, or seed tree.

Evolution—Modification of a species; the core theme of biology.

Exalbuminous—Descriptive of seeds that lack endosperm.

Excretion—Process of getting rid of waste materials.

Excurrent branching—Tree growth in which the main axis continues to the top of the tree and from which smaller, lateral braches arise (as in conifers). (See *Deliquescent branching.*)

Exergonic—Releasing energy.

Experiment—The conduct of a systematic, controlled test or investigation.

Facultative—Bacteria capable of growth under aerobic and anaerobic conditions.

Fastigate form—Strictly erect and more or less parallel vertical branches.

FBM—Foot, board measure.

Federal Water Pollution Control Act (Clean Water Act)—Act concerned with controlling and regulating the amount of municipal and industrial pollution fed into the nation's water bodies.

Fen—A bog with springs as a water source other than precipitation.

Feral goats—Goats that have escaped from domestication and become wild.

Fermentation—The decomposition of organic substances by microorganisms or enzymes. The process is usually accompanied by the evolution of heat and gas and can be aerobic or anaerobic.

Fertilizer—Substance that adds essential nutrients to the soil and makes the land or soil capable of producing more vegetation or crops.

Fine woody debris (FWD)—Dead branches, twigs, or wood splinters 0.1 to 2.9 inches (0.3 to 7.4 cm) in diameter.

First law of thermodynamics—In any chemical or physical change, any movement of matter from one place to another, or any change in temperature, energy is neither created nor destroyed but is merely converted from one form to another.

Fission product—The atomic fragments left after a large nucleus fissions.

Flagged krummholz—The tallest trees protrude from the protective snow pack and become wind-battered.

Floating leaf vegetation—Part of the littoral zone in a lake or pond where vegetation rooted under the surface allows stems to produce foliage that is able to reach and float on the water surface.

Fluvial—Produced by steam action.

Food—Nutritious substances needed by living things to grow, develop, and repair body parts.

Food chain—A sequence of transfers of energy in the form of food from organisms in one trophic level to organisms in another trophic level when one organism eats or decomposes another.

Food web—A complex network of many interconnected food chains and feeding interactions.

Forage—Herbage or browse that is accessible and acceptable as food for grazing and browsing animals.

Forest—An ecosystem characterized by tree cover.

Forest floor—The entire thickness of organic material overlying the mineral soil, consisting of the litter and the duff (humus).

Forest land—As defined by the U.S. Forest Service, the 749 million acres of forest land in 2002 consist of "land at least 10 percent stocked by trees of any size, including land that formerly had such tree cover and that will be naturally or artificially regenerated. Forest land includes transition zones, such as areas between heavily forested and nonforested lands that are at least 10 percent stocked with forest trees and forest areas adjacent to urban and built up lands. Also included are pinyon–juniper and chaparral areas in the West and afforested areas."

– Forest land that is producing or is capable of producing crops (in excess of 20 cubic feet per acre per year) of industrial wood and not withdrawn from timber utilization by statute or administrative regulation. Currently inaccessible and inoperable areas are included.

– Forest land withdrawn from timber utilization through statute, administrative regulation, or designation without regard to productive status. Wilderness areas and parks are included in this category. The definition changed slightly in 1997. Prior to 1997, the reserved forest land definition depended on the timberland designation. Reserved timberland was classed as "unproductive reserved" and included under the "other forest land" category (see below).

– Forest land other than timberland and reserved forest land. This includes available and reserved unproductive forest land, which is incapable of producing 20 cubic feet per acre per year of industrial wood under natural conditions because of adverse site conditions such as sterile soils, dry climate, poor drainage, high elevation, steep slopes, or rockiness. Urban forest land is also included. This definition changed slightly starting in 1997, when "other forest land" no longer included land classified as unproductive reserved. This area, amounting to about 12 million acres in 1997, is now included in the "reserved forest land" category.

Forest-use land—A major land uses category based on use of the forest land as opposed to the forest cover alone. The forest-use category includes both grazed and ungrazed forests but excludes an estimate of forest land in parks, wildlife areas, and similar special-purpose uses from the U.S. Forest Service's inventory of total forest land. Although it is impossible to eliminate overlap with other uses, this reduced area is a closer approximation of the land that may be expected to serve commercial forest uses as opposed to having forest cover; nevertheless, forest-use land may still be economically unsuited for timber harvests. In addition, private landowners may have objectives other than timber harvest; for example, it was found that only 29% of private forest owners reported managing their land primarily for timber production.

Formula weight—Sum of the atomic weight of all atoms that comprise one formula unit.

Frost rings—A zone of injured cambium tissue caused by frost.

Fuel bed—Down woody material fuel complex measured from the top of the duff layer to the highest piece of woody debris found at the transect point.

Fuels, 1-hour—Fine woody debris with a transect diameter less than 0.24 inch (0.6 cm).

Fuels, 10-hour—Fine woody debris with a transect diameter from 0.25 to 0.9 inch (0.6 to 2.3 cm).

Fuels, 100-hour—Fine woody debris with a transect diameter from 1 to 2.9 inches (2.5 to 7.4 cm).

Fuels, 1000+-hour—Coarse woody debris with a transect diameter of 3 inches (7.6 cm) or more.

Fungi—A saprophytic or parasitic organisms that may be unicellular or made up of tubular filaments and that lacks chlorophyll.

Funiculus—The basal stalk of an ovule arising from the placenta in the angiosperms.

Gabbro—A dark, coarse-textured, heavy rock composed of calcium feldspar and augite with a small amount of quartz.

Gaining stream—Typical of humid regions, where groundwater recharges the stream.

Gas—In the widest sense, minute particles that exhibit the tendency to fly apart from each other in all directions. Normally, gases are found in that state at ordinary temperature and pressure. They can only be liquefied or solidified by artificial means, either through high pressure or extremely low temperatures.

Gas laws—Physical laws concerning the behavior of gases. They include Boyle's law and Charles's law, which are concerned with the relationships between the pressure, temperature, and volume of an ideal (hypothetical) gas.

Gene—The smallest transmittable unit of genetic material consistently associated with a single primary genetic effect.

General biological succession—The process whereby communities of plant and animal species in a particular area are replaced over time by a series of different and usually more complex communities (also known as *ecological succession*).

Genet—A single sexually produced individual.

Genome—A complete haploid set of chromosomes.

Genus—A group of species with many common characteristics.

Geology—The science of the Earth and its origin, composition, structure, and history.

Geophysical testing—Used to evaluate the subsurface layers, locate the water table, and map contaminant contours using resistivity and conductivity meters.

Geosphere—Consists of the inorganic (nonliving) portions of the Earth which are home to the entire globe's organic (living) matter.

Geothermal energy—The use of the Earth's natural heat for human purposes; a form of alternative energy that is massive but difficult to tap.

Geothermal power—See *Geothermal energy*.

Germinative capacity—Percentage of seeds that germinate during the normal period of germination.

Germplasm—Within an individual or group, a collection of genetic resources that are the physical basis for inheritance.

Gibberellin—Plant hormone useful in regulating growth characteristics of many plants.

Glade—An open space in a forest.

Global dimming—The gradual reduction in the amount of global direct irradiance at the Earth's surface.

Global distillation (or grasshopper effect)—The geochemical process by which certain chemicals, most notably persistent organic pollutants (POPs), are transported from warmer to colder regions of the Earth.

Global positioning system (GPS)—A system using satellites to locate ground positions.

Global warming—The long-term rise in the average temperature of the Earth.

Gneiss—A metamorphic rock derived from either igneous or sedimentary formations.

Grab sample—An individual discrete sample collected over a period of time not exceeding 15 minutes.

Graft incompatibility—Refers to plants that, when grafted together, fail to form a lasting union.

Gram—The basic unit of weight in the metric system; equal to 1/1000th of a kilogram. Approximately 28.5 grams equal 1 ounce.

Granite—A very hard natural igneous rock formation.

Graticule—The spherical coordinate system based on lines of altitude and longitude.

Gravity—The force of attraction that arises between objects by virtue of their masses. On Earth, gravity is the force of attraction between any object in the Earth's gravitational field and the Earth itself.

Graze—The consumption of standing forage by livestock or wildlife.

Great circle—Intersection of a sphere (e.g., the Earth) and a plane that passes through the center point of the sphere. The Equator, each meridian, and each other full circumference of the Earth forms a great circle. The arc of the great circle shows the shortest distance between points on the surface of the Earth.

Greenhouse effect—The trapping of heat in the atmosphere. Incoming short-wavelength solar radiation penetrates the atmosphere, but the longer wavelength outgoing radiation is absorbed by water vapor, carbon dioxide, ozone, and several other gases in the atmosphere and is reradiated to Earth, causing an increase in atmospheric temperature.

Greenhouse gases—Gases present in the Earth's atmosphere that contribute to the greenhouse effect.

Grid—Used to locate places on a map. Global grids are formed by crossings of parallels (line of latitude) and meridians (lines of longitude). A letter or number coordinate grid system is used to locate places on maps of smaller places such as state, city, and highway maps.

Grood soils—Nut-structured soils characteristic of the transition zone between prairie soils and podzolic soils (i.e., prairie–forest soils).

Gross primary production—Total amount of organic matter in an ecosystem including aboveground (leaves and stems) and belowground (roots) biomass. Also sometimes referred to as *net primary production*.

Groundwater—Water collected underground in porous rock strata and soils; it emerges at the surface as springs and streams.

Group selection—The silvicultural system in which trees are removed periodically in small groups, resulting in openings that do not exceed 0.4 to 0.8 hectare (1 to 2 acres) in size. This leads to the formation of an uneven-aged stand in the form of a mosaic of age-class groups in the same forest.

Growing stock level (GSL)—A numerical index; the residual square meters of basal area per hectare (square feet of basal area per acre) when the average stand DBH is 25 cm (10 inches) or more. Basal area retained in a stand with an average DBH of less than 25 cm. (10 inches) is less than the designated level.

Habitat—The place or type of place where an organism or community of organisms naturally or normally thrives.

Haploid—An organism with one basic chromosome set symbolized by n.

Harden-off—The process of gradually reducing the amount of water and lowering the temperature for plants in order to toughen their tissues, making it possible for them to withstand unfavorable (usually cold) environmental conditions.

Hardness—A water-quality parameter. The scale found in pots, pans, and kettles is caused by the presence of certain salts of calcium and magnesium in the water supply.

Headwaters—The part of a stream or river proximate to its source.

Heat—A condition of matter caused by the rapid movement of its molecules. Energy has to be applied to the material in sufficient amounts to create the motion and may be applied by mechanical or chemical means.

Heat balance—The constant trade-off that takes place when solar energy reaches the Earth's surface, is absorbed, and then must return to space to maintain Earth's normal heat balance.

Hedging—Close-cropping.

Heptane—Any of several isometric hydrocarbons of the methane series.

Herb—Any flowering plant except those developing persistent woody stems.

Herbage—Aboveground biomass of herbaceous plants.

Herbicide—Used to kill unwanted plants.

Hermaphrodite (bisexual)—A flower with both functional male and female reproductive organs.

Heterotrophic—Refers to a category of organisms that obtain their energy by consuming the tissues of other organisms; they break down the other biological material using digestive enzymes and then assimilate the usable byproducts.

Heterozygote—An organism whose cells have one or more sets of unlike alleles.

High-lining—The underside of a forest canopy that is uniformly cropped by deer at the highest level they can browse; a browseline.

Hilum—The scar on a seed marking the point of attachment of the ovule.

Horizon—In soil, a layer of soil approximately parallel to the soil surface and differing in properties and characteristics from adjacent layers below or above it.

Humidity—The amount of water vapor in a given volume of the atmosphere (absolute humidity) or the ratio of the amount of water vapor in the atmosphere to the saturation value at the same temperature (relative humidity).

Humus—The more or less stable fraction of soil organic matter remaining after the major portions of added plant and animal residues are decomposed; usually dark in color.

Hybrid swarm—An extremely variable population derived from the hybridization of two different taxa and consisting of the products of subsequent segregation and recombination, backcrossing, and crossing between the hybrids themselves. It occurs where the ranges of interfertile species overlap.

Hydraulic gradient—The difference in hydraulic head divided by the distance along the fluid flow path. Groundwater moves through an aquifer in the direction of the hydraulic gradient.

Hydrocarbon—A chemical containing only carbon and hydrogen atoms. Crude oil is a mixture largely of hydrocarbons.

Hydrological cycle—The means by which water is circulated in the biosphere; cooling in the atmosphere and precipitation over both land and oceans counterbalances evapotranspiration from the land mass plus evaporation from the oceans.

Hydroponics—The cultivation of plants, without soil, in water solutions of nutrients required for growth.

Hydrosere—An ecological sere (plant community) originating in an aquatic habitat.

Hydrosphere—The portion of the Earth's surface covered by the oceans, seas, and lakes.

Hypanthium—A floral tube formed by the fusion of the basal portions of the sepals, petals, and stamens and from which the rest of the floral parts emanate.

Hypha (sing.), **hyphae** (pl.)— In fungi, a tubular cell that grows from the tip and may form many branches.

Hypocotyl—The part of an embryo or seedling below the cotyledon and above the radicle (but sometimes including it).

Hypogeal—Describes seed germination in which the colytedons remain beneath the surface of the soil. (See *Epigeal*.)

Hypogeous—Growing or developing below the soil surface.

Hypolimnion—The cold, relatively dense bottom layer of water in a stratified lake.

Ideal gas law—A hypothetical gas that obeys the gas laws exactly with regard to temperature, pressure and volume relationships.

Igneous rock—Formed by solidification of molten magma.

Imperfect flower—A flower that lacks either stamens or carpels.

Impoundment—A lake classification; an artificially manmade lake made by trapping water from rivers and watersheds.

Inbreeding—In plants, a breeding system in which sexual reproduction involves the interbreeding of closely related plants by self-pollination or backcrossing.

Indicator species—Any plant that by its presence, frequency, or vigor reflects a particular property of the site.

Individual tree selection—The silvicultural system in which trees are removed individually each year, here and there, over an entire forest or stand. The resultant stand usually regenerates naturally and becomes all-aged.

Indolebutyric acid (IBA)—A synthetic auxin widely used in horticulture to induce rooting of cuttings.

Indoor air quality—A term referring to the air quality within and around buildings and structures, especially as it relates to the health and comfort of building occupants.

Inference—A conclusion drawn based on data or observations.

Infiltration capacity—The maximum rate at which soil can absorb rainfall.

Ingestion—Taking in food or producing food.

Ingrowth—The volume of trees that have grown into the lowest inventoried size class between two measurements.

Inland Empire—A region in eastern Washington, northern Idaho, and western Montana, named for commercial purposes.

Inorganic compounds—May or may not contain carbon.

Inorganic substance—A substance that is mineral in origin that does not contain carbon compounds, except as carbonates, carbides, etc.

Insolation—The amount of direct solar radiation incident per unit of horizontal area at a given level.

Intergeneric—Existing or occurring between genera.

Intermediate—Crown class of trees of the middle canopy whereby the crown extends somewhat into the lower part of the main canopy. The crown intercepts direct sunlight only at a limited area on the top and none at the sides; it is narrow and short, with limited leaf surface area and a lower live–crown ratio. Tree diameter is within the lower range of those present, but not necessarily the smallest.

Intermountain—A U.S. Forest Service area that includes the states of Montana, Idaho, Utah, Nevada, and the western quarter of Wyoming.

Intraspecific—Refers to some relationship between the members of the same population or species.

Introgression—The entry or introduction of a gene from one gene complex to another.

Invasive species—Non-indigenous species (e.g., plants or animals) that adversely affect the habitats they invade economically, environmental, or ecologically.

Ionic bonds—A chemical bond in which electrons have been transferred from atoms of low ionization potential to atoms of high electron affinity.

Irrigation—Artificial water supply for dry agricultural areas created by means of dams and channels.

Isoline—Isogram; a line on a map or chart along which there is a constant vale (temperature, pressure, or rainfall).

Isozymes (isoenzymes)—Two or more chemically distinct but functionally similar enzymes.

Jackstrawed fuel—Trees that have fallen in tangled heaps.

Juvenile cuttings—The youngest parts of the branches are severed from the plant and rooted to produce new plants.

Karyotype—The character of the chromosomes as defined by their size, shape, and number.

Kelvin—Temperature scale used by scientists that begins at absolute zero and increases by the same degree intervals as the Celsius scale; that is, 0°C is the same as 273 K and 100°C is 373 K.

Knee—An abrupt bend in a stem or tree trunk or an outgrowth rising from the roots of some swamp-growing trees such as baldcypress.

Krummolz—Stunted growth habit characterized by crooked wood caused by wind; found in certain tree species at the upper limit of their distribution.

Lacustrine—Related to or growing in lakes.

Lake Agassiz Basin—A later glacial and early postglacial lake area in southern North Dakota and western Minnesota.

Lake states—States bordering the Great Lakes (Minnesota, Wisconsin, Illinois, Indiana, Michigan, Ohio, Pennsylvania, and New York).

Laminar flow (water)—Occurs in a stream where parallel layers of water shear over one another vertically.

Lammas—The part of an annual shoot that is formed after a summer pause in growth.

Land farming—Another name for land treatment whereby various contaminants are spread on soil and worked into the surface and subsurface to allow biodegradation to take place.

Lapse rate—The rate of temperature change with increasing height. On average, temperature decreases –65°C/100 m or –6.5°C/km, which is considered to be the *normal lapse rate*.

Latent heat of fusion—The amount of heat required to change one gram of a substance from the solid to the liquid phase at the same temperature.

Latent heat of vaporization—The amount of heat required to change one gram of a substance from the liquid to the gas phase at the same temperature.

Law of conservation of mass—In any ordinary physical or chemical change, matter is neither created nor destroyed but merely changed from one form to another.

Layering—The rooting of an undetached branch, laying on or partially buried in the soil, that is capable of independent growth after separation for the mother plant.

Leach liquors—Refers to liquid leached from a substance via water circulation through or over it.

Leaf area index (LAI)—Leaf surface area per unit of land surface area. For broadleaf forests, the index is calculated using only one side of the leaf blade; for needleleaf stands, the total leaf surface is used; for mixed broad- and needleleafed stands, a combination of the two is used.

Legend—Key included on a map to explain the meaning of colors and symbols used on the map; may include a key to elevation (distance above or below sea level).

Lentic—Refers to calm waters found in lakes, ponds, and swamps.

Life span—Maximum length of time an organism can be expected to live.

Light pollution—Excessive or obtrusive artificial light (photopollution or luminous pollution).

Lightwood (fatwood, lightered wood or stumps, stumpwood)—Coniferous wood having an abnormally high content of resin and therefore easily set alight.

Lignotubers—A woody swelling at ground level originating from the axils for the cotyledons from whose concealed dormant buds a new tree can develop if the old one is injured; characteristic of many eucalypts.

Limited—Limiting nutrients such as carbon, nitrogen, and phosphorous.

Limiting factor—Factor such as temperature, light, water, or a chemical that limits the existence, growth, abundance, or distribution of an organism.

Limiting nutrient—See *Limited*.

Limnetic—The open water surface layer of a lake through which reaches sufficient sunlight for photosynthesis.

Limnology—The study of lakes and other bodies of open freshwater in terms of their plant and animal biology and their physical properties.

Limonene—A component of pine turpentine with a density of approximately 0.84 and an index of refraction of about 1.47, both at 25°C (77°F).

Linear scale—The relation between a distance on a map and the corresponding distance on the Earth. Scale varies from place to place on every map; the degree of variation depends on the projection used when making the map.

Lipids—Energy-rich compounds made of carbon, oxygen, and hydrogen.

Liquid—A state of matter between a solid and a gas.

Liter—A metric unit of volume, equal to 1 cubic decimeter (1.76 pints).

Lithosphere—The layers of soil and rock that comprise the Earth's crust.

Litter—The intact and partially decayed organic matter (bark, twigs, flowers, and fruits) lying on top of the soil; discards thrown about without regard to the environment; the L layer of the organic portion of the soil profile.

Littoral—The shallow zone of water near the shore of a body of water.

Loam—The textural class name for soil with a moderate amount of sand, silt, and clay. Loam soils contain 7 to 27% clay, 28 to 50% silt, and 23 to 52% sand.

Loess—A uniform and unstratified fine sand or silt (rarely clay) deposit transported by wind (an aeolian soil). It is sometimes described as *rock flour*.

Log rule—A formula or table for estimating the volume (usually in board feet) of lumber that may be sawed from logs of different sizes.

Losing stream—Where streams can recharge groundwater; typical of arid regions.

Lotic—Running freshwater systems (e.g., rivers, streams).

Lumen—(1) Cell cavity. (2) Unit of luminous flux equal to the light emitted by a uniform point source of one candle intensity.

Lye—A strong alkaline solution of sodium hydroxide, potassium hydroxide, or the leachate of weed ashes that is rich in potassium carbonate.

Maceration—Removal of the fleshy tissue surrounding seeds, often by soaking in water.

Macronutrients—Includes the nutritional elements of nitrogen, phosphorus, potassium, calcium, magnesium, and sulfur, which are essential for normal plant growth, development, and production. They are usually derived from the soil.

Magma—Molten rock material within the Earth's core.

Map parts—Help users read maps and analyze the physical and human landscapes of the world. The title identifies the map and its contents. The direction indicator identifies directions or orientation; usually direction is shown on a map by a single arrow labeled "N" that points north. Other maps have a compass rose (directional indicator) symbol that indicates direction with arms that point to the cardinal and intermediate directions. The cardinal directions are north, south, east, and west, and the intermediate directions are northeast, southeast, northwest, and southwest.

Map projection—A systematic representation of a round body such as the Earth or a flat (plane) surface. Each map projection has specific properties that make it useful for specific purposes.

Map scales—Used to measure distances on maps. Different scales are used on different maps; they are necessary for developing map representations because sizes of maps in relation to the size of the real world differ; scale is shown by giving the ratio between distances on the map and actual distances on earth. Areas can be represented using a variety of scales. The amount of detail shown on a map is dependent on the scale used:

  – *Written or statement scale*—A statement that relates distance on the map to the distance it represents on Earth (e.g., 1 inch equals 4 miles).

  – *Representative fraction*—Fractions or ratios relate distance on the map to the distance it represents on Earth (e.g., 1:250,000, 1/250,000).

  – *Graphic or bar scale*—A short line that represents the number of miles or kilometers on the Earth's surface compared to the corresponding distance on the map; the line is divided into equal parts and labeled with miles or kilometers.

Mass—The quantity of matter and a measurement of the amount of inertia that a body possesses.

Mast—Fruits, nuts, and seeds produced by woody plants and used as food by animals.

Mature pond—A pond that reaches maturity; characterized by being carpeted with rich sediment, with aquatic vegetation extending out into open water, and a great diversity of plankton, invertebrates, and fishes.

Mean—The average value of a variable for all units in a population or sample.

Mean annual increment—The total increment of trees in a stand up to a given age divided by the age; usually expressed in annual cubic meters of growth per hectare (cubic feet of growth per acre).

Meandering—Stream condition whereby flow follows a winding and turning course.

Median—The value of a variable so that half of the values are larger and half are smaller than this value in a population or sample.

Megagametophyte—The female gametophyte that develops from the megaspore and produces female gametes.

Megasporangium—The sporangium in which megaspores are produced.

Megaspore—Heterosporous plants have two types of spores: microspores (male) and megaspores (female); megaspores germinate into female (egg-producing) gametophytes, which are fertilized by sperm produced by male gametophytes developed from microspores.

Meiosis—Reduction division resulting in the production of haploid gametes; a process consisting of two specialized nuclear divisions ultimately leading to the formation of eggs or sperm.

Meristem—Tissue primarily associated with protoplasmic synthesis and formation of new cells by division.

Mesic—Characterized by intermediate moisture conditions, neither decidedly wet nor decidedly dry.

Mesophyte—A plant whose normal habitat is neither very wet nor very dry.

Mesozoic—An era of geologic history marked by cycads, evergreen trees, dinosaurs, marine and flying reptiles, and ganoid fishes.

Era	Period	Millions of Years Before Present
Cenozoic	Quaternary	2.5–present
	Tertiary	65–2.5
Mesozoic	Cretaceous	135–65
	Jurassic	190–135
	Triassic	225–190
Paleozoic	Permian	280–225
	Pennsylvanian	320–280
	Mississippian	345–320
	Devonian	400–345
	Silurian	440–400
	Ordovician	500–440
	Cambrian	570–500
	Precambrian	4600–570

Metabolism—The chemical processes of living organisms; a constant alternation of building up and breaking down. Green plants, for example, build up complex organic substances from water, carbon dioxide, and mineral salts (photosynthesis). By digestion, animals partially break down complex organic substances ingested as food and subsequently resynthesize them in their own bodies.

Metamorphism—A pronounced change effected by pressure, heat, and water that results in a more compact and more highly crystalline condition.

Meter—The standard of length in the metric system, equal to 39.37 inches or 3.28 feet.

Methane (CH_4)—The simplest hydrocarbon of the paraffin series. Colorless, odorless, and lighter than air, it burns with a bluish flame and explodes when mixed with air or oxygen. Methane is a greenhouse gas.

Microbial community—The community of microbes available to biodegrade contaminants in the soil.

Microbial degradation—The natural process whereby certain microbes in soil can degrade contaminants into harmless constituents.

Microbiology—The study of organisms that can only be seen under the microscope.

Micronutrients—Nutritional elements (trace elements) necessary in minute quantities for normal plant growth, such as boron and manganese.

Micropyle—A minute opening in the integument of an ovule through which the pollen tube normally passes to reach the embryo sac, usually closed in the mature seed to form a superficial scar.

Microsporangium—In plants having two types of haploid spores (microspores and megaspores), the saclike structure in which microspores are produced.

Microspore—A haploid spore produced by meiosis of the microsporocyte and developing into the male gametophyte; the pollen grain of seed plants.

Mine spoil—Earth and rock excavated from a mine.

Mitochondria—A microscopic body found in the cells of almost all living organisms and containing enzymes responsible for the conversion of food to usable energy.

Mitosis—Normal distribution of a nucleus into two identical daughter nuclei by a process of duplication and separation of chromosomes.

Mixture—In chemistry, a substance containing two or more compounds that still retain their separate physical and chemical properties.

Mode—The value of a variable that occurs most frequently in a population or sample.

Modeling—The use of mathematical representations of contaminant dispersion and transformation to estimate ambient pollutant concentrations.

Molar concentration (molarity)—In chemistry, the amount of a constituent of a mixture divided by the volume of that mixture.

Mole—SI unit for the amount of a substance; the amount of a substance that contains as many elementary entities as there are atoms in 12 g of the isotope carbon-12.

Molecular weight—The weight of one molecule of a substance relative to ^{12}C, expressed in grams.

Molecule—The fundamental particle that characterizes a compound. It consists of a group of atoms held together by chemical bonds.

Monadnock—A hill or mountain of resistant rock surmounting land of considerable area and slight relief shaped by erosion.

Monoecious—Having staminate and pistillate flowers in separate places on the same plant.

Montane—Biogeographic zone made up of relatively moist cool upland slopes below timberline that is characterized by large evergreen trees as a dominant life form.

Morphogenesis—Evolutionary development of the structure of an organism or part.

Motility—An organism's mobility; its ability to move.

Movement—Nonliving material moves only as a result of external forces, whereas living material moves as a result of internal processes at cellular level or at organism level (locomotion in animals and growth in plants).

Muck—Highly decomposed organic material formed under conditions of waterlogging, with few recognizable remains of the original plants.

Mull—A soil whose upper mineral layer has become intimately mixed (mainly through the action of earthworms) with amorphous organic material, sometimes to a depth of 1.2 to 1.5 meters (4 to 5 feet).

Multilevel sampling—A sampling design that utilizes more than one phase or stage of sampling. The first levels are used to collect information on covariates useful for more efficient estimation of the ultimate parameters of interest, for which information is usually collected at the last phase or stage.

Mycelium—An interwoven mass of threadlike filaments or hyphae forming the main body of most fungi. The reproductive structures, or "fruiting bodies," grow from the mycelium.

Mycology—The branch of botany that deals with fungi.

Natural pruning—The freeing of the stem of a standing tree of its braches by natural death, disintegration, or fall; it is caused by decay, by a deficiency of light or water, or by snow, ice, and wind breakage.

Naval stores—Products of the resin industry. In the United States, they are turpentine, rosin, pine tar, and pitch (resin). Gum naval stores refer specifically to gum turpentine and gum rosin; wood naval stores to wood turpentine and wood rosin.

Necromass—The accumulated dead biomass or litter.

Neutron—Elementary particles that have approximately the same mass as protons but have no charge. They are one constituent of the atomic nucleus.

Niche—The functional role of an organism within its community; the complete ecological description of an individual species (including habitat, feeding requirements, etc.).

Nitrates—In freshwater pollution, a nutrient, usually from fertilizer, that enters the water system and can be toxic to animals and humans in high enough concentrations.

Nitrogen cycle—The natural circulation of nitrogen through the environment.

Nitrogen dioxide (NO₂)—A reddish-brown, highly toxic gas with a pungent odor; one of the seven known nitrogen oxides that participate in photochemical smog and primarily affect the respiratory system.

Nitrogen fixation—Nature accomplishes nitrogen fixation by means of nitrogen-fixing bacteria.

Nitrogen oxide (NO)—A colorless gas used an anaesthetic; soil bacteria form it from decomposing nitrogenous material.

Nonpoint source—Source of pollution in which wastes are not released at one specific, identifiable point but from a number of points that are spread out and difficult to identify and control.

Nonrenewable resources—Resources that exist in finite supply or are consumed at a rate faster than the rate at which they can be renewed.

Nonvolatile—A substance that does not evaporate at normal temperatures when exposed to the air.

Normal lapse rate—See *Lapse rate.*

Normal yield table—A table showing, for one or more species in a fully stocked stand, the growth pattern of a managed even-aged stand derived from measurements at regular intervals covering its useful life. It includes mean DBH and height, number of stems, and standing volume per unit area. The table may also contain a variety of other useful data; the data presented are averages derived from any stands considered to be fully stocked at the time they were sampled.

Northcentral—U.S. Forest Service area that includes Indiana, Illinois, Michigan, Wisconsin, Minnesota, Iowa, and Missouri.

Northeastern—U.S. Forest Service area that includes the New England states plus New York, New Jersey, Delaware, Maryland, Pennsylvania, Ohio West Virginia, and Kentucky.

Nucellus—The tissue of an ovule, in which the female gametophyte (embryo sac) develops; the megsporangium.

Nucleic acids—Biological molecules (RNA and DNA) that store information essential for life.

Nutrient cycles—See *Biogeochemical cycles.*

Nutrients—Elements or compounds necessary for the survival, growth, and reproduction of a plant or animal.

Nutrition—The process of nourishing or being nourished.

Old growth—Timber stands with the following characteristics: large mature and overmature trees in the overstory, snags, dead and decaying logs on the ground, and a multi-layered canopy with trees of several age classes. The specific attributes of an old-growth stand are primarily dependent on plant associations and forest cover type.

Oleoresin—The nonaqueous secretion of resin acids dissolved in a terpene hydrocarbon oil that is produced in, or exuded from, the intercellular resin ducts of a living tree or accumulated, together with oxidation products, in the dead wood of weathered limbs or stumps. Commonly called *pine gum, gum, pitch,* or *sap.*

Open-grown—Trees grown in the absence of woody competition.

Operculum—A caplike structure composed of used sepals and petals that suggest a lid.

Organelle—A specialized part of a cell that resembles and functions as an organ.

Organic chemistry—The branch of chemistry concerned with compounds of carbon.

Organic compounds—Found in living things; they contain carbon.

Organic matter—Includes both natural and synthetic molecules containing carbon and usually hydrogen. All living matter is made up of organic molecules.

Organic soil layers—L-layer, freshly fallen or only slightly decomposed leaves, twigs, flowers, fruit, and bark lying on the soil surface. F-layer, zone of active organic matter fermentation. H-layer, humidified zone; the more or less stable fraction from decomposed soil organic material that generally is amorphous, colloidal, and dark colored.

Organic substance—Any substance containing carbon.

Organism—Any living thing.

Ortet—An original plant from which a vegetatively propagated clone has been derived.

Overgrazing—Consumption of vegetation on rangeland by grazing animals to the point that the vegetation cannot be renewed or is renewed at a rate slower than consumption.

Overstory—Trees in a forest of more than one story that form the upper or uppermost canopy layer.

Ovulate—Bearing or possessing ovules.

Oxidation—The process by which electrons are lost.

Oxygen—An element that readily unites with materials.

Ozone—The compound O_3, which is found naturally in the atmosphere in the ozonosphere; a constituent of photochemical smog.

Ozone depletion—Ozone concentrations vary naturally with sunspots, the seasons, and latitude; these processes are well understood and predictable. Scientists have established records spanning several decades that detail normal ozone levels during these natural cycles. Each natural reduction in ozone levels has been followed by a recovery. Recently, however, convincing scientific evidence has shown that the ozone shield is being depleted well beyond changes due to natural processes.

Pacific Northwest—U.S. Forest Service area that includes the states of Washington and Oregon.

Pacific Southwest—U.S. Forest Service area that includes the states of California and Hawaii, plus Guam and the Trust Territories of the Pacific Islands.

Parameter—A characteristic or function of the values of the units in a population; the population characteristic of interest, such as average volume per ha or total volume of trees in a forest.

Parasite—Primary, secondary, or higher consumer that feeds on a plant or animal, known as a host, over an extended period of time.

Parent material—The unconsolidated and more or less chemically weathered mineral or organic matter from which pedogonic processes develop the solum in soils.

Parthenocarpy—The development of fruit without viable seeds. It may be induced artificially, as by some foreign pollen, or with hormones.

Particulate matter—Normally refers to dust and fumes; travels easily through air.

Pascal (Pa)—A unit of pressure equal to one newton per square meter.

Pathogen—Any disease-producing organism.

Peak standing crop—The maximum amount of standing crop observed during a given year.

Peat—Undecomposed or only slightly decomposed organic matter accumulated under conditions of excess moisture. Plant residues show little, if any, morphological change.

Pedologist—A person who study soils.

Peds—A unit of soil structure such as an aggregate, crumb, prism, block, or granule, formed by natural processes.

Peduncle—A stalk bearing a flower, flower cluster, or a fructification.

Perennial stream—A type of stream in which flow continues during periods of no rainfall.

Perfect flower—A flower having both stamens and carpels; may or may not have a perianth.

Perianth—A collective term for the floral envelope, usually comprised of the calyx and corolla of a flower.

Pericarp—The wall of a ripened ovary (fruit) that is homogeneous in some genera and in others is composed of three distinct layers: exocarp, mesocarp, and endocarp.

Periodic law—The properties of elements are periodic functions of the atomic number.

Permafrost—Permanently frozen ground; generally refers to a layer at some depth below the soil surface. Any layer above it that thaws in summer is termed the *active layer*.

Perpetual resource—A resource such as solar energy that comes from an essentially inexhaustible source and thus will always be available on a human time scale regardless of whether or how it is used.

Persistent substance—A chemical product with a tendency to persist in the environment for quite some time (e.g., plastics).

Pesticide—Any chemical designed to kill weeds, insects, fungi, rodents, and other organisms that humans consider to be undesirable. A substance or mixture of substances used to kill pests.

pH—A numerical designation of relative acidity and alkalinity; a pH of 7.0 indicates precise neutrality, high values indicate increasing alkalinity, and lower values indicate increasing acidity.

Phenotype—An organism's observable characteristics or traits; the product of the interaction of the genes of an organism (genotype) with the environment.

Phosphates—A nutrient substance obtained from fertilizers.

Phosphorous cycle—A biogeochemical cycle in which phosphorus is converted into various chemical forms and transported through the biosphere.

Photochemical reaction—A reaction induced by the presence of light.

Photochemical smog—A complex mixture of air pollutants produced in atmosphere by the reaction of hydrocarbons and nitrogen oxides under the influence of sunlight.

Photoperiodism—The physiological response of an organism to the periodicity and duration of light and darkness which affect many processes, including growth, flowering, and germination.

Photosynthesis—A complex process that occurs in the cells of green plants whereby radiant energy from the sun is used to combine carbon dioxide (CO_2) and water (H_2O) to produce oxygen (O_2) and simple sugar or food molecules, such as glucose.

Phyllodes—A flat expanded petiole that replaces the blade of a foliage leaf and fulfills the same functions in photosynthesis.

Physical change—The process that alters one or more physical properties of an element or compound without altering its chemical composition. Examples include changing the size and shape of a sample of matter and changing a sample of matter from one physical state to another.

Physical weathering—The physical changes produced in rocks by atmospheric agents (e.g., wind, precipitation, heat, cold).

Phytomass—Total weight of plant mass per unit of area in an ecosystem.

Pioneer—A plant capable of invading a newly exposed soil surface and persisting there until supplanted by successor species.

Pioneer community—The first successfully integrated set of plants, animals, and decomposers found in an area undergoing primary ecological succession.

Pistil—Ovule-bearing organ of an angiosperm composed of ovary, style, and stigma; collectively, the pistils are called the *gynoecium.*

Pistillate—Having only female organs; may apply to individual flowers or inflorescences, or to plants of a dioecious species in angiosperms.

Plankton—Microscopic floating plant and animal organisms of lakes, rivers, and oceans.

Ploidy—Degree of repetition of the basic number of chromosomes.

Plume—(1) The column of noncombustible products emitted from a fire or smokestack. (2) A vapor cloud formation having shape and buoyancy. (3) A contaminant formation dispersing through the subsurface.

Plus-tree—A phenotype judged, but not proven by test, to be unusually superior in some quality or qualities.

Plutonium in the environment—An article (part) of the actinides series in the environment.

Podzol—A soil characterized by a superficial layer of raw humus above a generally gray A horizon of mineral soil depleted of sesquioxides of iron and aluminum and of colloids and overlying a B horizon wherein organic matter or sesquioxides of iron have accumulated.

Point source—Discernable conduits, including pipes, ditches, channels, sewers, tunnels, or vessels, from which pollutants are discharged.

Point source pollution—Pollution that can be traced to an identifiable source.

Pole-size—A young tree with a DBH of not less than 10.2 cm (4 inches). A small pole has a maxim DBH of 20.3 cm (8 inches), and a large pole has a maximum DBH of 30.5 cm (12 inches).

Polygamodioecious—Bearing perfect and pistillate flowers on female trees and only staminate flowers on male trees.

Polygamous—Plants bearing both perfect and imperfect flowers.

Polymorph—One of several forms of an organism.

Pond—A still body of water, smaller than a lake, often of artificial construction.

Pond succession—Pond transformation process whereby a young pond is formed and develops over time to a mature pond and then a senescent pond.

Pool—In a stream or river, segment where the water is deeper and slower moving.

Population—An aggregate of items each with a common characteristic or common set of characteristics. In the statistical sense, a population is an assembly of individual units formed in order to describe the population quantitatively; for example, it might be all of the trees in a particular forest stand or all of the uses of a recreation area.

PPS sampling—A sampling design where sample units are selected with a probability proportional to a measure of size, usually a covariate such as DBH or basal area in the case of tree volume.

Precision—Relative freedom from random variation. In sampling, it is expressed as the standard error of the estimate and relates to the degree of clustering of sample values about their own average or the reproducibility of an estimate in repeated sampling. It is also used to indicate the resolving power of a measuring deice.

Prescribed fire—Any fire ignited by management actions to meet specific objectives prior to ignition; a written, approved prescribed fire plan must exist, and National Environmental Protection Act requirements must be met.

Pressure—Force per unit area.

Primary consumers—In the food chain, organisms that consume producers (autotrophs).

Primordium—An organ, a cell, or an organized series of cells in their earliest stage of differentiation (e.g., leaf primordium, sclereid primordium, vessel primordium).

Probabilistic sampling—Procedures in which samples are selected such that all units and each pair of units in the population have a positive probability of selection.

Procaryotic—A type of primitive cell lacking a membrane-delimited nucleus.

Producers—Organisms that use solar energy (green plants) or chemical energy (some bacteria) to manufacture their own organic substances (food) from inorganic nutrients.

Proembryo—Series of cells that are formed after fertilization within the ovule of a flowering plant, before formation of the embryo.

Propagule—A plant part such as a bud, tuber, root, or shoot used to reproduce (propagate) an individual plant vegetatively.

Protandry—The termination of shedding of pollen by a flower prior to the stigma of the same flower being receptive.

Proteranthous—Having flowers appearing before the leaves.

Protogyny—The termination of receptivity prior to the maturation of pollen on the same plant or flower.

Protozoa—Single-celled microorganisms; includes the most primitive forms of animal life.

Provenance—The original geographic source of seeds, pollen, or propagules.

Pumice—A volcanic glass full of cavities and very light in weight.

Pyrene—The pit or seed of a drupe which is surrounded by a bony endocarp.

Pyric—Resulting from, induced by, or associated with burning.

Ramet—An individual member of a clone, derived from an ortet.

Randomization—A deliberately haphazard arrangement of observations to simulate selection by chance.

Reactive—The tendency of a material to react chemically with other substances.

Receptivity—The condition of the female flower that permits effective pollination.

Recharge area—The area in which precipitation percolates through to recharge groundwater.

Relative humidity—The percentage of moisture in given volume of air at a given temperature in relation to the amount of moisture the same volume of air would contain at the saturation point.

Renewable resources—Resources that can be depleted in the short run if used or contaminated too rapidly but that normally are replaced through natural processes.

Representative sample—A sample of a universe or whole, such as a waste pile, lagoon, or groundwater, that can be expected to exhibit the average properties of the whole.

Residual biomass—Amount of vegetation remaining after grazing is completed.

Residue pile density—In reference to CWD residue piles, the density of the volume (as defined by the corresponding shape code dimensions) of the slash pile occupied by actual CWD.

Residue pile shape—The shape of each CWD residue pile coded as one of four shapes—half-section sphere, half-cylinder, half-frustum of cone, or irregular solid.

Residue piles—Piles and windrows of CWD created by human activity or natural causes in cases where CWD tally is physically impossible (i.e., large wind throws).

Reservoir—A large and deep human-created standing body of freshwater.

Resource—Something that serves a need, is useful, and is available at a particular cost.

Rhumb line—A line on the surface of the Earth cutting all meridians at the same angle. A rhumb line shows true direction. Parallels and meridians, which also maintain constant true directions, may be considered special cases of the rhumb line. A rhumb line is a straight line on a mercator projection. A straight rhumb line does not show the shorter distance between points that are on the Equator or on the same meridian.

Riffles—Shallow, high-velocity flow over a mixed-gravel–cobble substrate.

Rocky Mountains—U.S. Forest Service area that includes the Dakotas, Nebraska, Kansas, Oklahoma, and Texas west of the 100th meridian; New Mexico; Arizona; Colorado; and the eastern three-quarters of Wyoming.

Run—Somewhat smoothly flowing segment of the steam.

Runoff—Surface water that enters rivers, freshwater lakes, or reservoirs from land surfaces.

Saddle—A ridge connecting two higher elevations.

Samara—A dry, indehiscent (not split open), winged fruit; one-seeded, as in *Fraxinus* and *Ulmus*, or two-seeded, as in *Acer*.

Sample—A subset of a population used to obtain estimates of one or more of its parameters.

Sample surveys—The design and execution of surveys to provide estimates of characteristics (parameters) of well-defined finite populations.

Sample unit—A unit from a population, such as a tree or all trees located within a plot (fixed area, strip, or point samples).

Sampling design—A formalized method of selecting a sample from the population, or example simple random sampling.

Sampling frame—A list of all sample units used to represent a population.

Sampling strategy—Comprised of both the sampling design and estimators used, such as simple random sampling with estimators of the population mean.

Sapling—A tree more than 3 feet in height and less than 4 inches in DBH.

Saprophyte—An organism that uses enzymes to feed on waste products of living organisms or tissues of dead organisms.

Saturated zone—Subsurface soil saturated with water; the water table.

Savannah—Essentially, lowland tropical and subtropical grassland, generally with a scattering of trees and shrubs. If woody growth is absent it is termed a *grass savannah*; with shrubs and no trees, a *shrub savannah*; or with shrubs and widely irregularly scattered trees, a *tree savannah*.

Scarification (for seed)—Pregerminative treatment to make seed coats permeable to water and gases; usually accomplished by mechanical abrasion or by soaking seeds briefly in a strong acid or other chemical solution.

Schist—A metamorphic crystalline rock having a closely foliated structure divisible along approximately parallel planes.

Science—The observation, identification, description, experimental investigation, and theoretical explanation of natural phenomena.

Scientific method—A systematic form of inquiry that involves observation, speculation, and reasoning.

Scion—An aerial plant part, often a branchlet, that is grafted onto the root-bearing part of anther plant.

Sclerenchyma—A protective or supporting tissue in higher plants composed of cells with walls thickened and lignified and often mineralized.

Second law of thermodynamics—Natural law that dictates that in any conversion of heat energy to useful work some of the initial energy input is always degraded to a lower-quality, more dispersed, less useful form of energy, usually low-temperature heat that flows into the environment.

Sedimentary rock—Rock formed from materials deposited from suspension or precipitated from solution and usually being more or less consolidated. The principal sedimentary rocks are sandstones, shales, limestones, and conglomerates.

Sediments—Soil particles dislodged by rain drops that travel via runoff into streams, rivers, lakes, or oceans and are deposited there.

Seed coat (testa)—The outer coat of the seed derived from the integument.

Seed tree—The silvicultural system in which the dominant feature is the removal of all trees except for a small number of seedbearers left singly or in small groups, usually 20 to 25 per hectare (8 to 10 per acre). The seed trees are generally harvested when regeneration is established. An even-age stand results.

Seedling—A tree grown from seed that has not yet reached a height of 3 feet or does not exceed 2 inches in DBH.

Selfing—The self-pollution of an individual or biotype with its own pollen, the offspring being termed *self*.

Senescent pond—A pond that has reached old age.

Sere—A sequence of plant communities that successively follow one another in the same habitat from the pioneer stage to a mesic climax.

Serotinous—Late in developing; particularly applied to plants that flower or fruit late in the season and to fruit and cones that remain closed for a year or more after the seeds mature, but also to bud opening, leaf shedding, etc.

Serpentine—A mineral or rock consisting essentially of a hydrous magnesium silicate. It usually has a dull green color and often a mottled appearance.

Serpentinite—A rock consisting almost wholly of serpentine mineral derived from the alteration of previously existing divine and pyroxene.

Sessile—Without a stalk; sitting directly on its base.

Shade-tolerance classes—The classes are very intolerant, intolerant, intermediate, tolerant, very tolerant.

Shelterwood—The silvicultural system in which, in order to provide a source of seed or protection for regeneration, the old crop (the shelterwood) is removed in two to more successive shelterwood cuttings. The first cutting is ordinarily the seed cutting, though it may be a preparatory cutting, and the last is the final cutting. Any intervening cutting is termed a *removal cutting*. An even-age stand results.

Short-rotation woody crops—Tree crops grown primarily for their fuel value.

Sialic—Light rock rich in silica and alumina; typical of the outer layer of the Earth.

Silviculture—The science and art of controlling the establishment, composition, and growth of forests.

Silviculture system—A process whereby forests are tended, harvested, and replaced, resulting in a forest of distinctive form. Systems are classified according to the method of carrying out the fellings that remove the mature crop with a view to regeneration and according to the type of forest thereby produced. These are individual tree selection, shelterwood, seed tree, and clearcut.

Single-level sampling—A sampling design where units are selected directly from the sampling frame of the population.

Sinks—Areas, whether natural or artificial, where the products or effluents from production and consumption in one place are physically exported to another for storage or dispersal.

Sinuosity—The bending or curving shape of a stream course.

Site class—A measure of the relative productive capacity of a site based upon the volume or height (dominant, codominant, or mean) or the maximum mean annual increment of a stand that is attained or attainable at a given age.

Site index (SI)—A measure of a site class based upon the height of the dominant trees in a stand at an arbitrarily chosen age, most commonly at 50 years in the East and 100 years in the West.

Skep—A woven straw beehive.

Slope—A soil property in which the steepness of the soil layer is directly related to the degree of erosion that may occur.

Small diameter—Timber that is usually 4- to 8-inches in diameter that has not been economical to remove for traditional timber production.

Smog—Visible air pollution; a dense, discolored haze containing large quantities of soot, ash, and gaseous pollutants such as sulfur dioxide and carbon dioxide.

Soil—The word *soil* is derived from the Latin *solum*, which means floor or ground. Soil is a dynamic natural body in which plants grow composed of mineral and organic materials and living forms. Important soil terms and definitions include:*

- *Ablation till*—A superglacial coarse-grained sediment or till that accumulates as the subadjacent ice melts and drains away to finally be deposited on the exhumed subglacial surface.
- *Absorption*—Movement of ions and water into the plant roots as a result of either metabolic processes by the root (active absorption) or diffusion along a gradient (passive absorption).
- *Acid rain*—Atmospheric precipitation with pH values less than about 5.6, the acidity being due to inorganic acids such as nitric and sulfuric that is formed when oxides of nitrogen and sulfur are emitted into the atmosphere.
- *Acid soil*—A soil with a pH value of <7.0 or neutral. Soils may be naturally acid from their rocky origin or by leaching, or they may become acid from decaying leaves or from soil additives such as aluminum sulfate (alum). Acid soils can be neutralized by the addition of lime products.
- *Actinomycetes*—A group of organisms intermediate between the bacteria and the true fungi that usually produce a characteristic branched mycelium; includes many (but not all) organisms belonging to the order of Actinomycetales.
- *Adhesion*—Molecular attraction that holds the surfaces of two substances (e.g., water and sand particles) in contact.
- *Adsorption*—The attraction of ions or compounds to the surface of a solid.
- *Aeration, soil*—The process by which air in the soil is replaced by air from the atmosphere. In a well-aerated soil, the soil air is similar in composition to the atmosphere above the soil. Poorly aerated soils usually contain more carbon dioxide and correspondingly less oxygen than the atmosphere above the soil.
- *Aerobic*—Growing only in the presence of molecular oxygen, as aerobic organisms.

---

* This definition was compiled and adapted from several sources, including *Keys to Soil Taxonomy*, U.S. Department of Agriculture, Washington, D.C., 2010; *Resource Conservation Glossary*, Soil Conservation Society of America, Anheny, IA, 1982; *Glossary of Soil Science Terms*, Soil Science Society of America, Madison, WI, 1987.

– *Aggregates, soil*—Soil structural units of various shapes, composed of mineral and organic material, formed by natural processes, and having a range of stabilities.

– *Agronomy*—A specialization of agriculture concerned with the theory and practice of field crop production and soil management; the scientific management of land.

– *Air capacity*—Percentage of soil volume occupied by air spaces or pores.

– *Air porosity*—The proportion of the bulk volume of soil that is filled with air at any given time or under a given condition, such as a specified moisture potential; usually the large pores.

– *Alkali*—A substance capable of liberating hydroxide ions in water, measured at a pH of more than 7.0, and possessing caustic properties; it can neutralize hydrogen ions, with which it reacts to form a salt and water, and it is an important agent in rock weathering.

– *Alluvium*—A general term for unconsolidated, granular sediments deposited by rivers.

– *Amendment, soil*—Any substance other than fertilizers (e.g., compost, sulfur, gypsum, lime, sawdust) used to alter the chemical or physical properties of a soil, generally to make it more productive.

– *Ammonification*—The production of ammonia and ammonium-nitrogen through the decomposition of organic nitrogen compounds in soil organic matter.

– *Anaerobic*—Without molecular oxygen.

– *Anion*—An atom that has gained one or more negatively charged electrons and is thus itself negatively charged.

– *Aspect, slope*—The direction that a slope faces with respect to the sun.

– *Assimilation*—The taking up of plant nutrients and their transformation into actual plant tissues.

– *Atterburg limits*—Water contents of fine-grained soils at different states of consistency.

– *Autotrophs*—Plants and microorganisms capable of synthesizing organic compounds from inorganic materials by either photosynthesis or oxidation reactions.

– *Available water*—The portion of water in a soil that can be readily absorbed by plant roots; the amount of water released between the field capacity and the permanent wilting point.

– *Bedrock*—The solid rock underlying soils and the regolith in depths ranging from zero (where exposed by erosion) to several hundred feet.

– *Biological function*—The role played by a chemical compound or a system of chemical compounds in living organisms.

– *Biomass*—The total weight of living biological organisms within a specified unit (area, community, population).

– *Biome*—A major ecological community extending over large areas.

– *Blow-out*—A deflation depression, eroded by wind from the face of a vegetated dune.

– *Breccia*—A rock composed of coarse, angular fragments cemented together.

- *Calcareous soil*—Containing sufficient calcium carbonate (often with magnesium carbonate) to effervesce visibly when treated with hydrochloric acid.
- *Caliche*—A layer near the surface, more or less cemented by secondary carbonates of calcium or magnesium precipitated from the soil solution. It may occur as a soft, thin soil horizon; as a hard, thick bed just beneath the solum; or as a surface layer exposed by erosion.
- *Capillary water*—Water held within the capillary pores of soils; mostly available to plants.
- *Catena*—The sequence of soils that occupy a slope transect, from the topographic divide to the bottom of the adjacent valley.
- *Cation*—An atom that has lost one or more negatively charged electrons and is thus itself positively charged.
- *Chelate*—From the Greek word for "claw," a complex organic compound containing a central metallic ion surrounded by organic chemical groups.
- *Class, soil*—A group of soils having a definite range in a particular property such as acidity, degree of slope, texture, structure, land-use capability, degree of erosion, or drainage.
- *Clay*—A soil separate consisting of particles that are <0.0002 mm in equivalent diameter.
- *Cohesion*—Force holding a solid or liquid together due to attraction between like molecules; decreases with rise in temperature.
- *Colloidal*—Matter of very fine particle size.
- *Convection*—A process of heat transfer in a fluid involving the movement of substantial volumes of the fluid concerned; very important in the atmosphere and to a lesser extent in the oceans.
- *Denitrification*—The biochemical reduction of nitrate or nitrite to gaseous nitrogen, either as molecular nitrogen or as an oxide of nitrogen.
- *Detritus*—Debris from dead plants and animals.
- *Diffusion*—The movement of atoms in a gaseous mixture or ions in a solution, primarily as a result of their own random motion.
- *Drainage*—The removal of excess water, both surface and subsurface, from plants. All plants (except aquatics) will die if exposed to an excess of water.
- *Duff*—The matted, partly decomposed organic surface layer of forest soils.
- *Erosion*—The wearing away of the land surface by running water, wind, ice, or other geological agents, including such processes as gravitational creep.
- *Eutrophication*—A process of lake aging whereby aquatic plants are abundant and waters are deficient in oxygen. The process is usually accelerated by enrichment of waters with surface runoff containing nitrogen and phosphorus.
- *Evapotranspiration*—The combined loss of water from a given area, during a specified period of time, by evaporation from the soil surface and by transpiration from plants.
- *Exfoliation*—Mechanical or physical weathering that involves the disintegration and removal of successive layers of rock mass.

- *Fertility, soil*—The quality of a soil that enables it to provide essential chemical elements in quantities and proportions necessary for the growth of specific plants.
- *Fixation*—The transformation in soil of a plant nutrient from an available to an unavailable state.
- *Fluvial*—Deposits of parent materials laid down by rivers or streams.
- *Friable*—A soil consistency term pertaining to the ease of crumbling of soils.
- *Heaving*—The partial lifting of plants, buildings, roadways, fence posts, etc., out of the ground as a result of freezing and thawing of the surface soil during the winter.
- *Heterotroph*—An organism capable of deriving energy for life processes only from the decomposition of organic compounds and incapable of using inorganic compounds as sole sources of energy or for organic synthesis.
- *Horizon, soil*—A layer of soil, approximately parallel to the soil surface, differing in properties and characteristics from adjacent layers below or above it.
- *Humus*—More or less stable fraction of the soil organic matter (usually dark in color) remaining after the major portions of added plant and animal residues have decomposed.
- *Hydration*—The incorporation of water into the chemical composition of a mineral, converting it from an anhydrous to a hydrous form; the term is also applied to a form of weathering in which hydration swelling creates tensile stress within a rock mass.
- *Hydraulic conductivity*—The rate at which water is able to move through a soil.
- *Hydrolysis*—The reaction between water and a compound (commonly a salt). The hydroxyl from the water combines with the anion from the compound undergoing hydrolysis to form a base; the hydrogen ion from the water combines with the cation from the compound to form an acid.
- *Hygroscopic coefficient*—The amount of moisture in a dry soil when it is in equilibrium with some standard relative humidity near a saturated atmosphere (about 98%); expressed in terms of percentage on the basis of oven-dry soil.
- *Infiltration*—The downward entry of water into the soil.
- *Ions*—Atoms that have lost or gained one or more negatively charged electrons.
- *Land classification*—The arrangement of land units into various categories based on the properties of the land and its suitability for some particular purpose.
- *Leaching*—The removal of materials in solution from the soil by percolating waters.
- *Liebig's law*—The growth and reproduction of an organism are determined by the nutrient substance (e.g., oxygen, carbon dioxide, calcium) available in minimum quantity with respect to organic needs, the limiting factor.
- *Loam*—The textural class name for soil having moderate amounts of sand, silt, and clay.

- *Loess*—An accumulation of wind-blown dust (silt) that may have undergone mild digenesis.
- *Marl*—An earthy deposit consisting mainly of calcium carbonate, usually mixed with clay. Marl is used for liming acid soils, but it is slower acting than most lime products used for this purpose.
- *Mineralization*—The conversion of an element from an organic form to an inorganic state as a result of microbial decomposition.
- *Nitrogen fixation*—The biological conversion of elemental nitrogen (N_2) to organic combinations or to forms readily utilized in biological processes.
- *Osmosis*—The movement of a liquid across a membrane from a region of high concentration to a region of low concentration. Water and nutrients move into roots independently.
- *Oxidation*—The loss of electrons by a substance.
- *Parent material*—The unconsolidated and more or less chemically weathered mineral or organic matter from which the solum of soils is developed by pedongenic processes.
- *Ped*—A unit of soil structure such as an aggregate, crumb, prism, block, or granule, formed by natural processes.
- *Pedogenic and pedological processes*—Processes associated with the formation and development, respectively, of soil.
- *pH*—The degree of acidity or alkalinity of the soil; also referred to as *soil reaction*. On the pH scale 7.0 is neutral; values from 0.0 to 7.0 are acid; and values from 7.0 to 14.0 are alkaline. The pH of soil is determined by a simple chemical test where a sensitive indicator solution is added directly to a soil sample in a test tube.
- *Photosynthesis*—The process by which green leaves of plants, in the presence of sunlight, manufacture their own needed materials from carbon dioxide in the air and water and minerals taken from the soil.
- *Porosity, soil*—The volume percentage of the total bulk not occupied by solid particles.
- *Profile, soil*—A vertical section of the soil through all its horizons and extending into the parent material.
- *Reduction*—The gain of electrons and therefore the loss of positive valence charge by a substance.
- *Regolith*—The unconsolidated mantle of weathered rock and soil material on the Earth's surface; loose earth materials above solid rock.
- *Rock*—The material that forms the essential part of the Earth's solid crust, including loose incoherent masses such as sand and gravel, as well as solid masses of granite and limestone.
- *Rock cycle*—The global geological cycling of lithospheric and crustal rocks from their igneous origins through all of any stages of alteration, deformation, resorption, and reformation.
- *Runoff*—The portion of the precipitation on an area that is discharged from the area through stream channels.
- *Salinization*—The process of accumulation of salts in soil.

- *Sand*—A soil particle between 0.05 and 2.0 mm in diameter; a soil textural class.
- *Silt*—A soil separate consisting of particles between 0.05 and 0.002 mm in equivalent diameter; a soil textural class.
- *Slope*—The degree of deviation of a surface from horizontal, measured in a numerical ratio, percent, or degrees.
- *Soil air*—The soil atmosphere; the gaseous phase of the soil; that volume not occupied by soil or liquid.
- *Soil horizon*—A layer of soil, approximately parallel to the soil surface, with distinct characteristics produced by soil-forming processes. These characteristics form the basis for systematic classification of soils.
- *Soil profile*—A vertical section of the soil from the surface through all of its horizons, including C horizons.
- *Soil structure*—The combination or arrangement of primary soil particles into secondary particles, units, or peds. These secondary units may be, but usually are not, arranged in the profile in such a manner as to give a distinctive characteristic pattern. The secondary units are characterized and classified on the basis of size, shape, and degree of distinctness into classes, types, and grades, respectively.
- *Soil texture*—The relative proportions of the various soil separates in a soil.
- *Soluble*—Dissolves easily in water.
- *Solum* (sing.), *sola* (pl.)—The upper and most weathered part of the soil profile; the A, E, and B horizons.
- *Subsoil*—That part of the soil below the plow layer.
- *Till*—Unstratified glacial drift deposited directly by the ice and consisting of clay, sand, gravel, and boulders intermingled in any proportion.
- *Tilth*—The physical condition of soil as related to its ease of tillage, fitness as a seedbed, and its impedance to seedling emergence and root penetration.
- *Topsoil*—The layer of soil moved in cultivation.
- *Weathering*—All physical and chemical changes produced in rocks, at or near the Earth's surface, by atmospheric agents.

Soil boring—Using a boring tool (such as an auger) to take soil samples for analysis.

Soil factor—In *in situ* soil remediation, soil factors include water content, porosity/permeability, clay content, and adsorption site density.

Soil families and series—The family category of classification is based on features that are important to plant growth such as texture, particle size, mineralogical class, and depth. Terms such as *clayey*, *sandy*, and *loamy* are used to identify textural classes. Terms used to describe mineralogical classes include *mixed*, *oxidic*, and *carbonatic*. For temperature classes, terms such as *hypothermic*, *frigid*, and *cryic* are used. The soil series gets down to the individual soil, and the name is that of a natural feature or place near where the soil was first recognized. Familiar series names include Amarillo (Texas), Carlsbad (New Mexico), and Fresno (California). In the United States, there are more than 18,000 soil series.

Soil fertility—The quality of a soil that enables it to provide essential chemical elements in quantities and proportions for the growth of specified plants.

Soil forming process—The mode of origin of the soil, with special reference to the processes or soil-forming factors responsible for the development of the solum, or true soil, from the unconsolidated parent material.

Soil great groups and subgroups—Suborders are divided into great groups. They are defined largely by the presence or absence of diagnostic horizons and the arrangements of those horizons. Great group names are coined by prefixing one or more additional formative elements to the appropriate suborder name. More than 230 great groups have been identified. Great groups are divided into subgroups. Subgroup names indicate to what extent the central concept of the great group is expressed. A Typic Fragiaqualf is a soil that is typical for the Fragiaqualf great group.

Soil Guideline Values (SGVs)—A series of measurements and values used to measure contamination of the soil.

Soil horizon—A layer of soil, approximately parallel to the soil surface, differing in properties and characteristics from adjacent layers below or above it.

Soil orders—The 11 soil orders are*

- *Alfisol*—Mild forest soil with gray to brown surface horizon, medium to high base supply (refers to amount of interchangeable cations that remain in soil), and a subsurface horizon of clay accumulation.
- *Andisol*—Formed on volcanic ash and cinders and lightly weathered.
- *Aridsol*—Dry soil with pedogenic (soil forming) horizon; low in organic matter.
- *Entisol*—Recent soil without pedogenic horizons.
- *Histosol*—Organic (peat or bog) soil.
- *Inceptisol*—Soil at the beginning of the weathering process with weakly differentiated horizons.
- *Mollisol*—Soft soil with a nearly black, organic-rich surface horizon and high base supply.
- *Oxisol*—Oxide-rich soil principally a mixture of kaolin, hydrated oxides, and quartz.
- *Spodosol*—Soil that has an accumulation of amorphous materials in the subsurface horizons.
- *Ultisol*—Soil with a horizon of silicate clay accumulation and low base supply.
- *Vertisol*—Soil with high-activity clays (cracking clay soil).

Soil pollution—Contamination of the soil and subsurface by the addition of contaminants or pollutants.

Soil profile—A vertical section of the soil from the surface through all its horizons, including C horizons.

Soil remediation—The use of various techniques or technologies to decontaminate or dispose of contaminated soil.

Soil sampling—Conducted to determine through analysis the type, texture, and structure of a soil, as well as the degree and extent of contamination.

Soil separates—Soil particles classified into groups (sand, silt, and clay) based on their size by the International Soil Science Society System, the U.S. Public Roads Administration, and the U.S. Department of Agriculture. In this text, we use the

---

* Adapted from *Soil Classification: A Comprehensive System,* U.S. Department of Agriculture, Natural Resources Conservation Service, Washington, D.C., 1975.

classification established by the USDA. The size ranges in these separates reflect major changes in how the particles behave, and in the physical properties they impart to soils.

Soil structure—The combination or arrangement of primary soil particles into secondary particles, units, or peds. These secondary units may be, but usually are not, arranged in the profile in such a manner as to give a distinctive characteristic, pattern. The secondary units are characterized and classified on the basis of size, shape, and degree of distinctness into classes, types, and grades, respectively.

Soil suborders—Soil orders are further divided into 55 suborders, based primarily on the chemical and physical properties that reflect either the presence or absence of waterlogging or genetic differences caused by climate and vegetation, to give the class the greatest genetic homogeneity. Thus, the Aqualfs are formed under wet conditions, and Alfisols become saturated with water sometime during the year. The suborder names all have two syllables, with the first syllable indicating the order, such as Alf (Alfisol).

Soil texture—The relative proportions of the various soil separates in a soil.

Solid—Matter that has a definite volume and a definite shape.

Solidification—A stabilization technique used to convert hazardous waste from its original form to a physically and chemically more stable material. Accomplished by reducing the mobility of hazardous compounds in the waste prior to its land disposal.

Solubility—The ability of a substance to mix with water.

Solum—The upper and most weathered part of the soil profile (i.e., the A and B horizons).

Solute—The dissolved substance in a solution.

Solvent—The substance in excess in a solution.

Sorption—Process of adsorption or absorption of a substance on or in another substance.

Source—The spring from which the stream originates, or other point of origin of stream.

Southeastern—U.S. Forest Service area that includes Virginia, the Carolinas, Georgia, and Florida.

Southern—U.S. Forest Service area that includes Alabama, Tennessee, Mississippi, Arkansas, Louisiana, and Oklahoma and Texas east of the 100th meridian.

Southern pines—Within the United States, the ten species of hard pines with major portions of their ranges below the Mason–Dixon line: longleaf, shortleaf, slash, loblolly, spruce, Virginia, sand, pitch, Table Mountain, and pond pine.

Species—A group of individuals or populations potentially able to interbreed and unable to produce fertile offspring by breeding with other sorts of animals and plants.

Specific gravity—Ratio of the weight of the volume of liquid or solid to the weight of an equal volume of water.

Specific heat—The amount of heat energy in calories necessary to raise the temperature of 1 gram of a substance 1°C.

Sporangiospore—Spores that form within sacs called *sporangia*, which are attached to stalks called *sporangiophores*.

Sporangium—A hollow, unicellular or multicellular saclike, spore-producing structure.

Spore—Reproductive stage of fungi.

Sporophyll—A modified leaf or leaflike structure that bears sporangia (stamens and carpels of angiosperms).

Spring—The point at which a stream emerges from an underground course through unconsolidated sediments or through caves.

Staminate—Having pollen-bearing organs (stamens) only; may apply to individual male plants of a dioecious species or to flowers, inflorescences, or strobili.

Stand density—A measure of the degree of crowding of trees within stocked areas, commonly expressed by various growing-space ratios such as crown length to tree height, crown diameter to diameter at breast height, crown diameter to tree height; or stem (triangular) spacing to tree height.

Stand of trees—A tree community that possesses sufficient uniformity in composition, constitution, age, spatial arrangement, or condition to be distinguishable from adjacent communities.

Standard deviation—The square root of the variance.

Standard temperature and pressure (STP)—Because the density of gases depends on temperature and pressure, defining the pressure and temperature against which a volume of gas was measured is customary. The normal reference point is a standard temperature of 0°C and a standard pressure of 760 mmHg.

Standing crop—The amount of biomass at a given time; usually refers to the amount of aboveground plant biomass.

Statistical inference—Expressing the connection between the unknown state of nature and observed information in probabilistic terms.

Statistical survey—Design and execution of surveys to provide estimates of characteristics of well-defined finite populations.

Stemflow—Precipitation that is intercepted by vegetative cover and runs down the stem or major axes of such cover.

Steppe—Arid land with xerophilous vegetation usually found in regions of extreme temperature range and loess soil.

Stereome—A collective physiological term for all supporting tissues in a plant, such as sclerenchyma and collenchyma.

Sterigma—A peg-shaped projection to which the leaves of some conifers (as spruces) are attached on the twigs.

Stigma—The part of the pistil, usually the tip, often sticky, which receives the pollen and upon which the pollen germinates.

Stipe—A supporting stalk, such as the stalk of pistil, a gill fungus, or petiole of a fern leaf.

Stipule—A small structure or appendage found at the base of some leaf petioles, usually present in pairs. They are morphologically variable and appear as scales, spines, glands, or leaflike structures.

Stoma—A pore in the epidermis and the two guard cells surrounding it; sometimes applied only to the pore.

Stool—A living stump capable of producing sprouts.

Stratification—A pregerminative treatment to break dormancy in seeds and to promote rapid uniform germination accomplished by exposing seeds for a specified time to moisture at near-freezing temperatures, sometimes with a preceding exposure to moisture at room temperature.

Stratosphere—A region of the atmosphere based on temperature; located approximately between 10 and 35 miles in altitude.

Strobilus—The male or female fruiting body of a gymnosperm.

Style—The stalk of a pistil which connects the stigma with the ovary.

Subadiabatic—The ambient lapse rate when it is less than the dry adiabatic lapse rate.

Submerged vegetation—In a pond, the submerged plants that grow where light can penetrate the water surface and reach them.

Subsidence inversion—A type of inversion usually associated with high-pressure systems, known as *anticyclones*, which may significantly affect the dispersion of pollutants over large regions.

Subsoil—The part of the soil below the plow layer.

Substrate—The material or substance upon which an enzyme acts.

Sulfur cycle—The natural circulation of sulfur through the environment.

Sulfur dioxide—A primary pollutant originating chiefly from the combustion of high-sulfur coals.

Suppressed—Crown class of very slowly growing trees with crowns in the lower layer of the canopy and leading shoots not free; such trees are subordinate to dominants, codominants, and intermediates in the crown canopy.

Sustainability—The capacity to meet the needs of the present without compromising the ability of future generations to meet their own needs; integrates environmental, social, and economic concerns and outcomes.

Symbiotic—A close relationship between two organisms of different species; one where both partners benefit from the association.

Sympatric—Species or populations inhabiting the same or overlapping areas. (See *Allopatric*.)

Sympodial—A branching growth pattern in which the main axis is formed by a series of successive secondary axes, each of which represents one fork of a dichotomy.

Synthesis—The formation of a substance or compound from more elementary compounds.

Taungya method—The raising of a forest crop in conjunction with a temporary agricultural crop.

Taxon—Any formal taxonomic group such as genus, species, or variety.

Temperature—A measure of the average kinetic energy of the molecules.

Temperature inversion—A condition characterized by an inverted lapse rate.

Tepal—Perianth parts undifferentiated into distinct sepals and petals.

Terpene—Any of various isometric hydrocarbons found especially in essential oils (as from conifers), resins, and balsams.

Testa—The outer coat of the seed derived from the integument.

Tetraploid (polyploid)—A cell, tissue, or organism having four sets of chromosomes.

Thalweg—Line of maximum water of channel depth in a stream.

Thermal circulation—Atmospheric circulation caused by the heating and cooling of air.

Thermal inversion—A layer of cool air trapped under a layer of less dense warm air, thus preventing reversing to the normal situation.

Thermic soil temperature—The mean annual soil temperature is 15°C (59°F) or higher but lower than 22°C (72°F), and the difference between mean summer and mean winter soil temperature is more than 5°C (9°F) at a depth of 50 cm (20 in).

Thermocline—The fairly thin transition zone in a lake that separates an upper warmer zone from a lower colder zone.

Thermosphere—A region of the atmosphere based on temperature between approximately 60 and several hundred miles in altitude.

Throughfall—All the precipitation reaching the forest floor minus the stemflow (i.e., canopy drip plus direct precipitation).

Tilth—The physical condition of soil as related to its ease of tillage, fitness as a seedbed, and its impedance to seedling emergence and root penetration.

Top-to-root ratio or root-to-shoot ratio—The relative weights or volumes of the epicotyl and the hypocotyl of a tree seedling, expressed as a ratio.

Topsoil—The layer of soil moved during cultivation.

Tracheid—An elongated thick-walled, nonliving conducting and supporting cell found in the xylem of most vascular plants.

Transect diameter—The diameter of a piece of down woody material at the point of intersection with a transect.

Triploid—A cell, issue, or organism having three sets of chromosomes.

Trophic level—The feeding position occupied by a given organism in a food chain, measured by the number of steps removed from the producers.

TSI (Timber Stand Improvement)—A loose term comprising all intermediate treatments made to improve the composition, constitution, condition, and increment of a timber stand.

Tuff—A rock composed of the finer kinds of volcanic detritus usually fused together by heat.

Turbidity—Reduced transparency of the atmosphere caused by absorption and scattering of radiation by solid or liquid particles other than clouds and held in suspension.

Turbulent flow—Movement of water in a stream that results in complex mixing.

Turnover—The mixing of the upper and lower levels of a lake that most often occurs during the spring and fall; it is caused by dramatic changes in surface water temperature.

Umbo—A blunt or rounded projection arising from a surface, as on a pine cone scale.

Unconfined aquifer—An aquifer not underlain by an impermeable layer.

Unequal probability sampling—Sampling designs where units are selected with different probabilities; these probabilities need to be known for unbiased estimation.

Uneven-aged—A condition of forest or stand that contains intermingled trees that differ markedly in age. By convention, a minimum range of 10 to 20 years is generally accepted, although with rotations of not less than 100 years, 25% of the rotation may be the minimum.

Unit—The basic sample unit used in the last stage of multistage sampling.

Unsaturated zone—Lies just beneath the soil surface and is characterized by crevices that contain both air and water; water contained therein is not available for use.

Vadose water—Water in the unsaturated zone that is essentially unavailable for use.

Valence—The net electric charge of an atom or the number of electrons an atom can give up (or acquire) to achieve a filled-out shell.

Valley winds—At valley floor level, slope winds transform into valley winds that flow downvalley, often with the flow of a river.

Variable—A characteristic that varies from unit to unit (e.g., age of a tree).

Variance—The arithmetic mean of the squares of the deviations of all values in a set of numbers from their arithmetic mean; used to indicate how widely individuals in a group vary. Population variance and sample variance are defined slightly differently.

Variety—A subdivision of a species, usually separated geographically from the typical, having one or more heritable, morphological characteristics that differ from the typical, even when grown under the same environmental conditions; a morphological variant.

Volatile—When a substance (usually a liquid) evaporates at ordinary temperatures if exposed to the air.

Volatilization—When a solid or liquid substance passes into the vapor state.

Volume—Surface area multiplied by a third dimension (e.g., height).

Water content—In *in situ* volatilization, water content influences the rate of volatilization by affecting the rates at which chemicals can diffuse through the vadose zone. An increase in solid water content decreases the rate at which volatile compounds are transported to the surface via vapor diffusion.

Water table—The upper surface of the saturation zone below which all void spaces are filled with water.

Water vapor—The most visible constituent of the atmosphere (H_2O in vapor form).

Watershed—The region draining into a river, river system, or body of water.

Watershed divide—A ridge of high land dividing two areas drained by different river systems.

Weather—The day-to-day pattern of precipitation, temperature, wind, barometric pressure, and humidity.

Weathering—The chemical and mechanical breakdown of rocks and minerals under the action of atmospheric agencies.

Weight—The force exerted upon any object by gravity.

Wetland—A lowland area, such as a marsh or swamp, saturated with moisture and usually thought of as natural wildlife habitat.

Wetted perimeter—The line on which the stream's surface meets the channel walls.

Whole-tree harvesting—A harvesting method in which the whole tree (above the stump) is removed.

Woody biomass—The trees and woody plants, including limbs, tops, needles, leaves, and other woody parts, grown in a forest, woodland, or rangeland environment that are the byproduct of forest management.

Xerophyte—A plant that is adapted to dry or desert habitats.

Appendix. Answers to Chapter Review Questions

Chapter 1

1.1   Answers will vary

Chapter 2

2.1   8
2.2   1
2.3   25
2.4   8
2.5   25
2.6   2
2.7   2.5
2.8   22.5
2.9   4.28
2.10 −8
2.11 0.001 or 1/1000

Chapter 3

3.1   a.   0.103022 km
      b.   252.928 ha
      c.   150.22 km^2
      d.   8.77822 m^2
3.2   a.   24278.22 ft
      b.   170.5 acres
      c.   1765.7777 ft^3
      d.   8,046.21 ft^3 per acre
3.3   a.   Global positioning system
      b.   Basal area
      c.   Thousand board feet
      d.   Diameter at breast height

Chapter 4

4.1   Sample

4.2   Population

4.3   Parameter

Chapter 5

5.1   Higher heating value

5.2   Dry weight basis

5.3   Stem, bark, stump, branches, and foliage

Chapter 6

6.1   a. I, 13 plots; II, 16 plots; III, 21 plots; then randomly locate each plot within each cover type

   b. Locate plots randomly across the total land area and hope to get representation of all classes in proportion to size

6.2   Basal

6.3   Rule by which the sample is drawn; the care exercised in measurement; the degree to which bias can be avoided

6.4   Often

6.5   Variable

6.6   Regression coefficient

6.7   Double sampling

Chapter 7

7.1   Art

7.2   Log rule

7.3   Smalian Cubic Volume rule

7.4   Basal area

7.5   0.785

Chapter 8

8.1 Scientific term for living matter, but the word *biomass* is also used to denote products derived from living organisms—wood from trees, harvested grasses, plant parts, and residues such as twigs, stems, and leaves, well as aquatic plants and animal wastes

8.2 Forestland, agricultural land

8.3 Biopower

8.4 10^{18}

8.5 Cofiring

Chapter 9

9.1 Apical meristem

9.2 Dicot

9.3 Meristem

9.4 Monocots

9.5 Xylem

9.6 Sapwood

9.7 Vascular

9.8 Leaves

9.9 Chloroplasts

9.10 Cambia

9.11 Gibberellins

Chapter 10

10.1 Sugar, starch, vegetable oils

10.2 Glycerin

10.3 Switchgrass

10.4 Hybrid poplars

10.5 Quad

10.6 Algae

10.7 Thallus

10.8 Carbohydrates, protein, natural oils

10.9 Logging residues

10.10 GGE

Chapter 11

11.1   Answers will vary
11.2   Answers will vary
11.3   Hydrogen
11.4   Thermal, electrolytic, photolytic
11.5   Answers will vary

Chapter 12

12.1   Answers will vary
12.2   Answers will vary

Index